2013 Joint IMEKO TC1-TC7-TC13 Symposium: Measurement Across Physical and Behavioral Sciences

Journal of Physics: Conference Series Volume 459

Genoa, Italy
4-6 September 2013

Editors:

Giovanni Battista Rossi
Francesco Crenna
Vittorio Belotti

ISBN: 978-1-62993-097-8
ISSN: 1742-6588

Printed from e-media with permission by:

Curran Associates, Inc.
57 Morehouse Lane
Red Hook, NY 12571

Some format issues inherent in the e-media version may also appear in this print version.

Printed by Curran Associates, Inc. (2013)

For permission requests, please contact the Institute of Physics
at the address below.

Institute of Physics
Dirac House, Temple Back
Bristol BS1 6BE UK

Phone: 44 1 17 929 7481
Fax: 44 1 17 920 0979

techtracking@iop.org

Additional copies of this publication are available from:

Curran Associates, Inc.
57 Morehouse Lane
Red Hook, NY 12571 USA
Phone: 845-758-0400
Fax: 845-758-2634
Email: curran@proceedings.com
Web: www.proceedings.com

TABLE OF CONTENTS

012001 Measurement in Psychology: Conceptual Analysis and Synthesis Revisited 1
B Berglund

012002 A Gentle Introduction to Rasch Measurement Models for Metrologists 2
L Mari, M Wilson

012003 Urns and Chameleons: Two Metaphors for Two Different Types of Measurements 3
L Accardi

012004 New Scale Classification Within the Representational Theory of Measurement 5
E Benoit

012005 Probability As a Logic for Measurement Representations 10
G B Rossi, F Crenna

012006 Intransitivity in Multiple Solutions of Kemeny Ranking Problem 15
S V Muravyov, I A Marinushkina

012007 Quantification is Neither Necessary Nor Sufficient for Measurement 21
L Mari, A Maul, D T Irribarra, M Wilson

012008 On the Conceptual Foundations of Psychological Measurement 27
A Maul, M Wilson, D T Irribarra

012009 Aligning Physical Elements with Persons' Attitude: An Approach Using Rasch Measurement Theory 33
F R Camargo, B Henson

012010 Modular Applications with Smartphones and Smartpads in Shape, Color and Spectral Measurements for Industry, Biology and Medicine plus Science, Education and Training 39
D Hofmann, P-G Dittrich, E Düntsch, D Kraus, C Gärtner, R Klemm

012011 Incorporating Human Stress Measurements Into Biomedical Engineering Class 45
O Bruna, P Soucek, J Holub

012012 Innovation of Laboratory Exercises in Course Distributed Systems and Computer Networks 51
P Soucek, O Slavata, J Holub

012013 Delta-Notch Lateral Inhibition Within the Organ of Corti 55
R Summers, T Abdulla, R Luff

012014 Scaffold Characterization Using NLO Multimodal Microscopy in Metrology for Regenerative Medicine 59
L Mortati, C Divieto, M Boffitto, S Sartori, G Ciardelli, M P Sassi

012015 Extracting the Respiration Cycle Lengths from ECG Signal Recorded with Bed Sheet Electrodes 65
A Vehkaoja, M Peltokangas, J Lekkala

012016 Ex Vivo Time Evolution of Thrombus Growth through Optical and Electrical Impedance Data Fusion 71
A Affanni, R Specogna, F Trevisan

012017 Muscle Activity Characterization by Laser Doppler Myography 76
L Scalise, S Casaccia, P Marchionni, I Ercoli, E P Tomasini

012018 Design and Test of a Semi-automated System for Metrological Verification of Non-contact Clinic Thermometers 82
R Giannetti, M A Sáenz-Nuño, J M Valderrama, A Fernandez

012019 Methods for Measuring Risk-aversion: Problems and Solutions 88
P J Thomas

012020 Theory-Based Parameterization of Semiotics for Measuring Pre-literacy Development 94
N Bezruczko

012021 New Perspective on the Accessibility of Railway Transport for the Vulnerable Traveller 100
C Sundling, B Berglund, M E Nilsson, R Emardson, L Pendrill

012022 Measuring Study Habits in Higher Education: The Way Forward? 105
E D Fitkov-Norris, A Yeghiazarian

012023 The Measurement of Statistical Reasoning in Verbal-numerical and Graphical Forms: A Pilot Study 111
M Agus, M Peró-Cebollero, J Guàrdia-Olmos, M P Penna

012024 Overcoming the Invisibility of Metrology: A Reading Measurement Network for Education and the Social Sciences 117
W P Fisher Jr, A J Stenner

012025 Metrological Traceability in the Social Sciences: A Model from Reading Measurement 123
A J Stenner, W P Fisher Jr

012026 From Concrete to Abstract in the Measurement of Length 129
A Stephanou, W P Fisher Jr

012027 Merging Psychophysical and Psychometric Theory to Estimate Global Visual State Measures from Forced-Choices 135
R W Massof, K M Schmidt, D M Laby, D Kirschen, D Meadows

012028 Measurement As Cognitive and Applied Problem 141
V A Granovskii

012029 Pca-based Algorithm for Calibration of Spectrophotometric Analysers of Food 147
R Z Morawski, A Miekina

012030 Non-invasive Measurement of Cholesterol in Human Blood by Impedance Technique: an Investigation by 2D Finite Element Field Modelling 153
E Aristovich, S H Khan

012031 Implementation of Perceptual Aspects in a Face Recognition Algorithm 161
F Crenna, E Zappa, L Bovio, R Testa, M Gasparetto, G B Rossi

012032 Dynamic Investigation Test-rig on hAptics (DITA) 169
F Cannella, L Scalise, E Olivieri, M Memeo, D G Caldwell

012033 the Evaluation of Uncertainty in Mass Calibration: Possible Approaches in a Comparison Study 175
J A Sousa, A B Forbes, A S Ribeiro, P M Harris, F Carvalho, L Bacelar

012034 Uncertainty in Airflow Field Parameters in a Study of Shock Waves on Flat Plate in Transonic Wind Tunnel 182
M L C C Reis, J B P F Filho, M C C Araújo

012035 Evaluation of the Type A Uncertainty in Measurements with Autocorrelated Observations 191
Z L Warsza

012036 Why Should Correction Values Be Better Known Than the Measurand True Value? 197
F Pavese

012037 Consideration of the Measurement Uncertainty in the Assembling of Components Characterized by High Dimensional Variation 203
C Hernández, R Tutsch

012038 Observations on the Worst Case Uncertainty 209
L Fabbiano, N Giaquinto, M Savino, G Vacca

012039 Fit for Purpose Models for Metrology: A Model Selection Methodology 215
L Wright, T J Esward

012040 A Structured Strategy of Concept Definition in Measurement: the Case of 'Sensitivity' 221
L Mari, A Mencattini

012041 The Problem of False Discoveries from a Metrological Point of View 227
G E D'Errico

012042 Problems in Implementing Some of the New Measurement Unit Definitions of the SI 233
F Pavese

012043 Are Fluctuating Measurement Results instable, Drifting or Non-stationary? – A Survey on Related Terms in Metrology 239
K H Ruhm

012045 Wireless Sensing Node Network Management for Monitoring Landslide Disaster 245
S Takayama, J Akiyama, T Fujiki, N A B Mokhtar

012046 Problems in Choosing Methods and Tools for Conformity Assessment of Prepackages at the Manufacturer 251
J Dobiliene, A Meskuotiene, E Raudiene

012047 A Test Procedure for Energetic and Performance Analysis of Cold Appliances for the Food Industry 257
F Armani, A Boscolo

012048 Experimental Techniques for Ballistic Pressure Measurements and Recent Development in Means of Calibration 263
L Elkarous, F Coghe, M Pirlot, J C Golinval

012049 Protein Aggregation Measurement through Electrical Impedance Spectroscopy 275
A Affanni, A Corazza, G Esposito, F Fogolari, M Polano

012050 Study on the AFM Force Spectroscopy Method for Elastic Modulus Measurement of Living Cells 280
A Demichelis, S Pavarelli, L Mortati, G Sassi, M Sassi

012051 Uncertainty Analysis of Cell Counting by Metabolic Assays 286
C Divieto, L Revel, G Sassi, M P Sassi

012052 The Importance of Risk-aversion As a Measurable Psychological Parameter Governing Risk-taking Behaviour ... 292
P J Thomas
012053 Frequency Analysis of DC Tolerant Current Transformers ... 298
P Mlejnek, P Kaspar
012054 Modelling Adequacy of Calibration Curves and Dynamic Characteristics of Measuring Instruments ... 303
V A Granovskii, T N Siraia
012055 Conception and Development of an Optical Methodology Applied to Long-distance Measurement of Suspension Bridges Dynamic Displacement ... 309
L L Martins, J M Rebordão, A S Ribeiro
012056 Uncertainty Analysis of a Stereo System Performing Ego-motion Measurements in a Simulated Planetary Environment ... 315
M Pertile, S Debei, E Lorenzini
012057 Quantifying Human Response: Linking Metrological and Psychometric Characterisations of Man As a Measurement Instrument ... 321
L R Pendrill, W P Fisher Jr
012058 The Experimental Verification of Metrological Properties of Direct Current Watt-hour Meter ... 327
P Otomanski, P Zazula
012059 New Concept of Laboratory Exercise on Temperature Measurements Using Thermocouple ... 333
O Slavata, P Soucek, J Holub
012060 Selection of Excitation Signals for High-impedance Spectroscopy ... 339
M Kowalewski
012061 An Investigation on Probabilistic Relations in Measurement ... 345
F Crenna, G B Rossi, V Belotti
012062 Conductivity Measures Coupled with Treatment with Ion-exchange Resin for the Assessment of Sodium Concentration in Physiological Fluids: Analyses on Artificial Solutions ... 350
A Tura, S Sbrignadello, E Mambelli, P Ravazzani, A Santoro, G Pacini
012063 A Wearable Multipoint Ultrasonic Travel Aids for Visually Impaired ... 355
I Ercoli, P Marchionni, L Scalise
012064 Metrological Reliability of Optical Coherence Tomography in Biomedical Applications ... 361
C M Goloni, G P Temporão, E C Monteiro
012065 Non-contact Wearable Single Forearm Cardiac Biopotential Acquisition Device ... 367
S Gonçalves, R C Martins
Author Index

Journal of Physics: Conference Series **459** (2013) 011001 doi:10.1088/1742-6596/459/1/011001

Preface

The 2013 Joint IMEKO (International Measurement Confederation) TC1–TC7–TC13 was organised by the University of Genova – DIME/MEC, Measurement Laboratory, Italy, on 4–6 September 2013. The work of this symposium is reported in this volume.
The scope of the symposium includes the main topics covered by the above Technical Committees:

TC1 Education and Training in Measurement and Instrumentation
TC7 Measurement Science
TC13 Measurements in Biology and Medicine

This is in keeping with the tradition set by the previous events of this well established series. There has been a special focus on measurement across physical and behavioural sciences, with the aim of highlighting the interdisciplinary character of measurement science and of promoting constructive interactions with scientists in other disciplines. The discussion was introduced by keynote lectures on measurement challenges in psychophysics, psychometrics and quantum physics. The symposium was attended by experts working in these areas from 18 countries, including USA, Australia and Japan, and provided a useful forum for them to share and exchange their work and ideas.

In total over sixty papers are included in the volume, organised according to the presentation sessions. Each paper was independently peer-reviewed by two reviewers from a distinguished international panel.

The Symposium was held in Genova, which was the European Capital of Culture in 2004, and took place in Palazzo Ducale, an important historical building whose construction started in the 13th century, and that has been the house of the Duke of Genova from the 14th century. Genova, whose name comes from the Latin word 'Janua' (meaning 'door', as January is the door month of the year), has been regarded over the centuries as a door connecting Europe with the different countries and cultures of the Mediterranean basin and thus was an appropriate site for an international symposium involving different and new scientific visions and approaches to measurement, focused on a common objective: the human being.

We would like to take this opportunity to thank the members of the Symposium Steering and International Programme Committees, many of whom acted as reviewers of the papers presented here under a very short timescale, as well as our sponsor, National Instruments.

The editors hope that this volume will provide a useful contribution to enhancing the science, technology, education, and training in measurement and instrumentation.

Giovanni Battista Rossi, Francesco Crenna and Vittorio Belotti
Editors

Università degli Studi di Genova - DIME/MEC
Laboratorio di Misure
Via all'Opera Pia 15 a
I - 16145 Genova
Italy

Journal of Physics: Conference Series **459** (2013) 011002 doi:10.1088/1742-6596/459/1/011002

Peer review statement

All papers published in this volume of *Journal of Physics: Conference Series* have been peer reviewed through processes administered by the proceedings Editors. Reviews were conducted by expert referees to the professional and scientific standards expected of a proceedings journal published by IOP Publishing.

Journal of Physics: Conference Series **459** (2013) 012001 doi:10.1088/1742-6596/459/1/012001

Measurement in Psychology: Conceptual Analysis and Synthesis Revisited

Birgitta Berglund

Stockholm University and
Institute of Environmental Medicine of the Karolinska Institutet
Stockholm, Sweden

E-mail: Birgitta.Berglund@ki.se

Short CV.

Birgitta Berglund is professor at Stockholm University (SU) and the Institute of Environmental Medicine of the Karolinska Institutet. She has held full professor chairs in Environmental Psychology and in Perception and Psychophysics. In 1971, she received her PhD in psychology from SU and held a Sloan Foundation post doc at Smith College, USA. In 1972-1986, she was researcher at the Swedish Council for Social Sciences, in 1987-1994 she held a full-professor research chair donated by the Swedish Council for Building Research to SU, and in 1994-2010 a chair as full professor of perception and psychophysics at SU, where she is still active as re-employed professor. Birgitta Berglund is founding member and former president of the International Society for Psychophysics (ISP) and the International Academy of Indoor Air Sciences (IAIAS). She has served as Secretary, President and Past President for the International Commission on the Biological Effects of Noise (ICBEN) and frequently served as scientific advisor and expert for the World Health Organization and the European Commission on works in environmental noise, indoor air quality and odorous air pollution. She coordinated the European Commission's NEST research on "Measuring the Impossible" and was the leading researcher on the development of electronic noses for measuring odors.

Abstract

The imagined existences of "quantity" in psychology is a prerequisite for measurement. Any person (researcher or subject) may imagine homogeneous concepts as quantities, for example, (perceived) redness, loudness or heaviness of different kinds of objects or events. Conversely, heterogeneous concepts may be merged to imagined quantities such as intelligence or personality, which both are constructed from agreed upon sets of theoretical concepts. Measurement in physics, psychophysics and psychology will be contemplated, comparatively.

Journal of Physics: Conference Series **459** (2013) 012002 doi:10.1088/1742-6596/459/1/012002

A gentle introduction to Rasch measurement models for metrologists

Luca Mari [1] **and Mark Wilson** [2]

[1] Cattaneo University – LIUC, Italy
[2] University of California, Berkeley, USA

E-mail: lmari@liuc.it ; MarkW@berkeley.edu

Short CV.

Luca Mari is a full professor of measurement science at the Cattaneo University – LIUC (Italy), where he teaches courses on measurement science, statistical data analysis, system theory. He is currently the chairman of the TC1 (Terminology) and the secretary of the TC25 (Quantities and units) of the International Electrotechnical Commission (IEC), and an IEC expert in the WG2 (VIM) of the Joint Committee for Guides in Metrology (JCGM). He has been the chairman of the TC7 (Measurement Science) of the International Measurement Confederation (IMEKO).

Mark Wilson is a full professor of education at the University of California, Berkeley, where he teaches courses on measurement in the social sciences, multidimensional measurement and applied statistics. He was the president of the Psychometric Society for 2011-12, and also became a member of the US National Academy of Education in the same year. He is fellow of both the American Psychological Association, and the American Educational Research Association. He has chaired two US National Research Council committees, both on assessment of science achievement, and been a member of several others.

Abstract

The talk introduces the basics of Rasch models by systematically interpreting them in the conceptual and lexical framework of the International Vocabulary of Metrology, third edition (VIM3). An admittedly simple example of physical measurement highlights the analogies between physical transducers and tests, as they can be understood as measuring instruments of Rasch models and psychometrics in general.

From the talk natural scientists and engineers might learn something of Rasch models, as a specifically relevant case of social measurement, and social scientists might re-interpret something of their knowledge of measurement in the light of the current physical measurement models.

IMEKO 2013 TC1 + TC7 + TC13 IOP Publishing
Journal of Physics: Conference Series **459** (2013) 012003 doi:10.1088/1742-6596/459/1/012003

Urns and Chameleons: two metaphors for two different types of measurements

Luigi Accardi

Centro Vito Volterra, Facoltà di Economia,
Università di Roma "Tor Vergata", Roma , Italy
http://volterra.uniroma2.it

E-mail: accardi@volterra.uniroma2.it

Short CV.

Luigi Accardi Academic Experience: - Degree in Mathematics cum laude, University of Naples 31.10.70. - PhD (Candidat Nauk) May 1974, Faculty of Mathematics and Mechanics, Moscow University. - 1974--1978 Researcher at CNR and teaches at the Universities of Salerno and Milano. - 1980 Full professor. Leaves the CNR. From 1981--present: University of Roma Torvergata. - Director, since July 1988, of the Centro V. Volterra. - Plenary speaker in several international conferences.

Editorial activities

Journals: - Quantum Probability and Infinite Dimensional Analysis (Managing Editor since 2004) - Open Systems and Information Dynamics (co-editor-in-chief since 1992) - Random Operators and Stochastic equations (Editorial Board since 1990) - Probability and Mathematical Statistics (Editorial Board since 1991) - Reports of Mathematical Physics (Editorial Board since 1989) - Probability Theory and Related Fields (Editorial Board from 1986 to 1991).

Books: - Accardi L.: Urne e camaleonti, Dialogo sulla realtà, le leggi del caso e la teoria quantistica, Il Saggiatore (1997), Regular and Chaotic dynamics (2002), Japanese translation to appear (2014) - Accardi L., Lu Y.G., Volovich I.: Quantum Theory and its Stochastic Limit, Springer Verlag (2002) -Accardi L., Obata N.: Elementary algebraic Probability, (in japanese) University of Nagoya Press (1998); 2--d Edition, Makino Shooten (2003)

Abstract

The awareness of the physical possibility of models of space, alternative with respect to the Euclidean one, begun to emerge towards the end of the 19-th century.

At the end of the 20-th century a similar awareness emerged concerning the physical possibility of models of the laws of chance alternative with respect to the classical probabilistic models (Kolmogorov model).

In geometry the mathematical construction of several non-Euclidean models of space preceded of about one century their applications in physics, which came with the theory of relativity.
In physics the opposite situation took place. In fact, while the first example of non Kolmogorov probabilistic models emerged in quantum physics approximately one century ago , at the beginning of 1900, the awareness of the fact that this new mathematical formalism reflected a new mathematical model of the laws of chance had to wait until the early 1980's.

In this long time interval the classical and the new probabilistic models were both used in the description and the interpretation of quantum phenomena and negatively interfered with each other because of the absence (for many decades) of a mathematical theory that clearly delimited the respective domains of application.
The result of this interference was the emergence of the so-called the "paradoxes of quantum theory".

For several decades there have been many different attempts to solve these paradoxes giving rise to what K. Popper baptized "the great quantum muddle": a debate which has been at the core of the philosophy of science for more than 50 years.

However these attempts have led to contradictions between the two fundamental theories of the contemporary physical: the quantum theory and the theory of the relativity.

Quantum probability identifies the reason of the emergence of non Kolmogorov models, and therefore of the so-called the paradoxes of quantum theory, in the difference between the notion of passive measurements like "reading pre-existent properties" (urn metaphor) and measurements consisting in reading "a response to an interaction" (chameleon metaphor).

The non-trivial point is that one can prove that, while the urn scheme cannot lead to empirical data outside of classic probability, response based measurements can give rise to non classical statistics.

The talk will include entirely classical examples of non classical statistics and potential applications to economic, sociological or biomedical phenomena.

Journal of Physics: Conference Series **459** (2013) 012004 doi:10.1088/1742-6596/459/1/012004

New scale classification within the representational theory of measurement

Eric Benoit

LISTIC, Universite de Savoie, B.P. 80439, 74944 Annecy le Vieux Cedex, FRANCE

E-mail: eric.benoit@univ-savoie.fr

Abstract. Within the representational theory of measurement, Stevens presented a classification of scales based on the set of empirical operations preserved by each type of scale. In this classification, the group structure of admissible transformations of each scale type were exposed. In this paper, we present recent studies that focus the scale classification on the group structure of admissible transformations. Then the classification is extended to other types of scales.

1. Introduction

Introduced by Stevens in [1] , the representational theory of measurement formalizes the concept of scale. The scales are presented as structures that define the representation process of manifestations of empirical quantities by symbols defined on a representational abstract space. During this process, called measurement, an empirical relational structure is preserved and transposed into the representational abstract space. In his seminal paper, Stevens proposed to classify scales into 4 types: nominal, ordinal, interval and ratio scales depending on the set of relations preserved during the measurement process. In parallel, Stevens remarks that for each type of scale, the set of all possible admissible transformations is a group under the composition law. In recent studies, we shown that the classification can be driven by such groups and by the subgroup relation [2]. This paper specifically develops this approach in comparison with other classifications and will propose the inclusion of new scale types.

2. Presentation of the classification

The representational theory of measurement has inspired several studies, especially for ordinal scales that represent a central interest in psychophysics [3]. The link between scale types and group structures remained unexplored until 1987, when Luce and Narens studied measurement scales on continuous spaces [4]. More recently, Narens showed the importance of an approach based on group theory to analyse the representational theory of measurement [5]. We will see that the group theory is the central pivot of the scale classification. Within the representational theory of measurement, a scale $< X, S, M, R_X, R_S >$ is defined by an empirical relational system $< X, R_X >$, a representational relational system $< S, R_S >$, and an homomorphism M from X to S that preserves the relational structure. The representation theorem induces for each type of scale the existence of a set of admissible transformations [6]. An admissible transformation is a mapping f into the representation set S that transforms a scale $< X, S, M, R_X, R_S >$ into

Table 1. Scale classification given by Stevens.

Scale	Basis empirical operation	mathematical group structure	admissible transfromation
nominal	determination of equality	permutation group	$y = f(x)$, f is a bijection
ordinal	+ determination of greater or less	isotonic group	$y = f(x)$, f is a monotonic increasing function
interval	+ determination of equality of intervals	general linear group	$y = ax + b, a > 0$
ratio	+ determination of equality of ratio	similarity group	$y = ax, a > 0$

another scale $< X, S, f \circ M, R_X, R_S >$ of same type such that:

$$\forall r_s \in R_s, n = arity(r_s), \forall x_1, \ldots, x_n \in X, r_s(M(x_1), \ldots, M(x_n)) \Leftrightarrow r_s(f(M(x_1)), \ldots, f(M(x_n))) \quad (1)$$

According that $S = M(X)$, the preceding definition is defined on the representation set.

$$\forall r_s \in R_s, n = arity(r_s), \forall s_1, \ldots, s_n \in S, r_s(s_1, \ldots, s_n) \Leftrightarrow r_s(f(s_1), \ldots, f(s_n)) \quad (2)$$

2.1. Nominal scales

In the case of a nominal scale, the representational relational structure includes only the equality relation. $R_S = \{=\}$. The admissible transformations are then mappings that respect the following constraint.

$$\forall s_1, s_2 \in S, s_1 = s_2 \Leftrightarrow f(s_1) = f(s_2) \quad (3)$$

The set of all admissible transformations of a nominal scale is then the set of bijective mappings on S. Such set defines, with the operator of function composition, the group of all permutations on S also called the Symmetric group and denoted $Sym(S)$.

2.2. Ordinal scales

With the ordinal scales the relational structure preserved during the measurement process includes also an ordering relation. Then the set of admissible transformations are relations that preserve a relational system $< S, \{=, <\} >$.

$$\forall s_1, s_2 \in S, s_1 < s_2 \Leftrightarrow f(s_1) < f(s_2) \quad (4)$$

The isotonic group cited in the seminal paper is also called order-preserving group and is the group of monotonic increasing functions on S. This group is a subgroup of $Sym(S)$ and will be denoted $OA(S)$ for order-preserving automorphisms on S.

2.3. Interval scales

Stevens explicitly defined the interval scales by their admissible transformations:

$$\forall x \in \Re, f(x) = a.x + b, a > 0 \quad (5)$$

In the seminal paper, the group of the admissible transformations of interval scales is given as the General linear group but is actually a subgroup of the affine group $GA(1,\Re)$ on $\Re$. Indeed, If the preceding constraint is compared with the definition, given in (6), of a member of $GA(1,\Re)$, it is deduced that any admissible transformation of interval scales is an affine increasing mapping. The group of the admissible transformations of interval scales is $GA(1,\Re) \cap OA(\Re)$. By definition this group is a subgroup of $OA(\Re)$.

$$\forall x \in \Re, f \in GA(1,\Re) \Leftrightarrow f(x) = a.x + b, a \neq 0 \quad (6)$$

The relations preserved by the admissible transformations are given by: (3), (4) and (7).

$$\forall s_1, s_2, s_3, s_4 \in \Re, s_1 - s_2 = s_3 - s_4 \Leftrightarrow f(s_1) - f(s_2) = f(s_3) - f(s_4) \quad (7)$$

2.4. Ratio scales

Finally, the similarity group associated in the original classification with the ratio scales is the Linear group $GL(1,\Re)$. Actually, and for the same reason than for interval scales, the set of admissible transformations of ratio scales (see (8)) is the set $GL(1,\Re) \cap OA(\Re)$. It's a group, subgroup of $GA(1,\Re) \cap OA(\Re)$. The relations preserved by the admissible transformations are: (3), (4) and (9).

$$\forall x \in \Re, f(x) = a.x, a > 0 \quad (8)$$

$$\forall s_1, s_2 \neq 0, s_3, s_4 \neq 0 \in \Re, \frac{s_1}{s_2} = \frac{s_3}{s_4} \Leftrightarrow \frac{f(s_1)}{f(s_2)} = \frac{f(s_3)}{f(s_4)} \quad (9)$$

Table 2. Scale classification updated with recent group theory.

Scale	Basis empirical operation	mathematical group structure	admissible transformation
nominal	determination of equality	Symmetric group $Sym(S)$	$y = f(x)$, f is a bijection
ordinal	+ determination of greater or less	$OA(S)$	$y = f(x)$, f is a monotonic increasing function
interval	+ determination of equality of intervals	$GA(1,\Re) \cap OA(\Re)$	$y = ax + b, a > 0$
ratio	+ determination of equality of ratio	$GL(1,\Re) \cap OA(\Re)$	$y = ax, a > 0$

3. New scale types

According to this approach, other scales introduced in the literature can be added to the classification.

The first trivial example is the scale type which set of admissible transformations is the group $T(\Re)$ of translations on $\Re$: $f(x) = x+b$. Actually, by the composition with exponential function, the group of translations is isomorphic to the group of ratios. We can conclude that this scale type is identical to a ratio scale.

Journal of Physics: Conference Series **459** (2013) 012004 doi:10.1088/1742-6596/459/1/012004

3.1. Metrical scales

In [7] Coombs introduces the metrical scales as scales that preserve a distance from the empirical space of manifestations to the representational space. In this case, the admissible transformations are bijective mapping that maintain distances: isometries. The preserved relational structure is then $< S, \{=, r_d\} >$ such that r_d is the quaternary relation:

$$\forall s_1, s_2, s_3, s_4 \in S, (s_1, s_2) r_d (s_3, s_4) \Leftrightarrow d(s_1, s_2) = d(s_3, s_4) \tag{10}$$

The admissible transformations are then bijective mappings that respect:

$$\forall s_1, s_2, s_3, s_4 \in S, d(s_1, s_2) = d(s_3, s_4) \Leftrightarrow d(f(s_1), f(s_2)) = d(f(s_3), f(s_4)) \tag{11}$$

Both isometries and scaling functions respect this constraint. A subgroup including only isometries can be defined after reinforcement of the last constraint.

$$\forall s_1, s_2 \in S, d(s_1, s_2) = \alpha \Leftrightarrow d(f(s_1), f(s_2)) = \alpha \tag{12}$$

The set of admissible transformations is then reduced to a group of isometries. As an example, a metrical scale must be used for the measurement of angle over the full circle. Indeed, in this case nor ordinal or interval or ratio scales can be used due to the lack of ordering relation. But a distance can be defined on the circle. In recent studies we proposed the concept of fuzzy scales that are actually varieties of metrical scales.

3.2. Multidimensional scales

If we consider the affine group and the linear group in higher dimensions, we obtain new multidimensional scales that are not the simple combination of several ratio scales or interval scales. Let first have a look on the linear group $GL(n, \Re)$ and on the affine group $GA(n, \Re)$.

$$f \in GL(n, \Re) \Leftrightarrow \forall x \in \Re^n, f(x) = A.x, A \in \Re^{nxn}, detA \neq 0 \tag{13}$$

$$f \in GA(n, \Re) \Leftrightarrow \forall x \in \Re^n, f(x) = A.x + B, A \in \Re^{nxn}, B \in \Re^{nxn}, detA \neq 0 \tag{14}$$

These 2 groups can be used to define the set of admissible transformation of the extension to higher dimensions of interval scales and ratio scales. In this case the set of invariant properties define a large field of mathematics that we won't open in this paper. The study of multi-dimensional scale is directly linked with the analysis of the invariant properties of the admissible transformations.

4. Conclusion

Considering the groups of admissible transformations and the *subgroup of* relation between them to classify the scales is a way to change the viewpoint on scales and to analyse new types of scales. In the new representation of scales, appear the historical branch including the order preserving scales, the branch of multi-dimensional scales, and the branch of metrical scales. These last branches are not really new but must be studied within the group theory approach. The advantage of considering scales with the group theory is to benefit from the last advances of this field. In particular the analysis of invariant properties of each scale gives the tools for the computation of data derived from measurements.

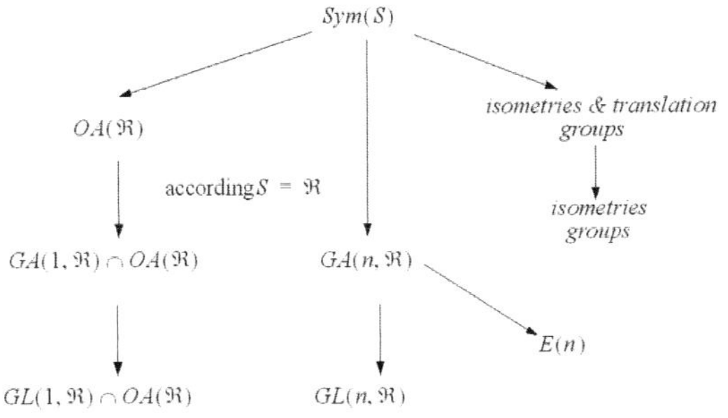

Figure 1. Classification of groups of admissible transformations.

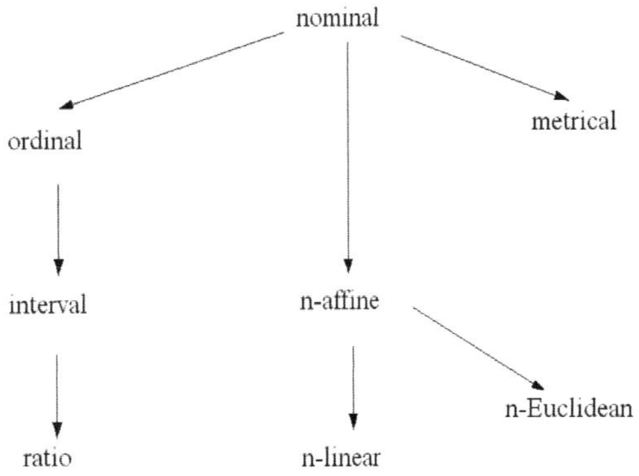

Figure 2. Proposal for a general classification of scales.

References

[1] S.S. Stevens, On the theory of the scales of measurement, *Science*, **103** (1946), 677680.
[2] E. Benoit, The role of fuzzy scales in measurement theory, *Measurement* (2013) 10.1016/j.measurement.2013.04.043.
[3] L. Finkelstein, M. S. Leaning, A review of the fundamental concepts of measurement, *Measurement* **2**(1) (1984) 25-34.
[4] R.D. Luce, L. Narens, Measurement scales on the continuum, *Science* **236** (1987) 1527-1532.
[5] L. Narens, A meaningful justification for the representational theory of measurement, *Journal of Mathematical Psychology* 46 (2002) 1527-1532.
[6] L. Finkelstein, Representation by symbol systems as an extension of the concept of measurement, *Kybernetes* **4** (1975), 215-223.
[7] C. H. Coombs, A theory of psychological scaling, Engineering Research Institute, University of Michigan, 1952.

Journal of Physics: Conference Series **459** (2013) 012005 doi:10.1088/1742-6596/459/1/012005

Probability as a logic for measurement representations

Giovanni Battista Rossi and Francesco Crenna

Laboratorio di Misure
Università degli Studi di Genova - DIME/MEC, Genova, Italy

Abstract. Measurement can be defined on the basis of the notion of empirical relation, which appears as a primitive term in the representational theory. Here a language is introduced for dealing with empirical relations and a probabilistic logic is applied to statements concerning them.

1. Introduction

According to Finkelstein, measurement can be understood as "a process of empirical, objective assignment of symbols to attribute of objects and events of the real world, in such a way as to represent them or to describe them" (1982) [4][1]. Such a representation involves the mapping of the "states" of objects, in respect of the quantity under consideration, into numbers, usually called *measure* or *measurement values*, so that empirical relations among objects correspond to similar relations among the associated values. For example, in the case of order structures and of their corresponding ordinal scales, if a and b are two objects, each expressing a specific state of the quantity under consideration, and $m(a)$ and $m(b)$ are the corresponding measure values, the claimed representational property can be concisely stated as

$$a \succsim b \Leftrightarrow m(a) \geq m(b), \tag{1}$$

where $\succsim$ denotes a weak order and $\geq$ is the familiar (simple) order among numbers. Representation results such (1) are good for clarifying the meaning of measurement, but they do not account for uncertainty [1, 10, 14]. For doing so, one, quite natural, possibility is to restate them under a logic able to express uncertainty. In classical logic a statement is either true of false. This means, for example, that, given two objects, a and b, the statement $a \succsim b$ is either true or false. If τ is a "truth" function, the result expressed in (1) can be equivalently stated as

$$\tau(a \succsim b) = \tau(m(a) \geq m(b)). \tag{2}$$

For properly expressing uncertainty, we can shift from a deterministic (classic) to a probabilistic logic, that is to a logic in which the category of truth is substituted by that of probability [12, 13]. So in the case of statement (1), we look for a reformulation as

$$P(a \succsim b) = P(m(a) \geq m(b)), \tag{3}$$

[1] References are listed in a chronological order in order to give a feeling of the historical development of the subject.

where the truth function has been replaced by the probability, P, function [11]. This approach is different from that of, say Falmagne, who identifies specific conditions for a probabilistic representation to hold true [2]. We rather look for a systematic translation of deterministic statements into their probabilistic counterparts. An approach similar to ours was rather developed, to some extent, by Regenwetter [6]. A noteworthy study on the logical aspects of probability is due to Fenstad [3]. In the following we will investigate this, new, for the field under consideration, approach (at least to some extent).

2. General probabilistic framework

Consider a finite universe A, a finite collection $\mathcal{E} = \{\mathcal{A}_1, \mathcal{A}_2, ...\mathcal{A}_M\}$ of structures over A and a probabilistic distribution over $\mathcal{E}$, $P(\mathcal{A}_i)$, such that

$$P(\mathcal{E}) = P\{\mathcal{A} \in \mathcal{E}\} = \sum_{i=1}^{M} P(\mathcal{A}_i) = 1. \tag{4}$$

Consider now any statement ϕ whose truth can be assessed on any element (structure) of $\mathcal{E}$. Then

$$P(\phi) = P\{\mathcal{A} \in \mathcal{E}|\phi\} = \sum_{\mathcal{A}_i|\phi} P(\mathcal{A}_i), \tag{5}$$

where $\mathcal{A} \in \mathcal{E}|\phi$ denotes a structure where the statement ϕ is true. Let us now apply this approach to order relations and to related order structures.

3. Probabilistic order and m-nary relations

Let us consider the statement

$$u \succsim v, \tag{6}$$

and let us discuss its truth. The symbol "$\succsim$" denotes a weak order relation, that is a relation such that for each u, v, z

- $u \succsim v \vee v \succsim u$,
- $u \succsim v \wedge v \succsim z \longrightarrow u \succsim z$.

Now the truth of the above statement may be not assessed in general but only by interpreting it in a specific context. Consider the case we have a set of objects A, some characteristic x and $\succsim$ is an empirical order relation among the elements of A, with respect to x. Suppose that one such relation exists and denote it by $\succsim_A$. Note that $\succsim_A$ may be regarded as a list of pairs of elements, $(a, b) \in A$ that satisfy $\succsim$. So we have a truth criterion: for each pair of elements $(a, b) \in A$,

$$a \succsim b \Longleftrightarrow (a, b) \in \succsim_A.$$

This "interpretation" of the above statement may be summarised by the *structure* $\mathcal{A} = (A, \succsim_A)$: so *we have a definite truth criterion if we interpret the statement in a proper well defined structure*[9][2].

Suppose now that $\succsim_A$ is not the only possible ordering on A, rather there is a finite collection of possible such orders. Then we can consider the finite collection of order structures, all referred to the same set (universe) A, that is $\mathcal{E} = \{\mathcal{A}_1, \mathcal{A}_2, ...\mathcal{A}_M\}$, where $\mathcal{A}_i = (A, \succsim_i)$. Now the previous

[2] When apparent from the context that we are referring to the realization of the relation in A we will drop the index A

Table 1. An illustrative example of a probabilistic order structure on $A = \{a, b, c\}$

Structure $\mathcal{A}_i$	Ordering $\succcurlyeq_i$	x_a	x_b	x_c	$P(\mathcal{A}_i)$
$\mathcal{A}_1$	$a \succ b \succ c$	3	2	1	0.1
$\mathcal{A}_2$	$a \succ c \succ b$	3	1	2	0.1
$\mathcal{A}_3$	$b \succ a \succ c$	2	3	1	0.0
$\mathcal{A}_4$	$b \succ c \succ a$	1	3	2	0.0
$\mathcal{A}_5$	$c \succ a \succ b$	2	1	3	0.0
$\mathcal{A}_6$	$c \succ b \succ a$	1	2	3	0.0
$\mathcal{A}_7$	$a \sim b \succ c$	2	2	1	0.0
$\mathcal{A}_8$	$a \sim c \succ b$	2	1	2	0.0
$\mathcal{A}_9$	$b \sim c \succ a$	1	2	2	0.0
$\mathcal{A}_{10}$	$a \succ b \sim c$	2	1	1	0.8
$\mathcal{A}_{11}$	$b \succ a \sim c$	1	2	1	0.0
$\mathcal{A}_{12}$	$c \succ a \sim b$	1	1	2	0.0
$\mathcal{A}_{13}$	$a \sim b \sim c$	1	1	1	0.0

truth criterion is no more applicable, since, say, $a \succcurlyeq b$ may be true in some structures and false in others. What can we do now is to associate a probability to each structure, $P(\mathcal{A}_i)$, in such a way that condition (4) above is satisfied. Then for any pair of objects, $a, b \in A$, we can assess, not the truth, but the probability of statement (6), as

$$P(a \succcurlyeq b) = P\{\mathcal{A} \in \mathcal{E} | a \succcurlyeq b\} = \sum_{\mathcal{A}_i | a \succcurlyeq b} P(\mathcal{A}_i). \tag{7}$$

In such a way it is thus possible to assess the truth of any "atomic" statement of the kind $u \succcurlyeq v$, as well as of any finite conjunction or disjunction of them. For example, the equivalence relation $a \sim b$ can be expressed by the statement $a \succcurlyeq b \wedge b \succcurlyeq a$, and its probability is provided by

$$P(a \sim b) = P(a \succcurlyeq b \wedge b \succcurlyeq a) = P\{\mathcal{A} \in \mathcal{E} | a \succcurlyeq b \wedge b \succcurlyeq a\} = \sum_{\mathcal{A}_i | a \succcurlyeq b \wedge b \succcurlyeq a} P(\mathcal{A}_i). \tag{8}$$

As an example, consider the probabilistic order in Table 1. We see, for example, that $a \succ b$ is true in all the ordering having a non null probability, and we thus obtain

$$P(a \succ b) = P(\mathcal{A}_1) + P(\mathcal{A}_2) + P(\mathcal{A}_{10}) = 1.$$

As another example, $b \succcurlyeq c$ is verified in $\mathcal{A}_1$ and in $\mathcal{A}_{10}$ and thus we have

$$P(b \succcurlyeq c) = P(\mathcal{A}_1) + P(\mathcal{A}_{10}) = 0.9$$

and so forth.

What so far expressed can be generalised, in a quite straightforward fashion, to n-nary relations, as well as to structures including more than one relation. Examples of m-nary relations[3], with $m > 1$, that occur in the representation theory are the *ternary* relation $a \sim b \circ c$ and the *quaternary* relation $\Delta_{ab} \succcurlyeq \Delta_{cd}$ [5, 7, 8]. So, if $R(u_1, ...u_m)$ is a m-nary relation and

[3] In what immediately follows the symbol m denotes a natural number. Do not confuse it with the measure function, also denoted by m. We apologise for this ambiguity of symbols, but their use is so common that substituting them with others would result into unfamiliar expressions.

Table 2. An illustrative example of a probabilistic (or random) variable from A to B.

Structure	Function	$f_i(a)$	$f_i(b)$	$f_i(c)$	$P(\mathcal{A}_i)$
$\mathcal{A}_1$	f_1	1	1	1	0.1
$\mathcal{A}_2$	f_2	1	1	2	0.4
$\mathcal{A}_3$	f_3	1	2	1	0.1
$\mathcal{A}_4$	f_4	1	2	2	0.2
$\mathcal{A}_5$	f_5	2	1	1	0.0
$\mathcal{A}_6$	f_6	2	1	2	0.1
$\mathcal{A}_7$	f_7	2	2	1	0.0
$\mathcal{A}_8$	f_8	2	2	2	0.1

$\mathcal{E} = \{\mathcal{A}_1, \mathcal{A}_2, ...\mathcal{A}_M\}$ is a finite collection of structures, each having a definite m-nary relation of the kind considered, $\mathcal{A}_i = (A, R_i)$, and a probability measure P is assigned to them, then the probability that the relation R holds for the n-ple $(a_1, ...a_m) \in A^m$ is

$$P(R(a_1, ...a_m)) = P\{\mathcal{A} \in \mathcal{E} | R(a_1, ...a_m)\} = \sum_{\mathcal{A}_i | R(a_1,...a_n)} P(\mathcal{A}_i). \quad (9)$$

This approach can be extended to functions, by remembering that a function can be considered as a relation as well.

4. Probabilistic variables

Consider now the case in which B is a set of numbers, as in the example in Table 2. Then each function $f : A \to B$ provides a numerical description of A in that it associates a number, x_a, to each element a in A. If x denotes in general the number that f assigns to a generic element of A, we can say that x is a *variable*, since *it can assume different values*. Yet for each element $a \in A$ there is just one "truc" value, x_a.

Conversely, in a probabilistic environment, as we already know, there are different possible descriptions and consequently different possible values describing each element a. We can thus calculate the probability that the variable x_a assumes values in B, with the rule

$$P(x_a = i) = P\{\mathcal{A} \in \mathcal{E} | f(a) = i\}, \quad (10)$$

for each $i \in B$. This rule defines a probability distribution for the probabilistic (or random[4]) variable x_a.

5. The logic of a probabilistic representation

Consider now a set of order structures and of the related ordinal scales. Consider a collection of structures of the following kind

$$\mathcal{S} = (A \times \mathcal{M}, \succsim, m),$$

where A is a finite set of objects, $\mathcal{M}$ is the set of the first M natural numbers, $\succsim$ is a weak order on A and m is a function, $m : A \to \mathcal{M}$, such that for each $a \in A$, x_a is the image of $a \in A$ in $\mathcal{M}$ through m, and for each $a, b \in A$,

$$a \succsim b \Leftrightarrow m(a) \geq m(b) \Leftrightarrow x_a \geq x_b.$$

[4] We prefer the term "probabilistic variable" to the more common "random variable" since when probability is considered from a logical standpoint, it does not necessarily implies randomness [12, 15].

Now consider a finite collection $\mathcal{Z} = \{\mathcal{S}_1, ...\mathcal{S}_M\}$, of such structures and assign a probability $P(\mathcal{S})$ to each of them. Then,

$$P(a \succsim b) = P\{\mathcal{S} \in \mathcal{Z} | a \succsim b\},$$
$$P(m(a) \geq m(b)) = P\{\mathcal{S} \in \mathcal{Z} | m(a) \geq m(b)\}, \text{ and}$$
$$P(m(a) \geq m(b)) = P\{\mathcal{S} \in \mathcal{Z} | x_a \geq x_b\}.$$

But since, for each S,

$$a \succsim b \Leftrightarrow m(a) \geq m(b) \Leftrightarrow x_a \geq x_b.$$

we conclude that

$$P(a \succsim b) = P(m(a) \geq m(b)) = P(x_a \geq x_a).$$

Proof is thus simpler and more straightforward than presently available ones [6, 10].

6. Final remarks

We have shown how probabilistic representations in measurement can be obtained by recognising that statements concerning empirical and numerical relations can be systematically turned into probabilistic. This provide a "logical" interpretation of probability, at least in this part of measurement theory [15]. The present discussion does not include a model of the measurement process and its probabilistic expression. Such development is necessary in order to treat the propagation of measurement uncertainty throughout measurement transformations and it could be the object of future studies.

References

[1] Leaning M S and Finkelstein L 1980 *ACTA IMEKO 1979* ed G Streker (Amsterdam: Elsevier) p 73
[2] Falmagne J C 1980 *Philosophy of Science* **47** 277
[3] Fenstad J E 1980 *Studies in inductive logic and probability Vol. 2* ed R C Jeffrey (Berkeley: University of California Press) p 251
[4] Finkelstein L 1982 *Handbook of measurement science Vol. 1* ed P H Sydenham (Chichester: Wiley) p 1
[5] Narens L and Luce R D 1986 *Psychological Bulletin* **99** 166
[6] Regenwetter M 1996 *Journal of mathematical psychology* **40** 219
[7] Luce R D and Suppes P 2002 *Stevens' Handbook of Experimental Psychophysics Vol. 4* (Wiley)
[8] Hand D J 2004 *Measurement - Theory and practice* (Chichester, UK)
[9] Rigamonti G 2005 *Corso di logica* (Torino: Bollati Boringhieri)
[10] Rossi G B 2006 *Measurement* **39** 34
[11] Accardi L 2008 La matematica: problemi e teoremi ed C Bartocci and P Odifreddi (Bologna: Einaudi) p 793
[12] Costantini D 2011 *Verso una visione probabilistica del mondo* (Padova: GEM)
[13] Haenni R, Romeijn J W, Wheeler G, Williamson J 2011 *Probabilistic Logics and Probabilistic Networks* (Dordrecht: Springer)
[14] Benoit E 2011 *Uncertainty in fuzzy scales based measurements* Int. IMEKO TC7 Symp. August 31th - September 2nd Jena, Germany
[15] Rossi G B 2012 *IEEE Transactions on Instrumentation and Measurement* **61** 2097

Journal of Physics: Conference Series **459** (2013) 012006 doi:10.1088/1742-6596/459/1/012006

Intransitivity in multiple solutions of Kemeny Ranking Problem

S V Muravyov and I A Marinushkina
Department of Computer-aided Measurement Systems and Metrology
National Research Tomsk Polytechnic University
Tomsk, 634050, Russia

E-mail: muravyov@camsam.tpu.ru

Abstract – Kemeny rule is one of deeply justified ways to solve the problem allowing to find such a linear order (Kemeny ranking) of alternatives that a distance from it to the initial rankings (input preference profile) is minimal. The approach can give considerably more than one optimal solutions. The multiple solutions (output profile) can involve intransitivity of the input profile. Favorable obstacle in dealing with intransitive output profile is that the intransitive cycles are lexicographically ordered what can help when algorithmically revealing them.

1. Introduction

In this paper we deal with multidimensional ordinal measurements. Their results are consensus preference relations being discovered for given m rankings, possibly including ties, of n alternatives. A classical problem of single consensus ranking determination for m voters of n candidates has been intensively investigated as a Voting Problem in the framework of Social Choice Theory, see, e.g. [1-3]. Modern control and measurement systems frequently investigate complicated objects characterized by multiple heterogeneous properties using multisensors and distributed networking. This is just the domain where the problem could be affectively applied; see, for example [4-6]. Also promising is a development of measurement systems based on mobile agents, collective purposeful work of which could be organized using consensus ranking determination. We use here a widely known and deeply justified distance-based approach to determination of the consensus relation, so called Kemeny rule. However, it may result in not single and considerably greater amount of consensus relations. A set of the relations may include intransitivities. Although (in)transitivity of preferences is a subject of many papers (e.g. [7,8]), non-transitive Kemeny rankings were ignored by researchers. Aim of this paper is to demonstrate examples of this phenomenon and draw up some ways how to deal with it.

2. Definitions

To find the consensus ranking we use the *Kemeny rule* [9] that is shortly described as follows. Let us have m rankings on set $A = \{a_1, a_2, ..., a_n\}$ of n alternatives and the relation set $\Lambda = \{\lambda_1, \lambda_2, ..., \lambda_m\}$, where each of m rankings (also called preference relations or weak orders) $\lambda = \{a_1 \succ a_2 \succ ... \sim a_s \sim a_t \succ ... \sim a_n\}$ may include $\succ$, a strict preference relation π, and $\sim$, an equivalence relation (or tie) ν, so that $\lambda = \pi \cup \nu$. The relation set Λ we will call an *input preference profile* for the given m rankings. Let Π be a set of all $n!$ linear (strict) order relations $\succ$ on A. Each linear order corresponds to one of permutations of first n natural numbers $\mathbf{N_n}$.

Kemeny rule allows to find a consensus relation as such a linear order (Kemeny ranking) $\beta \in \Pi$ of alternatives that a *distance* $D(\beta, \Lambda)$ from β to the profile Λ is minimal, that is

$$\beta = \arg\min_{\lambda \in \Pi} D(\lambda, \Lambda). \qquad (1)$$

Journal of Physics: Conference Series **459** (2013) 012006 doi:10.1088/1742-6596/459/1/012006

The problem (1) is called *Kemeny Ranking Problem (KRP)*.

The *distance* $D(\lambda, \Lambda)$ between arbitrary ranking λ and profile Λ is defined as

$$D(\lambda, \Lambda) = \sum_{i<j} \sum_{k=1}^{m} d_{ij}^{k}, \tag{2}$$

where $d_{ij}^k = \begin{cases} 0 & \text{if } a_i^k \succ a_j^k \\ 1 & \text{if } a_i^k \sim a_j^k \\ 2 & \text{if } a_i^k \prec a_j^k \end{cases}, \quad i, j = 1, \ldots, n.$

To describe all rankings of a preference profile we use the *profile matrix P*, rows and columns of which are labeled by the alternatives' numbers, where

$$p_{ij} = \sum_{k=1}^{m} d_{ij}^{k}, \quad i, j = 1, \ldots, n. \tag{3}$$

Then the KRP statement (1) is changed to

$$\beta = \arg\min \sum_{i<j} p_{ij}, \tag{4}$$

that means the determination of such a transposition of profile matrix rows and columns that the sum of elements of its upper triangle submatrix is minimal.

2. Profile intransitivity

Though each ranking of the preference profile is transitive, the corresponding matrix P (and, hence, the preference profile) may be both *transitive*, i.e. $p_{ik} \le p_{ki}$ if $p_{ij} \le p_{ji}$ and $p_{jk} \le p_{kj}$, for all $i \ne j \ne k = 1, \ldots, n$, and *non-transitive* if the condition fails. Well known *Condorcet paradox* (Condorcet rule: if some alternative obtains a majority of votes in pair-wise contests against every other alternative, the alternative is chosen as the *winner* in the consensus ranking β; however, for some β and a_i, a_j, $a_k \in \beta$, it can be that $a_i \succ a_j$ and $a_j \succ a_k$ while $a_k \succ a_i$) just takes place if the given preference profile is non-transitive.

The following profile

$$\begin{aligned} \lambda_1&: a_1 \succ a_2 \succ a_3 \\ \lambda_2&: a_2 \succ a_3 \succ a_1 \\ \lambda_3&: a_3 \succ a_1 \succ a_2 \end{aligned} \tag{5}$$

includes an evident circular ambiguity (shortly, cycle) that leads to the Condorcet paradox if we would apply the Condorcet rule to determine a consensus ranking. In this case a Condorcet winner does not exist. The Kemeny rule being applied in this case gives three optimal solutions (with $D(\beta, \Lambda) = 8$) coinciding with rankings of the initial profile.

Thus, in this case, the initial profile intransitivity seems to produce multiple solutions in the Kemeny rules. Reasonable way to resolve the situation is to declare all three alternatives to be equivalent to each other, that is the final solution is $a_1 \sim a_2 \sim a_3$. In fact, this weak order is in minimal feasible distance, equal to 9, from all the rankings (5).

Matrix P can be characterized by a *least distance* D_{least} from its preference profile Λ to some linear order [5,10]. To find the distance a lesser element of each pair (p_{ij}, p_{ji}) should be included into, that is

$$D_{\text{least}} = \sum_{i<j} \min(p_{ij}, p_{ji}), \quad i, j = 1, \ldots, n. \tag{6}$$

It can be easily shown that if matrix P is *transitive* (this means all initial rankings are consistent), then $D_{\text{least}} = D(\beta, \Lambda)$ and D_{least} is an accessible value. An inverse proposition is also valid. Thus, for any input profile, after determination of β by Kemeny rule (4), its transitivity can be determined using the following criterion:

$$\text{matrix } P \text{ is } \begin{cases} \text{transitive} & \text{if } D_{\text{least}} = D(\beta, \Lambda) \\ \text{intransitive} & \text{if } D_{\text{least}} < D(\beta, \Lambda) \end{cases}. \tag{7}$$

3. Monte-Carlo investigation of input profile intransitivity

We have experimentally investigated intransitivity of input profiles using our branch-and-bound (B&B) algorithm described in [5] to solve the Kemeny Ranking Problem. We used Monte-Carlo approach in order to simulate profile matrices calculated for weak orders that were obtained by uniting pseudo-random strict orders and ties generated separately. Strict orders were generated using the *uniform distribution* of integers in a specified range 1, ..., n.

For each combination (m, n), where number of rankings m = 4, 5, 6, and 15 and number of ranking elements n = 10, 15 and 20, 100 individual profiles (problems) were generated (solved). Thus, totally 1500 profiles were generated each served as input of the B&B algorithm.

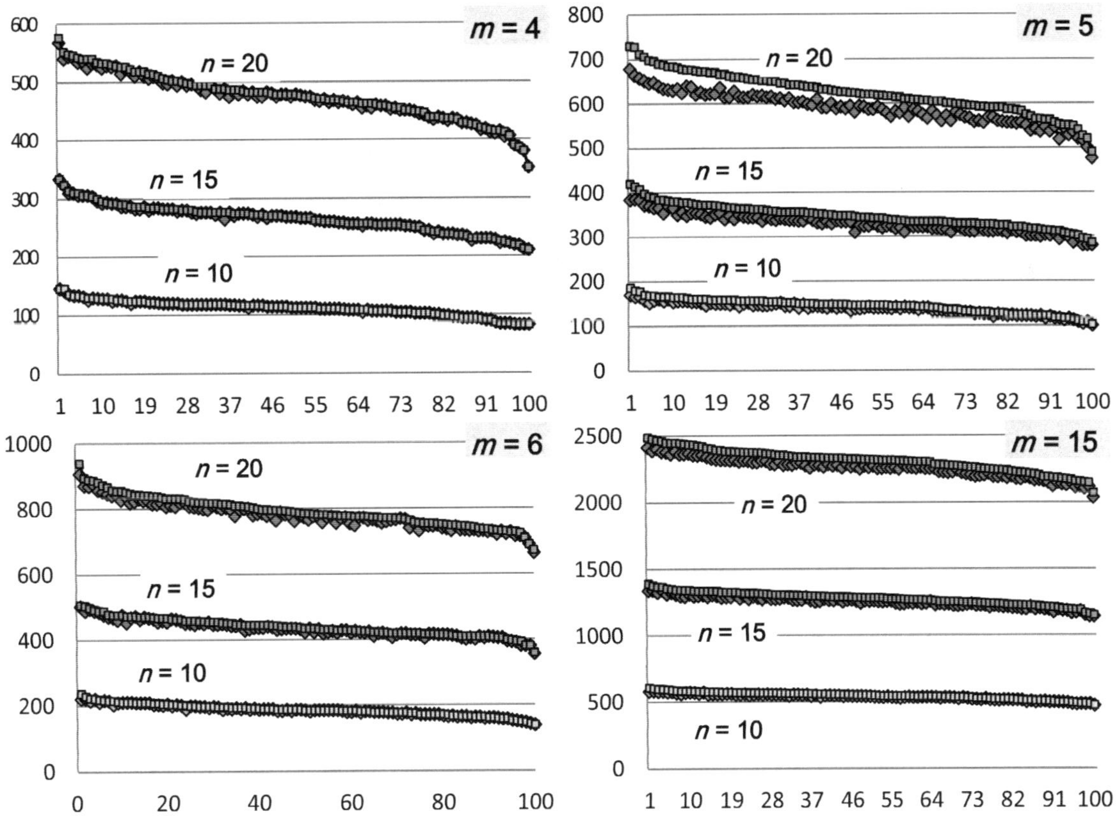

Figure 1. Distances D_{least} and $D(\beta, \Lambda)$ (vertical axis) for each of 100 individual problems (horizontal axis) for m = 4, 5, 6 and 15 and candidate numbers n = 10, 15 and 20; $D(\beta, \Lambda)$ curves are shown with boxes, D_{least} – with diamonds.

Figure 1 shows curves of distances D_{least} and $D(\beta, \Lambda)$ plotted by their values obtained for each of 100 individual problems and sorted by distance $D(\beta, \Lambda)$ in descending order.

One can see from figure 1 that, for the given n and m, profiles were frequently transitive (that is the condition $D_{least} = D(\beta, \Lambda)$ was valid) *only for even* values of m. Odd m have practically always resulted in intransitive input profiles. In all cases amount of transitive profiles decreases as the number n of alternatives increases.

Generally, the Kemeny rule produces a number N_{kem} of optimal solutions which shape *output profile* so that there can be $N_{kem} >> m$. Multiple solutions of the KRP is not a seldom phenomenon. Computing experiments [5,10-12] show that a probability of single KRP solution is in close agreement with the probability of Condorcet winner existence [13-15]. The latter decreases as the values of n increase. The same can be said of m, however, evenness of m seriously decreases chances of Condorcet winner comparatively to odd values of m [13].

Using the obtained experimental data presented in figure 1 it is possible to estimate probabilities of intransitive input profiles $P_{intr}(m, n)$ for fixed m and n, dividing a number of cases where $D_{least} \neq D(\beta, \Lambda)$ by 100. In this way the data in table 1 were calculated; they, in turn, were used to the curves in figure 2.

Table 1
Experimental estimations of probability $P_{intr}(m, n)$ of intransitive profile.

m	n		
	10	15	20
4	0.06	0.21	0.53
5	0.96	1	1
6	0.33	0.7	0,94
15	0.97	1	1

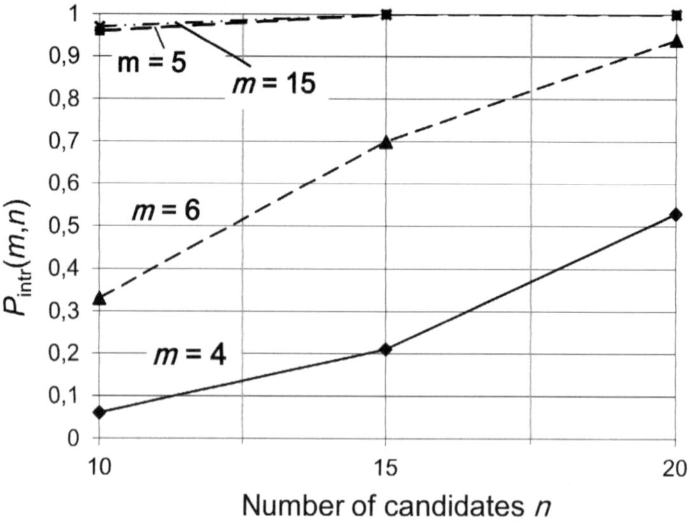

Figure 2. Experimental curves of probability $P_{intr}(m, n)$ of intransitive input profile.

One can see from figure 2 that probability $P_{intr}(m, n)$ of intransitive input profile is close to one for *odd* values of m in the range $10 \le n \le 20$. If m is even, the input profile can rather frequently be transitive (or intransitive) in dependence on particular values of m and n. On the whole, the probability increases as both the number n of alternatives and the number m of rankings increase. This observation is in satisfactory agreement with findings presented in papers [13-15].

4. Multiple solutions and their intransitivity

Necessary condition of intransitivity presence in the KRP multiple solutions is $N_{kem} \ge 3$. Also, if, in accordance with criterion (7), an initial profile is intransitive, an output profile is intransitive too but the cycles are lexicographically ordered and can be algorithmically revealed. To illustrate this let us consider two short examples.

In table 2 one can see that the initial profile is intransitive, the output profile consisting of three solutions has one cycle (shown by bold frame) attesting intransitivity. Following to approach described in Section 2 we have the final solution $a_1 \succ a_2 \sim a_4 \sim a_5 \succ a_3$.

Table 2
Examples of input and corresponding output profiles at $m = 5$, $n = 5$.

Input profile					Output profile				
1	4	5	2	~3	1	2	4	5	3
3	2	1	4	5	1	4	5	2	3
1	5	4	2	3	1	5	2	4	3
1	~5	2	3	4					
2	4	1	5	3					
$D_{least} = 30$		$D(\beta, \Lambda) = 32$					$N_{kem} = 3$		

An example of output profile having four cycles reduced in table 3. In this case the final optimal solution could be $a_1 \sim a_6 \sim a_9 \succ a_2 \sim a_{10} \succ a_8 \succ a_4 \succ a_3 \sim a_7 \sim a_5$.

Table 3
Example of output profile at $m = 5$, $n = 10$.

cycle 1→	1	6	9	2	10	8	4	3	7	5	
cycle 2→	1	9	6	2	10	8	4	5	3	7	←cycle 3
	6	1	9	2	10	8	4	7	5	3	
cycle 1→	6	9	1	10	2	8	4	3	7	5	
cycle 2→	9	6	1	10	2	8	4	5	3	7	←cycle 4
cycle 1→	9	1	6	10	2	8	4	7	5	3	
	$D_{least} = 129$			$D(\beta, \Lambda) = 131$				$N_{kem} = 6$			

5. Conclusion

Output profile of the Kemeny Ranking Problem can involve intransitivity of an input preference profile. Computing experiment had shown that most likely intransitivities of input profiles are in case of odd values of m in the range $10 \leq n \leq 20$. Favorable obstacle in dealing with intransitive output profile is that the intransitive cycles are lexicographically ordered and can be algorithmically revealed.

References

[1] Nurmi H 1999 *Voting Paradoxes and How to Deal with Them* Berlin-Heidelberg-New York: Springer-Verlag.

[2] Gehrlein W V 2003 Condorcet Efficiency and Proximity to Single-Peaked Preferences *Proc. 3rd Intern. Conf. on Logic, Game Theory and Social Choice (September 2003, Siena, Italy)* pp 282-3.

[3] Elkind E, Lang J, Saffidine A 2011 Choosing Collectively Optimal Sets of Alternatives Based on the Condorcet Criterion *Proc. 22nd Intern. Joint Conf. on Artificial Intelligence (IJCAI'11, 16–22 July 2011, Barcelona, Spain)* vol 1 pp 186-91.

[4] Muravyov S V and Savolainen V 2001 Special interpretation of formal measurement scales for the case of multiple heterogeneous properties *Measurement* **29** 209-23.

[5] Muravyov S V 2013 Ordinal measurement, preference aggregation and interlaboratory comparisons *Measurement* In press DOI: 10.1016/j.measurement.2013.04.044.

[6] Tao S, Chan M C, Muravyov S V and Tarakanov E V 2010 Pritrans: A Prioritized Converge-cast Scheme using Consensus Ranking in Wireless Sensor Networks *Proc. IEEE Sensors Applications Symp. (SAS-2010, 23-25 February 2010, Limerick, Ireland)* pp 251-6.

[7] Regenwetter M, Dana J and Davis-Stober C P 2011 Transitivity of Preferences *Psychological Review* **118** 42–56.

[8] Vidu L 2000 The minimal quota for a complete and transitive majority relation to be transitive *Group Decision and Negotiation* **9** 531-4.

[9] Kemeny J G and Snell J L 1962 *Mathematical Models in the Social Sciences* New York Ginn.

[10] Muravyov S V and Tarakanov E V 2011 Multiple solutions of an exact algorithm for determination of all Kemeny rankings: preliminary experimental results *Proc. Intern. Conf. on Instrumentation, Measurement, Circuits and Systems (ICIMCS 2011, 13-14 December 2011, Hong Kong)* ASME Press New York vol 1 pp 17-20.

[11] Muravyov S V 2012 Chaotic results of multidimensional ordinal measurements *Proc. XX IMEKO World Cong. (September 9-14, 2012, Busan, Republic of Korea).*

[12] Muravyov S V Dealing with chaotic multidimensional ordinal measurement results *Measurement* submitted.

[13] Maassen H and Bezembinder T 2002 Generating random weak orders and the probability of a Condorcet winner *Social Choice and Welfare* **19** 517-32.

[14] Jones B, Radcliff B, Taber C and Timpone R 1995 Condorcet Winners and the Paradox of Voting: Probability Calculations for Weak Preference Orders *The American Political Science Review* **89** 137-44.

[15] Van Deemen A M A 1999 The probability of the paradox of voting for weak preference orderings *Social Choice and Welfare* **16** 171-82.

Journal of Physics: Conference Series **459** (2013) 012007 doi:10.1088/1742-6596/459/1/012007

Quantification is Neither Necessary Nor Sufficient for Measurement

Luca Mari[1#], Andrew Maul[2], David Torres Irribarra[3], Mark Wilson[3]

[1] School of Engineering, Università Cattaneo – LIUC, Castellanza (VA), Italy
[2] School of Education, University of Colorado, Boulder, CO, USA
[3] Graduate School of Education, University of California, Berkeley, CA, USA

E-mail: lmari@liuc.it

Abstract. Being an infrastructural, widespread activity, measurement is laden with stereotypes. Some of these concern the role of measurement in the relation between quality and quantity. In particular, it is sometimes argued or assumed that quantification is necessary for measurement; it is also sometimes argued or assumed that quantification is sufficient for or synonymous with measurement. To assess the validity of these positions the concepts of measurement and quantitative evaluation should be independently defined and their relationship analyzed. We contend that the defining characteristic of measurement should be the structure of the process, not a feature of its results. Under this perspective, quantitative evaluation is neither sufficient nor necessary for measurement.

1. Introduction

Being an infrastructural, widespread activity, performed by human beings for millennia, measurement is laden with stereotypes, some of which concern the role of measurement in the relation between quality and quantity (or: between qualitative and quantitative properties). As an authoritative source, consider, for example: "The primary aim of a given theory of measurement is to show in a precise fashion how to pass from qualitative observations to the quantitative assertions needed for more elaborate theoretical stages of science." [1]. Contained in this statement are two basic suppositions. First, qualities such as beauty and wisdom are basic elements of human knowledge, but only on quantities such as length and weight can objective information be obtained, and only on quantity values can mathematical equations be computed; thus, the transition from qualities to quantities is a worthy endeavor. Second, measurement is the privileged tool to achieve the desired quantitative representations. The path leading to the recognition that a property is in fact a quantity, or can be represented as a quantity (that quantities are particular properties is a position accepted by the International Vocabulary of Metrology (VIM) [2]), is an evolutionary one, implying an advancement of knowledge related to the scale type on which the property can be evaluated [3] and therefore of the semantics of the property itself [4]. Against this background, and given the further acknowledgment that measurement is the tool that historically enabled physics to represent properties quantitatively, the conclusion is sometimes reached that there is a strict tie between quantities and measurement. Such a relation is differently instanced, as follows.

In terms of measurement as a process: each quantity evaluation is a measurement (i.e., the only way to assign a value to a quantity is by means of measurement: quantitative evaluation is sufficient

---

[#] To whom any correspondence should be addressed.
One of the authors is a member of the Joint Committee on Guides in Metrology (JCGM) Working Group 2 (VIM). The opinion expressed in this paper does not necessarily represent the view of this Working Group.

for measurement); alternatively, each measurement is a quantity evaluation (i.e., only quantities can be measured: quantitative evaluation is necessary for measurement).

In terms of the ontology of the measurability of properties: each quantity is measurable (i.e., for a property to be a quantity is a sufficient condition for it to be measurable); alternatively, each measurable property is a quantity (i.e., if a property is not quantitative, then it cannot be measured: for a property to be a quantity is a necessary condition for it to be measurable).

These are the positions we intend to challenge here.

2. Framing measurement

Our claim is that measurement is not a natural entity, existing independently of human beings and discovered by them, but it is a designed-on-purpose process. Hence, some conventionality in the definition of measurement is unavoidable, and at least partially different concepts of measurement can be maintained in different contexts. On the other hand, one may safely assume that, for example, measurement results and subjective opinions are not the same. The relation between measurement and quantitative evaluation is a critical component of an appropriate characterization of measurement. It can be studied along three orthogonal dimensions (for the sake of simplicity, each of them will be considered as a Boolean feature). Is the evaluated property: 1. a quantity? 2. evaluated as a quantity? 3. evaluated by means of a process conveying good-quality information (in a sense to be specified)?

Issue 1 relates to claim on what the property *is*, and is thus *ontological*: is *to be* a quantity sufficient, and/or necessary, to be measurable? A positive answer would rule out, in particular, nominal properties as potentially measurable entities. Issue 2 acknowledges that a property that is a quantity might be nevertheless evaluated in a non-quantitative way. For example the length of rigid objects may be evaluated according to the condition "is it longer than 1 m and shorter than 2 m?", resulting in a pure, binary classification, even though at the ontological level length remains (interpreted as) a quantity. This issue is then *operative*, and adds a specification to the ontological one: is *to be evaluated* quantitatively sufficient, and/or necessary, for measurement? Issue 3 focuses on the information obtained in the evaluation, and questions whether measurement should be characterized in terms of desirably epistemic properties, e.g., in terms of objectivity and inter-subjectivity [5, 6]. This issue is *epistemic*: is *to be evaluated so to convey objective and inter-subjective information* (in a sense to be specified) sufficient, and/or necessary, for measurement?

With the explicit position that measurement is primarily an epistemic process, we argue that the relation between measurement and quantitative evaluation is loose: *not all quantitative evaluations are measurements* and *not all measurements are quantitative evaluations*; furthermore, *being a quantity is neither sufficient nor necessary for being measurable*. Our argument develops along two complementary lines. First quantities are, historically and conceptually, tied with *measures*, not measurement, and 'measure' and 'measurement' are different concepts: the results that hold for measures should not uncritically applied also to measurement. Second, the structure of measurement is generally independent of the possible quantitative structure of the property under consideration: that only quantities are considered to be measurable is just a matter of convention. In this paper we explore the first topic, thus mainly developing the *pars destruens* of the work.

3. A historical overview

In Euclid's *Elements*, sometimes thought of "the earliest contribution to the philosophy of measurement available in the historical record" [7], the concept 'measure' (as a noun) is often exploited, whereas 'measurement' is never used, because "in the geometrical constructions employed in the Elements [...] empirical proofs by means of measurement are strictly forbidden" (from the introductory notes of [8]). This highlights that 'measurement' and 'measure' should not dealt with as synonyms: even though they are terms so entangled with daily activities and speaking that their meaning and relations are often spurious, in the scientific context the former refers specifically to the *process* of measuring, not to the structure of its input or output entities – properties and property values respectively. For example, the VIM defines 'measurement' as "the process of experimentally obtaining one or more quantity values that can reasonably be attributed to a quantity". Interestingly, the VIM avoids using "measure" as noun (but in the technical, idiomatic term "material measure") to reduce ambiguity, by preferring "measurement result" to denote the *outcome* of the process. Indeed, while "measurement" appears to have a relatively well-defined meaning even in everyday use – "the action of measuring", according to the *Oxford English Dictionary* [9] – "measure" is remarkably

Journal of Physics: Conference Series **459** (2013) 012007 doi:10.1088/1742-6596/459/1/012007

polysemic, with several distinct meanings. One of them is particularly important here: a measure is "a quantity contained in another an exact number of times; a divisor". This is indeed the definition that can be traced back to the Elements. The oft-quoted first two definitions of Book 5 are [8] "A magnitude is a part of a(nother) magnitude, the lesser of the greater, when it measures the greater" and "the greater (magnitude is) a multiple of the lesser when it is measured by the lesser". Given that the Greek verb for 'to measure' contains the root "metr-" one might conclude that there is a conceptual continuity (or even identity) between the Euclidean 'to measure' and what nowadays is the object of metrology, i.e., (physical) measurement. But let us consider another definition by Euclid, now from Book 7: "A number is part of a(nother) number, the lesser of the greater, when it measures the greater". Definitions 3.1 and 7.3 have the same structure: "An x is a part of a(nother) x, the lesser of the greater, when it measures the greater", and in both cases 'to measure' is used. But while in the first case x = magnitude, in the second case x = number ("numbers *measure* one another"), thus showing that here *'to measure' does not necessarily have an empirical connotation*. These definitions are followed by several others in which divisibility between numbers is presented in terms of their "measurability": *the Euclidean 'to measure' is actually a shortcut for 'to be (integer) part of'*.

The generalization of measures as additive functions is immediate, via the hypothesis that such parts are pairwise disjoint. Indeed, the mathematical literature proposes definitions where "the study of measures and their application to integration is known as *measure theory*." [10]. This shows the conceptual continuity (even though surely not an identity) between the traditional, Euclidean concept of measure and the contemporary measure theory, which – now the emphasis should be obvious – is *not* a measurement theory. This highlights that sentences such as "Euclid's concept [...] explained the place of numbers in measurement [...] and what it is that is being estimated in measurement" [7], and then, as apparently a direct consequence, "to understand measurement theory, it is necessary to revisit the theory of integration and, particular, Lebesgue measure theory" [11], are plainly false.

With a path including the contributions by, e.g., Newton and Maxwell, and more recently Campbell, what in the classical world was a feature of a measure – numerical ratios – has become a definitional characteristic of measurement. Even more critically, the emphasis on measurement as representation has generated a shift from the original ontological claim to much weaker representability conditions. For example, according to [1] an appropriate representation theorem "makes the theory of finite weak orderings a theory of measurement, *because of its numerical representation*" (emphasis added): even though weakly ordered entities surely do not satisfy the Euclidean conditions on a measure, they are considered measurable because they can be represented by means of quantity values. From this point of view, Michell's call to stronger conditions is appropriate: unfortunately it just misses the point.

4. Measurement and quantities

The structure of the position that identifies measurement and quantitative evaluation is transparent:

(a) physical laws are written as mathematical equations whose variables represent quantities;

(b) properties can be mutually related via physical laws only if they are quantities;

(c) only quantities are measurable.

While (a) and (b) are patent facts, they do not entail (c), which is explained in the light of the equation:

(d) quantity = (Euclidean) magnitude

together with the stipulation that:

(e) objects of measurement are (Euclidean) magnitudes.

While (d) can be assumed as a definition of 'quantity', or just the acknowledgment of a synonymy, as we have just noted (e) cannot be justified by the Euclidean tradition, which does not deal with any (empirical) "action of measuring". It seems that (e) is simply a matter of convention or tradition, and thus the question arises: why should measurement be only related to quantities?

From the Joint Committee for Guides in Metrology comes a hint that the current conception of measurement is a moving target, instead of being stably bound to the Euclidean tradition. The already mentioned current version of the VIM maintains that only quantities are measurable (hypothesis (c) above), but at the same time introduces the concept of (measurable) 'ordinal quantity', "quantity, defined by a conventional measurement procedure, for which a total ordering relation can be established, according to magnitude, with other quantities of the same kind, but for which no algebraic operations among those quantities exist", thus breaking hypothesis (d): in the Euclidean perspective, the very term "ordinal quantity" would be an oxymoron.

The position (d) + (e) hints at a possible direction towards establishing the measurability of non-physical properties, and in fact it may partially explain the activities aimed at guaranteeing (i.e., discovering or constructing, depending on one's philosophical orientation) a quantitative structure to non-physical properties, such as those typically derived from (simultaneous) "conjoint measurement" [12]. In the lack of a well-established theoretical and metrological structure [13, 14], to study the structure of a property (and, possibly, to discover that it is quantitative) is a worthwhile endeavor, but in principle this should be acknowledged as a distinct task from measuring the property itself. As an example of the critical relation between quantification and measurement, consider the sentence "a demonstration that intelligence, say, satisfies (or fails to satisfy) the axioms of conjoint *quantification* would complete the scientific task of quantification for intelligence", taken from [15] and where the original term "conjoint measurement" has been replaced by "conjoint quantification". Hence [7] correctly notes that "conjoint measurement" refers "not so much [to] a method of measurement [but to] a context within which indirect evidence for quantitative structure could be collected". Were instead the mentioned replacement accepted, the sentence would be more or less tautological. The original sentence is meaningful only if 'measurement' and 'quantification' are assumed as (i) in principle distinct concepts and (ii) actually coincident. Nevertheless, we have shown that *this position is wrong*.

5. The concept of quantity and philosophies of measurement

Up to this point we have attempted to formulate both the positions that quantification is necessary and/or sufficient for measurement, and our own comments, in as philosophically agnostic a manner as possible. However, given that these claims are often stated in terms of or influenced by particular philosophical stances, and given also that our own position on measurement inevitably involves certain philosophical commitments, it is worth exploring the implications of two major philosophical schools of thinking about measurement. Our treatment here is necessarily brief (but see [16]).

Empiricist philosophical stances generally emphasize a commitment to direct observation as the basis for knowledge. Representational measurement theory, operationalism, and the writings of S.S. Stevens may all be considered empiricist approaches to measurement, in that they characterize measurement in terms of the manner in which numerical assignments are derived from observable relations. The idea that quantification is sufficient for measurement can be formulated as a direct corollary to operationalism, insofar as operationalism is commonly interpreted as holding that measurement is nothing more than the the results of applying a particular procedure ('measurement by fiat'); on one reading, Stevens' [17] declaration that "measurement, in the broadest sense, is defined as the assignment of numerals to objects or events according to rules" is consistent with this position as well, insofar as anything other than random assignment can be considered a rule. Others have argued, and we agree, that this trivializes the concept of measurement; certainly, if one accepts that measurement is an epistemic activity, it can easily be seen that there are instances of rule-based number-assignments that have nothing to do with the acquisition of knowledge, and thus cannot count as instances of measurement. For example, if a group of persons were each to flip a coin a given number of times, they could be assigned numbers based on how many times their coin landed heads-up – and further, these numbers or transformations thereof could be shown on some criteria to possess quantitative structure [18] – but one would be hard-pressed to argue that such numerical assignments should be considered to be values of a property of persons.

By contrast, *realist* philosophical stances on scientific inquiry emphasize the commitments that (a) there is a (single) natural world, which exists regardless of what any conscious being thinks or perceives; (b) scientific claims about the world are to be taken at face value, as possessing truth-values, and (c) so interpreted, true scientific claims constitute knowledge of the world. Michell's writings are consistent with (but do not explicitly require) *metaphysical realism,* which entails the *correspondence theory of truth*: statements are true if they directly correspond to facts in the world. In terms of measurement, the implication is that whether or not a property is a quantity is entirely an empirical matter – that is, a mind-independent fact about the way the world really is.

A number of challenges have been raised to metaphysical realism. In our view, the most compelling of these invoke the observation that there are simply too many ways in which beliefs and symbols can be mapped onto the world for it to be plausible that there is a single complete and true description of the way the world really is [19]. A commitment to scientific realism need not entail the belief that the world consists of a fixed totality of mind-independent objects and their properties;

Journal of Physics: Conference Series **459** (2013) 012007 doi:10.1088/1742-6596/459/1/012007

rather, it is possible to maintain a realist view of measurement while acknowledging that *knowledge* is constructed by humans, and can be constructed in multiple ways depending in part on pragmatically determined frames of reference.

For example, we may wish to acquire knowledge about the extent to which students have mastered a set of concepts related to statistical reasoning. We may design an assessment of this property of students, based on our best available theories of learning and cognition, and we may employ statistical models such as the Rasch model [20], which treat the measured property of students as a quantity, to test hypotheses regarding both our cognitive theories and our assessment (see [21] for an example of how this might be done), and to represent differences amongst students in their degrees of mastery of statistical concepts. Doing so may (a) advance our collective understanding of how learning works, (b) communicate information about students to educators and other stakeholders in an efficient manner, and (c) suggest further avenues of exploration both for educational practice and educational psychology. Thus, in this example, information has been acquired and represented using the logic of measurement, to scientifically and practically productive ends, without it ever being necessary to assume that the measured property *is* a quantity in a mind-independent sense.

6. What, then, is measurement all about?

In this paper we have argued that the defining characteristic of measurement should be the structure of the process, rather than a particular feature of the results of the process (such as whether information can be represented on a ratio scale) or a feature of the input of the process (such as whether the property being evaluated is a quantity). Alternatively stated, measurement is an epistemic activity, and may be applied to a variety of properties of parts of the natural world. In our view, when one says that scientists are engaging in 'measurement activities', one is saying they are attempting to develop methods of obtaining objective and inter-subjective information about instances of a property. It may turn out that the property under investigation is truly quantitative (or can be treated as quantitative within a specified frame of reference and according to the available knowledge of it), in which case the particular method of obtaining objective and inter-subjective information about the property may indeed be to discover ratios of magnitudes of quantity relative to a standard unit. However, this may not be known from the outset (for example, it was only discovered after centuries of study that temperature was a quantitative property [15]; are we then to say that all pre-1760 work on discovering differences in temperatures do not count, *a priori* and independently of how they were performed, as measurement activities?); further, it may turn out that the property is not quantitative at all, and in our view it seems arbitrary and unnecessary to then disallow the use of the term 'measurement' when referring to the study of that property.

We certainly agree with Michell [20] that the *a priori* assumption that a given property is quantitative is problematic (one might even say 'pathological'). Part of the point of measurement activities, in our view, is to *test* hypotheses about properties, and in so doing learn more about how facts about them can be (accurately *and* usefully) represented. There are two points on which we depart from Michell: (a) we do not think that *quantitivity* is necessarily a mind-independent feature of properties, and (b) we do not think that *quantity* – mind-independent or not – should be considered a necessary condition for measurement.

As argued previously, empiricism is motivated by the intuition that the preferred method of acquiring knowledge is through observation and experience, while realism is motivated by the intuition that that scientific inquiry seeks to gain knowledge about a natural world. Neither of these intuitions contradicts the other. An example of a philosophic framework that is consistent with both is found in Putnam's recent (e.g., [19]) writings on *pragmatic realism*, which acknowledge that conceptual relativity is not at odds with realism, but rather, it is the interface between the world and the rich fabric of our concepts and linguistic schemes that jointly determines what we see. On this account, the existence of natural reality is not denied, but neither is it seen as directly presented to our senses; instead, our various substantive and methodological models and theories and pragmatic concerns cause us to organize, prioritize, and make sense of experience in a particular way. Further, the connection between natural reality and the outcomes of a measurement procedure is not in itself compromised by the fact that we choose to privilege certain contrast classes, levels of explanation, methods of summarization, and modes of description.

Measurement is a complex and challenging endeavor. In our view, the claims that quantification is sufficient or necessary for measurement both trivialize the concept of measurement: the former position in effect denies that measurement is an epistemic activity, and the latter arbitrarily ties an empirical activity (measurement) to a specific mathematical concept (Euclidean magnitude). We have argued, instead, that locating the defining characteristics of measurement in the structure of the activity – rather than a specific feature of its subject matter or output – pays respect to the fact that measurement is an infrastructural, dynamic, designed-on-purpose epistemic activity.

References

[1] P. Suppes, Representation and invariance of scientific structures, CSLI Publications, 2002.
[2] JGCM 200:2012, International Vocabulary of Metrology – Basic and General Concepts and Associated Terms (VIM), Joint Committee for Guides in Metrology, 2012 (2008 editions with minor corrections), http://www.bipm.org/en/publications/guides/vim.html.
[3] A. Giordani, L. Mari, Property evaluation types, Measurement, 45, 437–452, 2012.
[4] L. Mari, Notes towards a qualitative analysis of information in measurement results, Measurement, 25, 3, 183–192, 1999.
[5] A. Frigerio, A. Giordani, L. Mari, Outline of a general model of measurement, Synthese, 175, 123–149, 2010.
[6] L. Mari, P. Carbone, D. Petri, Measurement fundamentals: A pragmatic view, IEEE Trans. Instr. Meas, 61, 8, 2107–2115, 2012.
[7] J. Michell, The logic of measurement: A realist overview, Measurement, 38, 285–294, 2005.
[8] Euclid's Elements of geometry, the Greek text of J.L. Heiberg (1883–1885) edited, and provided with a modern English translation, by Richard Fitzpatrick, 2008, ISBN 978-0-6151-7984-1, http://farside.ph.utexas.edu/euclid.html.
[9] Oxford Dictionaries, Oxford University Press, 2010, http://oxforddictionaries.com.
[10] A. Cortzen, E.W. Weisstein, Measure, in: MathWorld – A Wolfram Web Resource, http://mathworld.wolfram.com/Measure.html.
[11] K. Sawyer, H. Sankey, R. Lombardo, Measurability invariance, continuity and a portfolio representation, Measurement, 46, 89–96, 2013.
[12] R.D. Luce, J.W. Tukey, Simultaneous conjoint measurement: A new type of fundamental measurement, J. Mat. Psych. 1, 1–27, 1964.
[13] L. Finkelstein, Widely, strongly and weakly defined measurement, Measurement, 34, 39–48, 2003.
[14] L. Finkelstein, Problems of measurement in soft systems, Measurement, 38, 267–274, 2005.
[15] D. Sherry, Thermoscopes, thermometers, and the foundations of measurement, Studies in History and Philosophy of Science, 42, 509–524, 2011.
[16] A. Maul, D. Torres Irribarra, M. Wilson, On the conceptual foundations of psychological measurement. Paper presented at Dimensions of Measurement, Bielefeld, Germany, March 14-16, 2013.
[17] S.S. Stevens, On the theory of scales of measurement, Science, 103, 2684, 677–680, 1946.
[18] R. Wood, Fitting the Rasch model—A heady tale. British Journal of Mathematical and Statistical Psychology, 31, 27-32, 1978.
[19] H. Putnam. The threefold cord: Mind, body, and world, Columbia University Press, 1999.
[20] G. Rasch. Probabilistic models for some intelligence and attainment tests, Danish Institute for Educational Research, 1960.
[21] M. Wilson, Constructing measures: An item response theory approach, Routledge Academic, 2004.
[22] J. Michell, Measurement in psychology: A critical history of a methodological concept, Cambridge University Press, 2005.

IMEKO 2013 TC1 + TC7 + TC13 IOP Publishing
Journal of Physics: Conference Series **459** (2013) 012008 doi:10.1088/1742-6596/459/1/012008

On the conceptual foundations of psychological measurement

Andrew Maul[1], Mark Wilson[2] ,David Torres Irribarra[2]

[1] School of Education, University of Colorado, Boulder, CO, USA
[2] Graduate School of Education, University of California, Berkeley, CA, USA

E-mail: andrew.maul@colorado.edu

Abstract. Measurement has long been an important element of epistemology in the physical sciences and natural philosophy. More recently, the psychological sciences have developed a variety of techniques that purport to be instances of measurement as well. However, it is not clear how the understanding of measurement invoked in psychological science applications accords with the understanding of measurement found in other scientific disciplines. A sharper focus on conceptual clarity and coherence across the psychological and physical sciences has the potential to add a great deal to efforts to improve such practices. In this paper, we argue that it is possible to formulate a philosophically coherent account of how measurement works in both the physical and the human sciences.

1. Introduction

Measurement has long been an important and prominent concept in the physical sciences and natural philosophy. More recently, the psychological sciences have developed a variety of techniques that purport to be instances of measurement as well. However, it is not clear how the way in which the concept of measurement is understood in these disciplines accords with the way it is understood in science more generally, or by the professional and general publics.

In recent years, a number of scholars (e.g., [1-5]) have subjected the conceptual and philosophical foundations of psychological measurement to vigorous investigation and critique. On one reading, this body of work seems to suggest that the way in which measurement is understood by psychological scientists is entirely dissimilar to the way in which it is understood by physical scientists and philosophers of science, and, by the criteria of the latter group, there may not have yet been any instances of successful measurement of psychological attributes, or at least no such instances that can be described in a philosophically coherent manner.

In our experience, psychological scientists who are aware of this body of work tend to react to it one of two distinct ways. The first is to accept the premise that there is but one definition of measurement throughout all of science, and that the activities and conceptual vocabulary of psychological scientists are out of touch with this definition. The second is to deny that the concept of measurement has or needs to have a consistent definition across scientific disciplines, and thus conclude that differences in understandings of measurement are not necessarily problematic. In the first case, the word 'measurement', when used by psychological scientists, would be a metaphor at best and a conceptual error at worst; in the second case, the word would be merely a homonym. We argue that it is not necessary for psychological scientists to react in either of these ways. Instead, we claim that it is possible to formulate a philosophically coherent account of how psychological measurement could work, in a manner consistent with mainstream positions in philosophy of science.

The structure of the paper is as follows. First, we review the ways in which three major intellectual strands have influenced thinking about both psychological measurement: first, *empiricism,* second, *pragmatism*, and third, *scientific realism*. We then argue that it is possible to formulate a coherent account of how measurement works that is consistent with certain important and relevant motivations of each of these strands of thought, but that also avoids pitfalls and contradictions associated with more severe formulations of each of them; furthermore, such a new formulation clarifies that there are foundational ideas about measurement that apply in both physical and psychological scientific disciplines, and thus that different fields' conceptions of measurement need not and should not be thought of as incompatible.

2. Empiricism and psychological measurement

A wide range of philosophical positions falls under the broad heading of *empiricism*; they share in common a commitment to direct observation as the basis for knowledge. In the twentieth century, the movement known as *logical positivism* synthesized many ideas from classical empiricism along with then-current advances in the philosophy of language and mathematics. Logical positivism was associated with the position that statements regarding unobservable (theoretical) entities or forces should only be regarded as meaningful if such statements can be linked to observations in a clear and consistent manner. In the fledging psychological sciences, *behaviorism* captured many of the same intuitions as those behind positivism. At this time, the concept of the human mind was regarded as too metaphysical and unobservable to be a proper object of scientific inquiry.

The positivist project is now widely regarded to have failed, and behaviorism as a logical doctrine has been abandoned [6]; modern philosophies of science typically include a much greater degree of acceptance of the inclusion of unobservable phenomena—such as the human mind—in scientific theories. However, positivism and behaviorism have left a significant legacy on methodological thinking in the psychological sciences [6, 7], including thinking about measurement. Notably, two major strands of thinking about measurement emerged in the early-to-mid 20th century that are consistent with many positivist principles. The first is *operationalism* [8], which is characterized by the stance that the meaning of any theoretical concept is exhausted by the operations undertaken to measure instances of the concept. The second is *representational measurement theory* (RMT; [9]), which is characterized by the stance that measurement is the construction of morphisms between empirical and numerical relations. The most widely-used definition of measurement in psychology, namely Stevens' view that "measurement, in the broadest sense, is defined as the assignment of numerals to objects or events according to rules" ([10], p.667), is consistent with the principles of both operationalism and RMT.

Operationalism had a strong influence on psychology through the influence of early behaviourists [7] such as Boring [11] and his student Stevens [10]. One of the reasons for this is surely the difficulty of precisely defining psychological attributes; stating that 'intelligence is what the tests test' ([11], p.35) neatly sidesteps the issue, and also gives at least the appearance of rigor and objectivity by anchoring abstract ideas in observables.

Operationalism has been almost uniformly rejected as irreconcilable with general scientific practice and vocabulary [6]. To see why, note that operationalism has the consequence that each unique set of operations must be associated with a distinct theoretical term; thus, for example, the outcome of the application of an alcohol thermometer and the outcome of the application of a mercury thermometer cannot refer to the same theoretical property. Additionally, the concept of measurement error—a central concept in psychological measurement—is ill-fitting with operationalism: if the results of applying a procedure are by definition a measurement of the theoretical term, what is there to be in error about?

Representational measurement theory shares in common with operationalism an empiricist emphasis on the exhaustion of the meaning of measurement claims by observable facts. More specifically, RMT holds that to measure is to construct a representation of an empirical relational system via a numerical relational system. On this view, the starting point for measurement is the determination of empirical relations amongst objects (e.g., X is greater than Y and less than Z). This requires that empirical relations be directly observable, or "identifiable" ([12], p.7), though it is not always obvious what this means (c.f. [1, 3]).

In its unadorned form, RMT suffers from many of the same weaknesses as operationalism. RMT cannot account for the role of measurement error except via the introduction of metaphysics [3], and again conflates ontology and epistemology by denying that a measured attribute has an existence beyond the manner in which (relations between) instances of it are observed. Thus, it is very difficult to make sense of how RMT accords with both scientific and lay discourse about measurement and knowledge acquisition more generally [1].

Following the collapse of logical positivism and an associated general retreat from extreme forms of empiricism, many scholars became increasingly willing to accept that the interpretation of concepts like temperature and knowledge outrun their associated measurement procedures or observable relations.

3. Pragmatism and psychological measurement

Philosophically *pragmatist* orientations generally share in common an orientation towards action (i.e., practice) and utility: "The pragmatic method ... is to try to interpret each notion by tracing its respective practical consequences. What difference would it practically make to anyone if this notion rather than that notion were true? If no practical difference whatever can be traced, then the alternatives mean practically the same thing, and all dispute is idle" ([12], p.14).

From the pragmatist's perspective, the purpose of beliefs and theories is to facilitate successful engagement with the world. Thus, a pragmatic orientation towards measurement might de-emphasize concerns such as the ontology of the measured attribute or the manner in which numerical assignments are formally constructed out of empirical relations, and ask instead whether the results of the measurement procedure can be usefully applied. The concept of 'usefulness' is, of course, relative to the motivations and values of individuals and society as a whole, and thus as such motivations and values change, so might the usefulness of the measure.

Psychological and educational tests and assessments are generally developed with a certain purpose or set or purposes in mind and are deeply embedded in a complex social structure; thus, a focus on the practical consequences of test use seems sensible. Such a focus is reflected in, for example, the concept of *validity* (in psychological testing contexts), crucial aspects of which are defined with reference to the actions taken on the basis of testing procedures as well as the direct interpretation of the results of such procedures (e.g., [13]).

However, one possible danger of the emphasis on the uses of tests and the consequences of such use is that the concept of *measurement*, as a knowledge-acquisition enterprise, can get buried. Testing and measurement are not the same thing, and testing procedures that are successful by many important pragmatic criteria without any measurement taking place (for example, when they are used as devices for prediction, as formalizations of values, as social signals, and as deterrents of certain behaviors or prods to certain forms of action); however, it seems common in the psychological sciences to speak of testing and measurement as if they were synonymous [14]. Moreover, the lack of a distinction between these concepts makes it more difficult to recognize when genuine claims about the measurement of psychological attributes are implicit in larger claims about the use of tests. For example, it may be claimed that it is appropriate to use the results of educational tests to guide decisions about the retention or firing of teachers. It is very difficult to make sense of this claim without the additional claims that (a) educational tests measure important attributes of students, (b) teachers have causal potency over these student attributes, and so (c) educational tests can be used to measure the effectiveness of teachers. Articulation and evaluation of such measurement claims are completely missing from most discussions of the 'validity' of the use of educational tests for the purposes of teacher evaluation (e.g., [15]).

Thus, a great deal of thinking about measurement in psychology is characterized by the views that tests should be evaluated with reference to their intended purposes, and that testing is synonymous

Journal of Physics: Conference Series **459** (2013) 012008 doi:10.1088/1742-6596/459/1/012008

with measurement. The first view we take to be quite reasonable; the second, however, is clearly false, and surely contributes both to the seeming incompatibility of conceptions of measurement in the physical and psychological sciences.

4. Realism and psychological measurement

As with empiricism and pragmatism, there are a broad range of positions that fall under the broad heading of *scientific realism;* they share in common the commitments that (a) there is a (single) natural world, which exists regardless of what any conscious being thinks or perceives (the 'metaphysical commitment'); (b) scientific claims about the world are to be taken at face value, as possessing truth-values (the 'semantic commitment'), and (c) so interpreted, true scientific claims constitute knowledge of the world (the 'epistemological commitment').

Michell has argued that the 'classical' understanding of the concept of measurement—the discovery of ratios of magnitudes of quantity—entails realism. More specifically, to the extent that "scientists see themselves as investigating independently existing natural systems and see their theories and hypotheses as attempts to capture something of the structure and ways of working of such systems" [2, p.286] the common understanding of the act of scientific discovery—and, *a fortiori*, measurement—invokes all three of the aforementioned realist commitments.

Michell's account is consistent with (but does not explicitly require) *metaphysical realism*. Metaphysical realism entails the *correspondence theory of truth* (the position that statements are true if they directly correspond to facts in the world), and implies that scientific objectivity depends on the subject matter of science itself being mind-independent, or *ontologically objective* [16]. In the context of psychological measurement, the implication that measured attributes of objects must possess mind-independent existence in order to be measurable is often interpreted as implying that they must possess physical (perhaps specifically neurophysiological) identity, and in some contexts a genetically-determined biological basis for variation in the attribute.

Further, for attributes to be measureable in the relevant population, there must be differences in the ways in which the attribute is instantiated in the relevant population. Whether these differences must themselves posses a certain sort of structure is a point of some contention in the literature. Michell argues that, by definition, measurement requires that the measured attribute possess a quantitative structure; other sources do not impose this restriction: for example, in the International Vocabulary of Metrology [17] it is stated that ordinal attributes are measurable, and some realists (e.g., [3]) use the concept of measurement in reference to attributes with purely nominal differences, such as gender.

A number of challenges have been raised to metaphysical realism. In our view, the most compelling of these invoke the observation that there are simply too many ways in which beliefs and symbols can be mapped onto the world for it to be plausible that there is a single complete and true description of the way the world really is (e.g, [18]). A commitment to scientific realism need not entail the belief that the world consists of a fixed totality of mind-independent objects and their properties; rather, it is possible to maintain a realist view of measurement while acknowledging that *knowledge* is constructed by humans, and can be constructed in multiple ways depending the pragmatically determined frames of reference.

5. Synthesis: A pragmatic-realist view of measurement

Empiricism is motivated by the intuition that the preferred method of acquiring knowledge is through observation and experience. Pragmatism is motivated by the intuition that theories are made useful via their links with practice. Realism is motivated by the intuition that that scientific inquiry seeks to gain knowledge about a natural world. None of these intuitions contradicts the others; however, severe formulations of each position may wind up saying something false by denying the correct intuitions of the other views. In our view, extreme proponents of operationalism and representational measurement theory overreach in denying that attributes exist independently of observations; extreme advocates of pragmatism overreach in denying that epistemic tools such as measurement seek to gain knowledge of a natural world; and finally, extreme supporters of metaphysical realism overreach in denying that knowledge is constructed by humans, and that it can be constructed in multiple correct ways depending on the observer's pragmatic concerns.

An example of a philosophic framework that is consistent with each of these intuitions is found in Putnam's recent writings on *pragmatic realism* [18], which acknowledge that conceptual relativity is not at odds with realism, but rather, "to use a Wittgensteinian idiom, *seeing* is always *seeing as*, and it is the interface between the world and the rich fabric of our concepts that jointly determines what we see." Thus realism does not necessarily entail a correspondence view of truth, nor does it deny that our conceptual schemes, models, and linguistic frameworks actively shape our experience of the world and frame our knowledge of it. Our application of the pragmatic-realist view of measurement is illustrated in Figure 1. The dotted boxes show how severe formulations of each of the aforementioned philosophical stances see a smaller part of a bigger picture. In this account, the existence of natural reality is not denied, but neither is it seen as directly presented to our senses; instead, our various substantive and methodological theories and pragmatic concerns cause us to organize and prioritize experience in a particular way.

Figure 1: A pragmatic-realist view of measurement

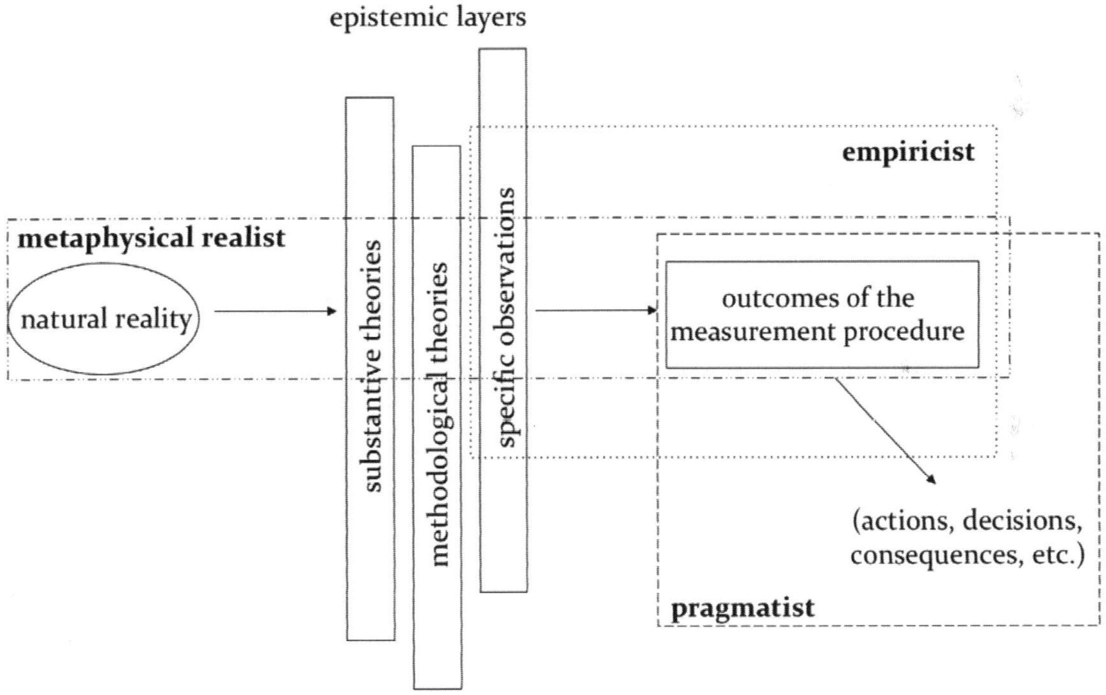

With regards to psychological attributes, contrary to the claims of metaphysical realism, it is not necessary for such attributes to be ontologically objective for them to be real components of the natural world. Elsewhere [19], we have discussed how the distinction between ontological and epistemic subjectivity and objectivity [16], and the recognition of the existence of intentionality-dependence of objects and attributes of objects, permit the formulation of a coherent realist account of the ontology of psychological attributes. Briefly, psychological attributes can (a) involve conscious phenomena with subjective first-person ontology, and (b) have conceptual boundaries delineated by contextually and pragmatically-driven frames of reference, rather than being natural kinds (or attributes) in the classic sense. Further, the connection between natural reality and the outcomes of a measurement procedure is not in itself compromised by the fact that we choose to privilege certain contrast classes, levels of explanation, methods of summarization, and modes of description. Notably, scientific models (including statistical models) serve precisely the purpose of organizing experience. This implies that scientists have the responsibility to be aware of and acknowledge the role that their conceptual frameworks play in shaping their investigations, and to explicate and empirically investigate the hypothesized connections between the psychological realities under investigation and the outcomes of purported measurement procedures.

We think the account of measurement given here shows that measurement *is* concerned with gaining knowledge about the real world, that it *is* an empirical process, and that it *is* shaped by the pragmatic concerns of the agents involved. Further, there is nothing logically distinct about the structure of this process when applied in the psychological sciences and the physical sciences.

References

[1] J. Michell, An introduction to the logic of psychological measurement. Lawrence Erlbaum Associates, 1990.
[2] J. Michell, The logic of measurement: A realist overview, Measurement, 38, 285–294, 2005.
[3] D. Borsboom, Measuring the mind: Conceptual issues in contemporary psychometrics. Cambridge University Press, 2005.
[4] G. Trendler, Measurement theory, psychology and the revolution that cannot happen. Theory & Psychology, 19, 579-599, 2009.
[5] N. Cliff, Abstract measurement theory and the revolution that never happened. Psychological Science, 1992.
[6] M. Bickhard, The tragedy of operationalism. Theory and Psychology, 11, 35–44, 2001
[7] C. Green, Of immortal mythological beasts: Operationism in psychology. Theory and Psychology, 2, 291–320, 1992.
[8] P. Bridgman, The logic of modern physics. New York: Macmillan, 1927.
[9] D.H. Krantz, R. D. Luce, P. Suppes, A. Tversky. Foundations of measurement volume I: additive and polynomial representations. Vol. 1. Dover Publications, 2006.
[10] S.S. Stevens, The operational basis of psychology. The American Journal of Psychology, 1935.
[11] E. Boring, Intelligence as the tests test it. New Republic, 35-37, 1923.
[12] W. James, Pragmatism: a new name for some old ways of thinking. Cambridge MA: Harvard University Press, 1907.
[13] S. Messick, Validity. In R. L. Linn (Ed.), Educational Measurement (3rd ed., pp.13-103). New York: American Council on Education/Macmillian, 1989.
[14] D. Borsboom, Educational measurement, 4th edition: Book review. Structural equation modeling, 16, 702-711, 2009.
[15] L. Shepard, Evaluating test-based teacher evaluation: Validity as a theory-of-action framework. Applied Measurement in Education (forthcoming).
[16] J. Searle, The rediscovery of the mind. MIT Press, 1992.
[17] JGCM 200:2012, International Vocabulary of Metrology – Basic and General Concepts and Associated Terms (VIM), Joint Committee for Guides in Metrology, 2012 (2008 editions with minor corrections), http://www.bipm.org/en/publications/guides/vim.html.
[18] H. Putnam. The threefold cord: Mind, body, and world. Columbia University Press, 1999.
[19] A. Maul, The ontology of psychological attributes. Paper presented at the annual meeting of the National Council for Measurement in Education, Vancouver, April 13-16, 2012.

Aligning physical elements with persons' attitude: an approach using Rasch measurement theory

F R Camargo[1] **and B Henson**

School of Mechanical Engineering, University of Leeds
Woodhouse Lane, Leeds, UK, LS2 9JT

[1]E-mail: mnfrc@leeds.ac.uk

Abstract. Affective engineering uses mathematical models to convert the information obtained from persons' attitude to physical elements into an ergonomic design. However, applications in the domain have not in many cases met measurement assumptions. This paper proposes a novel approach based on Rasch measurement theory to overcome the problem. The research demonstrates that if data fit the model, further variables can be added to a scale. An empirical study was designed to determine the range of compliance where consumers could obtain an impression of a moisturizer cream when touching some product containers. Persons, variables and stimulus objects were parameterised independently on a linear continuum. The results showed that a calibrated scale preserves comparability although incorporating further variables.

1. Introduction

Research in affective engineering has focused on converting the information obtained from people's feelings (e.g., emotions, feelings, sentiments and moods) stimulated by the physical properties of materials or product features (e.g., roughness, temperature, odour and colour) into an improved design. However, current methods in the domain have not in many cases met measurement assumptions. The relation amongst variables in a measurement structure ought, in this sense, to present additivity, invariant comparisons and a constant unit [1]. Some sources of violation to those assumptions have been associated with linguistic ambiguity, redundancy, misinterpretation, variance emerged from bias, reliability of qualitative dimensions and the difficulties of assuming equal intervals in categorical scales [2]. Additionally, some approaches have not embodied metrological rules, such as those of traceability and variance control [3]. As a consequence, one cannot infer beyond the considered sample and results from different studies cannot reliably be compared, limiting the understanding of a more general human–product interaction.

This paper proposes overcoming the problem by establishing a novel theory-driven approach in the domain of affective engineering using probabilistic models underpinned by Rasch measurement theory (RMT). The Rasch model (RM) provides theory and mechanisms to examine how well the data fit together and cooperate to define the attribute being measured. Accordingly, the RM's procedures verify whether the observations meet the assumptions necessary for quantifying the numerical validity of the data employing the tools of standard statistics.

The research demonstrates through an empirical study that if data from affective responses to physical elements fit the model, further variables can be added to the calibrated core of measurement whilst preserving comparability. The study established a scale to compare affective responses indicating different degrees of endorsement associated with the tactile sensory information when squeezing a collection of product containers with different characteristics of compliance. The aim was to determine the range of compliance where consumers could obtain an impression of a moisturizer cream. Persons, variables (i.e., statements in a questionnaire, called items henceforth) and stimulus objects were independently parameterised on a sole linear continuum. This alignment allowed the inclusion of further items in the scale, re-calibrating it through the responses from a different group of individuals.

2. Separation of parameter estimates

As a result of the additive correspondence obtained from a calibrated metric, comparisons can be made by the difference between the numbers associated with the responses, where a particular difference has the same interpretation across the scale continuum.

From RMT's perspective, separation of parameters estimates is a determinant condition for achieving invariant comparisons. The Rasch model, named after the Danish mathematician George Rasch, expresses the probability that a person will endorse an item with two-category responses (e.g., yes or no) as a logistic function of the difference between two mathematically independent parameters on a linear continuum: the persons' parameter and the items' parameter [4]

$$\Pr\{X_{ni} = 1 | \beta, \delta\} = \frac{e^{(\beta_n - \delta_i)}}{1 + e^{(\beta_n - \delta_i)}} \tag{1}$$

where $\Pr\{X_{ni} = 1 | \beta, \delta\}$ is the probability that a person n will endorse an item i, β is the persons' parameter and δ is the items' parameter. A number of models have extended the dichotomous Rasch model for accommodating items with more than two categories [5][6]. We have adapted a derivation of the Rasch model [7], namely the many-facet Rasch model (MFRM) developed by Linacre [8], which embodies one or more parameters called facets to the RM' structure. In the domain of affective engineering we have used the facet denoted stimulus, associating it with physical elements of design, such that

$$\Pr\{X_{nis} = x_{nis} | \beta, \delta, \zeta\} = \frac{\exp\left[\sum_{k=0}^{x} (\beta_n - \delta_i + \zeta_s - \tau_{isk})\right]}{\psi_{nis}} \tag{2}$$

given that $\psi_{nis} = \sum_{k=0}^{m} \exp\left[\sum_{k=0}^{x} (\beta_n - \delta_i + \zeta_s - \tau_{isk})\right]$ is a normalising factor and $\Pr\{X_{nis} = x_{nis} | \beta, \delta, \zeta\}$ is the probability of a respondent n to give a rating of $k, k \in (1,\ldots,m)$, on item i for stimulus s; β_n is the inclination of a person n to endorse the item i for stimulus s; ζ_s is the level of the attribute fulfilment for stimulus s; δ_i is the difficulty of endorsement of item i and τ_{isk} is the threshold parameter given a rate k on item i for stimulus s.

3. Empirical study - Method

3.1. Preliminary pool of statements and sampling

Initially, documented verbatim statements to express affective requirements were collected from a focus group along with other publicly available sources. Sixteen preliminary statements related to the packaging of everyday products were identified and selected according to the context of the study (Table 1). Subsequently, five available products in the market with different characteristics associated with the compliance of their containers were presented to 120 respondents. They were neither able to see the containers nor able to make contact with the product inside them. After squeezing each container participants expressed their degree of endorsement on a five-point Likert-style scale using computer-base self-report questionnaires. The metric was then calibrated. Finally, five additional items were introduced in the calibrated metric (Table 2). A different sample of 66 participants gave their ratings on the set of statements with the additional items. The same containers presented to the previous sample were used in this administration. A new calibration was then carried out and compared with the first test.

Table 1. Preliminary pool of statements for the first administration

Code	Item	Code	Item
1	The product in this container would give me a heavy, greasy film on my skin.	9	There is a lightweight cream in this container.
2	The product in this container is likely to look and smell delightful.	10	It is easy to know how much is left in the packaging.
3	I might get a bit watery product in this container.	11	The product inside this container could be sticky.
4	I feel the product in this container would hydrate my skin.	12	The product in this packaging is likely to flow easily.
5	The product in this packaging might be pricey.	13	The product in this packaging might seem more medicinal than anything else.
6	The container feels only half filled when squeezing it.	14	It is quite hard to explain the product when touching its packaging.
7	The container makes me feel like I would be buying a great product.	15	The product in this container could give me a refreshing sensation.
8	The product inside the container would spread easily.	16	The product in this packaging could be a bit boring.

Table 2. Additional statements for the second administration

Code	Item	Code	Item
17	I could get just the right amount of the product when I squeeze its container.	20	It's too soft for a creamy product.
18	I've got a pleasant touch with this container.	21	I could find no consistency in the product inside this container.
19	I feel this container as a skin care product.		

3.2. Rasch analysis

The investigation using Rasch analysis determined whether the data fitted the model. The analysis' procedures, therefore, contrasted with the practice in statistical modelling, which attempts to fit a model to data. The residuals were examined to determine the difference between the observed score and the expected score for each person for each item. Adequate fit to the model was established within the theoretical interval of ±2.50, which represents approximately 99% of the confidence interval (CI) [9].

Items were tested for misfit associated with the respondents' inconsistent use of the response categories. This involved examining whether the transitions between categories, called thresholds, were consistent, i.e., whether each response category had a point along the scale continuum where the most probable response was located. Another procedure was to examine item bias for sex and age, denoted as differential item functioning (DIF), i.e., whether a person subgroup demonstrated consistently greater inclination to endorse an item than another subgroup. Violation of the assumption of local independency, which underpins the RM, was examined through of the degree of correlation in which the response to an item determined the response to another one. In addition, test for trait independence, also called unidimensionality, determined whether after the calibration there was no significant pattern in the residuals resultant from relationships between items except random associations. The person separation index (PSI) is another indication of quality control of the calibration associated with the reliability of the metric. However, in RMT the precision of the individual estimates is emphasised and therefore, the index was solely used as an element of a comprehensive interpretation of the data set.

3.3. Measurement of the containers' compliance

The testing system to measure compliance consisted of a force platform (MiniDyn, multi-component dynamometer Type 9256C2, Kistler), an X–Z motion table (Series 1000 Cross Roller, Motion link), a steel ball of radius 10mm, a controller and a computer program. The containers were positioned between the steel ball and the force platform. The ball was pressed against the surface of each stimulus and the ball's displacement D_y with increasing load F_y was recorded. The measure of compliance was empirically taken to be the value of D_y (mm) when F_y was 3N [10].

4. Results

4.1. Calibration of the metric using the first sample

A preliminary analysis using the software package RUMM2030®, professional edition [11], indicated high residual correlations and evidence of multidimensionality in the structure. Disordered thresholds were identified in every item. The score system was then recoded into four categories and Items 1, 11, 14 and 16 were given reversed scores. Examination of the correlation matrix of person-item residuals identified items with indication of dependency. Misfit to the model was identified for Items 3, 6, 9, 10 and 13, being removed from the analysis. No DIF was observed. The remaining items fitted the model, indicating relative independence and meeting the model's assumption of unidimensionality (Table 3).

The metric presented in Figure 1 is the representation of the relative locations of all facets on the same logit scale. Person locations are plotted on the scale represented in the first column. On the top of the second column are indicated items that obtained endorsement by fewer persons. The location of stimuli on the continuum demonstrated that the container with lower compliance was located at the bottom of the scale in relation to the containers with higher compliance, indicating higher degree of endorsement to the affective attribute for the latter.

Table 3 - Fit statistics of the scale for the first sample.

Stimulus	Location	*SE*	Fit-residual	Items	Mean location	*SE*	Fit-residual	χ^2	df	*p*	PSI
St1	0.03	0.14	0.04	1	-0.04	0.14	0.37	121.64	110	0.21	0.7
St2	-0.07	0.13	0.07	2	-0.25	0.15	-0.08				
St3	0.34	0.15	-0.04	4	0.00	0.13	-0.06				
St4	0.28	0.15	0.09	5	-0.22	0.15	-0.03				
St5	-0.58	0.14	0.27	7	0.04	0.15	0.02				
				8	-0.50	0.13	0.04				
				11	0.43	0.14	0.31				
				12	-0.31	0.14	0.12				
				14	0.27	0.13	0.44				
				15	-0.02	0.14	-0.05				
				16	0.60	0.16	-0.06				

4.2. Calibration of the metric with additional items using the second sample

The calibrated items from the first administration were used as the core of a second calibration with the five incorporated items. In a preliminary analysis significant item-trait interaction and high residual correlations were identified, giving evidence of misfit to the model. Items were then rescored into four categories and reversed for Items 20 and 21. Items 18 and 20 were removed from the preliminary set. After calibration, a non-significant item-trait interaction (p=0.96) indicated invariance across the structure (Table 4). Figure 2 is the representation of the relative locations of the calibrated 14-item set.

Table 4 - Fit statistics of the scale for the second sample.

Stimulus	Location	*SE*	Fit-residual	Items	Mean location	*SE*	Fit-residual	χ^2	df	*p*	PSI
St1	-0.25	0.19	0.29	1	-0.17	0.21	0.38	112.27	140	0.96	0.8
St2	-0.17	0.20	0.46	2	-0.55	0.22	0.26				
St3	0.59	0.21	0.14	4	0.31	0.20	0.25				
St4	1.23	0.23	0.18	5	0.96	0.21	0.32				
St5	-1.40	0.18	0.46	7	-0.10	0.21	0.25				
				8	-0.88	0.20	0.19				
				11	0.44	0.19	0.33				
				12	0.03	0.19	0.22				
				14	0.54	0.18	0.63				
				15	-0.18	0.21	0.16				
				16	0.41	0.23	0.15				
				17	-0.46	0.20	0.26				
				19	0.02	0.18	0.33				
				21	-0.38	0.18	0.58				

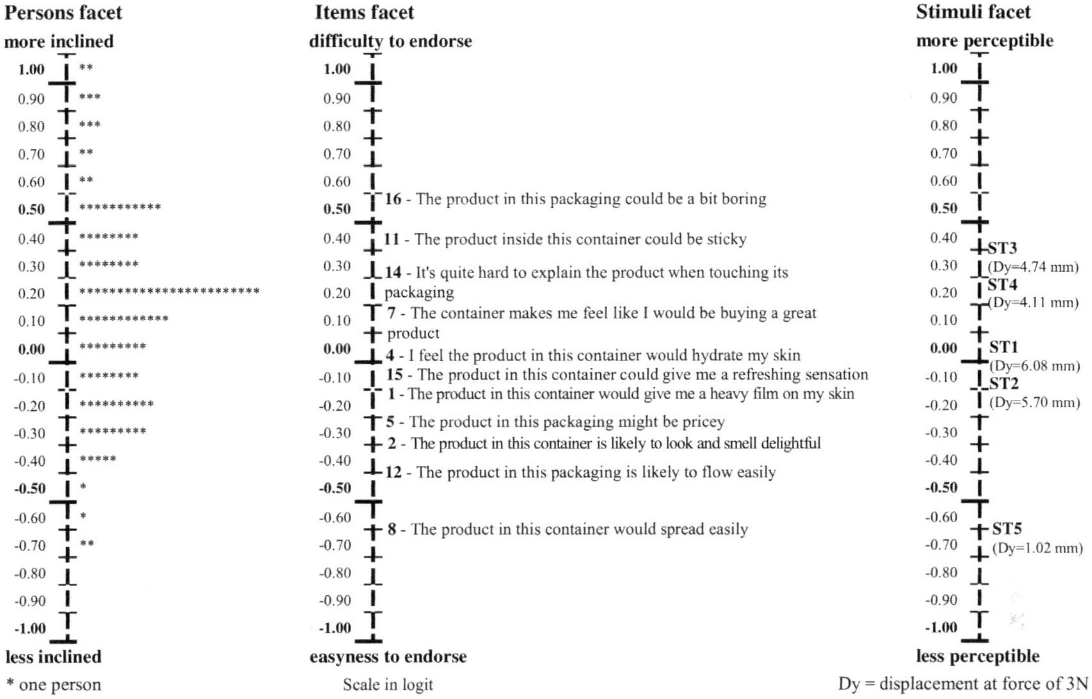

Figure 1. Rasch-calibrated metric with 11 items for the impression of a moisturizer cream associated with the compliance of a range of product containers.

5. Discussion

In Rasch analysis items that do not fit the model are not automatically rejected; rather they are investigated to identify their source of misfit and to what extent they corrupt measurement. For example, responses to Item 3, “I might get a bit watery product in this container”, presented dependence on Item 12 and Item 18, “I’ve got a pleasant touch with this container”, can have been ambiguous originating an inconsistent responses pattern.

The inclusion of further items in the scale yielded a better discrimination of the effects of the compliance of the containers for that particular sample. This was not unexpected because more information was aggregated to the measurement structure.

Although item locations on the continuum did not achieve identical estimates, these differences can be associated with the measurement error in the compared calibrations. The modelled standard error was estimated to be lower than 0.5 logit interval in a two-tailed 95% confidence interval of ± 2.0, based on the calibration stability for the sample size adopted in the study. An exception was Item 5 that extrapolated its measurement error and therefore, it shall be further examined. Nevertheless, the measurement precision of the calibration will normally improve as more data become available.

The metrics have suggested that the degree of affective responses does not follow the order of the physical element compliance for the containers. The metrics indicate, however, that according to participants’ impression there is a higher likelihood of the intermediate range of compliance being associated with a container of a moisturizer cream. It is possible to draw this conclusion because of the alignment of persons, items and stimuli on the same scale continuum.

6. Implications

The major outcome from this empirical study is, perhaps, the potential of incorporating new variables in calibrated structures for the measurement of an underlying attribute of a product. A practical value from the inclusion of further variables is the construction of an item bank. As a consequence, analysts can develop a bespoken structure with additional items without losing the properties of the core of the original, off-the-shelf calibrated scale to make whatever general comparisons they require. We envisage, therefore, that future applications could incorporate computer-aided assessment of the effects of different characteristics of physical components and product features on persons' attitude.

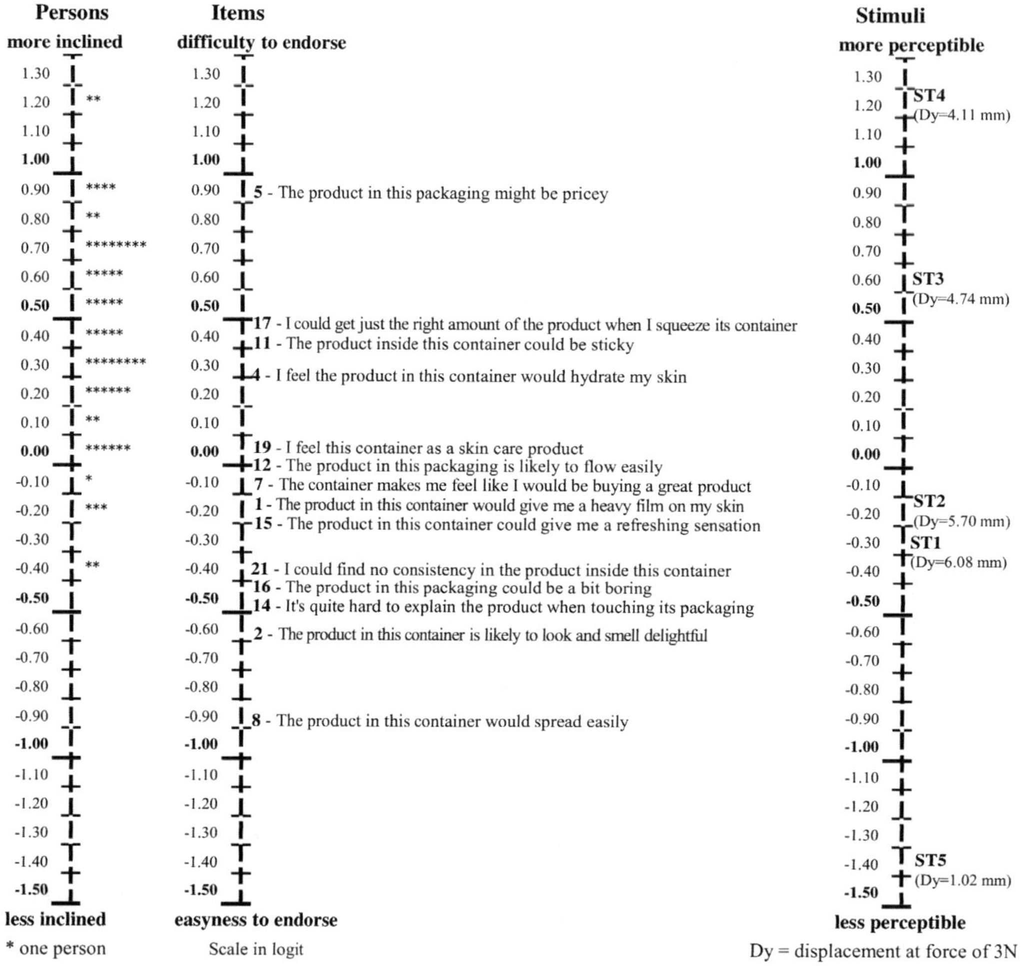

Figure 2. Rasch-calibrated metric including with 14 items for the impression of a moisturizer cream associated with the compliance of a range of product containers.

References

[1] Andrich D 1988 *Rasch models for measurement* No. 68 (London: Sage Publications)
[2] Camargo F R and Henson B 2011 *J. Design. Res.* **9** pp 360 – 75
[3] Rossi G B 2007 *Measurement* **40** pp 545 – 62
[4] Rasch G 1960, 1980 *Probabilistic models for some intelligence and attainment tests* (Chicago: University of Chicago Press)
[5] Andrich D 1978 *Appl. Psych. Meas.* **2** pp 581 – 594
[6] Masters G N 1982 *Psychometrika*, **47** pp 149 – 174
[7] Camargo F R and Henson B 2012. Invariant comparisons in affective design *Advances in Affective and Pleasurable Design* ed Y G Ji (Boca Raton: CRC Press) pp 490 – 499
[8] Linacre J M 1989 *Many-facet Rasch measurement* (Chicago: MESA Press)
[9] Pallant J F and Tennant A 2007. *Brit. J. Clin. Psychol.*, **46** pp 1 – 18
[10] Shao F, Chen X, Barnes C J and Henson B 2010. *J. Eng. Med.*, **224** pp 97 – 105
[11] Andrich D, Sheridan B, Lyne A and Luo G 2012. *RUMM laboratory*. Perth.

Journal of Physics: Conference Series **459** (2013) 012010 doi:10.1088/1742-6596/459/1/012010

Modularized Applications with Smartphones and Smartpads in Shape, Color and Spectral Measurements for Industry, Biology, Medicine, Science, Education and Training

Prof. Dr. Dietrich Hofmann[1], B.Eng. Paul-Gerald Dittrich, B.Eng. Eric Düntsch, Daniel Kraus, Dr. Claudia Gärtner[2], Dipl.-Ing. Richard Klemm
[1]SpectroNet c/o Technology- and Innovation Park Jena GmbH, Hans-Knöll-Straße 6, 07745 Jena, Germany
[2]microfluidic ChipShop GmbH, Stockholmer Str. 20, 07747 Jena, Germany

E-Mail: d.hofmann@spectronet.de, pg.dittrich@spectronet.de, e.duentsch@spectronet.de, d.kraus@spectronet.de, claudia.gaertner@microfluidic-chipshop.com, richard.klemm@microfluidic-chipshop.com

Abstract. Aim of the paper is the demonstration of a paradigm shift in shape, color and spectral measurements in industry, biology and medicine as well as in measurement science, education and training. Laboratory applications will be supplemented and replaced by innovative in-field and point-of-care applications. Innovative functional modules are smartphones and/or smartpads supplemented by additional hardware apps and software apps. Specific examples are given for numerous practical applications concerning optodigital methods. The methodological classification distinguishes between different levels for combinations of hardware apps (hwapps) and software apps (swapps) with smartphones and/or smartpads. These methods are fundamental enablers for the transformation from conventional stationary working places in industry, biology, medicine, science, education and training towards innovative mobile working places with in-field and point-of-care characteristics. The innovative approach opens so far untapped enormous markets for measurements in industry, biology and medicine as well as in science, education and training. These working conditions will be very common due to their convenience, reliability and affordability. The fundamental enablers are smartphones and/or smartpads. A highly visible advantage of smartphones and/or smartpads is the huge number of their distribution, their worldwide connectivity via cloud services and the experienced capabilities of their users for practical operations. Young people are becoming the pioneers.

1. Mobilization of Computers

The computer market is changing. Miniaturization and mass production are in progress. The special advantages of smartphones and/or smartpads are their mobility and their multitouch user interfaces. Market leaders are Apple with iOS operation system and Google with Android operation system. Microsoft with Windows 8 operation system is a late but powerful and experienced follower (Fig. 1-01). Windows 8 smartphones and/or smartpads combine the typical Windows working style with the leisure look and feel of iOS and Android life style (Fig. 1-02). The innovative computers are equipped with consumerized interfaces for wired and/or wireless communications (Fig. 1-03). Smartphones and/or smartpads are additional equipped with powerful internal hardware apps for image acquisitions that are light sources, cameras, camcorders and scanners and with software apps for image processings.

Smartphones and smartpads are mobile because they are light weight and battery powered. In addition, the number, quality and diversity of external hardware apps for smartphones and/or smartpads are increasing. In the next chapters selected examples will be given for optodigital hardware apps and digital software apps which are already on the market. They are applicable for industrial, biological and medical diagnostics as well as for scientific investigations or mobile education and training purposes.

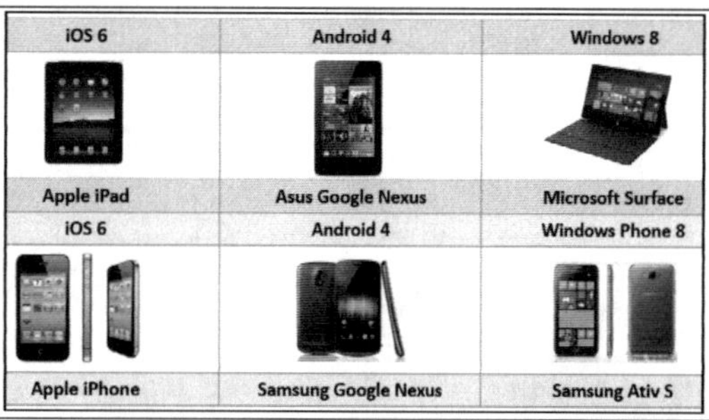

Fig. 1-01 Innovative Mobile Computers

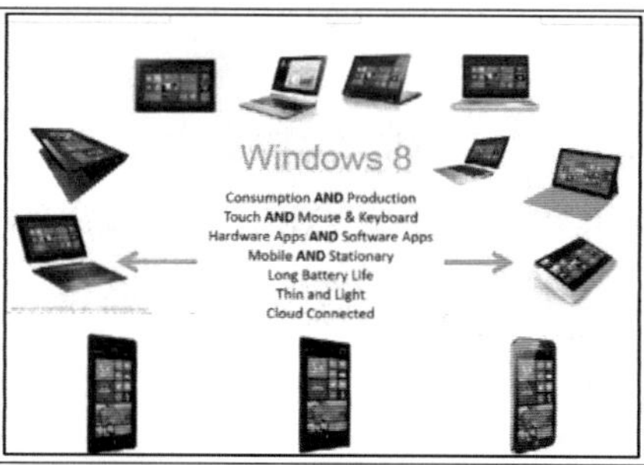

Fig. 1-02 Innovative Windows Mobile Computers

	USB 2.0	USB 3.0	NFC	Bluetooth	WiFi	LTE
Transfer Rate	480 Mbit/s	2400 Mbit/s	0,424 Mbit/s	24 Mbit/s	600 Mbit/s	1000 Mbit/s
Range	5 m	4,5 m	4 cm	100 m	300 m	10 000 m
Wired	✓	✓	-	-	-	-
Wireless	-	-	✓	✓	✓	✓

Fig. 1-03 Innovative Mobile Interfaces

The Ozcan Group at the University of California, Los Angeles (UCLA) is working since years successfully on this topic. The research results were recently summarized in a very informative overview [1]. A fundamental purpose of the present paper is to show that many innovative mobile smart solutions can be designed with already existing functional modules according to the LEGO principle. Due to standardized operation systems (see Fig. 1-01 and Fig 1-02) and standardized interfaces (see Fig 1-03) the existing hardware modules and software modules must **not** be modified in their original structure and functionality [2]-[4]. The disadvantage of multi-hybrid solutions can sometimes be that they are more complex and also more complicated in its structures and in its functions as specialized solutions. But the advantage of these multi-hybrid solutions is that they are available quickly, because they are composed of commercially available components.

2. Mobile Smart Optodigital Hardware and Software Apps for Shape Measurements

Shape measurements are the analysis of geometric shapes in 1, 2, 2.5 and 3 dimensions. Typical optodigital hardware apps are cameras and microscopes preferably with USB-interfaces (Fig. 2-01). Very numerous software packages for image processing [5] are considered software apps. Normally the real objects are compared with fundamental master objects in algorithms or data bases. Most commonly the boundary representations are used. Once the objects are represented by extracting the shapes from images of the object, the data clouds have to be simplified before a comparison with standardized masters can be achieved. The simplified representation is often called a shape descriptor (or fingerprint, or signature) of the real object. The simplified representations try to carry most of the important information, while being easier to handle, to store and to compare than the shapes directly [6].

Optodigital shape measurement apps	Manufacturer / Distributor
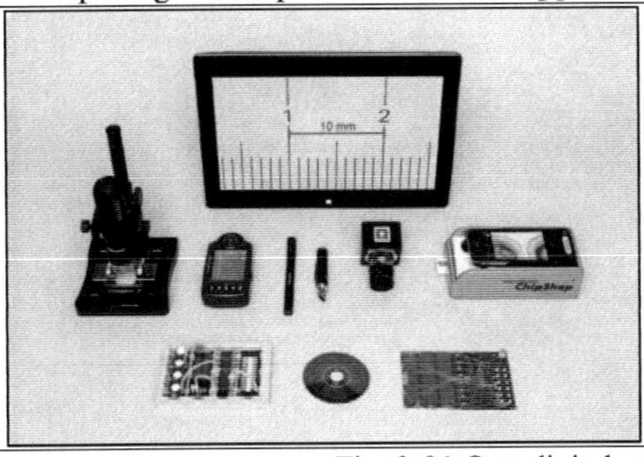	www.spectronet.de www.tip-jena.de www.dnt.de www.easypix.eu www.supereyes.cc www.quick-image.de www.weidmann-plastics.com www.fotofinder.de www.cda.de

Fig. 2-01 Optodigital modules for shape measurements

3. Mobile Smart Optodigital Hardware and Software Apps for Color Measurements

Color measurements are the analysis of light and body colors near to the human color perception. Typical optodigital hardware apps are color sensors preferably with USB interfaces (Fig 3-01). With colorimeters an unknown color is evaluated in terms of known colors near to a standardized human color perception. Color measurements are of great importance for the analysis of raw materials and finished products in industry. Color measurements are also critical in biology and medicine. For color measurements the quantification of colors is based on the three-component theory of true color vision. It states that the human eye possesses receptors for three primary colors red, green and blue. All colors are seen by human as mixtures of these primaries. In colorimetry, these components are referred to as X-Y-Z coordinates. Colorimeters, based on this theory of color perception, employ three optosensors as receptors to see colors in much the same way as the human eyes (Fig 3-01) [7].

Optodigital color measurement apps	Manufacturer / Distributor
	www.spectronet.de www.tip-jena.de www.mathai.de www.colorstriker.com www.oceanoptics.com www.seachangeronline.com www.jencolor.de www.mazet.de www.osram.de www.xrite.com

Fig. 3-01 Optodigital modules for color measurements

4. Mobile Smart Optodigital Hardware and Software Apps for Spectral Measurements

Spectral measurements are the analysis of unknown spectra in comparison with evaluated terms of known spectra. Typical optodigital hardware apps are spectrometers preferably with USB interfaces (Fig. 4-01). With spectrometers an unknown spectrum is evaluated in terms of known spectra concerning its wavelengths and amplitudes. Spectral diagnostics is of great importance in chemical industry as well as in biology and medicine. Real objects are becoming visible by emitting or reflecting photons. The wavelengths of these photons depend on the object's composition, along with other attributes such as temperature. The human eyes perceive the presence and absence of photons with different wavelengths as different colors. For example, photons with a wavelength of 620 to 750 nanometers are perceived as red. An object that primarily emits or reflects photons in that range looks red [8].

Optodigital spectral measurement apps	Manufacturer / Distributor
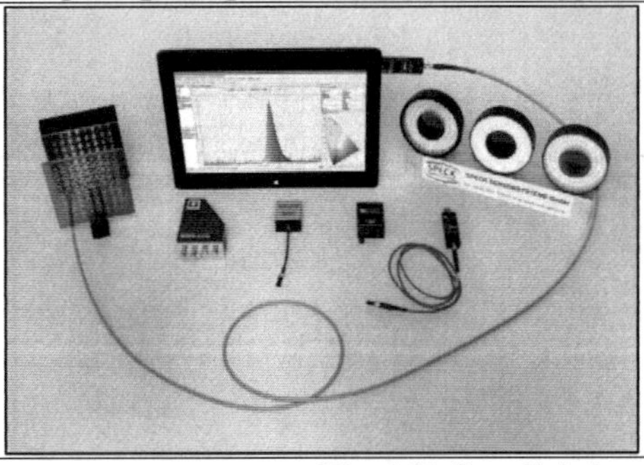	www.spectronet.de www.tip-jena.de www.rgb-laser.com www.zeiss.de www.stz-ilmenau.de www.quick-image.de www.oceanoptics.com

Fig. 4-01 Optodigital modules for spectral measurements

5. Modularized Mobile Smart Applications for Shape, Color and Spectral Diagnostics

Modularized mobile smart applications are realizable by the combination of the contents of chapter 1, 2, 3 and 4 (Fig 5-01, Fig 5-02 and Fig 5-03). It is of great methodical importance that modularized mobile smart solutions are harmonizing the present barriers between industry, biology and medicine.

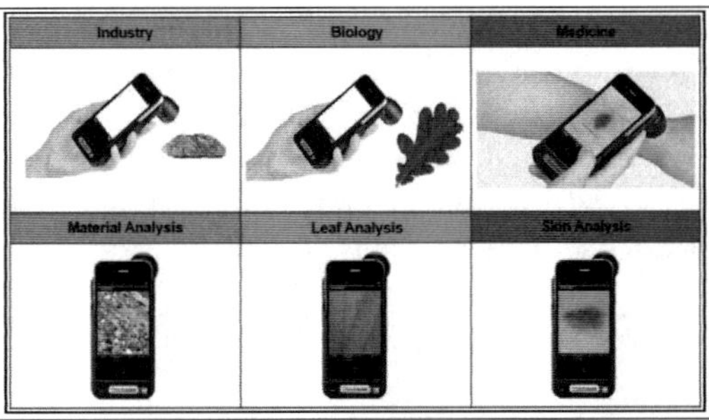

Fig. 5-01 Mobile smart shape diagnostics

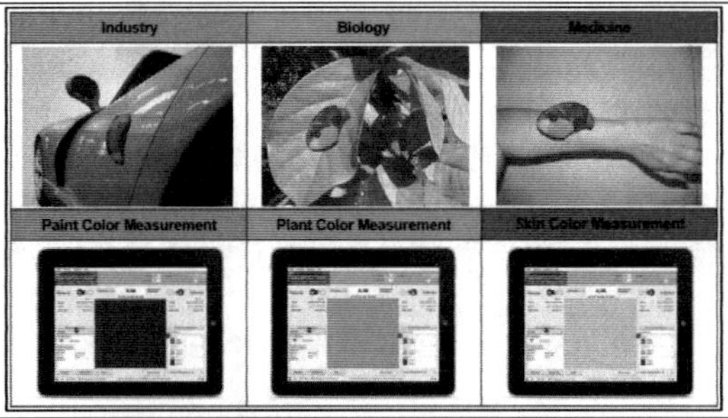

Fig. 5-02 Mobile smart color diagnostics

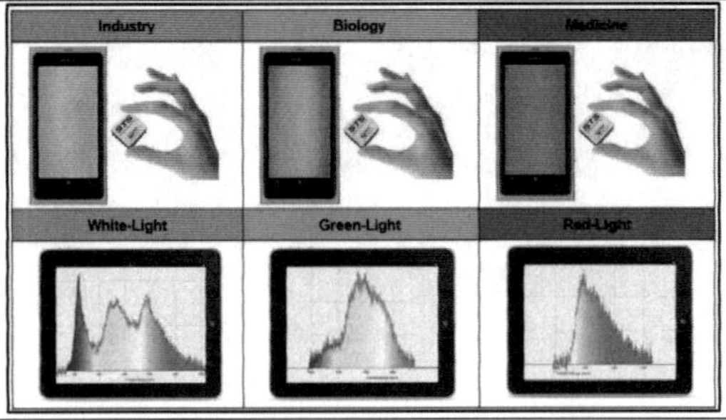

Fig. 5-03 Mobile smart spectral diagnostics

6. An example application for mobile smart biomedical diagnostics

The actual challenges for biomedical diagnostics lie primarily in creating miniaturized systems in which the presence of low concentrations of target biomarkers can be reliably detected in small volumes of biological samples. The multi hybrid combination of common mobile smart instruments like smartphones and smartpads with microfluidic analytical chips will lead to a new class of in-field or point-of care diagnostic systems. The availability and advantages of diagnostic instruments with commercial smart phones and/or smartpads combined with commercial microscopic hardware and software apps and combined with commercial miniaturized microfluidic chips generate awareness on the public market for mobile smart POC systems [9], [10]. An example application for this combination was developed for different immunological assays like Francisella tularensis as pathogen detection example and Borrelia burgdorferi as diagnostic example [11]. The setup of the analytical diagnostic instrument contains three components: A smartphone iPhone 4S from Apple, a microscope adapter with software from FotoFinder Systems (see Fig. 2-01) and a microfluidic chip from microfluidic ChipShop (Fig. 6-01). The microscope adapter developed by FotoFinder Systems is a hardware app which uses an optical lens setup to realize a 20x focusing of the already integrated camera inside the smartphone combined with a software app for image analysis and processing. Around this setup another hardware setting was realized to introduce the microfluidic chip aligned to the optical setup (Fig. 6-02).

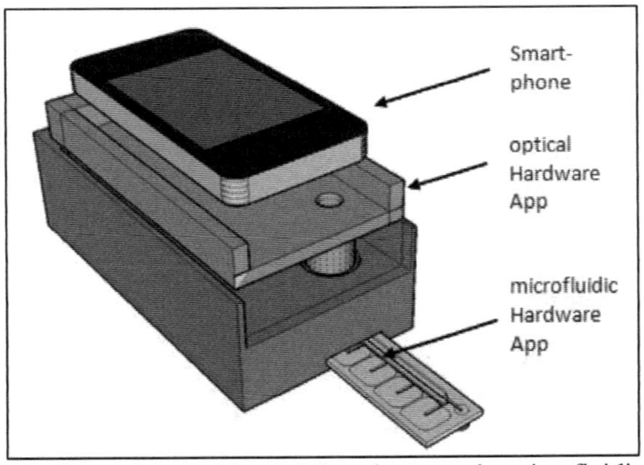

Fig. 6-01 Smartphone with modular adapter and a microfluidic chip

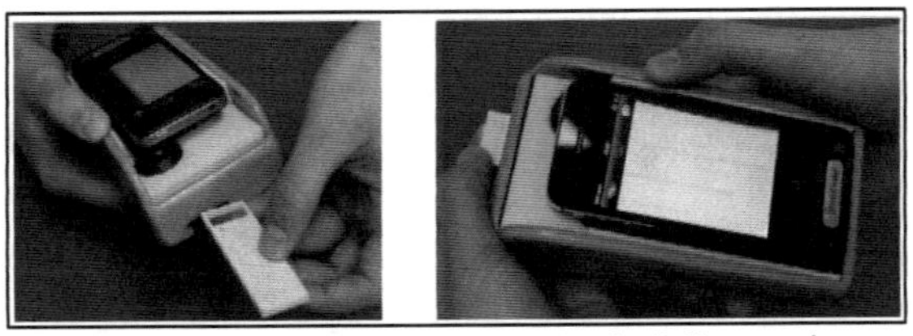

Fig. 6-02 Colorimetric immunological diagnostics

7. An example application for mobile smart education and training materials

STEMMER IMAGING is Europe´s largest imaging technology provider with subsidiaries in Germany, UK, France, Switzerland and The Netherlands. STEMMER IMAGING offer an industry leading service, a comprehensive support programme and competitively priced products. An innovative service is the education and training material of the STEMMER IMAGING ACADEMY, very special in its combination of training material and product catalogue (Fig. 7-01 and Fig. 7-02).

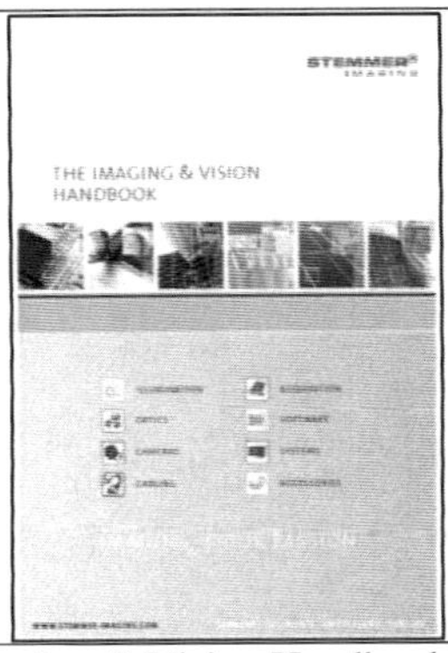

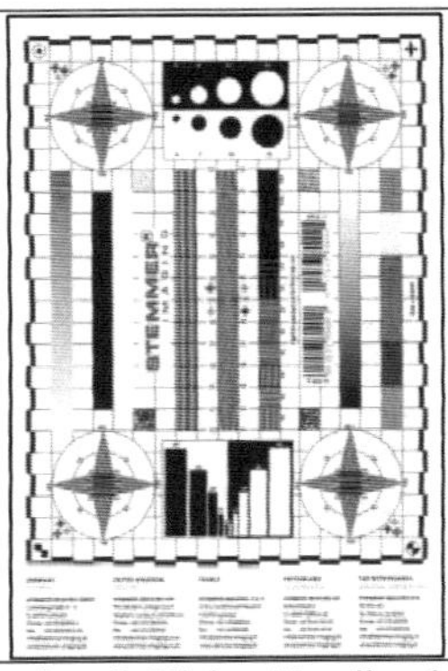

Fig. 7-01 Imaging & Vision Handbook - front cover

Fig. 7-02 Imaging & Vision Handbook - back cover

ISBN 978-3-00-039674.8

An optodigital version of the handbook is available via www.stemmer-imaging.de

8. Conclusions

Smartphones and Smartpads became global mass products with great convenience, high reliability and affordable prices. These market developments enable a paradigm shift in industrial, biological and medical measurements and diagnostics. With modular hardware apps and software apps new classes of hybrid instruments can be created. Modular hardware apps and software apps are reducing development time, cost & risk. Smartphones & smartpads are revolutionary game changers in lifestyle and workstyle. Miniaturization & consumerization inspire each other with increasing speed. Field instrumentations will be based on re-invented laboratory instrumentations. First steps on innovative routes are already gone successfully. An interesting convergence of measurements and diagnostics in industry, biology and medicine can be expected.

9. Sources

[1] H. Zhu; S.O. Isikman; O. Mudanyali; A. Greenbaum; A. Ozcan: Optical imaging techniques for point-of-care diagnostics. Lab chip, 2013, 13, 51-67

[2] D. Hofmann; P-G. Dittrich; E. Duentsch: Smartphone measurement engineering – Innovative challenges for science & education, instrumentation & training. Journal of Physics: Conference Series Vol. 238 (2010) p 012006 jpcs.iop.org doi:10.1088/1742-6596/238/1/012006

[3] D. Hofmann; R. Margull; P-G. Dittrich; E. Duentsch: Smartphone green vision at dawn of industry 4.0, Advanced Materials Research Vols 403-408 (2012) pp. 4079-4083

[4] D. Hofmann; R. Margull; P-G. Dittrich; E. Duentsch: Smart instrumentations for mobile diagnostics in industry, biology and medicine. In: D. Beckmann; M. Meister (Hrsg.): Technische Systeme für die Lebenswissenschaften. 2012. S. 7-13. ISBN 978-3-00-039458-3

[5] http://www.vdmashop.de/mv_emag/

[6] http://en.wikipedia.org/wiki/Shape_analysis_(digital_geometry)

[7] http://sensing.konicaminolta.us/ KMSA-Colorimetry-White-Paper.pdf

[8] http://www.wisegeek.com/what-is-optical-spectroscopy.htm

[9] N. Hlawatsch; R. Klemm; C. Carstens; T. Brandstetter; H. Becker; R. Elbracht; C. Gaertner: A lab-on-achip system for the development of complex assays using modular microfluidic components. Proc. SPIE 8251. Microfluidics, BioMEMS, and Medical Microsystems X, 82510D (February 9, 2012) doi:10.1117/12.910269; http://dx.doi.org/10.1117/12.910269

[10] R. Klemm: Mobile hybrid microlab with microfluidic chips and smartphone for analytical diagnostics. http://spectronet.de/de/vortraege_bilder/vortraege_2012/15-spectronet-collaboration-forum-puchheim-near-mu_has9upce.html

[11] D.Hofmann, P-G. Dittrich, C.Gärtner, R.Klemm: Multi Hybrid Instrumentations with Smartphones and Smartpads for Innovative In-Field and POC Diagnostics. Microfluidics, BioMEMS, and Medical Microsystems XI, edited by Holger Becker, Bonnie L. Gray, Proc. of SPIE Vol.8615, 861508 - 2013

Journal of Physics: Conference Series **459** (2013) 012011 doi:10.1088/1742-6596/459/1/012011

Incorporating human stress measurements into biomedical engineering class

Ondřej Bruna, Pavel Souček, Jan Holub

Dept. of Measurement, FEE CTU Prague, Technicka 2, CZ 166 27 Prague 6, Czech Republic

E-mail: brunaond@fel.cvut.cz, soucepa3@fel.cvut.cz, holubjan@fel.cvut.cz

Abstract: The topic of this paper is to describe the current course at Czech Technical University in Prague X38KLS (Construction of medical systems) and to describe proposed improvements and differences between novel and old approach. The changes and in a state of preparation and have not been fully implemented. The new course should take over the old in September 2013.

1. Introduction

The field of biomedical engineering is fast growing and expanding. Medical experts are equipped with sophisticated tools to examine the patient and to process patient's data [1], [2] to decide on diagnosis. Because doctors are more often asked to interact with complex devices, there is a need for experienced technician, who would be able to understand both the data and the patient. At this level the technician is not a competition to a doctor, but partner, who should help with data, analyses and who should be able to identify corrupted data or signal affected by a technical issue to exclude wrong diagnosis. To achieve this new training including failures and person in the training loop is necessary. In this paper we describe the approach taken at our university in order to ensure the students will become this type of specialists. Current course is focused on electronics only and the aim is to expand its scope beyond this point and to describe the electronics in context of patients and process of setting diagnose.

2. Current structure

Current course is designed in a following way:

a) Lectures: the students are optionally attending lectures (organized in a lecture hall capable of accommodating 40 people at once), lectures teach students about the circuits in medical instruments and explain their operation.

b) Laboratory classes.

    a. Practical measurements in the laboratory serve other purpose, students are supposed to validate the knowledge gained during the lectures by measuring real hardware which incorporates instruments from lectures. This segment takes 7 weeks. When students are finished with the first part, they are considered eligible to proceed to second stage.

    b. Project of their own choice. This last and final step is supposed to increase students' creativity by letting them solve a problem themselves. There were many different things designed and tested by our students.

Lectures are focused on knowledge about electronics and measurements of biological signals. Lectures do not deal with signal processing. Measurement part of the course lasts for a first half of a semester (7 weeks) while the whole semester has 14 weeks. Practical measurement takes place in a laboratory equipped with standalone training electronic circuits (training box/training board) and the students are supposed to perform measurements on each of the parts separately to study its function. They are also supposed to simulate circuits in a National Instruments MultiSim and generate results for same tasks as in case of real circuits. When all the data are obtained, students are expected to compare both results in order to conclude with the advantages and disadvantages of simulation. Simulation also allows working with the circuit in more ways and students can try things, which they could not with real application.

The individual boxes used in current curriculum cover measurements of all important electronic parts. Students would see the electronic circuit on the lecture and during the exercise it is possible to examine real circuit. Topics of measurements are:

- Band filter.
- Instrumentation amplifier.
- Modelling of parasite capacities of active and passive parts.
- Function generator.
- Mean value convertor.

The second half of the semester is dedicated to design of their own device or circuit related to biomedical engineering. Students are supposed to design, implement and test in MultiSim (verification phase), and afterwards to assemble it in a breadboard and measure its properties, which will be then compared with original design in computer program. The goal is to compare these two approaches and to realize differences, advantages and disadvantages of both of them. As a result students write a report.

Some examples of what students did in last two semesters: volume unit meter, voltage controlled oscillator, alfa-wave monitor, defibrillation pulse protection circuit, metal detector, correction amplifier, ECG amplifier and so on.

During the class some students reported some problems and provided feedback to the teacher regarding the class set-up. Some students felt the class is too software oriented and would appreciate more circuit oriented approach. Some students complained they did not see the whole picture and could not relate the knowledge from lectures to exercises. Some students also complained about virtual oscilloscopes, which were used in class – Velleman PCSU1000USB. In order to satisfy all collected data, new curriculum was designed and is discussed in the next chapter.

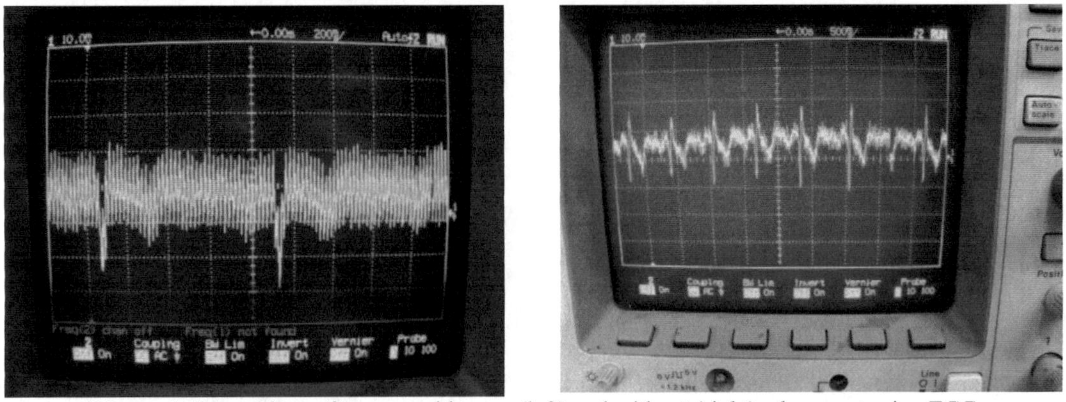

Figure 1: Screenshots of screens with error (left) and without (right) when measuring ECG.

3. Proposed structure of the course

New curriculum is taking into account the feedback provided from students. New approach introduces modules for measurement of galvanic skin resistance, alfa wave monitor and ECG monitor. These systems include most of the blocs, which were presented in original curriculum and are connected together to create a working instrument for measurement. So the students can understand single circuits as well as the whole system. Volunteers in the class can have measured their ECG (electrocardiogram). Training boards are made of several blocks, which can be measured. These blocks are substitutions to boxes used in former approach.

Table 1: Segments on new training boards.

ECG, EEG, EMG Board	GSR Board
Instrumentation amplifier	Inverting amplifier
Notch filter	Band pass filter
Alpha wave filter	Visualisation and signalling segment
Output amplifier	
Optocoupler	

First training box ECG, EEG and EMG – electromyography is a method for measurement of activity of skeletal muscles. The board allows measurement of forearm, biceps and triceps EMG. Second training box is for measurement of galvanic skin resistance. Table 1 shows segments of bought training boards.

Approach proposed in this paper is based on a similar platform as the original one. Students are still expected to attend exercises (mandatory) and lectures (optional). Exercises are divided into three sections.

a) Measuring and calibrating every single blocks of the device.
c) Measuring on a real subject.
d) Diagnosis of system failure.

The second part of semester includes measurement on students. The students are divided into groups of three and two and perform measurements on each other. When they are able to operate and measure with the device, the teacher will use a switch button to simulate one or more failures of the device – example Figure 1. Students are then asked to measure, detect and remove the error individually without any interference from the teacher. The teacher then evaluates their approach to solution of the problem. In this case, students do not have any instructions on how to proceed and they have to deal with the problem on their own. This should support their self-confidence in knowledge they have learned throughout the semester and help them improve analysis skills, which are essential for an engineer in today's world.

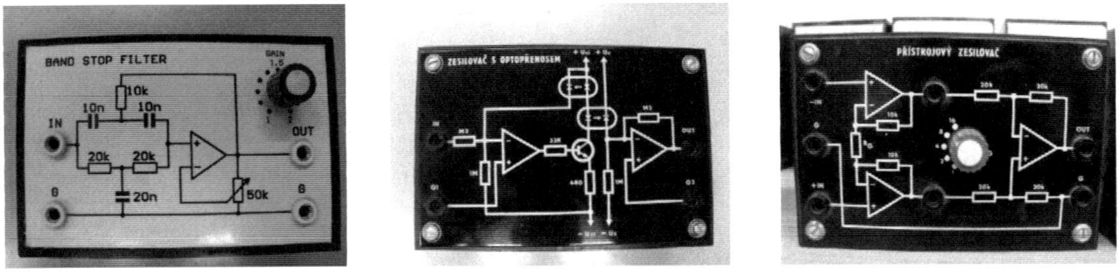

Figure 2: Original boards for measurement.

New approach is based on instruments, which incorporate the same circuits as did the boards from original approach. The difference is that here are circuits implemented on one board. Boards DL3155BIO8 and DL3155BIO5 from company DeLorenzo have been ordered for this purpose.

DL3155BIO8 is a board for measurement of galvanic skin resistance. The comparison of both can be seen on Figure 1 and Figure 2. Galvanic skin resistance has been used as an indicator of perspiration. In some cases it is considered, that increased perspiration is an indicator not only of increased physical load, but also psychical load or stress [3],[4]. This can be one of input variables for determining people's stress level. Board DL3155BIO5 is designed for measurement of EEG (electroencephalograph, alpha waves only), ECG (electrocardiogram) and EMG (electromyography). Both boards have special switches creating artificial defects in the system and corrupting the measurement. The DL3155BIO5 board is also equipped with board for simulation of two proper heartbeats. These two boards have been chosen for the following reason: the ECG and EEG are nowadays the most commonly measured data and students are supposed to be able to detect and remove any problems related to this electronics. When combined with the board DL3155BIO8 it can create a simple polygraph. Other equipment used was an oscilloscope HP54600, Velleman PCSU1000 digital oscilloscope and computer.

Table 2: Comparison of former and current exercises.

Old	Status in new curriculum
Convertor to mean value	Discontinued
Instrumentation amplifier	Continued
Modelling of parasite capacities of active and passive parts	Discontinued
Function generator	Discontinued
Band-pass filter	Continued
Notch-amplifier	New
Optocoupler	Continued
Alpha wave filter	New
Visualisation and signalling segment	New

How will the new training boxes replace the new ones is shown in table 2. During the first section of the class are students supposed to perform similar measurements as in the original plan, but this time the circuits are implemented in one measured board and there is clear purpose of this board. Students measure parameters of all blocks and make simulation using MultiSim as well. These measurements are performed with step by step instructions. After measuring and calibrating both boards, the students can proceed to measurements with subjects. There will be several tasks for them to accomplish. One class will be focused on measurement of human stress, next on assembling polygraph and using it as a simple lie detector, and identifying and dealing with several errors of teacher's choice (set-up on the board control panel).

Journal of Physics: Conference Series **459** (2013) 012011 doi:10.1088/1742-6596/459/1/012011

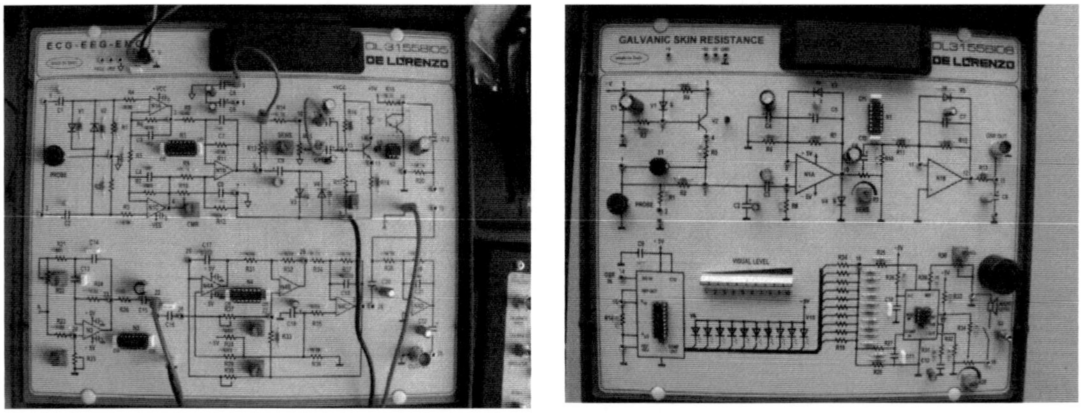

Figure 3: New boards for measurement DL3155BIO5 (left) and DL3155BIO8(right).

Teacher can use the boards to simulate a failure of certain blocks. Students are then supposed to use their knowledge about the boards perform measurements. Based on measurements they state a diagnosis. The diagnosis tells what the problem is and now to fix it. Part of the task is to explain, how they came to conclusion. GSR board can simulate 4 and ECG, EEG and EMG board 5 faults. Fault is for example failure of a resistor in a feedback of amplifier, disconnected output amplifier and more. Determining the failure would be the last exercise of the semester.

4. Summary

The proposed approach has not been implemented in university's curriculum yet and is still in a preparation phase. Despite this have been already performed some experiments with students, picture taken is on Figure 4. From last semester was chosen one group of volunteers, who went through partly modified semester plan, which included new boards. Their task was to measure on new devices instead of measurement on old standalone boards. Students were asked to assess new exercise – Stress factor. Stress factor exercise will be also part of new semester plan. Its goal is to show that people under stress react differently. To achieve it, special programs were used and ECG with GSR was measured. Human factors and stress play an important role in many manned systems from flight decks to control station of power plants.

Figure 4: New measurement station with students.

5. Conclusions

Students finished measurements successfully and reported, that they would recommend them to become a permanent part of this class. Test with students showed, that the goal – implement all three phases will probably not be doable. During the semester students had more problems with setting up the board than anticipated. Therefore it is planned to reduce the amount of exercises and give more space to measurement on students and fault diagnosis. This approach is still under development and will be revised before launching it in classrooms. Despite some drawbacks – very time consuming and some students might not keep up with their colleagues, it has been decided to continue with implementation and to start new approach the next semester.

References

[1] Hunt DL, Haynes R, Hanna SE, Smith K. "Effects of Computer-Based Clinical Decision Support Systems on Physician Performance and Patient Outcomes: A Systematic Review" *JAMA*. 1998;280(15):1339-1346. doi:10.1001/jama.280.15.1339.

[2] Taehwan Roh; Kyeongryeol Bong; Sunjoo Hong; Hyunwoo Cho; Hoi-Jun Yoo; , "Wearable mental-health monitoring platform with independent component analysis and nonlinear chaotic analysis," Engineering in Medicine and Biology Society (EMBC), 2012 Annual International Conference of the IEEE , vol., no., pp.4541-4544, Aug. 28 2012-Sept. 1 2012 doi: 10.1109/EMBC.2012.6346977

[3] Jing Zhai; Barreto, A.B.; Chin, C.; Chao Li; , "Realization of stress detection using psychophysiological signals for improvement of human-computer interactions," SoutheastCon, 2005. Proceedings. IEEE , vol., no., pp. 415- 420, 8-10 April 2005 doi: 10.1109/SECON.2005.1423280

[4] Ogorevc, J.; Podlesek, A.; Gersak, G.; Drnovsek, J.; , "The effect of mental stress on psychophysiological parameters," Medical Measurements and Applications Proceedings (MeMeA), 2011 IEEE International Workshop on , vol., no., pp.294-299, 30-31 May 2011 doi: 10.1109/MeMeA.2011.5966692

Journal of Physics: Conference Series **459** (2013) 012012 doi:10.1088/1742-6596/459/1/012012

Innovation of laboratory exercises in course Distributed systems and computer networks

Pavel Souček, Oldřich Slavata, Jan Holub

Dept. of Measurement, FEE CTU Prague, Technicka 2, CZ 166 27 Prague 6, Czech Republic

E-mail: soucepa3@fel.cvut.cz, slavao@fel.cvut.cz, holubjan@fel.cvut.cz

Abstract. This paper is focused on innovation of laboratory exercises in course Distributed Systems and Computer Networks. These exercises were introduced in November of 2012 and replaced older exercises in order to reflect real life applications.

1. Introduction

Exercises wireless network coverage prediction and TCP/IP network analysis are part of Distributed Systems and Computer Networks course. These lessons are compulsory for students Sensors and Instrumentation and optional for other students.

Distributed Systems and Computer Networks lessons are successor of long time taught course Information Transfer. This course is focused on basic principles used in order to transfer information in area of electrical engineering. It covers area from serial line to USB and TCP/IP networks.

The main purpose of this course is to show main approaches used in way data are transferred and to compare their advantages and disadvantages. Through this course students become familiar with basic ways to transfer data, from physical principles, for example difference between twisted and flat cable, to the ways of securing data delivery on higher layers of OSI model, like CRC.

This knowledge will help students in following lectures, so they can faster get to solving specialized problems and don't get stuck at basics of data communication, like how serial communication works.

2. Wireless network coverage

Laboratory exercise Wireless Network Coverage is extension of former version, which employed only wireless modems working on 868 MHz frequency and I-Prop software (figure 1) used for simulation of signal coverage. The main problem of original exercise was, that it was too theoretical and students had problems understanding the connection between measured data and real life application.

The main task is to simulate coverage of wireless signal in real environment and to demonstrate how the signal distribution is affected by obstacles and type of antenna used on both sides of connection.

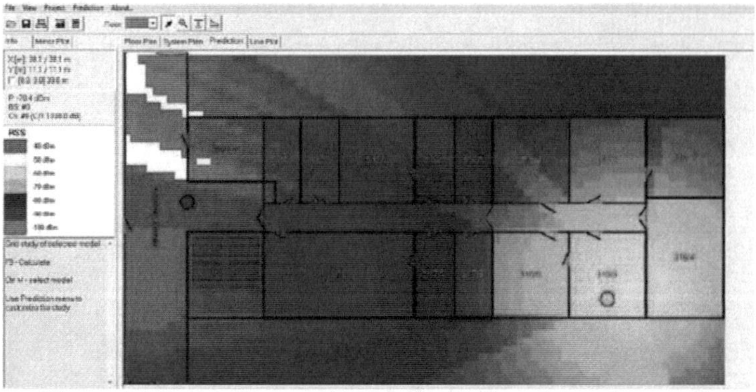

Figure 1 I-Prop software used for simulation of signal strength

After simulation students evaluate predicted signal coverage with real devices – 868 MHz modems, wi-fi access point Mikrotik RouterBoard and notebook. One of modems and RouterBoard serve as base station (figure 2), the other one with notebook serve as mobile client, which students use to measure real signal strength in several points of their choosing.

Figure 2 Mikrotik RouterBoard RB411U and 868 MHz modem

The main advantage of hardware used in this exercise is its versatility and robustness. Both modems and Mikrotik RouterBoard are practically indestructible and can be easily reconfigured after reset, in case of misconfiguration made by students. To determine the signal strength they use special query command in case of 868 MHz modems and visualization in inSSIDer program in case of wifi signal (figure 3).

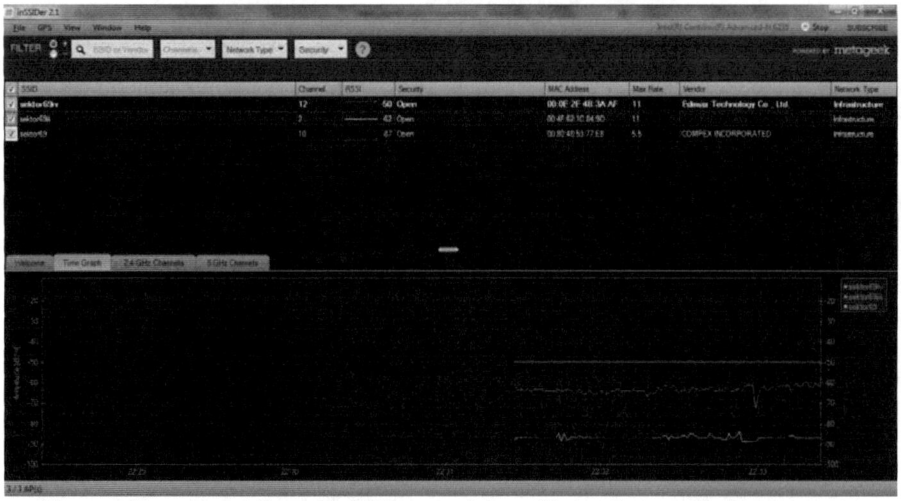

Figure 3 Freeware software inSSIDer used for wifi signal strength measurement

Another innovation is usage of several antennas, which help students to understand basic aspects of signal transmission like polarization. Students have one antenna with known gain and they have to determine the gain of the others (figure 4).

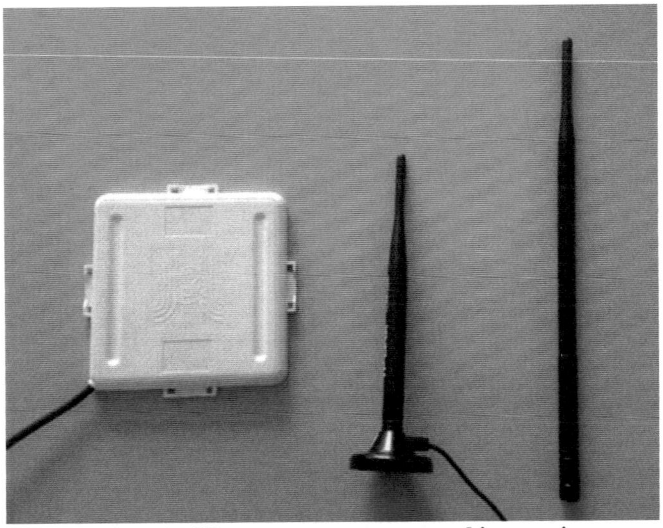

Figure 4 Different antenna types used in exercise

When students gather all the data they need they compare measured and simulated data.
The last part of exercise is focused on change in wi-fi signal strength depending on antenna used on side of base station. For this part we prepared three types of antenna. Using these antennas we can easily demonstrate the difference between them and with piece of dry and wet cloth we can demonstrate problems that wi-fi has with air humidity in real life.

3. TCP/IP network analysis

TCP/IP Network Analysis is brand new laboratory exercise focused on basic principles used in TCP/IP networks including OSI model and basic protocols.
In this laboratory exercise students will find out how TCP/IP networks work. They will gain understanding of principles used in networks. In simplified form they will acquire knowledge from routing, through DHCP and DNS system, to basics of network monitoring and analysis.
For network monitoring students use simple tools, which are part of operating system (both Windows and Linux) or are licensed under GNU/GPL. With these tools students are able to discover network topology and use this knowledge for localization of problems in this network. Of course these tools can be used to gain illegal access to network resources, but they can be also used to detect such attack or prevent it.
In first part of this laboratory exercise students have to find out how they are connected to into the internet and from publicly accessible databases they find out how DNS system works.
In second part of this laboratory exercise they are monitoring network communication using packet sniffer Wireshark (figure 5). Than students analyze collected data in order to solve problem defined in following parts of exercise.
Among other tasks they have to capture DHCP request and lease and to find out lease time or to analyze how web page is composed from several servers. Last but not least is capture of unsecured message transmission like smtp or Facebook chat.

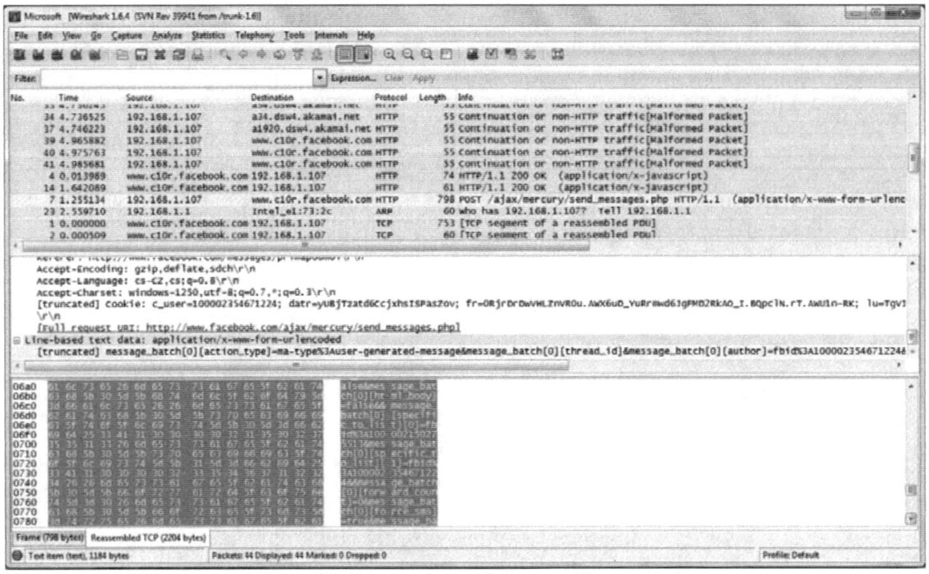

Figure 5 Wireshark used for traffic analysis

One of advantages of this laboratory exercise is that every part of it reflects real world conditions. As a method of line interception for we chose port mirroring, which can be easily set up on better switches.

4. Comparisoin

It is not surprising, that exactly same laboratory exercises don't exist. In case of Wireless network coverage similar exercise is taught at KTH Royal Institute of Technology [3]. The main difference is that they employ GSM base station and don't simulate signal coverage, only measure signal strength in defined distances.

Better situation is in case of TCP/IP network analysis where we can easily find very similar laboratory exercises, which often use Wireshark. These exercises are usually taught on Computer science departments and in many cases they reveal the subject more deeply [4], but that exceeds intention for this course, which is mainly to show possibilities and trends.

5. Conclusion

Innovation of two laboratory exercises in Distributed systems and computer networks course led to significant modernization and made the course more interesting. The course reflects real life application of principles demonstrated in course.

References

[1] Kocourek P, Novák J: Přenos informace, CTU in Prague, Prague, 2004

[2] Distributed Systems and Computer Networks lecture notes, 2013, available from: http://measure.feld.cvut.cz/en/education/courses/AE3B38DSP

[3] Wireless indoor coverage and path loss , KTH, Stockholm, 2013, available from: https://www.kth.se/social/upload/511ba910f276542ef42c4697/ik1330lab1.pdf

[4] Computer networks laboratory exercise no. 2, Halmstad University, 2013, available from: http://www.hh.se/download/18.70cf2e49129168da015800096178/1249984542542/Laboration_2.pdf

Delta-Notch Lateral Inhibition within the Organ of Corti

R. Summers, T. Abdulla, R. Luff

SEESE, Loughborough University, Ashby Road, Loughborough, LE11 3TU, UK.

R.Summers@lboro.ac.uk

Abstract. Lateral inhibition is described as an emergent property of the Delta-Notch signalling network. Two separate model representations of lateral inhibition are proposed for different purposes. One provides information about bioenergetics while the other has the capability to produce a physical representation. It is proposed that both can be used in further studies of the sensory pathways in the human connectome model of brain function.

1. Introduction

Over the last 60 or so years there have been many advances in the understanding of life at the level of scale at which microbiology provides the underpinning science. These advances have allowed several streams of thought to be intertwined to what might conveniently be termed here, 'systems biology'. In the study of systems biology: the gene provides the basis of heredity; biological physics and chemistry gives the details of cellular bioenergetics that drives the various cellular functions; the cell itself can be considered as the basic functional unit of life; and evolution is via natural selection. Each of these four processes operate at different spatial and temporal scales simultaneously, from which a multi-scale approach to functional analysis might provide an enhanced systemic understanding of behaviour.

For microbiology to advance further it may require further thinking in 'informatics and systems', where here there is a subtle difference in the meaning of the term 'system' than used in the previous term introduced – 'systems biology'. 'Systems science' [1] and more latterly 'systems engineering' [2] are themselves distinct disciplines that concern the study of organisational entities and the relations between them. From this simple definition it is possible to conduct numerous types of intervention that link functions to other functions, or functions to behaviour. So, using appropriate methods it is possible, for example, to study complex adaptive systems and their information flows which together may represent a regulatory system (including positive and negative feedback), an oscillatory system, and a network system. Clearly such systems occur in finance and economics, industry and commerce etc., but it is their use and interpretation in biological systems that are of concern in this paper. However, whatever the application, there are some systems-based terms that can be used to describe systemic behaviour that is of wider interest. Such a term is 'emergence' that leads to a need to define an 'emergent property' of a system. For the purposes of this paper an emergent property can be defined as behaviour or function that is exhibited at one level of scale that can not be determined from behaviour or function at a lower level of scale. Thus, the behaviour of a group of cells will not be deducible from the behaviour of individual cells.

Previous work by the research team at Loughborough University has investigated the link between multi-scale functions and behaviour in the development of heart tissue using an ontological approach [3] and more specifically epithelial to mesenchymal transition in cardiac cushion growth in the development of heart valves [4, 5]. In the latter studies, the role of the trans-membrane protein Delta-

Notch cell signalling network is particularly important. Further studies of the Delta-Notch protein network reveal examples of emergent properties of the system – in the patterns of activity at a cellular level obtained as a consequence of lateral inhibition and lateral induction of the Delta-Notch protein signals. This paper investigates an occurrence of Delta-Notch lateral inhibition by producing two model representations of the process applied to the development of hair cells within the sensory end-organ of hearing – the Organ of Corti.

2. Biological Context

The Delta-Notch protein signalling network is an example of juxtacrine signalling, meaning that the cell-cell signals are restricted to neighbouring cells in physical contact with one another. Both Delta and Notch are trans-membrane proteins. Juxtacrine signalling occurs when a Delta-type ligand binds to a Notch receptor (see Figure 1a, b). Binding causes a conformational change in the Notch protein which has the effect of breaking off the intra-cellular portion of the protein – termed the 'Notch Intra-cellular Domain' (NICD) component. The NICD component passes through the intra-cellular boundary to the nucleus of the cell where it interacts with a DNA-binding protein called CSL. The resultant change caused by the reception of this NICD 'signal' depends on a second intra-cellular Notch protein component termed $Notch^{intra}$. When $Notch^{intra}$ is absent the target cell is repressed which leads to lateral inhibition (Figure 1a); but in its presence the target cell is activated which leads to lateral induction (Figure 1b).

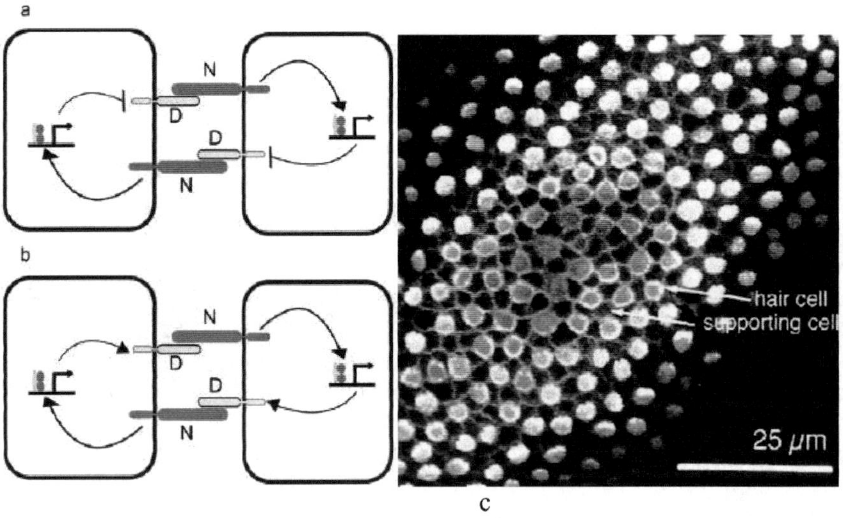

Figure 1: a) schematic diagram of lateral inhibition [6]; b) schematic diagram of lateral induction [6]; c) Sensory hair patches in a chick's inner ear [7].

Whereas the emergent pattern behaviour of lateral induction can take many forms, the behaviour of cells that exhibit lateral inhibition has a typical chequer-board pattern where each activated cell is surrounded by repressed cells. The sensory end-organs are examples of where lateral inhibition of neighbouring cells can be found, which can also contribute towards physiological function. In the eye, lateral inhibition of the rod cells yield a greater visual acuity between shades of grey; and in the organ of Corti in the inner ear, each hair cell is surrounded by a so-called 'supporting' cell (Figure 1c refers). The model representations below are sufficiently rich to demonstrate the tell-tale chequer-board patterns found in lateral inhibition. They can therefore be used in simulations to test hypotheses in behaviour of the model that maps to behaviour that may be found in the development of hair cells within the organ of Corti.

3. Outline Methods

3.1. Cellular Potts Models
Cellular Potts models (CPMs) provide simulations in which biological cells are represented by multiple sites on a pixelated lattice. The simulation is driven by updating the lattice one pixel at a time based on rules that reduce the entropy of the system as a whole. Finding the lowest energy configuration can be achieved in CPMs by using the Metropolis Monte Carlo algorithm. For this, a Hamiltonian Energy, H, is defined as a measure of entropy for the whole lattice. In each simulation step a random copy attempt is made at each lattice site that represents a cell surface. The resulting change in energy, ΔH, is calculated and accepted with a probability of $min(1, e^{\Delta H/T})$, where the T parameter is an intrinsic measure of cell motility.

Though CPMs can model cell behaviours such as adhesion, surface area, chemotaxis, and apoptosis, in this study cell-cell signalling is highlighted.

3.2. Biological Circuits
To represent Delta-Notch lateral inhibition as a biological circuit takes the programming environment used – National Instruments LabVIEW ® (LabVIEW) - into a novel domain and thereby stretches the use cases available to users across the globe. In essence, LabVIEW is a graphical programming environment for hardware and software integration to design and deploy measurement and control systems. It is employed as the program of choice here, as juxtacrine signalling can be considered as a biological measurement and control system, exhibiting regulatory cycles that are composed primarily of negative feedback control loops. The LabVIEW program component at the heart of the 'develop-and-test cycle' is termed the VI (for Virtual Instrument). The VI comprises three components: the block diagram, the front panel (user interface); and the connector panel (for sub-system integration).

4. Outline Results

4.1 Cellular Potts Models
Multi-scale cellular Potts models have been used to represent Delta-Notch lateral inhibition and lateral induction. This was achieved by assigning a SBML model to each cell in a simulation which had previously been solved dynamically. Cell level parameters were set as a function of sub-cellular variables, and the sub-cellular models took input values calculated from neighbouring cells in the Potts simulation environment. This multi-scale implementation gives the freedom to investigate developmental processes that involve both pattern formation through signalling, and changes in the physical shape, architecture and arrangement of cells in tissue. The development of the organ of Corti is an excellent candidate for this kind of modelling and simulation. It has been investigated thoroughly via experimental means, and involves juxtacrine signalling, growth and remodelling simultaneously.

4.2 Biological Circuits
A holistic systems approach was adopted to generate requirements for a simulation model to represent cell-cell communication. A 2-D array of cells (30 x 30 array in Figure 2a, b) was constructed with each cell possessing a random value between zero and 10. If the value of the cell was greater than 5, Delta is said to be expressed; otherwise the fate of the cell is Notch-expressed. The simulation was driven by averaging a 4-neighbour cell configuration; should this average exceed 5, then 0.5 is subtracted from the value of the target cell, otherwise 0.5 is added. Clearly, this may alter the fate of cell expression. This process is repeated for each cell in the array, and when all the cells in one iteration have been compared they adopt their new values. Figure 2a shows a typical simulation start state. After 55 whole array iterations, the tell-tale chequer-board pattern of lateral inhibition can be determined (Figure 2b refers).

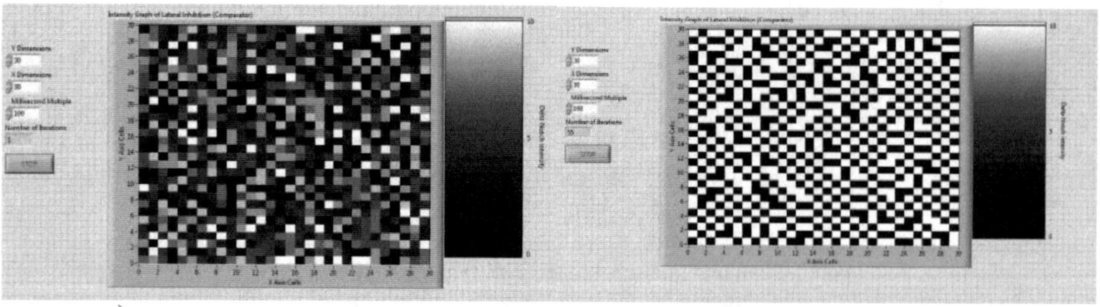

a) b)

Figure 2 A LabVIEW VI front panel for a simulation of lateral inhibition a) at the start of a simulation, b) at the end of a simulation.

To improve the verification and validation of the simulation, the simple arithmetic comparison rule that drives the simulation will be replaced by linked differential equations. A 6-neighbour configuration will also be adopted where each cell has a hexagonal representation.

5. Conclusions

It is hoped that model-based representations and simulations of sensory systems will provide deep insights into the study of the neuronal wiring of the brain. Though results thus far from mapping the connectome have produced stunning visualisations, it would be interesting to map the effect of lateral inhibition of the Delta-Notch signalling network of one of the sense end-organs to their locations in either the visual or auditory cortex. A multi-scale understanding of the signalling pathway, helped via further model-based interventions of cell signal transmission other than juxtacrine (e.g. by the use of neuristors that comprise a new electrical circuit component termed a 'memrister' [8] that can mimic the spikes seen in neurone models) might then yield new kinds of therapeutic interventions for blindness and deafness.

References

[1] von Bertalanffy, L., *General Systems Theory*, New York: George Braziller, 1968.

[2] Jenkins, G., The systems approach, In: Beishon, J. and Peters, G. (eds.), *Systems Behaviour*, (2nd Ed.), London: Harper Row, 1969.

[3] Summers, R., Abdulla, T., Houyel, L., Schleich, J-M. Progress with a Multiscale Systems Engineerng Approach to Cardiac Development, *Automatika*, 52, 2011, pp. 49-57.

[4] Abdulla, T., Luna-Zurita, L., de la Pompa, J. L., Schleich, J-M., Summers, R. Epithelial to Mesenchymal Transition – the roles of cell morphology, labile adhesion and junctional coupling, *Computer Methods and Programs in Biomedicine*, in press.

[5] Abdulla, T, Imms, R, Dillenseger, J-L, Schleich, J-M, Summers, R. Computational modelling of epithelial to mesenchymal transition, *IRBM*, 32(5), 2011, pp.306-310.

[6] Abdulla, T. Advances in Modelling of Epithelial to Mesenchymal Transition, *PhD Thesis*, SEESE, Loughborough University, 2013.

[7] Adam, J., Myat, A., Le Roux, I., Eddison, M., Henrique, D., Ish-Horowicz, D., Lewis, J. Cell fate choices and the expression of Notch, Delta and Serrate homologues in the chick inner ear: parallels with Drosophila sense-organ development, Development, 125, 1998, pp.4645-4654.

[8] Pickett, M. D., Medeiros-Ribeiro, G., Stanley Williams, R. A scalable neuristor built with Mott memristors, *Nature Materials*, Advance on-line publication, 16 December, 2012, DOI:10.1038/NMAT3510.

Journal of Physics: Conference Series **459** (2013) 012014 doi:10.1088/1742-6596/459/1/012014

Scaffold characterization using NLO multimodal microscopy in metrology for regenerative medicine

Leonardo Mortati[1], Carla Divieto[1], Monica Boffitto[2], Susanna Sartori[2], Gianluca Ciardelli[2], Maria Paola Sassi[1]

1Istituto di Ricerca Metrologica (INRIM), strada delle cacce, 91 – Torino(Italy)
2Politecnico di Torino, corso Duca degli Abruzzi, 24 – Torino (Italy)

m.sassi@inrim.it

Abstract. Metrology in regenerative medicine aims to develop traceable measurement technologies for characterizing cellular and macromolecule behaviour in regenerative medicine products and processes. One key component in regenerative medicine is using three-dimensional porous scaffolds to guide cells during the regeneration process. The regeneration of specific tissues guided by tissue analogous substrates is dependent on diverse scaffold architectural properties that can be derived quantitatively from scaffolds images. This paper discuss the results obtained with the multimodal NLO microscope recently realized in our laboratory in characterizing 3D tissue engineered (TE) scaffolds colonized from human Mesenchimal stem cells (hMSC), focusing on the study of the three-dimensional metrological parameters.

1. Introduction

Metrology in regenerative medicine addresses a field of research novel at International level, a field where the convergence of nanotechnology and biotechnology will provide a significant contribution to current and future progress. It aims to develop traceable measurement technologies for characterizing cellular and macromolecule behaviour in regenerative medicine products and processes. One key component in regenerative medicine is using three-dimensional porous scaffolds to guide cells during the regeneration process.[1,2] These scaffolds are intended to provide cells with an environment that promotes cell attachment, proliferation, and differentiation. After sufficient tissue regeneration using in vitro culturing methods, the scaffold/tissue structure is implanted into the patient, where the scaffold will degrade away, thereby leaving only regenerated tissue; on a different approach, noncellularised scaffolds are inserted into the patient to elicit in vivo cell recruitment, growth and tissue regeneration. Scaffolds need to meet both the biological goals of tissue formation and the stresses and loading conditions present in the human body. For this reason, any design approach must ensure that the mechanical properties of the resulting scaffold structure are compatible and optimally match the requirements from the environment, that, respectively, are the cell adhesion transmembrane protein, the cytoskeleton structure, the cell population. The need to design scaffold structures, the need for precision control during their fabrication and for determining the metrological indices and the need to characterize their structural behaviour at different scales have lead to numerous experimental and computational challenges. The regeneration of specific tissues guided by tissue analogous substrates is dependent on diverse scaffold architectural properties that can be derived quantitatively from scaffolds images. Over the last few years a number of imaging techniques have started to be developed which could be applied profitably to tissue engineering. Many of these techniques require labeling of the cells or scaffolds in order to accurately make measurements within the 3D tissue engineered product. However, recently a number of label free technologies have been developed which offer the potential for rapid information rich imaging without compromising or altering the product composition or interfering with the behavior of the cells. In particular, label free imaging technologies such as coherent anti-Stokes Raman scattering (CARS) and other Non-linear Optics (NLO) techniques offer a great deal of potential. A multimodal approach in which different NLO techniques are implemented in a same microscope is particulary powerful [3,4,5].

This paper discuss the results obtained with the multimodal NLO microscope recently realized in our laboratory [6,7,8,9] in characterizing 3D tissue engineered (TE) scaffolds colonized from human Mesenchimal stem cells (hMSC) [6], focusing on the study of the three-dimensional metrological parameters.

2. Materials and methods

2.1. Multimodal microscopy

The multimodal microscope integrate three NLO techniques CARS/SHG/TPEF on a single platform. It is based on a microscope scan unit (Fluoview FV300, Olympus) combined with an upright microscope (BX51WI, Olympus). Near infrared picosecond pulse trains are used to excite the non-linear optical processes in the samples. The optical source system is based on a passively mode-locked Nd:YVO_4 laser (Picotrain, HighQLaser) at 1064 nm and equipped with a SHG unit emitting at 532 nm. The frequency doubled output synchronously pumps an optical parametric oscillator (OPO - Levante Emerald, APE Berlin) with repetition rate 76 MHz and temporal width pulses about 6 ps. The OPO signal and idler waves are simultaneously generated by an LBO crystal in the resonant cavity, tunable in the wavelength range 700 nm - 1020 nm and 1110 nm - 2200 nm respectively. Both waves, exiting the system time and space overlapped and collimated, provides CARS stokes and pump signals. The OPO signal wave excites both SHG and TPEF processes, its photon energy having in the non-linear processes a higher efficiency with respect to the idler photon. The system was implemented with optics and mechanical systems suitable for the simultaneous detection of the different NLO techniques and to provide a z axis combined movement. *Ad hoc* algorithms have been developed for an automatic acquisition of 3D images [6].

2.2. Cell culture

Mesenchymal stem cells (hMSCs; Lonza) have been cultured in cell culture flasks with cell culture medium (DMEM from Sigma Aldrich with 10% FBS, 100 U/mL penicillin, 100 g/mL streptomycin) until they reach the 80% of confluence. Cells have been detached from culture flasks by using a trypsin/EDTA solution (Invitrogen) and counted by using a hemocytometer. 10^5 cells have been fluorescently stained with Calcein AM (Invitrogen), a vital staining specific for living cells only: the cell suspension (in cell culture medium) has been centrifuged at 1000 g for 5 minutes in a centrifuge tube to obtain a pellet of cells and 500 µl of PBS containing calcein AM 4µM has been used to re-suspend the pellet. After 10 minutes at 37 °C the cell suspension (in PBS/Calcein AM), stained cells have been washed in PBS to eliminate the excess of Calcien AM: cell suspension has been centrifuged again at 1000 g for 5 minutes. The pellet of stained hMSCs has been re-suspended in 100 µl of cell culture medium and this volume (a drop) has been seeded on the top surface of the scaffold PHBHV. The scaffold, with the drop containing cells, in a Petri dish has been incubated at 37 °C and 5% CO_2 for 1 hour to leave the cells in the drop to invade the scaffold. Then, the scaffold with cells has been attached to a Petri dish to clamp the sample during the imaging. To this purpose it has been used the fibrin gel made by mixing thrombin (12,5 µg/ml) and fibrinogen (5 mg/ml) at 1:2 ratio. This solution is liquid at 4°C but gradually became gel at room temperature. A small drop (10 µl) of fibrin gel is added on a dry Petri dish and very quickly the scaffold with cells is placed on the fibrin gel drop. After

5 minutes at 37 °C the scaffold is clamped to the Petri dish. Then, 10 ml of cell culture medium has been added to the Petri dish and incubated at 37 °C and 5% CO_2 over night to allow cells to adhere to the scaffold and to migrate into the scaffold internal structure.

2.3. Scaffolds
Degradable linear polyurethanes from non toxic building blocks were prepared. Bioactive peptides were incorporated in polyurethane chains, in order to create a synthetic equivalent of the Extracellular Matrix, able to mimic natural tissue. Both cell adhesive peptide (RGD sequences) and enzymatic proteolitically degradable peptides, containing the Ala-Ala sequence, were used as chain extender. Furthermore amino-derived diamines and cyclic diols were used as alternative chain extenders [10,11]

3. Results and discussion
Two complementary experiments have been done with the aim of studying with a non invasive and new methodology one of the most important features in regenerative medicine processes, the interaction of the cells with the scaffold. The metrological approach, studying the uncertainty components in measuring some of the scaffold's characteristic parameters, allows an estimate of the confidence achievable in characterizing those products and processes.

In the first experiment a dry fragment of the polymeric structure was imaged with CARS technique. The Coherent Antistokes Raman spectrum of the polymeric structure was detected and recorded in the wavelength range between 2750 cm^{-1} and 3150 cm^{-1}. Figure 1 shows this spectrum. Two peaks are evident at 2850 cm^{-1} and at 2920 cm^{-1}. This last wavelength was chosen to image the structure.

CARS three-dimensional reconstruction of a dry fragment of the structure architecture was done in a region size of 1414 x1414 x350 μm^3 with a Z-stack of 141 images with resolution 1024 x 1024 pixel and a Z step 2.5 µm. The 3D image was recorded and visualized importing the stack in the free software ImageJ, setting up the voxel size parameters and using 3D viewer plug-in.

Figure 2 shows the 3D reconstruction of the polymeric structure from the CARS signal.

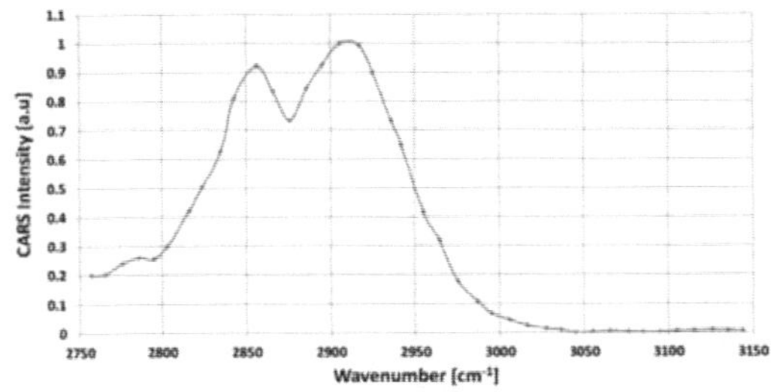

Figure 1 - Coherent Antistokes Raman spectrum of the polymeric structure in the wavelength range 2750 cm^{-1} and 3150 cm^{-1}

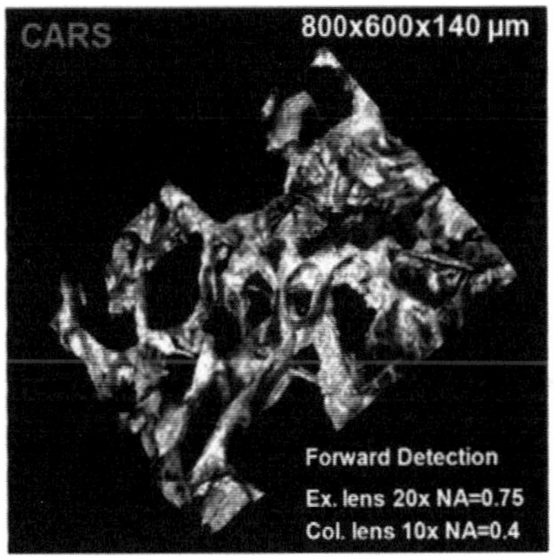

Figure 2 – CARS three-dimensional reconstruction of a dry fragment of the structure architecture

Recent literature and the standard guides [2,12] define specific geometrical parameters characterizing the micro-structure of scaffolds, fundamental to the cell attachment and in-depth penetration (and tissue vascularization), in conjunction with biological factors as cell seeding and nutrients supply. In particular, being an adequate transport of nutrient through the scaffold fundamental for culturing cells, the capability of scaffolds to be permeated by nutrients and metabolites, i.e., the ease with which fluids/species move through the scaffold, has a marked impact in

Tissue Engineered processes. As a consequence a metrological measurement of parameters as porosity, pore size and distribution, surface area, and tortuosity is the key argument for Tissue Engineered product knowledge and quality. This immediate translate in safety of regenerative medicine processes.

Different techniques are typically used to determine and to quantify these parameters; a drawback of the used techniques is their invasiveness with respect to the sample [2,13]. CARS imaging could bypass this limit through non-invasive measurements of samples, allowing the samples after the measurements to be used for cell culture and possibly for implants.

With this in mind a preliminary metrological study of the CARS imaging of the scaffold has been done, key parameters have been measured and an estimate uncertainty have been defined.

The 3D CARS images have been processed using an ad hoc developed computational algorithm [6] to extract a number of morphological parameters. The parameters are: the Pore Circularity (a value between 0 and 1 indicating a thin long shape when approaches 0 and a perfect circle when approaches 1), the Pore Roundness (a value between 0 and 1 indicating an elongated shape when approaches 0 and a perfect circle when approaches 1), the Pore Feret'sDiametre (the maximum distance between two points along the pore boundary), the Average Porosity (the ratio between the total pores area and the total scaffold area) and the Void Space Ratio (the ratio between the total pores volume and the total scaffold volume), following the standard guide [12] terminology and definitions.

With the aim of measuring uncertainty components due to measurement artifacts lied to (i) CARS technique itself, (ii) optical attenuation and distortion phenomena depending on the sample characteristics (iii) uncertainties in software threshold definition, different measurement have been performed of the same sample, changing optical parameters and software threshold analysis. Uncertainties have been estimated as reproducibility over these parameters.

Table 1 shows the results obtained by the measurement of morphological parameters of the polymeric scaffold in analysis, with associate estimated uncertainties.

Parameter	Measured value from CARS Images
Average Porosity	56.5%
Void Space Ratio	89.6%
Pore Circularity	0.16 ± 0.1
Pore Roundness	0.55 ± 0.2
Pore Feret's Diameter	(179 ± 180) µm

Table 1- Measured values of the shape descriptor parameters obtained by image processing of the CARS 3D reconstruction of the polymeric structure

A second study has been done on the ability of the specific polymeric structure to act, for human Mesenchimal stem cells (hMsc), as a scaffold; that means to be a suitable artificial matrix where cells could survive, proliferate, and colonize the structure in 3D. This study, novel for its capacity to image living samples without perturbing its behaviour, has the potential to quantify two main parameters: proliferation of cells and localization with respect to the 3D scaffold. Nevertheless this quantification is out of the scope of this paper.

A 3D polymeric structure 3 mm x 3 mm x 0,350 mm have been cultured as previously described. The construct was observed under the multimodal microscope for three days in a time-lapse experiment [6, 14] with the aim of studying the interaction between scaffold and hMsc cells.

Figure 3 shows the cells colonizing the structure at different times during three days in culture. Fluorescent calcein, metabolized by the cells was detected by the multimodal microscope in Two Photon Emission Fluorescence (TPEF) EPI detection modality, showing the 3D cell cytoplasma morphology. Simultaneously the 3D structure of the scaffold was detected by the multimodal microscope in Coherent Antistokes Raman Scattering (CARS) Forward modality.

In each measurement a stack of 66 images of 512x512 pixels averaged 2 times with a Kalman filter, with a Z step of 1 μm were collected in around six minutes. All the Z stacks were converted in a 3D view using 3D viewer imageJ plug-in.

The analysis of the 3D images shows the position of the single cells with respect to the static position of the scaffold. It is possible to observe the number of cells per area increasing during the time in culture, showing a strong proliferation. After the first 24 hours cells start to colonize the scaffold pores from the pore's external surface to its centre. Some of the cells disappeared, due to different death processes (apoptosis). These results demonstrate that this specific construct is a good example of regenerative medicine product, useful for metrological studies of regenerative medicine processes.

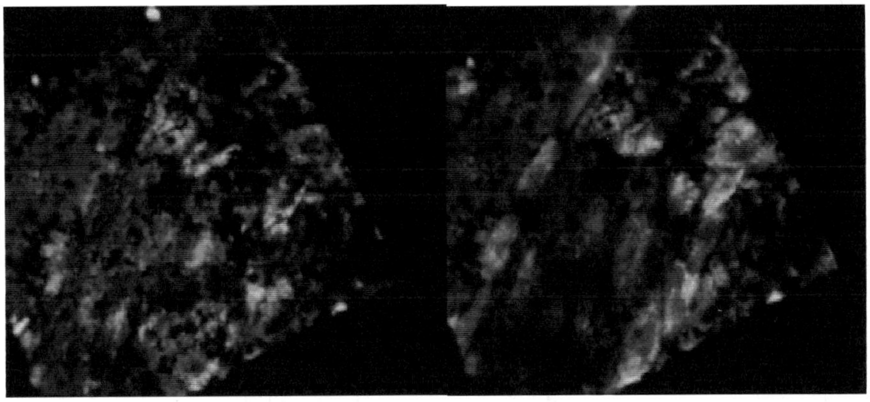

Figure 1 – NLO multimodal microscope images of living hMsc cells (green) colonizing the 3D scaffold (blue) at different times (24h, 30h) in a three day experiment.

With respect to the measurement method used for this analysis, it seems extraordinary promising for its intrinsic image resolution, for its ability to a deep 3D imaging, for the independent but simultaneous acquisition of scaffold and cells without any perturbation of their native processes.

4. Conclusion

The results obtained with the multimodal NLO microscope recently realized in our laboratory in characterizing 3D tissue engineered (TE) scaffolds colonized from human Mesenchimal stem cells (hMSC), have been focused on the study of the three-dimensional scaffold parameters and of the real time cell behaviour when cultured on a specific polymeric structure. Such studies demonstrate NLO microscopy as a powerful, not invasive and reproducible instrument for research in regenerative medicine.

These results show the possibility to develop traceable measurement technologies for characterizing cellular behavior in regenerative medicine products and processes. Further studies have to be done in order to better define eventual artifacts in measuring scaffold and cells morphology with the different imaging techniques.

Time-lapse experiments allowed by multimodal NLO microscopes are a real novelty in studying one of the most important features in regenerative medicine processes, the interaction of the cells with the scaffold.

References

[1] Nerem, R. M., and A. Sambanis. Tissue engineering: from biology to biological substitutes. Tissue Eng. 1(1):3–13, 1995; F. Consolo, F. Mastrangelo, G. Ciardelli , F.M. Montevecchi , U. Morbiducci, M. Sassi and C. Bignardi Multilevel experimental and modelling techniques for bioartificial scaffolds and matrices (2010) Applied Scanning Probe Methods, Vol. 14 Edited by B. Bhushan Springer-Verlag, Heidelberg.
[2] Pennella F, Cerino G, Massai D, Gallo D, Falvo D'Urso Labate G, Schiavi A, Deriu MA, Audenino A, Morbiducci U. A Survey of Methods for the Evaluation of Tissue Engineering Scaffold Permeability. Ann Biomed Eng. 2013 Apr 24. [Epub ahead of print] DOI:10.1007/s10439-013-0815-5
[3] W.R. Zipfel, R.M. Williams, and W.W. Webb, "Nonlinear magic: multiphoton microscopy in the biosciences," Nature Biotech. 21, 1369-1377 (2003)].
[4] S. Yue, M. N. Slipchenko, J. X. Cheng, "Multimodal nonlinear optical microscopy" Laser Photonics Rev. 5(4), 496–512 (2011).
[5] Potma EO, Cheng JX, Xie XS "Coherent Raman imaging techniques and biomedical applications" J Biomed Opt. 2011 Feb;16(2):021101. doi: 10.1117/1.3558735.
[6] Leonardo Mortati Ultrasensitive Coherent Anti-Stokes Raman Scattering Microscopy for Metrology in Regenerative Medicine (2013), PhD Thesis SCUDO Politecnico di Torino Dottorato di Ricerca in Metrologia;
[7] Guido Sassi, Marco Bernocco, Mariapaola Sassi (2011). Uncertainty of the Diffusion Measurements on Scaffolds for Cell Growth . Diffusion And Defect Data, Solid State Data. Part A, Defect And Diffusion Forum, vol. 312 315, p. 770-775, ISSN: 1012-0386
[8] Mortati L, Divieto C, Sassi M (2012). CARS and SHG microscopy to follow collagen production in living human corneal fibroblasts and mesenchymal stem cells in fibrin hydrogel 3D cultures. Journal of Raman Spectroscopy, vol. 43 , p. 675-680, ISSN: 0377-0486, doi: 10.1002/jrs.3171
[9] Sassi M, Mortati L (2011). Chemical selective in vitro bioassay for cells/EC matrix detection in 3D nonlabelled cell cultures. Regenerative Medicine, vol. Vol. 6 No.6 (Suppl. 2), p. 147, ISSN: 1746-0751
[10] V. Chiono, E. Descrovi, S. Sartori, P. Gentile, M. Ballarini, F. Giorgis, G. Ciardelli, NanoScience and Technology, 2011, Part 3, 645-689
[11] G. Ciardelli, S. Sartori, P. Serafini, M. Boffito, A. Caporale, A. Silvestri, E. Bernardi, F. Boccafoschi Nanotechnology 2011: Bio Sensors, Instruments, Medical, Environment and Energy, Chapter 3: Bio Nano Materials, 155 - 158 (2011)
[12] ASTM F2150 – 07 "Standard Guide for Characterization and Testing of Biomaterial Scaffolds Used in Tissue-Engineered Medical Products" January 2008
[13] F. Gelain, D. Bottai, A. Vescovi, S. Zhang, PLoS ONE 1(1): e119. doi:10.1371/journal.pone.0000119; B. D. Boyan, C. H. Lohmann, D. D. Dean, V. L. Sylvia, D. L. Cochran, Z. Schwartz, Ann Rev Mater Res 2001; 31: 357-71; N. R. Washburn, K. M. Yamada, C. G. Jr Simon, S. B. Kennedy, E. J. Amis, Biomaterials 2004; 25(7-8): 1215-24; T. P. Kunzler, T. Drobek, M. Schuler, N. D. Spencer, Biomaterials 2007; 28(13): 2175-82
[14] L. Mortati, M. P. Sassi, European Conference on Nonlinear Optical Spectroscopy (ECONOS 2012), 31st European CARS Workshop (ECW), 8-11 July 2012, University of Aberdeen, UK

Journal of Physics: Conference Series **459** (2013) 012015 doi:10.1088/1742-6596/459/1/012015

Extracting the respiration cycle lengths from ECG signal recorded with bed sheet electrodes

A Vehkaoja, M Peltokangas and J Lekkala
Department of Automation Science and Engineering,
Tampere University of Technology, Tampere, Finland

E-mail: antti.vehkaoja@tut.fi

Abstract. A method for recognizing the respiration cycle lengths from the electrocardiographic (ECG) signal recorded with textile electrodes that are attached to a bed sheet is proposed. The method uses two features extracted from the ECG that are affected by the respiration: respiratory sinus arrhythmia and the amplitude of the R-peaks. The proposed method was tested in one hour long recordings with ten healthy young adults. A relative mean absolute error of 5.6 % was achieved when the algorithm was able to provide a result for approximately 40 % of the time. 90 % of the values were within 0.5 s and 97 % within 1 s from the reference respiration value. In addition to the instantaneous respiration cycle lengths, also the mean values during 1 and 5 minutes epochs are calculated. The effect of the ECG signal source is evaluated by calculating the result also from the simultaneously recorded reference ECG signal. The acquired respiration information can be used in the estimation of sleep quality and the detection of sleep disorders.

1. Introduction and related work

Respiration is one of the most important physiological signals in sleep studies where the data is collected using polysomnography (PSG). It is used for the detection of breathing-related sleep disorders by following the amplitude of the breathing signal and comparing the breathing signals recorded with respiratory inductance plethysmography and temperature sensors. Respiration rate (RR) or respiration cycle length (RCL) can also be used in sleep staging because the variation of respiration rate is higher when a person is awake or during the rapid eye movement sleep stage and is decreased during deep sleep stages [1]. The respiration related features are even more important in novel unobtrusive systems developed for long term monitoring of sleep, like the one presented in [2]. The reason is that smaller amount of physiological signals are available through unobtrusive monitoring techniques and e.g. electroencephalogram and electrooculogram, which are normally used in sleep staging in PSG recordings cannot be measured unobtrusively.

Many authors have investigated the detection of the RR or RCL using ECG signal. In these studies, the ECG has been recorded using conventional electrodes attached to certain locations on the torso. The most commonly used ECG related features for the RR detection are respiratory sinus arrhythmia (RSA) and the modulation of the R-peak amplitude (RPA). The RSA is a result of the variation in the instantaneous heart beat intervals caused by the respiration related control of the autonomous nervous system, whereas the RPA is caused by the changes in the lead field and the heart orientation due to the breathing movement. [3–6]

We have earlier presented a system for recording night-time ECG unobtrusively using textile electrodes sewn on a bed sheet and used the data for detecting heart rate and heart rate variability parameters [7]. The current work investigates the usability of the recorded ECG in the detection of respiration cycle lengths. While other studies have investigated the detection of RR or RCL using conventional electrodes in fixed locations, the electrode locations do not stay constant when using the bed sheet integrated electrodes but change according to the sleeping posture and the measurement channel used. Also the quality of the ECG signal is not always as good as if using the regular gel electrodes.

2. Materials and Methods

2.1. ECG measurement system and R-peak detection algorithm

Our ECG measurement system consists of eight electrodes that are made by embroidering using silver coated polyamide yarn and are sewn on a bed sheet in 5 cm distances. The ECG is measured bipolarly between each adjacent electrode pair, which enables also combining the recorded ECG signals thus forming new ECG leads with wider inter-electrode distance. When recognizing the R-peaks and heart rate from the ECG, the ECG lead that produces the best signal is first selected and then used until the quality of the signal decreases too much. A detailed description of the channel selection and the R-peak recognition procedure is presented in [7] and the effect of combining the ECG channels in [8].

2.2. ECG derived respiration signals and RCL detection

We selected two commonly used respiration related features that are present in the ECG signal to determine the RCL: RSA and RPA. We also tested filtering the ECG signal with low, 0.6 Hz cut-off in order to reveal the baseline wandering caused by the breathing. The resulting signal however suffered from frequent interferences and did not provide as good results as the RSA and RPA signals.

The raw RSA and RPA signals contain one sample for each heartbeat. Both signals are first interpolated into constant 0.1 s sample interval using cubic spline interpolation. After this, the signals are filtered with different pass-bands in order to bring out the respiration related variation in the signals. Theoretically, both signals should contain a sinusoidal kind of variation at the frequency of the respiration but they also include interference components. Different pass-bands are used because the RCL usually varies between 2 s and 10 s, corresponding to the frequency range of 0.1–0.5 Hz. Depending on the respective value of the RCL, different pass-band will produce the best filtering result. The selected pass-bands are 0.1–0.22 Hz, 0.1–0.33 Hz, and 0.1–0.5 Hz. Forward-backward-filtering was used in order to eliminate the phase delay, which in this case could cause a significant distortion in the signals. The order of the Butterworth response filters was set as two. Using the forward-backward filtering then produces the overall filter order of four.

Next, the repetition intervals of each signal are calculated by searching local maxima and minima as well as rising and falling zero crossings and then calculating the repetition intervals of these features. This way, altogether 24 (respiration) interval signals are formed (2 ECG derived signals × 3 filters × 4 features). The interval signals are then linearly interpolated into 1000 ms sample interval. Finally, for each time instant a cluster of six RCL suggestions that are closest to each other is searched for. If the square sum of the distances is less than an empirically set threshold, the center of the cluster is chosen as the RCL for that time instant. If the smallest square distance of any six suggestions is higher than the threshold, no RCL value is chosen. Earlier we used a similar method for detecting the RCLs from the signals that were recorded using force sensors placed under a bed post [8].

2.3. Per epoch mean RCL detection

In addition to the instantaneous RCLs, we also tested how the presented method performs when the RCL values are used for calculating the mean RCL for epochs of certain length. The mean RCL is simply calculated by averaging all the RCLs detected during e.g. 1 or 5 minute epochs. If the mean RCL is calculated for all epochs that contain at least one detected instantaneous RCL, good coverage is achieved but the error of the result is probably higher. The error can be decreased by setting the minimum number of RCLs that must have been detected during an epoch before the mean RCL for the epoch is reported. The results were calculated using the overlaps of 50 and 240 seconds for the 60 and 300 second epoch lengths, respectively.

2.4. Reference respiration measurement
The reference respiration signal was measured using an NTC thermistor placed inside a breathing mask. The resistance of the thermistor was measured, converted into the temperature, and filtered. The respiration rate was calculated from the band-pass filtered temperature signal by finding the maximum from each separate segment that is above the zero level. These points correspond to the end of the inhale phase. Finally, the signals were visually inspected and false or missing detections were corrected. Short sections from the recordings of four test subjects were found, where the amplitude of the reference breathing signal was significantly decreased or the waveform distorted. These parts were discarded from the analysis because reliable reference was not available.

2.5. Test subjects and data collection
We made test measurements with eight male and two female subjects (subjects 6 and 10). The subjects were 23–33-year-old, normal weight and had no history of cardiovascular problems or sleeping related breathing disorders. The measurements were made in laboratory settings using an 80 cm wide spring mattress bed. The subjects were allowed to change their sleeping posture freely. The length of the recordings was approximately one hour and most of the subjects fell asleep during this time. In order to evaluate the effect of the ECG signal source to the performance of the proposed RCL detection algorithm, we also used the reference ECG signal that was recorded at the same time with disposable Ag/AgCl electrodes. The reference electrodes were placed under both clavicles so that they did not disturb the skin contact of the sheet electrodes in any posture.

3. Results

3.1. Instantaneous RCL detection results
Table 1 shows the results of instantaneous RCL detection for each subject. As seen in table 1, there is a significant variation in the performance of the method between the subjects. The best detection coverage was obtained with the data of subject 5, 83.2 % of the total recording time or 91.9 % of time when the source ECG signal was available. The average detection coverage for all the subjects is approximately 40 % of the whole recording time, which is nearly 50 % of the time the bed sheet ECG signal has been available. The ECG signal of the test subject 7 was so low in amplitude that the RPA signal was practically unusable and even when the R-peak detection algorithm was able to find the R-R-interval signal with 74 % coverage, the RCL detector found results that were reliable enough only 3 % of the time. The data of this subject was therefore excluded from the calculation of the average results.

The last two rows in table 1 present the instantaneous RCL detection results calculated using the reference ECG data. The coverage of the detected RCLs is higher when using the reference ECG than with the sheet ECG but this is explained by the fact that the RSA and RPA signals produced by the reference ECG are available almost 100 % of the time whereas the coverage of the sheet ECG based RSA and RPA is 82 % in average. Also the error figures for the reference ECG based RCL are slightly better than for the sheet ECG. The results of the subject 7 were now included because the reference ECG did not suffer from the same problems as the sheet ECG signal.

Table 1. Respiration recognition performance characteristics. The largest and smallest value of each parameter is boldfaced. The last two rows show the minimum, maximum, and the average of the results calculated for each test subject using the reference ECG signal.

	RRI Coverage (%)	*Respiration Coverage (%)*	*Concordance Correlation*	*RMSE* [a] *(s)*	*MAE* [b] *(s)*	*MAE* [b] *(%)*	*e < 0.5 s* [c] *(%)*	*e < 1 s* [c] *(%)*
1	93.3	32.1	0.770	**0.782**	**0.527**	**11.135**	**64.1**	**85.3**
2	64.0	42.4	0.948	0.367	0.176	3.744	96.1	99.1
3	**61.4**	28.6	0.791	0.307	0.183	4.714	97.1	99.6
4	69.3	28.3	**0.273**	0.579	0.398	9.871	73.3	93.7
5	90.5	**83.2**	0.929	0.194	0.148	**3.251**	98.1	**100.0**
6	83.6	47.1	0.902	0.224	0.145	3.492	97.5	99.3
7	*74.5*	*3.0*	*0.003*	*0.654*	*0.328*	*10.369*	*86.4*	*90.0*
8	**97.5**	24.4	0.655	0.434	0.239	5.863	88.9	98.0
9	91.5	**19.0**	**0.973**	0.356	0.212	4.621	93.2	98.3
10	94.5	51.2	0.749	**0.169**	**0.131**	3.689	**99.2**	**100.0**
Mean	82.0	39.6	0.777	0.379	0.240	5.598	89.7	97.0
Min-Max Ref ECG	98.5–100.0	13.0–83.6	0.184–0.991	0162–0.588	0.116–0.241	2.678–5.683	94.7–99.3	97.2–100.0
Mean Ref ECG	99.6	47.8	0.777	0.309	0.165	4.144	97.1	98.9

a Root-mean-square error of the instantaneous and interpolated RCL values.
b Mean absolute error of the instantaneous and interpolated RCL values.
c Percentage of the RCL values that are within 0.5 second and 1 second from the reference.

Figure 1 shows as an example, the histogram of the RCL error for the test subject 5. As seen in figure 1, the error produced by the presented method has a Gaussian distribution with zero mean, which shows that the method does not produce any bias to the result. Figure 2 shows the scatter plot of the sheet ECG based instantaneous RCL values and the thermistor reference values for the same person. The concordance correlation is a measure of the deviation of the variable pairs from the 45-degree line drawn through the origin (the grey line in figure 2).

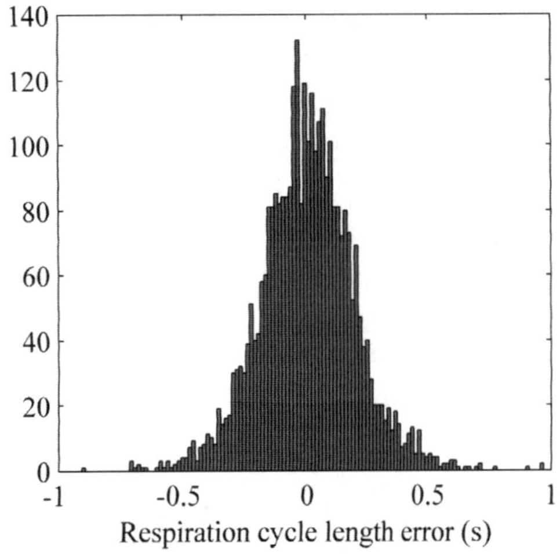

Figure 1. Histogram of the error of the detected respiration cycle lengths (subject 5).

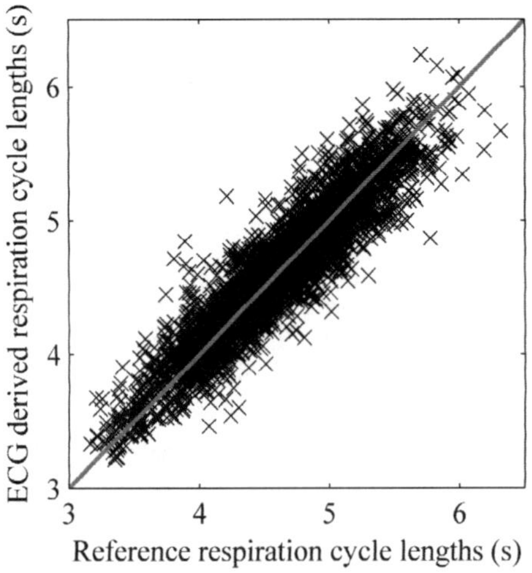

Figure 2. Scatter plot of the respiration results (subject 5).

3.2. Mean RCL detection results

Figure 3 shows the coverage and the mean absolute error (MAE) of the mean RCL values as a function of the minimum required number of the instantaneous RCLs in an epoch. The results in figure 3 are calculated for 1 minute epoch length and using 50 seconds overlap between epochs. 88.4 % of all epochs calculated using the sheet ECG data contained at least one RCL value and the average relative MAE for the mean RCL was 6.6 %. Somewhat better mean RCL coverage is achieved by using the reference ECG signal because the reference signal is always available. The average MAE is slightly smaller for the reference ECG when all possible epochs are considered but when the minimum number of the required RCLs in an epoch is increased, both signals produce similar errors.

Orphanidou *et al.* received 4.8 % average relative MAE for 1 minute mean respiration rate with young subjects when they combined the RSA and RPA signals by selecting as the result the one that has the higher magnitude of the pole in the autoregressive spectrum [3]. They did not calculate the results for all the available data but used a criterion for choosing the parts where a high quality reference signal was available. In this way they used 71 % of the available data. Similar coverage is achieved by the sheet ECG data when setting a 10 % minimum value for the within epoch RCL coverage. The average relative MAE is then approximately the same as in [3]. The reference ECG yields approximately 3.3 % MAE when using the same criterion.

Cysarz *et al.* used the concordance correlation coefficient (CCC) to describe the agreement between the respiration rate calculated in 5 minute time windows from a single ECG lead and the reference air flow sensor [4]. The regular Pearson correlation only considers the linear dependency of two variables but concordance correlation also takes into account the difference of the exact values of the tested variables. Cysarz *et al.* received the night-time CCC values of 0.79 and 0.73 younger group of subjects when using only RSA and RPA signals, respectively. They did not consider the epochs where the valid R-peaks were available less than 75 % of the time.

Table 2 shows the CCCs between the 1 and 5 minute mean RCLs and the thermistor based reference RCL. The average CCC of 0.63 was achieved with both sheet and reference ECG when all available epochs were considered. The CCCs are considerably improved (0.82 and 0.86 for the sheet and the reference ECG) when only the epochs that contain at least 25 % of the RCLs are considered. The mean RCL coverages are however then decreased to 64 and 79 %.

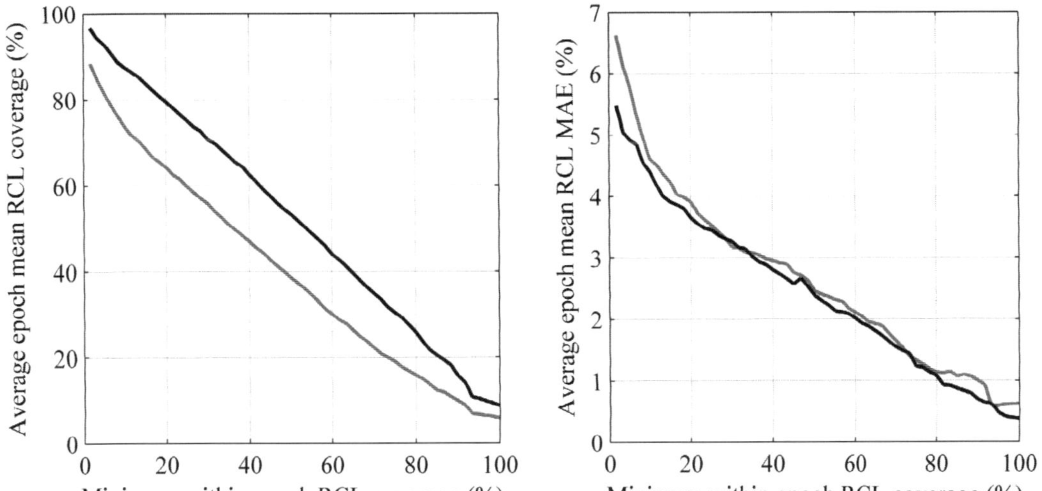

Figure 3. The average dependency of the coverage (left) and the MAE of the mean RCL values (right) for one minute epochs for all subject. The results calculated from the reference ECG are shown in black and the results calculated from the sheet ECG in grey.

Table 2. Average per epoch mean RCL recognition coverage and concordance correlation coefficients for 1 and 5 minute epoch lengths.

		Coverage Sheet (%)	*Coverage Reference (%)*	*Concordance Correlation Sheet*	*Concordance Correlation Reference*
1 minute epoch length	**All epochs**	88.4	96.6	0.69	0.59
	Epochs > 25 %	58.3	73.7	0.89	0.87
	Epochs > 50 %	37.4	51.9	0.94	0.95
	Epochs > 75 %	17.7	29.2	0.96	0.98
5 minute epoch length	**All epochs**	99.8	100	0.63	0.63
	Epochs > 25 %	64.2	79.2	0.82	0.86
	Epochs > 50 %	36.1	56.6	0.85	0.87
	Epochs > 75 %	13.3	23.8	0.96	0.87

Epochs > 25 %, 50 % or 75 %: only epochs that include more than the specified amount of recognized instantaneous RCL values are considered.

4. Conclusions

A method for detecting instantaneous respiration cycle lengths from an ECG signal recorded with bed sheet integrated textile electrodes or regular ECG electrodes has been presented. Also mean RCLs are calculated for 1 and 5 minute epochs and the results compared with methods found in the literature. The proposed method performs well with both sheet ECG and reference ECG signals at least with the test subjects who were younger than 35 years.

The future work includes the evaluation of the proposed method with people of wider demographics. It has been concluded that the RSA phenomenon attenuates along with aging and that parameter is not as efficient for estimating the RCL with older people [4].

References

[1] Chung G, Choi B, Lee J, Lee J, Jeong D and Park K S 2009 REM sleep estimation only using respiratory dynamics *Physiological Measurement* **30** 1327–40

[2] Paalasmaa J, Waris M, Toivonen H, Leppäkorpi L and Partinen M 2012 Unobtrusive online monitoring of sleep at home *Proc. 34th Ann. Int. Conf. IEEE Engineering in Medicine and Biology Society* (San Diego, CA, USA, 28 August–1 September 2012), pp 3784–88

[3] Orphanidou C, Fleming S, Shah S A and Tarassenko L 2013 Data fusion for estimating respiratory rate from a single-lead ECG *Biomedical Signal Processing and Control* **8** 98–105

[4] Cysarz D, Zerm R, Bettermann H, Frühwirth M, Moser M and Kröz M 2008 Comparison of Respiratory Rates Derived from Heart Rate Variability, ECG Amplitude, and Nasal/Oral Airflow *Annals of Biomedical Engineering* **36** 2085–94

[5] O'Brien C and Heneghan C 2007 A comparison of algorithms for estimation of a respiratory signal from the surface electrocardiogram *Computers in Biology and Medicine* **37** 305–14

[6] Boyle J, Bidargaddi N, Sarela A and Karunanithi M 2009 Automatic Detection of Respiration Rate From Ambulatory Single-Lead ECG *Transactions on Information Technology in Biomedicine* **13** 890–6

[7] Peltokangas M, Verho J and Vehkaoja A 2012 Night-Time EKG and HRV Monitoring With Bed Sheet Integrated Textile Electrodes *Transactions on Information Technology in Biomedicine* **16** 935–42

[8] Vehkaoja A, Peltokangas M., Verho J and Lekkala J 2012 Combining unobtrusive electrocardiography and ballistography for more accurate monitoring of sleep *Proc. IEEE 12th Int. Conf. on Bioinformatics and Bioengineering* (Larnaca, Cyprus, 11–13 November 2012) pp 202–7

Journal of Physics: Conference Series **459** (2013) 012016 doi:10.1088/1742-6596/459/1/012016

Ex vivo Time Evolution of Thrombus Growth through Optical and Electrical Impedance data fusion

A. Affanni, R. Specogna, F. Trevisan

University of Udine, Dept. of Electrical, Management and Mechanical Engineering, via delle Scienze 206, 33100 Udine (Italy)

antonio.affanni@uniud.it

Abstract. We designed a novel sensor specifically aimed at *ex vivo* measurements of white thrombus volume growth; a white thrombus is induced within an artificial micro-channel where hemostasis takes place starting from whole blood under flow conditions. The advantage of the proposed methodology is to identify the time evolution of the thrombus volume by means of an original data fusion methodology based on 2D optical and electrical impedance data simultaneously processed. On the contrary, the present state of the art optical imaging methodologies allow the thrombus volume estimation only at the end of the hemostatic process.

1. Introduction

Cardiovascular diseases and thrombotic complications are among the major causes of morbidity and mortality of the western world. Herein, we propose a novel methodology for the concurrent evaluation of platelet aggregate formation and activation of the coagulation cascade (fibrin formation) under *ex vivo* conditions that closely mimic the *in vivo* ones.

Current instruments estimate the thrombus growth through image processing of data acquired by inverted or confocal microscopes. For the imaging of white thrombus formation a fluorescent marker for platelets is used (e. g. quinacrine). The inverted microscope estimates the volume of white thrombi growth from a digital bi-dimensional (based on pixel luminance) data image of a whole blood sample and matematically extrapolates the 3D size, therefore the result may not reflect the actual volume of thrombi. On the other hand, the confocal microscope is a very expensive instrument and its use cannot be extended to a clinical screening; moreover it does not analyze the kinetics of thrombus formation and/or the thrombus stability since the volume estimation is performed at the end of the process.

In the paper, we present a novel methodology to measure the volume of white thrombus during its formation, by simultaneously monitoring electrical impedance and 2D optical image of pixel luminance of an aggregating blood sample.

Electrical Impedance Spectroscopy (EIS) on blood [1 – 2] has received a considerable attention because of non–invasive and real–time monitoring of the clotting process. The measurement bench is composed by an impedance meter, an optical microscope with a CCD camera, a control panel implemented on a Personal Computer and a developed sensor containing the microfluidic connections and electrodes to perform EIS. The control panel communicates via USB interface with the impedance meter and the CCD camera and acquires simultaneously the image and impedance information. At a time instant and at a specific frequency, the acquired impedance data are post processed and compared to a corresponding computed impedance obtained from an electromagnetic numerical model of the overall system formed by the microchannel, containing the thrombus, and the gold electrodes. The geometry of the thrombus is represented by a solid model deduced from the 2D optical image of pixel luminance. For each pixel, the height of the thrombus with respect to surface of the slide, is assumed proportional to the luminance according to a scale factor. In order to determine such a scale factor, a number of 3D electrical simulations are performed by modifying the scale factor in order to match the computed impedance with the measured one; the white thrombus volume is then evaluated by the thrombus solid model. The 3D electrical analysis tool is based on the Discrete Geometric Approach developed by the authors [3].

2. System Description

The developed microfluidic chip is connected to an LCR meter and posed under an optical microscope with a CCD camera; both LCR meter and CCD camera communicate with a calculator through USB interface and a software control panel simultaneously acquires optical and impedance data. Blood flows inside the sensor at a controlled temperature of 37 °C and at a controlled flow rate in order to simulate the *in vivo* conditions.

The sensor is composed by a Poly-Dimethyl-Siloxane (PDMS) substrate where the microchannel with microfluidic connections is molded and a quartz slide forming the upper wall of the channel where gold electrodes and electrical connections are sputtered. At every experiment a collagen is spread on the quartz slide to induce the thrombus formation and the microchannel is then assembled as shown in Figure 1 (a) and (b).

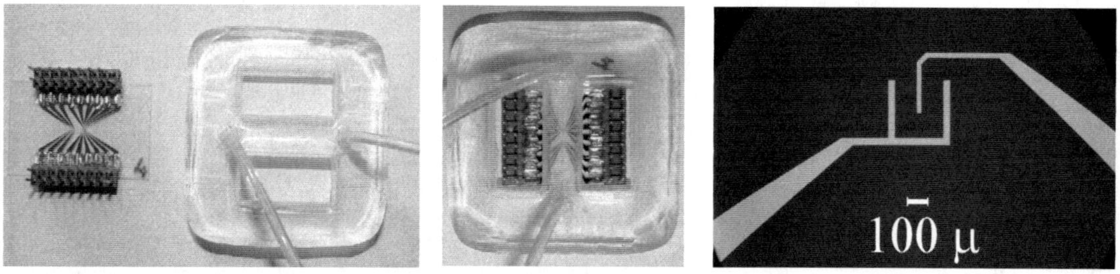

Figure 1: (a) view of PDMS substrate with fluidic connections and glass with sputtered gold electrodes; (b) assembly of substrate and slide; (c) microscope view of the electrodes disposition

Whole blood flows for a specific time interval (3 min) and thrombi grow from the surface of the slide where collagen has been spread. In order to characterize the accuracy of the estimated volumes with the actual volumes of the white thrombus, at the end of the experiment a 3D scan of the investigation area is performed with a confocal microscope.

2.1. Sensor Description and Measurement Setup

The developed microfluidic chip is composed by a PDMS substrate with fluidic connections and a quartz slide with 200 nm thick sputtered gold electrodes.

On the PDMS substrate it is molded an open microchannel without the upper wall whose dimensions are 400 μm in width, 250 μm in depth and 25 mm in length; also the fluidic connections have been molded on this substrate. The upper wall of the microchannel is formed by a quartz slide with 200 nm thick sputtered gold electrodes, arranged as shown in Figure 3(c).

The impedance spectroscopy measurements have been performed using the high precision LCR meter Agilent E4980A in the frequency range 1-300 kHz with eight logarithmic spaced steps, a two-wire configuration and a drive voltage of 100 mV; such drive voltage is by far lower than the standard half cell potential for gold (that is 1.5 V) in order to avoid artifacts due to redox reactions taking place between electrodes and salts dissolved in plasma. The LCR meter has been connected to the assembled empty device and an "open" calibration has been performed in order to eliminate the capacitive parasitic effects of connections. In these conditions the impedance measurement accuracy is 0.1 %.

Blood flows in the microchannel at a flow rate of Q = 750 μl/min in order to obtain a target shear rate of 3000 s^{-1}.

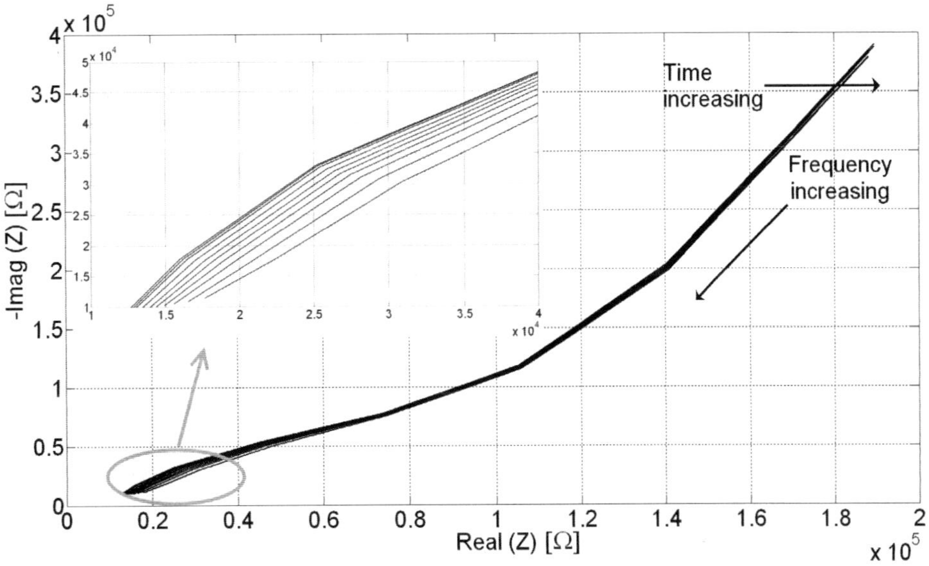

Figure 2: Impedance Spectroscopy during during thrombus growth; curves are 17 s time spaced

Figure 2 shows the Real - Imaginary plot of the impedance over a time interval of three minutes; the time resolution of acquired data is 1.7 s but for graphical reasons in Figure 2 each plotted curve is acquired every 17 s. As it can be seen in Figure 2, at low frequencies the interface electrode-blood (double layer) dominates the impedance behaviour masking the effects of thrombus formation. Since we already demonstrated that thrombus behaves like an insulator for frequencies below 300 kHz [3], the best frequency interval for the thrombus growth measurement is [100-300] kHz, in order to by-pass the double layer effects but not the cells membranes forming the thrombus. In this frequency range blood behaves mainly like a resistor whose resistance is increased by the thrombus growth, since the resistive cross section of blood is reduced proportionally to the thrombus height and width.

3. Experimental results

The measurements have been performed automatically acquiring data from the LCR meter and the CCD camera; the control panel has been set up in order to perform a two-wire measurement avoiding redox reactions and compensating the parasitic effects of cables and connections, as described in previous Section.

Figure 3 shows some significant frames acquired from the microscope during an experiment. The blood flows from the top to the bottom of each frame; the frames shown in Figure 3 (a) to 3 (f) are acquired 40 s, 60 s, 80 s, 100 s, 120 s, 140 s, 160 s and 180 s after the start of the experiment. The growing thrombus is represented by the white spots inside the investigation area. It is possible to see that, increasing time, single platelets are activated and start adhering to the collagen; then the aggregates increase both in base area and volume to form stripes in the direction of blood flow. With standard measurement systems (e. g. confocal microscope) the evaluation and measurement of the time evolution of growth is not possible, since the confocal scans can be performed only at the end of the experiment.

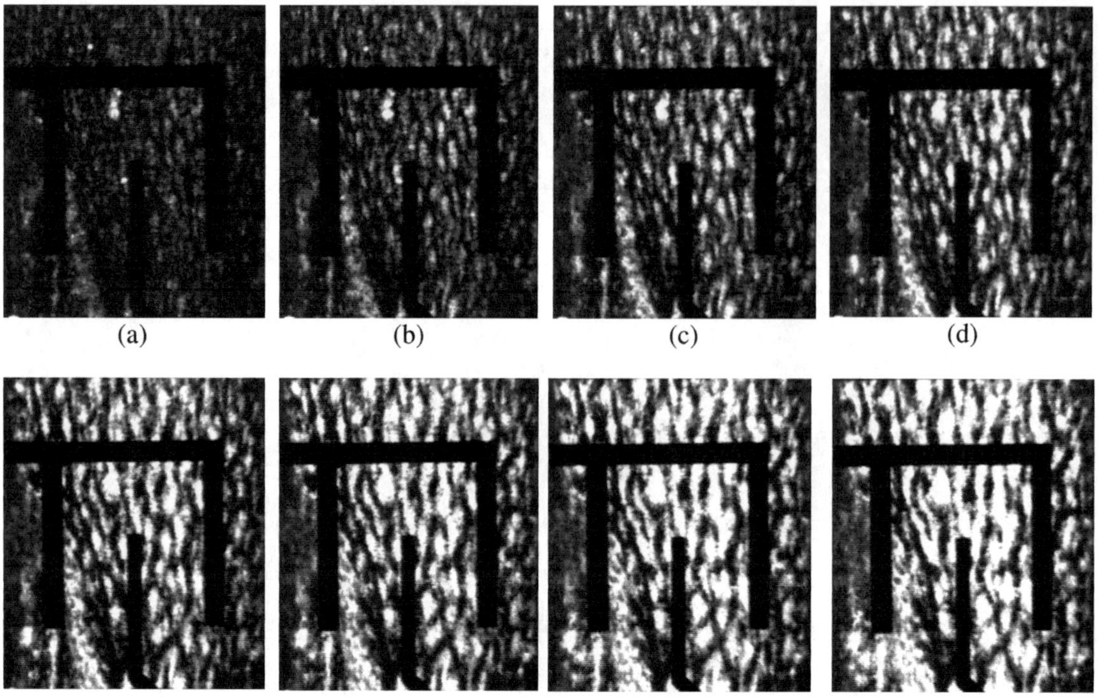

(a) (b) (c) (d)

(e) (f) (g) (h)

Figure 3: Acquired frames from microscope during thrombus growth, blood flows from the top to the bottom of each frame. Frames are acquired from the start of the experiment: (a) after 40 s; (b) after 60 s; (c) after 80 s; (d) after 100 s; (e) after 120 s; (f) after 140 s; (g) after 160 s; (h) after 180 s.

In Figure 4 (a) instead impedance data and reconstructed thrombus volumes are shown for all the frames depicted in Figure 3; the continuous line represents the impedance value (scale on left axis) and the circular markers represent the reconstructed thrombus volume (scale on right axis).

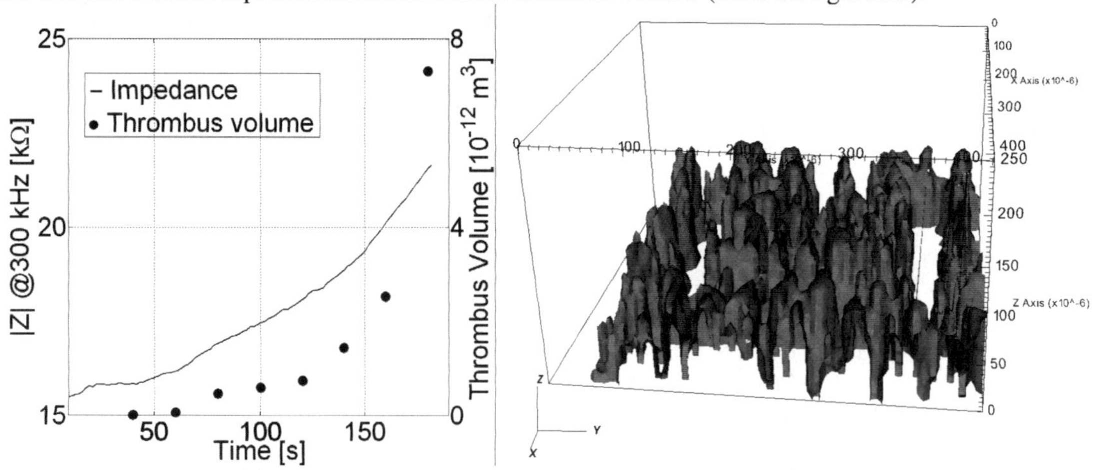

(a) (b)

Figure 4: (a) Impedance change over time measured at 300 kHz (line) and thrombus volume estimation (circles); (b) thrombus 3D model reconstruction

Journal of Physics: Conference Series **459** (2013) 012016 doi:10.1088/1742-6596/459/1/012016

The uncertainty on the volume reconstruction can be overestimated approximating each reconstructed thrombus with a cylinder; this hypothesis is pejorative since thrombus shape looks like a paraboloid. The uncertainty of each thrombus base diameter D_i is related to quantization pixel of microscope and can be quantified in $u(D_i) = 5$ µm, while the uncertainty of the height h_i is related to the simulations steps and can be quantified in $u(h_i) = 5$ µm. Since each thrombus volume can be overestimated as:

$$V_i = \pi D_i^2 h_i / 4 \tag{1}$$

Thus, propagating the uncertainty and considering height and diameter uncorrelated quantities since derived from two different measuring instruments, we obtain

$$u(V_i) = \sqrt{\left(\pi D_i\, h_i u(D_i)/2\right)^2 + \left(\pi D_i^2 u(h_i)/4\right)^2} \tag{2}$$

The uncertainties relevant to each volume reconstruction in Figure 4 are estimated less than 20%.

4. Discussions and conclusions

Here we presented a novel methodology, based on data fusion between optical and electrical information, which provides rapid information on the individual predisposition to form thrombi. This methodology will develop a quantitative mean to estimate the volume of thrombus during its formation, by monitoring the global electrical impedance of an aggregating whole blood sample. The perspective of the proposed device is the treatment of arterial thrombotic complication in cancer patients, due to the increased risk of haemorrhage induced by antithrombotic agents during the administration of prophylactic antiplatelet/anticoagulant therapies.

References

[1] F. G. Hirsch, E. C. Texter, L. A. Wood, W. C. Ballard, F. E. Horan, I. S. Wright, C. Frey, D. Starr, "The electrical conductivity of blood: 1. Relationship to erythrocyte concentration", Blood Journal, vol. 5 pp. 1017-1035,1950.

[2] H. Pauly, H. P. Schwan, "Dielectric properties and ion mobility in erythrocytes", Biophysical journal, vol. 6 pp. 621-639, 1966.

[3] A. Affanni, R. Specogna, F. Trevisan "A Discrete Geometric Approach to Cell Membrane and Electrode Contact Impedance Modeling", IEEE Transactions on Biomedical Engineering, vol. 59, No 9, pp. 2619-2627.

[4] A. Affanni, R. Specogna, F. Trevisan "Measurement Bench for Impedance Tomography during Hemostasis Process in whole Blood", IEEE International Symposium on on Medical Measurements and Applications (MeMeA 2011) Proceedings, Bari (Italy), 30-31 May 2011, pp. 80-83.

[5] A. Affanni, R. Specogna, F. Trevisan "Electrical Impedance Spectroscopy on flowing blood to predict white thrombus formation in artificial microchannels ", IEEE International Instrumentation and Measurement Technology Conference (I2MTC2012) Proceedings, Graz (Austria), 13-16 May 2012, pp. 1477-1480.

[6] H. Fricke, S. Morse, "The electric resistance and capacity of blood for frequencies between 800 and 4.5 million cycles", J. Gen. Phys. pp. 153 - 167, 1925.

[7] R. D. Fomekong, U. Pliquett, F. Pliquett, "Passive electrical properties of RBC suspensions: changes due to distribution of relaxation times in dependence on the cell volume fraction and medium conductivity", Bioelectrochemistry and Bioenergetics, vol. 47 pp. 81-88, 1998

[8] Agilent E4980A Precision LCR Meter Users Guide, 7th Edition, Agilent Technology, Santa Clara, CA, Manufacturing No. E4980-90070, Jun. 2008.

[9] E. Barsoukov, J. R. Macdonald, "Impedance Spectroscopy Theory, Experiment, and Applications", 2nd ed. J. Wiley and Sons, 2005 pp. 30 – 90.

[10] H. Berney, J. J. O'Riordan, "Impedance Measurement Monitors Blood Coagulation", Analog Dialogue, 42-08, August 2008.

Journal of Physics: Conference Series **459** (2013) 012017 doi:10.1088/1742-6596/459/1/012017

Muscle activity characterization by laser Doppler Myography

Lorenzo Scalise, Sara Casaccia, Paolo Marchionni, Ilaria Ercoli, Enrico Primo Tomasini

Dipartimento di Ingegneria Industriale e Scienze Matematiche (DIISM)
Università Politecnica delle Marche, 60131, Ancona, ITALY

E-mail: l.scalise@univpm.it

Abstract. Electromiography (EMG) is the gold-standard technique used for the evaluation of muscle activity. This technique is used in biomechanics, sport medicine, neurology and rehabilitation therapy and it provides the electrical activity produced by skeletal muscles. Among the parameters measured with EMG, two very important quantities are: signal amplitude and duration of muscle contraction, muscle fatigue and maximum muscle power. Recently, a new measurement procedure, named Laser Doppler Myography (LDMi), for the non contact assessment of muscle activity has been proposed to measure the vibro-mechanical behaviour of the muscle.

The aim of this study is to present the LDMi technique and to evaluate its capacity to measure some characteristic features proper of the muscle. In this paper LDMi is compared with standard superficial EMG (sEMG) requiring the application of sensors on the skin of each patient. sEMG and LDMi signals have been simultaneously acquired and processed to test correlations. Three parameters has been analyzed to compare these techniques: Muscle activation timing, signal amplitude and muscle fatigue. LDMi appears to be a reliable and promising measurement technique allowing the measurements without contact with the patient skin.

1. Introduction

Electromyography (EMG) is the most common biomedical instrumentation used for the evaluation of muscles activity [1]. EMG uses needle electrodes (invasive measurement of EMG) or adhesive skin electrodes which are placed in contact with skin (non-invasive measurement of EMG) to measure the myoelectric activity. In this work, we use the latter method to measure the EMG signal that is commonly known as surface electromyography (sEMG) [1,2]. sEMG is not always easy to be performed because it requires the patient collaboration and the possibility to apply the electrodes to the skin; it is therefore required an accurate skin preparation (removing eventual hair, skin detersion, etc.) and particular attention to the placement of the electrodes and cable positioning.

The aim of this study is demonstrate that it is possible to evaluate some parameters related to the muscle activity without contact by means of a novel optical measurement method (Laser Doppler myography, LDMi [3]). This is possible because the muscle contraction is characterized by two component: the electrical component measured by sEMG and the mechanical component measured by Laser Doppler myography. So, the Laser Doppler vibrometer measures the muscle vibrations [3,4] during an isometric muscle contraction. In recent years, the Laser Doppler vibrometer has been widely used for biomedical applications and it has already been demonstrated to be a valid instrument for the assessment of the vibrational behaviour of the human skin[3-7].

Journal of Physics: Conference Series **459** (2013) 012017 doi:10.1088/1742-6596/459/1/012017

2. Materials and methods

The study is focused on the detection of the muscle contraction from different muscles: flexor carpi ulnaris and tibialis anterior. The test involved the simultaneous acquisition of the sEMG signal and the LDMi signal during an isometric contraction of the muscle. In order to record the sEMG signal it is necessary to apply two Ag-AgCl electrodes on the skin in correspondence to the right and left flexor carpi ulnaris/tibialis anterior while a reference electrode is fixed on the wrist for the flexor carpi ulnaris and on the ankle for the tibialis anterior (bipolar configuration, fig.1). Before electrodes application, the skin must be shaved and cleaned and cables were fixed in order to limit the noise. During the test on the forearm carpi ulnaris, the subject was asked to seat on a chair with the elbow flexed at the right angle and the dorsal side of the forearm in a horizontal downwards position and for the muscle contraction he must contract the hand holding an object. While for the test on the tibialis anterior the subject was seated with the leg stretched and he brought the tip of the foot upward without extension of the great toe. The LDMi signal is measured with a single point system (PDV100; calibration accuracy ±0.05 mm/s, bandwidth 0.05 Hz- 22 kHz, spot dimension < 1 mm, working wavelength He-Ne 632.8 nm). The velocity of vibration of the measurement point was measured. The laser beam was pointed perpendicular to the skin surface at a distance of about 30 cm and the laser spot was pointed between the skin electrodes. sEMG and LDMi signals are acquired by a 12 bit acquisition board (ML865 PowerLab 4/25T). The surface EMG signal is filtered by a band-pass filter (10-200 Hz) and by a Notch filter to eliminate 50 Hz noise. Both the acquired signals are sampled at 1kHz, synchronously. The experimental setup is reported in fig.1.

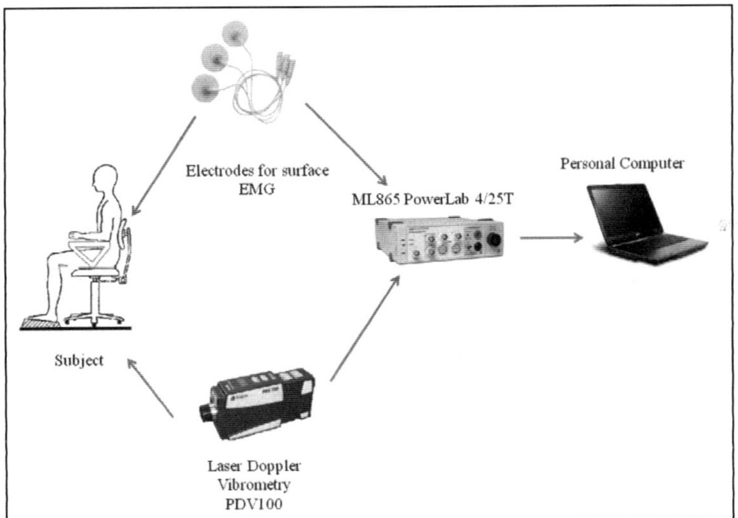

Figure 1. Experimental setup used for the tests.

Tests have been performed on 20 subjects (10 females and 10 males), of mean weight 65 ± (10) kg, height 170 ± (10) cm, and 24 ± (5) years. All subjects have been measured on flexor carpi ulnaris and tibialis anterior (left and right). They were trained with the experimental protocols before the test. To investigate that the LDMi method is valid to determine the muscle contraction, we have calculated the RMS (root mean square) of the signals which can be considered as an index of the signal sensitivity and an index of muscle fatigue [8].

3. Results

3.1 Timing of muscle activity

The detection of the muscle activation has been performed using a threshold algorithm. This method allows to evaluate activation events analyzing the RMS of the signals. This algorithm is applied to both the acquired signals (sEMG and LDMi) to evaluate the differences of timing of the signal activations measured by the sEMG and the LDMi (tab.1). The threshold to discriminate the muscle contraction was based on the rest condition. A muscle activation is detected when the RMS of the signal reaches values higher then the threshold for a at least 100 samples. Each test was characterized by 3s of rest, 10s of muscle activation and of 3s of rest after the contraction and each subject was asked to repeat the test 4 times.

Table 1. Mean and standard deviation of the duration of the muscle contraction, activation and deactivation time for sEMG and LDMi

	Deviation between sEMG and LDMi values		
	t_2-t_1	t_1	t_2
Right flexor carpi ulnaris (mean±SD)	0.12±0.09	0.01±0.05	0.13±0.09
Left flexor carpi ulnaris (mean±SD)	0.23±0.44	-0.008±0.04	0.22±0.06
Right tibialis anterior (mean±SD)	0.18±0.13	-0.07±0.04	0.12±0.11
Left tibialis anterior (mean±SD)	0.15±0.12	-0.05±0.03	0.10±0.09

The activation on the sEMG signal is characterized by a clear increase of the RMS signal amplitude respect to the amplitude of the RMS of the noise. The deactivation phase is clearly visible by the decreasing of the RMS signal amplitude. For the LDMi signal, the activation starts with the first group of peaks and it finishes before the last group of peaks. An example of the events detection (activation/deactivation) is shown in fig.2 for both the signals (sEMG and LDVi):

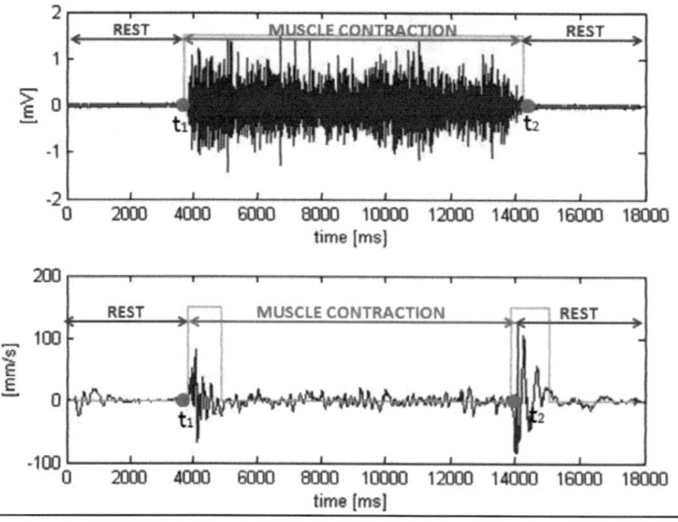

Figure 2. Timing of muscle contraction: sEMG signal and LDMi signal (t_2-t_1=total muscle activation)

3.2 Signal amplitude

The signal amplitude was evaluated by the calculation of the RMS value of the acquired signal during the activation and rest phases on sEMG signal and LDMi signal (fig.2). The ratio of the RMS values of each of two signals (sEMG and LDMi) during activation (t_2-t_1) and rest (t_1) was then calculated. This ratio is called Signal to Noise Ratio (S/N). With this parameter is possible to evaluate the sensitivity of the signal on the noise.

$$S/N = \frac{\sqrt{\frac{1}{t_2 - t_1}\sum_{i=t_1}^{t_2} musclecontraction_i^{\,2}}}{\sqrt{\frac{1}{t_1}\sum_{j=0}^{t_1} rest_j^{\,2}}} \quad \text{(eq. 1)}$$

LDMi signal was filtered in order to remove noise; a wavelet filter was applied using a Sym4 mother for the case of the flexor carpi ulnaris acquisition, while the Sym2 was used for the case of the tibialis anterior. Correlation between S/N of the sEMG signals and S/N LDMi signals - calculated using eq. 1 – are reported in fig.3, fig.4 and fig.5.

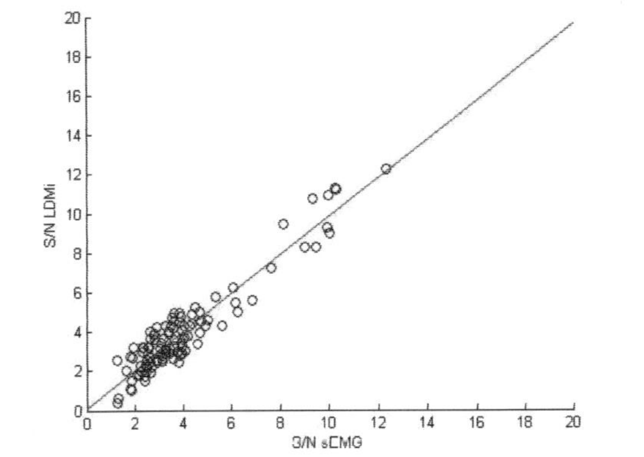

Figure 3. Correlation between S/N of sEMG signals and LDMi signals for the right and left flexor carpi ulnaris (Pearson's coefficient: 0.95).

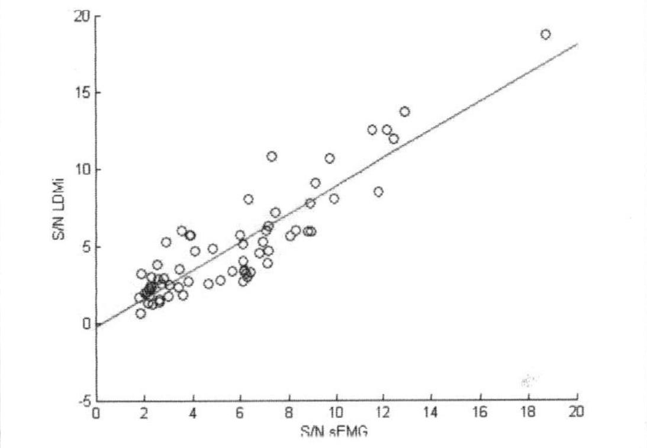

Figure 4. Correlation between S/N of sEMG signals and LDMi signals for the right and left tibialis anterior (Pearson's coefficient: 0.89).

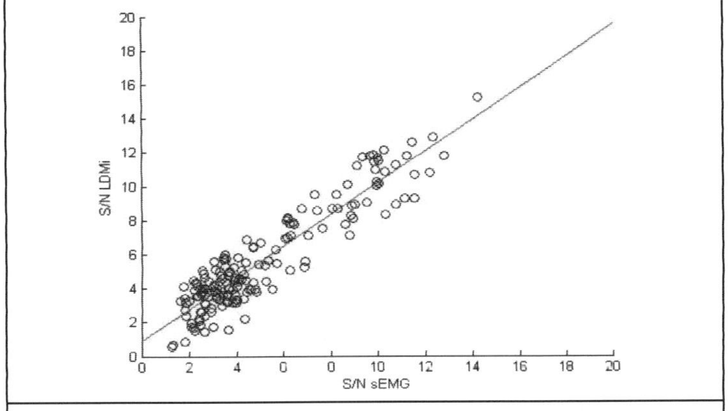

Figure 5. Correlation between S/N of sEMG signals and LDMi signals for both the flexor carpi ulnaris and the tibialis anterior (Pearson's coefficient: 0.93).

3.3 Relationship between sEMG and LDMi signals

A second test was carried out aiming to analyze the correlation between surface EMG and LDMi amplitudes. The subject contracted the flexor carpi ulnaris muscle with two different force levels (minimum and maximum). For this test the subjects were 3 and each subject repeated the test 4 times for the maximum level of force and 4 times for the minimum level of force. The parameter that was calculated is S/N (according eq.1) and the LDMi signal was filtered with a Wavelet (sym4).

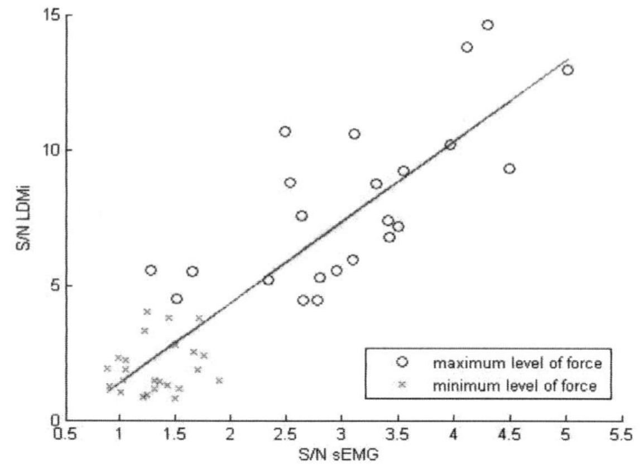

Figure 6. Scatter plot of sEMG and LDMi S/Ns for isometric contraction of the right and left *flexor carpi ulnaris* muscle (Pearson's coefficient: 0.88).

In correspondence to a force increase, it's possible to observe an increase of the S/N sEMG signal amplitude as well as of the LDMi signal which allows to cluster the data in terms of force level.

3.4 Muscle fatigue

The RMS value of the time signals and the slope of the interpolating line are known parameters for the evaluation of the muscle fatigue. The RMS value, according eq.1, is calculated during the muscle contraction for sEMG and LDMi signals. For this test, acquisitions stands for 60s. In fig. 8 and 9, an example of time series of the RMS of the acquired signals are reported; it is possible to observe a negative slope on both the interpolating lines testifying the effect of the muscle fatigue.

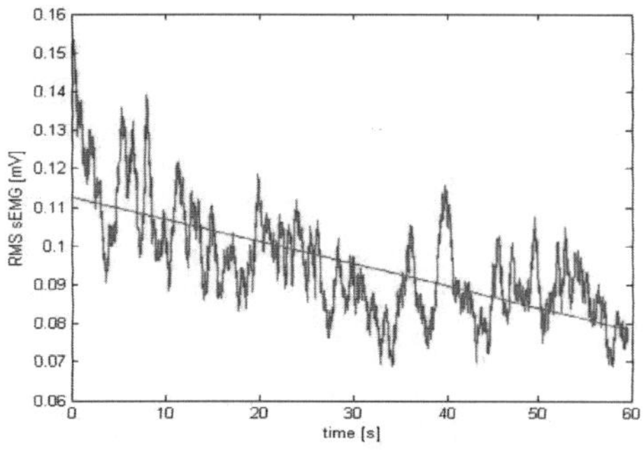

Figure 8. RMS of EMG signal activation.

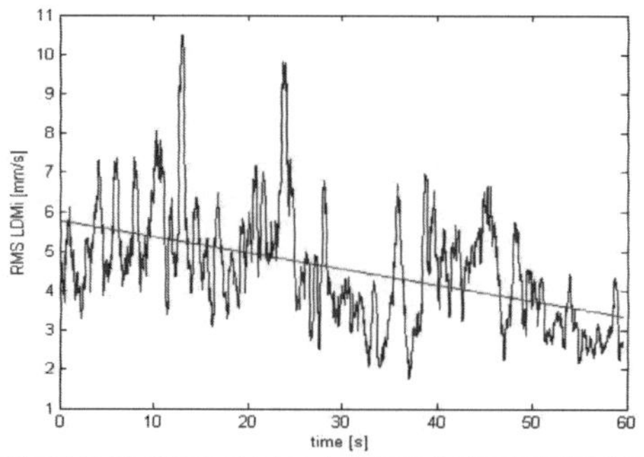

Figure 9. RMS of LDVi signal activation.

4. Discussion

The aim of this research is to present a non-contact measurement procedure for the determination of the muscle characteristics related to the muscle contraction. Comparative tests with sEMG and the proposed method have been carried out on a population of 20 subjects on the flexor carpi ulnaris and tibialis anterior (left and right) muscles. The RMS value of the acquired signals and the Signal to Noise ratio (S/N) have been calculated and results show that the detection timing is sufficiently accurate (maximum standard deviation: 440 ms). A good correlation between the S/N of sEMG signals and the S/N of LDMi signals is also reported (mean Pearson coefficient: 0.93). This result is also shown for the test at minimum and maximum level of force. Finally, it was also possible to evaluate the muscle fatigue.

Such results highlight the possibility to measure such parameters, related to the muscle activity, using the proposed method without contact, obtaining results which are in accord with the ones measured with the gold standard (sEMG).

References

[1] J.G Webster (ed.), *Medical instrumentation: application and design.* 3rd ed. New York: John Wiley & Sons, 1998.

[2] Medved, Cifrek (2011). Kinesiological Electromyography, Biomechanics in Applications, Vaclav Klika (Ed.), InTech, Available from: http://www.intechopen.com/books/biomechanics-in-applications/kinesiological-electromyography

[3] J W Rohrbaugh, E J Sirevaag, E J Richter, *Laser Doppler Vibrometry measurement of the mechanical myogram.* 10th Int. Conference on Vibration Measurements by Laser and Noncontact Techniques. AIP Conf. Proceedings, Vol. 1457, pp. 266-274 (2012).

[4] M. De Melis, U. Morbiducci , L. Scalise, E.P. Tomasini, D. Delbeke, R. Baets, L.M. Van Bortel, P. Segers, *A non contact approach for the evaluation of large artery stiffness: a preliminary study.* American Journal of Hypertension, Vol. 21, pp. 1280-1283 (2008).

[5] L. Scalise, U. Morbiducci, *Non contact cardiac monitoring from carotid artery using optical vibrocardiography.* Medical Engineering & Physics, Vol. 30(4), pp. 490-497 (2008).

[6] L. Scalise, P. Marchionni, I. Ercoli. *A non-contact optical procedure for precise measurement of respiration rate and flow,* Proc SPIE 7715, 77150G (2010).

[7] L. Scalise, F. Rossetti, N. Paone, *Hand vibration: non-contact measurement of local transmissibility.* Int Arch Occupational and Environmental Health. Vol. 81(1), pp. 31-40 (2007).

[8] A. Georgakis, L.K. Stergioulas, G. Giakas. *Fatigue analysis of the surface EMG signal in isometric constant force contractions using the averaged instantaneous frequency.* IEEE Transactions on biomedical engineering, 50, 2, 2003.

Journal of Physics: Conference Series **459** (2013) 012018 doi:10.1088/1742-6596/459/1/012018

Design and test of a semi-automated system for metrological verification of non-contact clinic thermometers.

Dr. R Giannetti[1], Dr. M.A. Sáenz-Nuño[1], J.M. Valderrama[1], A. Fernandez[2]

[1]School of Engineering (ICAI), Comillas Pontifical University, Alberto Aguilera 25, 28015 Madrid, Spain

[2]Quality Assurance Direction, Fremap Hospital, 28220, Majadahonda, Madrid, Spain; e-mail alfonso_fernandez@fremap.es

Corresponding author: Romano Giannetti, romano@dea.icai.upcomillas.es

Abstract. Clinic thermometers are, probably, the most used measurement instrument in the medical facilities (hospitals, clinics, etc.) all around the world. A good part of the assessment the physician does on the patient's health status will depend on the result of such a measurement. In this work, a system to assess the quality of non-contact clinic thermometers is developed and presented; the accuracy of the system is designed to be a useful tool in the phase of the instrument verification and as a base for a future automated calibration facilities.

1. Introduction

It is in some way surprisingly that there is commonly no habit, nor protocols, for metrological calibration and verification of the instruments used in hospitals in most parts of the world in its daily use. In Spain, our group is trying to create a committee of the standard metrological bodies with the aim to suggest "good practice" guidelines in this field and, in the long run, a possible regulation on metrological assessment of medical instrumentation used in hospitals. The newly founded committee is working on a directive of medical instrumentation systems, the related IEC or ISO standards, and on an assessment of the calibration and/or verification procedures implanted at the user facilities (clinics, hospitals, daily care centers).

The high turnover of some of the instruments (like, for example, thermometers) or the high cost of others require a time-effective procedure for the verification, easy to be carried out by the available personnel, and that will not incur in a strong added cost to the instrument (cost that, one way or another, would be passed on the final user, the patient).

To start with a simplified set of instruments to study, the group has decide to approach three kind of medical instruments and to analyse them with respect to metrological characteristics and verification need: the clinic thermometers, the ECG systems, and the cardiovascular pressure measurement systems. The work on the clinical thermometers is oriented to study two kinds of devices: direct contact (skin) thermometers, and no-contact instant thermometers.

One of the authors, as director of quality control in the rehabilitation hospital of FREMAP, Majadahonda, Madrid, set up a verification system for the calibration of the skin-contact clinical thermometer used daily in the hospital facilities [1]; a system for verification of the calibration point of

the devices has been designed and built; this system is currently in use in the aforementioned hospital and is part of the periodic verification procedure of the facility's quality control.
On the other hand, the usage of no-contact thermometers is on the rise, giving the much more practical handling, especially in the case of forehead thermometers (as opposed to in-ear versions), which can be used without any kind of physical contact with the patient and therefore safer under the cross-contamination risk.
In this paper we will present the design of a non-expensive, easy-to-use, verification system for the "instant" no-contact clinic forehead thermometers, thought to be added to clinical facilities scheduled quality control procedures in a straightforward way.

2. System design and implementation

2.1. Stating the problem

The available literature on calibration and verification of near-IR no-contact thermometers is somehow biased towards the analysis of Infra-red Ear Thermometers (IET), for which analysis of different calibration setups are available [2],[3]. Infra-red Forehead Thermometers (IFT) seems to be less studied. Indeed the measurement system used in the case of IETs has clear advantages versus the IFTs method: the measurement is performed in a closed space, with a lot less variability introduced by external factors, and with a black-body approximation of the emitting surface much more at hand.
In the case of IFTs, the measurement is performed by approximating the sensing end of the device (typically an IR diode with or without some optic contraption) to the skin of the patient's forehead, and slightly moved --- to average or maximize the radiation received, depending on the algorithm implemented on the specific brand). The device must compensate for different skin thermal characteristics (that although studied quite extensively present a high variability [4]), radiation differences [5] and, in addition, ambient temperature variations [6]. This is the reason why IFT are widely considered less reliable and "accurate" than skin-contact of IET replacement.
Nevertheless, IFT devices are easier to use, do not require (if well handled) any contact with the patient skin (so that they can be freely used in consecutive manner on a high number of patients without any hygienic concern), and allow the personnel to scan a big number of patients in short periods --- useful for example for massive scanning in school or airports. The daily usage of these devices in medical facilities will be helped by an effective, fast and economically viable system able to periodically verity the correct behaviour of the thermometer.

2.2. System design

In Fig. 1 the system under development is sketched up as it will be when in final production. The whole system will be highly portable, with an approximate size of 30x10x20 cm, possibly operated by battery. It will have a USB computer connection for remote control and data dump and firmware upgrades. The controlling unit is based on a microcontroller unit PIC30F3010, which has deemed sufficient for the task of controlling the "target" part of the device.
The target is the real heart of the system. It is basically a sloping surface which will simulate the skin emissivity at various temperatures under the control of the CPU unit. The structure of the target is shown in more detail on the left side of the same figure.

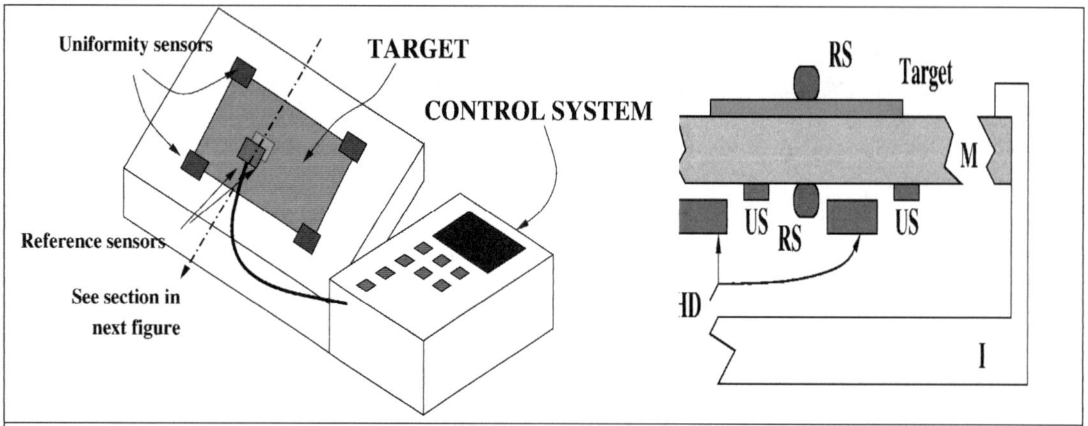

Figure 1. Structure of the proposed system (right), detail of the target (left, detailed description in the main text).

The target is composed by an highly thermally conductive metallic sheet, M, enclosed by one side with a thermal insulator (styrofoam or similar). The metallic sheet is heated by a set (four in the first prototype) of power transistors, controlled with pulse width modulation (PWM) directly by the microcontroller. On the central zone of the target, several low-cost temperature sensors (LM35 or similar) are bond to the metallic sheet; in order to check the uniformity of temperature on the target with the help of a pre-calibration made with a thermal IR camera. On the upper side of the metal sheet, where the thermometer under test (TUT) will be pointed to, a material simulating the reflectance of human skin will be layered upon the surface. Two reference sensors, one fixed on the lower side of the target sheet, and another one movable on the upper side, will be used as the source of the reference body temperature. The selected component is an AD592 sensor, calibrated in the 35-45 °C range with the help of the CEM (Spanish Center for Metrology, Tres Cantos, Spain).

The system will be calibrated in two phases. In the first phase, the low-precision sensors will be cross-calibrated to teach the system when the metal sheet can be considered at uniform temperature, using a IR thermal camera to check it in the first pass, and the movable reference sensor after that.

In the second phase, the system will be fully calibrated at the CEM facilities with a pyrometer system.

The finished system is, at the date of the writing of this article, in the state of advanced prototype; just the case and the (optional) batteries operated power supply is missing. The user interface is still rudimentary, as the analysis of the best skin-simulating material to be used in the final prototype, but the system has reached the point where a full set of measurements can be performed with meaningful results.

3. Measurements

3.1. Devices under test

To check the performance of the system, we have verified three different brand's IFT. The three instrument characteristics are described in Table 1. The brands of the three instrument are kept private at this stage, given that the verification board has not been properly calibrated nor the verification procedures has passed any kind of standardization. The data reported in the table has been obtained by the instrument data-sheet.Notice that the data easily available on the instruments' data sheets is the resolution, and not the accuracy.

Table 1. Devices under test.

DEVICE	Brand	*Aproximated Cost (EUR)*	Bought at	*Range (°C)*	Declared resolution (°C)
TIF1	B1	40	Pharmacy	32-42.9	0.1
TIF2	B2	80	Pharmacy	34-42.2	0.1
TIF3	B3	15	Internet (e-bay)	N/A	0.1

To verify the correct behaviour of the devices, a set of 5 tests have been performed for each one of the thermometers:

1. A hysteresis test: measurements has been taken with a ramp-like temperature pattern: the target was set to temperatures from 35 to 45 °C; then the temperature was dropped back from 45 to 35 °C and, finally, it was raised again to 45 °C. During the ramps, measurements were obtained for every integer temperature. (TEST. 1)
2. A repeatability test: the target was stabilized at 36.5°C and each thermometer was tested several times, switching it to off state between each measurement point. (TEST. 2)
3. An inclination sensitivity test in which it was tested the dependency of the measurement with the inclination between the optical axis of the thermometer and the normal to the skin. (TEST. 3)
4. A distance sensitivity test to find the dependence around the usual work conditions of the thermometers. (TEST. 4).
5. An ambient temperature sensitivity test to evaluate the influence of it on the measurement of each thermometer.(TEST. 5)

Figure 2. Photo of the prototype (second incarnation). The casing is missing, and the insulation is substituted by a simple Styrofoam layer.

3.1 Results.

Due to space limitation, a summary of the finding will be reported here.

TEST 1 was shown up to be the best verification procedure to be done in the hospital plant for the interim checks, once the other uncertainty contributions were evaluated.

TEST 2 allowed to evaluate the "drift" of the displayed value in each TIF for the same reference temperature on the prototype.

Journal of Physics: Conference Series **459** (2013) 012018 doi:10.1088/1742-6596/459/1/012018

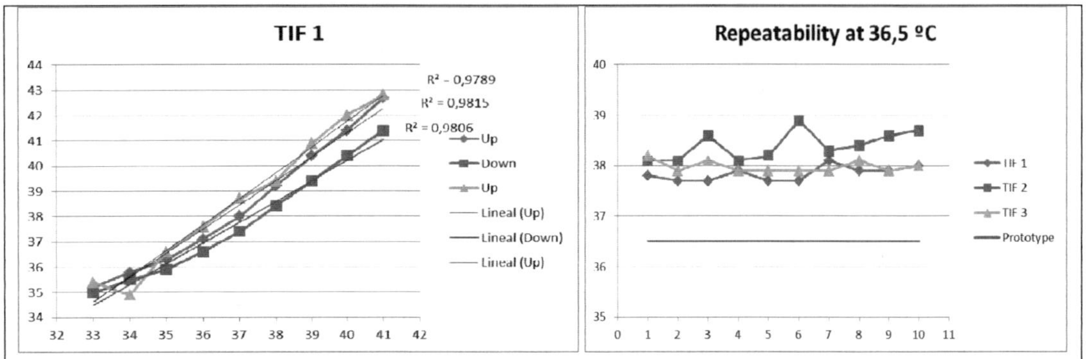

Figure 3. Results of TEST 1 with TIF 1, and TEST 2 with the three TIF's.

TEST 3 showed how the angle introduces a high distortion on the measurement, dealing with a rising contribution to the uncertainty, from TIF 1 to TIF 3.

TEST 4 did show how important is to keep the thermometer not further than 2 cm from the skin, as it has a big influence on TIF 2 and TIF 3. If the measurement is done carefully into the range of 1 cm approximately from the skin, the answer of each thermometer is comparable.
TEST 5 resulted in a small contribution to be taken into account if the ambient conditions may vary more than 5°C, dealing with a much smaller uncertainty, as shown in the following budget.

Table 2. Drift and verification Uncertainty evaluated for each thermometer with the developed prototype

DEVICE	Uncertainty (k=2)	Drift	Main contribution	Second contribution	Uncertainty (no amb. Infl.)
TIF1	± 2,4	1,3	Ambient. Cond.	Linearity of the TIF	± 1.9
TIF2	± 3	1,9	Ambient. Cond.	Angle influence	± 2,7
TIF3	± 3,1	1,5	Angle	Ambient. Cond.	± 2,8

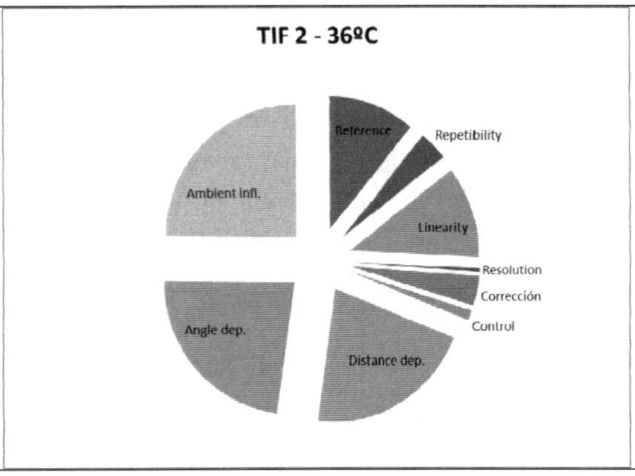

Figure 4. Example of the detailed graphical uncertainty budget for TIF 2.

4. Conclusions

In order to verify the correct operation of infrared clinical thermometers, we developed and implemented a system suitable to carry over a routine evaluation of the instruments in a simple and effective way. To test the system, we analyzed three thermometers for their metrological characteristics and presented the results here.

The results obtained in terms of uncertainty are quite on line on these found in literature [7] if only measurements carried over in "perfect" conditions are analyzed --- where the accuracy estimated with k=1 can be considered as low as ±0.5 ºC; however it is quite evident (see table 2) that in practical use, it is easy to find uncertainties of the order of ±2 ºC. The influence of the ambient conditions and the angle of sight of the optic are the main source of errors in our tests, so they have to be taken into account in field usage.

Summarizing, the system can be effectively used for a fast, routinely verification of the infrared thermometers. A couple of the tests (especially TEST 2) can be performed in little time by unspecialized or nursing personnel and can give a faithful hindsight on the correct operation of the instruments.

5. Acknowledgements

The authors would like to acknowledge the help of FREMAP, s.a., for the help in the organization of the metrology in medicine Spanish group. Moreover, we have to thank the technical staff of the Department of Electronics and Control Systems (Mr. Antonio Martín and José María Bautista) for the help in the development of the prototype.

6. References

[1] F.A. Fernández, 2011, *private communication,* in process of publication
[2] S. F. Tsai, "Comparison Measurements of Infrared Ear Thermometers Against Three Types of Blackbody Sources", Int J Thermophys (2010) 31:1821–1831
[3] Igor Pušnik et al, "Comparison of Blackbodies for Calibration of Infrared Ear Thermometers", Int J Thermophys (2011) 32:127–138
[4] M. L. Cohen, "Measurement of thermal properties of human skin, a review", The Journal of Investigative Dermatology, 69:333-338, 1977
[5] Kimio Otsuka et al, "Imaging of Skin Thermal Properties with Estimation of Ambient Radiation Temperature", IEEE Eng. Med. Bio, Nov/Dec 2002, 49-55.
[6] M. J. Martín et al, "Influencia de la temperatura ambiente y del material de las aperturas en calibraciones de termómetros de radiación de infrarrojo (8-14 μm) de 0 °C a 30 °C" (in Spanish: Influence of ambient temperature and openings material in calibrations of infrared radiation thermometers (8-14 μm) from 0 ° C to 30 ° C), Congreso Español de Metrología, junio 2009.
[7] J. A. Kistemaker et al., "Reliability of an infrared forehead skin thermometer for core temperature", Measurements Journal of Medical Engineering & Technology, Vol. 30, No. 4, July/August 2006, 252 – 261

Methods for measuring risk-aversion: problems and solutions

P. J. Thomas

Risk Management, Reliability and Maintenance Group, School of Engineering and Mathematical Sciences, City University London, Northampton Square, London EC1V 0HB, United Kingdom

p.j.thomas@city.ac.uk; pjt3.michaelmas@gmail.com

Abstract. Risk-aversion is a fundamental parameter determining how humans act when required to operate in situations of risk. Its general applicability has been discussed in a companion presentation, and this paper examines methods that have been used in the past to measure it and their attendant problems. It needs to be borne in mind that risk-aversion varies with the size of the possible loss, growing strongly as the possible loss becomes comparable with the decision maker's assets. Hence measuring risk-aversion when the potential loss or gain is small will produce values close to the risk-neutral value of zero, irrespective of who the decision maker is. It will also be shown how the generally accepted practice of basing a measurement on the results of a three-term Taylor series will estimate a limiting value, minimum or maximum, rather than the value utilised in the decision. A solution is to match the correct utility function to the results instead.

1. Introduction

Risk-aversion is a well established parameter in the field of economics and decision science, and the fact that it can be defined precisely in mathematical terms means that it offers the potential for accurate measurement. The paper will examine how far risk-aversion may be deduced from the individual's spending either (i) on insurance or physical protection to avoid an already-imposed risk or (ii) on a ticket for a lottery. A derivation of the Taylor series method of estimation will be presented first, followed by the more generally valid process of matching the correct utility function to the observed data in order to measure risk-aversion.

2. When the individual seeks to avoid an already-imposed risk: the protection/insurance method

Let us consider an individual of starting wealth, w_0, facing a loss, z, with probability, p, in a given period and use the maximum insurance premium, t, that he is prepared to pay to avoid any possibility of this loss to estimate his dimensionless risk-aversion, ε. (Alternatively he may be prepared to spend the same sum, t, on some other form of physical protection.)

If he does not buy insurance, his expected utility at the end of the period will be:

$$E[U_1] = pu(w_0 - z) + (1-p)u(w_0) \qquad (1)$$

whereas if he buys the insurance at a premium, t, his utility at the end of the period will be $u(w_0 - t)$, with certainty. The individual will be content to pay the premium, t, provided:

$$u(w_0 - t) \geq pu(w_0 - z) + (1-p)u(w_0) \qquad (2)$$

The term, $u(w_0 - t)$, may expanded using the first three terms of a Taylor series as

$$u(w_0 - t) \approx u(w_0) - tu'(w_0) + t^2 u''(w_0)/2 \qquad (3)$$

under the assumption that $t << w_0$. The term, $u(w_0 - z)$ may be expanded in a similar fashion, provided it is possible to assume that $z << w_0$:

$$u(w_0 - z) \approx u(w_0) - zu'(w_0) + z^2 u''(w_0)/2 \qquad (4)$$

The necessity of assuming that $z << w_0$ means that this analysis can apply to the insurance of, for example a washing machine costing £250, with an annual insurance premium of tens of pounds, but cannot be used to analyse the insurance of the average person's house, with a value of £150,000, for example. This point will be returned to later.

Substituting from equations (3) and (4) into inequality (2) gives:

$$u(w_0) - tu'(w_0) + \frac{1}{2} t^2 u''(w_0) \geq pu(w_0) - pzu'(w_0) + \frac{1}{2} pz^2 u''(w_0) + u(w_0) - pu(w_0) \qquad (5)$$

Hence

$$u''(w_0)\left(\left(\frac{t}{z}\right)^2 - p\right) \geq \frac{2}{z}\left(\frac{t}{z} - p\right) u'(w_0) \qquad (6)$$

Both the probability, p, of loss and the ratio of the premium to the potential loss, t/z, will be strictly fractional, which implies that

$$(t/z)^2 < t/z \text{ and } p < \sqrt{p} \qquad (7)$$

Now the expected loss is given by

$$E(Z) = pz + (1 - p).0 = pz \qquad (8)$$

(c.f. equation (1)). For the case where t is set less than or equal to the expected loss, $t \leq pz$ or $t/z \leq p$ so that, from equation (7), $(t/z)^2 - p < 0$. The same condition holds when an individual is prepared to set his premium, t, above the expected loss but in the range: $p < t/z < \sqrt{p}$. Since utility rises with wealth, it follows that $u'(w_0) > 0$. Hence dividing inequality (6) by first $u'(w_0)$ and then $(t/z)^2 - p$ gives

$$\frac{u''(w_0)}{u'(w_0)} \leq \frac{2}{z} \frac{(t/z) - p}{(t/z)^2 - p} = \frac{2}{z} \frac{p - (t/z)}{p - (t/z)^2} \quad \text{for } t/z < \sqrt{p} \qquad (9)$$

Thus

$$-\frac{u''(w_0)}{u'(w_0)} \geq \frac{2}{z} \frac{p - t/z}{p - (t/z)^2} \quad \text{for } t/z < \sqrt{p} \qquad (10)$$

In other words:

$$-\frac{u''(w_0)}{u'(w_0)} \geq \frac{2(pz - t)}{(t^2 - pz^2)} \quad \text{for } t/z < \sqrt{p} \qquad (11)$$

Using the definition of risk-aversion, $\varepsilon_{\min} = -w_0 u''(w_0)/u'(w_0)$ we may now calculate the minimum value for risk-aversion at which a premium, t, will be sanctioned (which is not the same thing as the actual value of risk-aversion at which the decision would be made):

$$\varepsilon_{\min} = \frac{2w_0(pz - t)}{(t^2 - pz^2)} \qquad (12)$$

The fact that equation gives a minimum value becomes plain when one considers a more risk averse individual, with $\varepsilon > \varepsilon_{\min}$, with the same wealth, w_0, and facing the same potential loss, z: it is intuitively obvious that he will be even keener to buy insurance at this premium, t.

The validity of equation (12) is restricted to cases where

- the possible loss, z, is much less than the individual's starting wealth, w_0: $z << w_0$
- the ratio of the premium, t, to the possible loss, z, is less than the square root of the probability of loss, p: $t/z < \sqrt{p}$

Journal of Physics: Conference Series **459** (2013) 012019 doi:10.1088/1742-6596/459/1/012019

3. Where the individual is invited to take a risk: a lottery

Consider now a lottery offering a prize, *z*, with probability of winning, *p*. The maximum amount, *t*, that an individual of starting wealth, w_0, is prepared to pay for a ticket may be used to deduce information about the individual's risk-aversion. A Taylor series analysis similar to that given in Section 2 may be used to find the maximum risk-aversion consistent with buying a lottery ticket at price, *t:*

$$\varepsilon_{\max} = \frac{2w_0(pz-t)}{pz^2 - 2pzt + t^2} \tag{13}$$

Again only a limit is derived. $\varepsilon_{\max}$ is not the value of risk-aversion at which the decision takes place, but an upper bound on that value. This fact is not always made plain [1], but becomes intuitively obvious when one considers a more risk confident individual, with $\varepsilon < \varepsilon_{\max}$, and the same wealth, w_0: he will be even keener to take the risk of betting on the lottery.

The conditions for equation (13) to be valid are that $t << w_0$ and $z - t << w_0$, implying that the possible prize, *z*, must be very much smaller than the person's starting assets, w_0. This excludes many lotteries in common use, such as national lotteries where the prize can be orders of magnitude greater than the starting wealth of the person buying the lottery ticket.

4. Difficulties in measuring risk-aversion by the insurance and lottery methods

It is clear from equations (12) and (13) that the limiting values of a person's risk aversion depends on his starting wealth, w_0. This dependence occurs not only explicitly but also via his choice of maximum insurance premium or maximum ticket price, *t*, which it is reasonable to assume will be affected by how much money the person has: $t = t(w_0)$. While a casual reading of equations (12) and (13) might suggest that risk-aversion will rise with wealth, in fact the reverse is more likely to be true, a fact that is explicable in the two cases considered above in terms of the variability of *t* with starting wealth. For example, a poor person might decide to insure a washing machine for more than the expected loss, indicating a positive risk-aversion, while a rich person might decide that he can stand the loss, thus setting his maximum acceptable premium at zero. Putting $t = 0$ into equation (12) shows the rich person's minimum risk-aversion is negative in this case. His greater wealth allows him to be risk confident in this case, and his risk-aversion is less than that of the poor person considered.

The individual may consider several insurance premia, *t*, and carry out a comparison of utilities in each case. Considering the terms on the left-hand side of equations (12) and (13) namely, *t*, *z*, *p* and w_0, it is clear than *z*, *p* and w_0 will remain constant for all such comparisons. The price or premium, *t*, will change before each successive comparison of utilities, but it will stay constant during the comparison process itself. Thus, if the actual risk-aversion used at the point of decision is governed by the same factors as its limiting values, then risk-aversion will be constant during the process of each utility comparison. This adds to the arguments set out in [2] for risk-aversion being constant during the process of any single decision. Such is, in any case, the universal assumption made in practice when utility functions are used to examine a financial decision under uncertainty.

But although a person's risk-aversion will not change during the decision process, it is likely to vary between individuals with different starting wealths and may also vary between individuals possessing the same starting wealth. Moreover, it has been shown elsewhere, [3], [4], [5], that it is rational for a person to experience a different risk-aversion when he is considering different investment decisions. It is rational for those decisions having only a small effect on his economic state either way to incur a lower risk-aversion than those where a poor outcome could threaten his economic survival. Thus any exercise to measure risk-aversion needs to take account of both variations in starting wealth between individuals and the financial importance of the decision to individuals, if the results are to be interpreted correctly.

The approach based on a Taylor series expansion requires the excursions from the starting wealth to be small, so that, for example, the possible win from a lottery must not be comparable with the starting wealth and therefore of low importance to all the individuals in the sample. This makes it rational for them all to employ a low risk-aversion in their assessment of insurance against the loss, rendering any study trying to pick out differences between groups problematic if the Taylor series is the chosen method of interpretation. Thus it is not surprising that Cramer et al. [1] reported difficulties in differentiating between the (maximum) risk-aversions of entrepreneurs and those in paid employment when they considered how the maximum acceptable price for a ticket in a lottery with a 1000 guilder prize (about €450).

Measurement of minimum risk-aversion from the maximum acceptable insurance premium suffers from a similar limitation. It can be used only for potential losses that are relatively unthreatening to the individuals' wealth, in which circumstance it is rational for the individuals all to use a low risk-aversion. The situation will be significantly different for most people faced with the problem of insuring their houses against fire: the potential loss will in many cases be almost all their wealth, and they would be expected to display high risk-aversions in considering house insurance as a consequence.

To measure risk-aversion in the general case it is necessary first to derive a general utility function with risk-aversion as its parameter and then to estimate risk-aversion by matching the function to the characteristics of the decision through adjusting this parameter.

5. Measuring risk-aversion from the maximum acceptable insurance premium using a utility function

A reason for the popularity of estimating risk-aversion via Taylor series expansions of utility rather than through a utility function is that the former avoids the need to specify a form of the latter. While Pratt [6] reduced the number of feasible utility functions, the number can be reduced much further to just one family. It has been shown [6] that, if an individual's risk-aversion stays constant, not over all time and decisions, but merely during the course of each decision, then only one general family of utility functions can be representative of human decision making. These are the Power utility functions, governed by risk-aversion as their sole parameter.

Using the Atkinson utility function [7], [2] (one of the Power family), it is possible to define the reluctance to invest, R_{120A}, as the difference in utility before and after investment, normalised to the starting utility relative to the utility of one unit of money [8], which is dependent on the reluctance to invest based on the Power utility, R_{120P}:

$$R_{120P} = 1 - p + p(1-c)^{1-\varepsilon} - (1-b)^{1-\varepsilon} \quad (14)$$

where b is the normalised premium: $b = t/w_0$, while c is the normalised potential loss: $c = z/w_0$.

For any given premium that a respondent is prepared to pay, it is required to find the value of risk-aversion, ε, that minimises the reluctance to invest, equivalent to maximising the desire to invest. The curve of R_{120A} versus ε will be smooth, and, if care is taken near $\varepsilon = 1.0$, it is possible to use calculus to find the valley minimum of R_{120A}:

$$\left(w_0^{1-\varepsilon} - 1\right)R'_{120P} - R_{120P}\log w_0 = 0 \quad (15)$$

where $R'_{120P} = d\,R_{120P}/d\varepsilon$ is given by:

$$R'_{120P} = p(1-c)^{1-\varepsilon}\log(1-c) - (1-b)^{1-\varepsilon}\log(1-b) \quad (16)$$

Solving equations (15) and (16) gives the risk-aversion associated with the minimum of R_{120A}, the value of risk-aversion at which the decision to invest is made, known as "the permission point" [4].

Not only is the utility function method not dependent on small-perturbation analysis but it has the advantage that it can identify the unique value of risk-aversion associated with the decision to invest in addition to the minimum value that could allow that decision to stand, which occurs when $R_{120A} = 0$.

Fig. 1 shows the behaviour of the reluctance to invest for the case of washing machine insurance, where the individual has wealth of £1,000 and the probability of complete loss of £250 is 0.1. The individual's maximum acceptable premium is assumed here to be £20. The reluctance to invest is zero, *viz.* $R_{120A} = 0$, meaning that the utility difference is zero at a risk-aversion of – 1.85, but the individual will be keener to buy the insurance at higher risk-aversions, with his minimum reluctance to invest, – 5.28 x 10^{-3}, occurring at a risk-aversion of – 0.32. The criterion for accepting the insurance premium is that the person paying it must experience at least a minimum desire to invest, a variable that depends on his risk-aversion, $-R_{120A\min} = -R_{120A\min}(\varepsilon)$. This minimum level of desire to invest will fall as risk-aversion increases (Fig.2). It is suggested that it is this process, which is rather more complex than simply finding the point where the utility difference is zero (*viz.* $R_{120A} = 0$), that provides the true explanation how higher premia will be sanctioned at higher levels of risk-aversion.

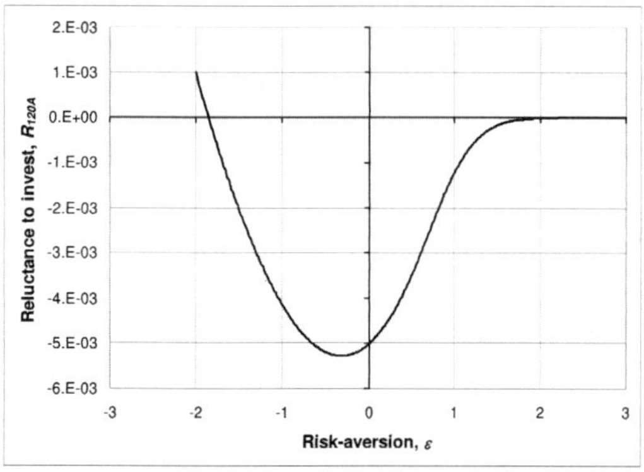

Figure 1 Reluctance to invest vs. risk-aversion

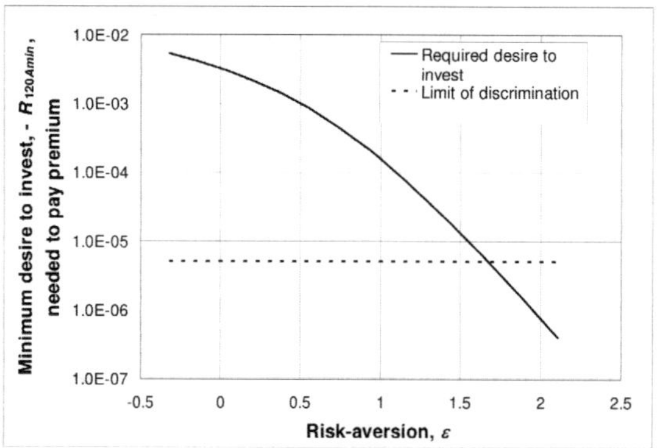

Figure 2 Minimum desire to invest needed to pay premium vs. risk-aversion

Fig. 3 shows a comparison between the methods for a range of insurance premia. The Taylor series expansion method matches the utility-function calculation of risk-aversion well near $\varepsilon = 0$, but it begins to fail as the risk-aversion moves away from zero. Only the utility function method is able to calculate the actual risk-aversion used by the subject when deciding to accept the insurance premium. While the actual risk aversion and the minimum risk-aversion begin to converge for high, positive values, it be noted that $|R_{120A}| < 5 \times 10^{-6}$ for a premium of £31 and above, meaning that it would not be sensible to authorise such a premium, as it lies beyond the point of indiscriminate decision [3]. This limit, only 20% up on the expected loss of £25 is thus very tight in this example where rather small sums of money are at stake. This reveals a further problem arising from the requirement of the Taylor series approach that experimental studies are restricted to risks that are small compared with the individual's assets.

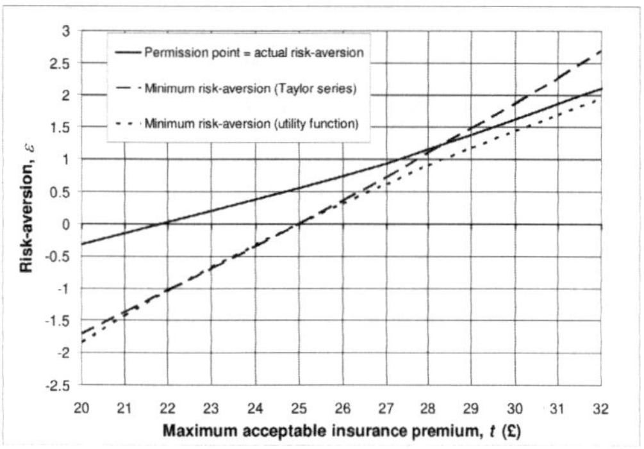

Figure 3. Risk-aversion versus premium

The capability of the utility function method to find the risk-aversion in general, insurance situations, including where the potential loss is large compared with the starting wealth allows a larger range of risk-aversions to be analysed. This improves the accuracy of risk-aversion observation.

6. Conclusions

Risk-aversion is a key parameter characterising human decision making in the necessarily uncertain environment in which we exist. The value of risk-aversion is likely to differ among different decisions for the same person, and among different people facing the same decision [9], [10]. It is possible to calculate a rational risk-aversion for a given insurance/protection situation, but it is likely that there may be differences between individuals.

Methods of estimating risk-aversion based on a Taylor series expansion suffer from two major weaknesses. First, they estimate only a limiting value, maximum or a minimum, of risk-aversion that could be compatible with the subject's economic choice, not the actual value used by the subject in making his decision. Secondly, their validity is restricted to situations where no outcome of the decision is going to make a great difference to the subject's economic well-being. By contrast, methods based on matching a utility function to the observed economic behaviour allow the estimation of not only the minimum or maximum compatible risk-aversion but also the actual value of risk-aversion used by the subject in deciding which economic course to take when a lot is at stake.

7. Acknowledgement

This work was carried out as part of the NREFS project, Management of Nuclear Risk Issues: Environmental, Financial and Safety, led by City University London and carried out in collaboration with Manchester, Warwick and Open Universities as part of the UK-India Civil Nuclear Power Collaboration. The authors acknowledge gratefully the support of the Engineering and Physical Sciences Research Council (EPSRC) under grant reference number EP/K007580/1. The views expressed in the paper are those of the author and not necessarily those of the NREFS project.

10. References

[1] Cramer, J. S., Hartog J., Jonker N., Van Praag C. M., 2002, "Low risk aversion encourages the choice for entrepreneurship: an empirical test of a truism", *J. of Economic Behavior & Org.* 48, 29–36

[2] Thomas, P. J., 2010, "An absolute scale for measuring the utility of money", *Proc. 13th* IMEKO *TC1-TC7 Joint Symposium Without Measurement no Science, without Science no Measurement* , September 1-3, City University, London, UK, IOP Publishing, Journal of Physics: Conference Series **238** (2010) 012039 doi:10.1088/1742-6596/238/1/012039.

[3] Thomas, P. J., Jones, R. D. and Boyle, W. J. O., 2010, "The limits to risk aversion: Part 1. The point of indiscriminate decision", *Process Safety and Environmental Protection*, Vol. 88, No. 6, November, pages 381 – 395.

[4] Thomas, P. J., Jones, R. D. and Boyle, W. J. O., 2010, " The limits to risk aversion. Part 2: The permission point and examples", *Process Safety and Environmental Protection*, Vol. 88, No. 6, November, pages 396 – 406.

[5] Waddington, I., Boyle, W. J. O. and Kearns, J., 2012, "Computing the Limits of Risk Aversion", *Process Safety and Environmental Protection*, http://dx.doi.org/10.1016/j.psep.2012.03.003

[6] Pratt, J. W., 1964, "Risk aversion in the small and in the large", *Econometrica*, **32**, No. 1 – 2 (Jan. – Apr.), 122 – 136.

[7] Atkinson, A. B., 1970, "On the measurement of inequality", Journal of Economic Theory, 2, 244 – 263.

[8] Thomas, P., Boyle, W. and Kearns, J., 2010, "The quantum of wealth", *Measurement + Control*, Vol. 43, No. 5, June, pages 156 – 158.

[9] Atkinson, J. W., 1957, "Motivational determinants of risk-taking behaviour", *Psychological Review*, Vol. 64, No. 6, 359 – 372.

[10] Thomas, P. J., 2013, " The importance of risk-aversion as a measurable psychological parameter governing risk-taking behaviour", Proc. 2013 Joint IMEKO TC1-TC7-TC13 Symposium, *Measurement across physical and behavioural sciences*, 4-6 September 2013, Genova, Palazzo Ducale – ITALY

Theory-Based Parameterization of Semiotics for Measuring Pre-literacy Development

N Bezruczko
Measurement and Evaluating Consulting
1524 E. 59th Street, Chicago, USA

Email: Nbezruczko@msn.com

Abstract. A probabilistic model was applied to problem of measuring pre-literacy in young children. First, semiotic philosophy and contemporary cognition research were conceptually integrated to establish theoretical foundations for rating 14 characteristics of children's drawings and narratives ($N = 120$). Then ratings were transformed with a Rasch model, which estimated linear item parameter values that accounted for 79 percent of rater variance. Principle Components Analysis of item residual matrix confirmed variance remaining after item calibration was largely unsystematic. Validation analyses found positive correlations between semiotic measures and preschool literacy outcomes. Practical implications of a semiotics dimension for preschool practice were discussed.

1. Introduction

Semiotics is the study of symbols and signs, which is directly linked to linguistics and has implications for language development, literature, and culture [1]. Relations between linguistic signs and knowledge have been studied since antiquity, and the Greeks considered signs instrumental to mental activity [2]. Augustine and Roger Bacon [3] continued an interest in semiotics in Middle Ages.

In this research, ideas of Charles Peirce, a 19th century American Pragmatist, were conceptual foundations for developing a statistical semiotic construct that models early literacy of three and four year old children. Peirce emphasized fundamental relations between semiotics and logic, while his concepts of icons, indices, and symbols were inter-related in a process of semiosis [4]. Many of Peirce's ideas re-appear in Luquet [5], and again in Piaget and Inhelder [6], as well as Bruner [7]. Semiotics, however, is not considered an integrated body of knowledge and efforts to measure language growth and learning does not appear in its literature. Despite ancient Western foundations, semiotics does not have a prominent empirical tradition or recognized body of cumulative knowledge.

Dual Coding Theory (DCT) [8, 9] is a cognition model that complements semiotics by emphasizing two symbolic systems, language (verbal) and mental imagery (nonverbal), which influence literacy development. Moreover, DCT provides a perspective on cognitive mechanisms that may underlie semiotic development in young children, and Figure 1 presents a schematic model. A key aspect of this model is referential linking between verbal and nonverbal systems, which has implications for comprehension and memory. A central DCT hypothesis in this research is semiotic function develops in conformity with a hierarchy defined by sensorimotor, iconic, and abstract symbolic components. DCT predicts children enrolled in preschools that emphasize simultaneous verbal and nonverbal processing will show higher semiotic function, as well as higher early literacy outcomes.

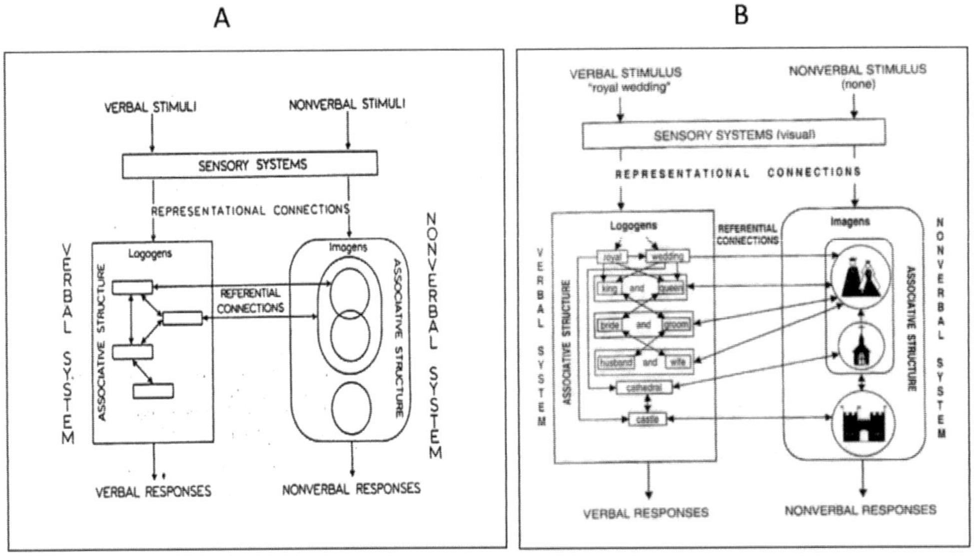

Figure 1. In Dual Coding Theory verbal and nonverbal symbol systems are linked by referential links during cognition. Panel A presents an overall schematic, while Panel B presents simultaneous activation of both systems for the concept "royal wedding" [10].

2. Problem

Researchers have long suspected early childhood language is associated with a semiotic process [11]. A problem has been lack of an empirical construct for measuring pre-literacy on a linear scale because semiotics has a weak empirical tradition, and its conceptual foundations are notoriously fragmented.

3. Objective

The objective of this research was to measure young children's acquisition of pre-literacy with a probabilistic model that is grounded in semiotics developed by Peirce and extended by others. This goal faced two major challenges. First, disparate semiotic ideas about signs and symbols needed to be consolidated into a coherent overarching construct. Then a measurement model needed to be implemented to demonstrate linear properties of this construct for measuring pre-literacy.

4. Method

The following steps were implemented to accomplish above objectives:

- standardize a procedure for collecting drawings and narratives from young children
- identify characteristics of child samples that represent semiotic function
- develop a rubric for quantifying characteristics of child drawings and narratives
- collect and rate child samples at Time 1 (fall) and Time 2 (spring)
- parameterize ratings with a probabilistic model and conduct validation analyses

4.1 Sample

Thirty preschools were selected from Chicago Public Schools (CPS), and at least four children were randomly selected from each preschool ($N = 120$). These preschools are demographically multi-ethnic, socio-economically disadvantaged, and many children were English language learners.

4.2 Data

Drawings, narratives, and standardized preschool outcomes were collected of children, while classrooms and teachers were evaluated with standardized observation and interview protocols.

4.2.1 Child samples and standardized assessments. A standard procedure was developed to collect authentic samples of child drawings and narratives. More specifically, teachers led a thematic storybook presentation, which emphasized "sharing" followed by whole group discussion. Instructions then were given for an "imagery" exercise, which requested children to close their eyes and imagine how they might experience sharing at home, in their neighborhoods, or at school. Finally, children formed small groups to talk about sharing and were given materials to draw about it.

4.2.2 Data collection. Authentic drawings and narrations were collected in preschools at Time 1 (fall) and Time 2 (spring). An adult transcribed child narratives. CPS provided standardized preschool outcomes.

4.2.3 Scoring rubric for drawings and narratives. Fourteen rating scale items with six categories (0, 1, 2, 3, 4, & 5) were derived from Peirce's triadic model (icons, indices, and symbols), as well as cognitive and developmental theory. Each child sample was rated by at least three trained evaluators.

4.2.4 Parameterization, analysis, and validation. Child ratings were transformed with a Rasch model for rating scales, which is summarized below and was implemented with Winsteps software [12].

$$\Pi_{nix} = \frac{\exp \sum_{j=0}^{x} [\beta_n - (\delta_i + \tau_j)]}{\sum_{k=0}^{m} \exp \sum_{j=0}^{k} [\beta_n - (\delta_i + \tau_j)]}$$

where β = observations, δ = item difficulties, and τ = rating scale thresholds. Π_{nix} is probability any item δ, will be rated X by participant β_n where X takes a value from a fixed range (j = 1, 2, 3, 4, & 5), m = number of steps for an item, and k = ith step.

5. Results

5.1 Principle Components Analysis

Thirty evaluators provided ratings and their agreement was high (>.90). Figure 2 presents examples of child drawings and narratives. Preliminary Principle Components Analysis (PCA) of raw data identified a single prominent factor (eigen value = 7.0), which accounted for 47 percent of raw data. Aggregation of three largest remaining factors only accounted for ~20 percent of variance.

5.2 Model fit, measurement properties, and residual analysis

With exception of two items, model fit was excellent. Likewise, all rating scale category thresholds were ordered, and conventional psychometric reliability was 0.84 (alpha). Some drawings, however, were unusual because children included both sophisticated and primitive characteristics such as Scribbling, Stick figures, and Tadpoles, which led to inconsistent ratings. In order to clarify dimensionality, PCA was conducted of item residuals. Results indicated calibrated dimension accounted for 79 percent of overall ratings, while only 3.2 percent of residual variation was systematic hence associated with unexpected factors that could threaten dimensionality. Figure 3 presents a plot of parameterized semiotic dimension with model residuals in eigen value units. These results support construct dimensionality. Validity was further investigated by decomposing item difficulties into sensorimotor, iconic, and abstract symbolic processing components, which showed semiotic components were associated with 70 percent of item difficulty variance ($R^2 = 0.71$, $F = 13.32$, $p<.001$).

Figure 2. A hierarchy of 14 characteristics was identified in child samples collected in fall and spring, then parameterized with a Rasch model. Oral narration was generally easier than drawing, while spatial relations, human figure details, and coherent graphic expression were more difficult.

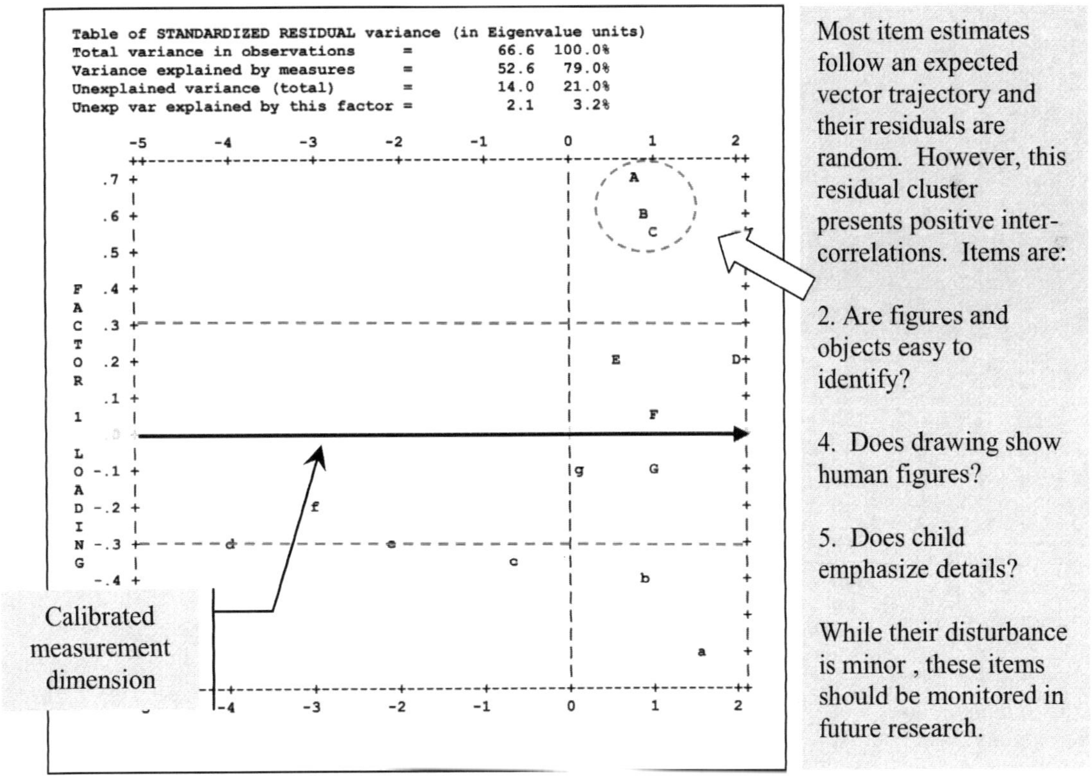

Figure 3. Residuals falling outside the measurement zone (factor loadings > +/- 0.3) indicate violation of local independence, which may threaten unidimensionality. PCA, however, indicated only 3.2 percent of residual variance was associated with nonrandom residual structures.

Journal of Physics: Conference Series **459** (2013) 012020 doi:10.1088/1742-6596/459/1/012020

5.3 Semiotic construct map

Figure 4 presents a calibrated construct map of items and children after ratings were transformed to linear units. Items on right of hierarchy are ordered from low to high difficulty, while linear measures of child ability appear on left side in common units. Drawings with stick figures, tadpoles, primitives, and scribbling on this hierarchy established lowest level of pre-literacy performance. Slightly higher were drawings with human figures and details, as well as those that demonstrated narrative clarity and consistency between narrative and drawing. Drawings at highest level emphasized spatial organization and tended to include letters and names. This progression from low to high is consistent with Peirce's triadic hierarchy of signs, icons, and symbols. Of particular significance, children between fall and spring tended to progress sequentially through this probabilistic structure. Item parameter invariance was confirmed across several criteria: gender, economic status, and home language background.

5.4 Differences between Time 1 and Time 2

Semiotic functioning of most children in this sample was sensorimotor when they enrolled in preschool, and, in general, most children advanced to iconic processing by spring assessment. Very few children reached abstract symbolic functioning. Overall fall mean was -1.04 logits (SD = 2.21, SE = 0.79), while spring mean was -.93 logits (SD = 1.42, SE = 0.60). Parameterized semiotic measures were significantly correlated with early literacy outcomes such as Beginning Sounds and Prints Letters, which supports construct validity. Standardized semiotic gains were largest for younger children in more disadvantaged neighborhoods, mainly because of very low performance at fall enrollment. Surprisingly, three and four year old children showed greater semiotic gain than five year olds, and early literacy outcomes did not significantly differ between older and younger children.

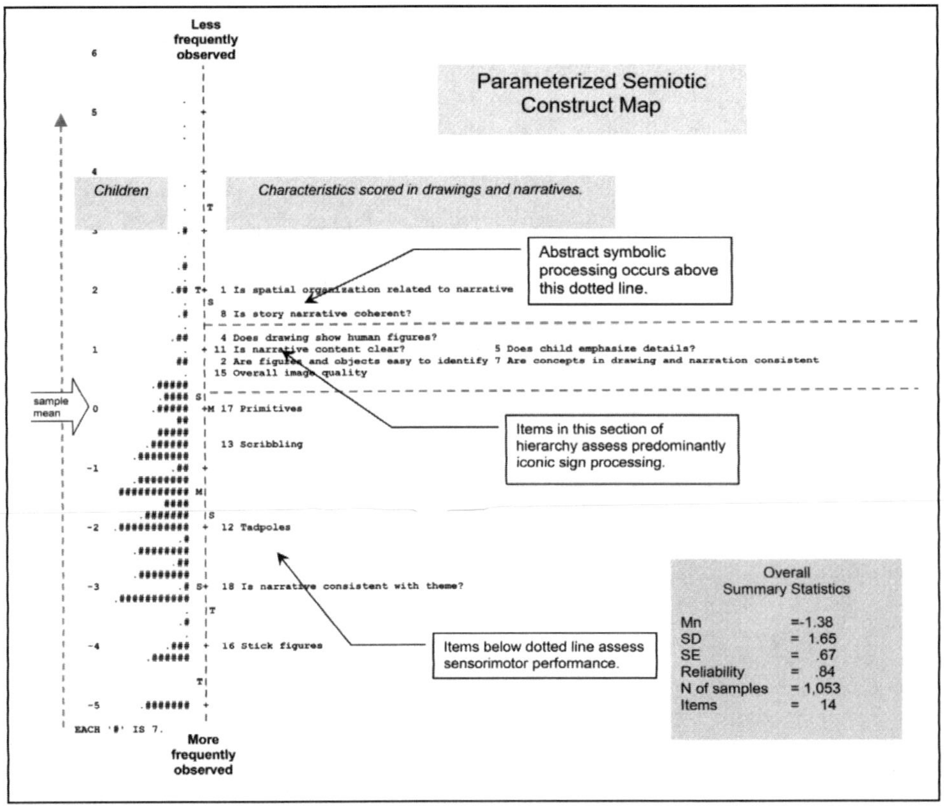

Figure 4. Transformed ratings present items and children in common logits.

6. Discussion

Semiotic hierarchy presented in this research suggests early literacy in preschool involves at least three progressive operations, namely, incoherent scribbling or sensorimotor functioning, physical integration of iconic signs in drawings, and advancement to abstract symbolic expression, which is largely consistent with Peirce's model of semiosis. Speculation suggests several cognitive mechanisms underlie child performance on this construct, that is, a) iconic matching, b) recognizing arbitrary relations between meaning and symbols, and, c) grasping relations between signs and language. Validation analyses supported relations of semiotic performance with kindergarten readiness, as well as DCT predictions that verbal and non-verbal language processing in preschool is related to both semiotic development and higher preschool outcomes. Measurement properties of the parameterized semiotic construct are generally adequate for most practical preschool applications. Further scale development should emphasize addition of more items at highest level of abstract symbol processing, which would extend validity and improve overall practical usefulness.

6.1 Limitations

Five year olds were already at top of parameterized construct at fall enrollment, a formulation that could not capture their gains. Restriction of range was also problematic because this sample was drawn from lower performing urban preschools in economically disadvantaged Chicago neighborhoods. More research is needed with a broader range of pre-literacy development and socio-economic background. Longitudinal studies are also needed to investigate long term effects of semiotic development, as well as influence on child attitude and motivation for learning.

6.2 Implications

These results support convergence of loosely organized semiotic ideas and cognitive science. Practical implication is preschools should exercise caution about emphasizing exclusively verbal learning.

7. Conclusion

Measurement results in this research provide theoretical semiotic foundations for measuring pre-literacy. This construct should be useful for identifying children who have reached abstract symbolic functioning related to early literacy and ready for advancement, as well as performance level of other children. This capacity to identify children who differ in semiotic functioning should have implications for diversifying early literacy pedagogical practices to accommodate developmental needs.

8. References

[1] Caesar M 1999 *Umberto Eco: Philosophy, Semiotics* (Hoboken, NJ: Wiley-Blackwell)
[2] Romeo L 1977 *Semiosis* **6** 37
[3] Maran T 2003 *Sign Systems* **31** 192
[4] Peirce C S 1931-35 *Collected Papers* (Cambridge, MA: Harvard University Press)
[5] Luquet G H 1927/2001 *Le Dessin Enfantin* (Trans. by Alan Costall) (London: Free Association)
[6] Piaget J and Inhelder R 1956 *Child's Conception of Space* (London: Routledge)
[7] Bruner J 1964 *American Psychologist* **19** 1
[8] Paivio A 1986 *Mental Representations: A Dual Coding Approach* (NY: Oxford University Press)
[9] Paivio A 1989 A dual coding perspective on imagery and the brain *Neuropsychology of Visual Perception* ed J.W. Brown (Hillsdale, NJ: Erlbaum) pp 203-216
[10] Sardoski M and Paivio A 2001 *Imagery and Text (*Mahwah, NJ: Erlbaum)
[11] Dyson A H 1986 *Research in the Teaching of English* **20** 379
[12] Linacre M J 2006 *Rasch Measurement Software and Manual* (Chicago: MESA Press)

Acknowledgments
I am indebted to Ava Haji and coaches at Doolittle School, as well as Barbara Bowman and Paula Cottone in Office of Early Childhood Education, Chicago Public Schools for supporting this research.

Journal of Physics: Conference Series **459** (2013) 012021 doi:10.1088/1742-6596/459/1/012021

New perspective on the accessibility of railway transport for the vulnerable traveller

C Sundling[1], B Berglund[1,2], M E Nilsson[1,2], R Emardson[3] and L Pendrill[3]

[1]Department of Psychology, Stockholm University, SE-106 91 Stockholm
[2]Institute of Environmental Medicine, Karolinska Institutet, P.O. Box 210, SE-171 77 Stockholm, Sweden
[3]Measurement Technology, SP-Technical Research Institute of Sweden, P.O. Box 857, SE-501 15 Borås, Sweden

E-mail: catherine.sundling@psychology.su.se

Abstract. Vulnerable travellers experience various problems in the transport environment. These may reduce public travel confidence and consequently lead to decreased mobility. A goal of our research is to find out how to improve the accessibility to railway travelling, especially, for persons with functional limitations. By reducing barriers, the ability of travelling would be improved, and consequently allow for more flexible travel behaviors. In order to develop a model and a method of measurement for accessibility, we (a) constructed a reference group of representative *'typical older persons'* (65-85 years) from questionnaire data, and (b) developed an accessibility measure for persons with functional limitations. In this measure barriers have different weights for the different persons depending on their functional ability and travel behavior. This gives the probability of facing a certain barrier when travelling to a certain destination; that is, a measure of accessibility for the individual. The more weight placed on a certain barrier, the less probable it is that the particular journey will take place. These weights will be obtained in forthcoming research on the perception of a set of various travel scenarios representing barriers.

1. Introduction

An overall goal of our research is to find out how we may improve the accessibility to railway travelling for persons with functional limitations. For the group of "vulnerable travellers" different kinds of functional limitations lead to different problems in the transport environment; covering the whole trip, from start to arrival at the intended place of destination. Traditionally, transportation research has been conducted with "mainstream" travellers who are viewed as interchangeable. But, the experience of the travel environment and its impact will differ among individuals, depending on where and when travel takes place. For the group of "vulnerable travellers" different kinds of functional limitations will lead to various problems in the transport environment. For instance, persons with cognitive impairment, involving stroke, dementia and brain injuries, often face specific accessibility problems such as coping difficulties in an increasingly changing travel environment. Also the elderly may have problems with our computerized information society. This can reduce public travel confidence and consequently lead to decreased mobility. Moreover, the severity of each type of functional limitation as well as other factors such as earlier travel experiences and factors related to personality, would certainly affect the interpretation of the travelling situation. Therefore, a focus on the individual with his/her specific or manifold of functional limitations would reveal needs that are disguised in the larger population. Moreover, it seems wise to include persons that are *not* travelling, because they already encountered insuperable barriers in travel environments. Without these would-be passengers, the most important factors might be missed among those that ought to be improved in the travelling system.

2. The concept of accessibility

There is no definition of accessibility that is uniform or generally accepted. Gould [1] describes it as "a slippery notion", that is a common concept widely used until faced with the problem of defining and measuring it. Accessibility research is found in i.a. geography, economics, occupational therapy and psychology, and its definition differs depending on the goal of the research. For example, in the United Kingdom [2], the concept of accessibility for persons with impairments is used in two different contexts: either to describe design characteristics that improve (or prohibit) the travelling or to describe the ease of reaching an intended destination. 'Accessibility' is used as a quantity in planning for measuring how the travel chain in a transport network performs for various groups of passengers. Even though its definition has not become uniform and generally accepted, the main distinction in recent definitions is that accessibility is an attribute of people *and* of places [3].

3. Accessibility measures

Not only are there many different definitions of accessibility, there are also many different ways of measuring it. Depending on scientific discipline, different ways of measuring accessibility have developed that capture different aspects or dimensions of the concept [4]. Thus, there is no consistent terminology for describing the various methods of measuring accessibility. However, it is fruitful to distinguish between two accessibility constructs: *place* accessibility and *individual* accessibility.

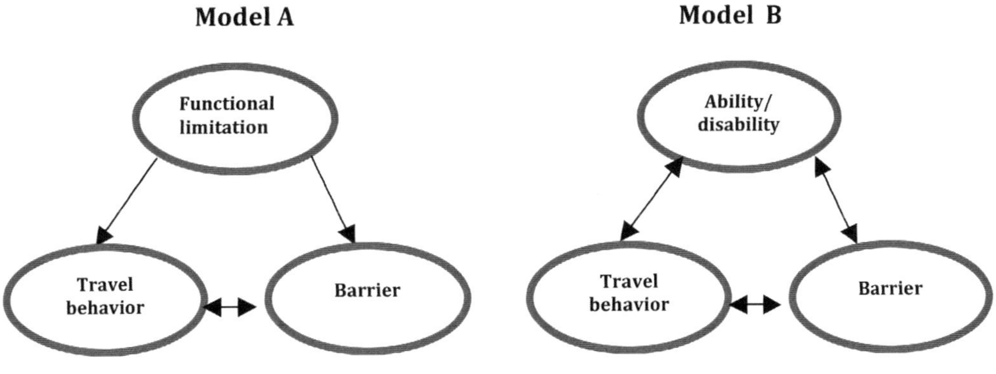

Figure 1. Accessibility may be modeled for travellers' functional limitations (Model A) or for travellers' abilities and/or disabilities (Model B). Please note that functional limitations are invariant relative to travel behavior and the barriers, whereas abilities/disabilities instead are variant owing to reciprocal interactions.

The measurement model for place accessibility is primarily based on various characteristics of the transport and geographical (physical) place, whereas individual accessibility is based on a set of features of each traveller in the transport system [2]. In our research, we focus on the measurement of *individual accessibility* (Figure 1).

A 'functional limitation' constitutes a physical, psychological or intellectual reduction in human function. It is a temporary or permanent condition in individuals. Alternatively, 'disability' is defined as the product of interactions among individual features and environmental characteristics [5]. Disability may involve anything from inaccessible information to physical design characteristics like staircases or doorsteps and/or psychological features like attitudes of fellow passengers or staff. The same functional limitation may thus, in interaction with one environment, but not another, become a disability.

A 'barrier' constitutes i.a. a physical, psychological, social or economic constraint in the travel chain, experienced by the individual passenger. Examples of barriers are design deficiencies of stations, trains and time-table information as well as delays. Over time, the set of barriers will vary for one traveller as well as among travellers. For example, if one traveller has already encountered a barrier during a trip, a subsequent barrier (the same kind or a new one) might well be perceived even worse than the previous one.

In the present research context, 'travel behavior' refers to the way persons use different means of transport. Here, we focus mainly on travel frequency and travel mode. In Sweden and in many other countries, car is the most frequently used travel mode. Thus, measured as number of movements per day, the car stands for 38% (drivers) and 15% (passengers). Car use declines with increasing age as does mobility in general. Among the Swedish car users in the age group 65-84 years old, men and high-income earners have more often a driver's license and own a car. Only 0.7% of all journeys are by railway [6]; see also [7] and [8]. The most common reasons for old persons' travels are shopping, physical exercise and recreational activities [9].

Rimmer et al [10] developed the AIMFREE set of psychometric measuring instruments for persons with functional limitations. They developed and validated 16 survey instruments that measured accessibility to recreational and fitness environments. Interestingly, they utilized the Rash model for demonstrating good psychometric properties of their instrument. Also a Swedish research team has developed a supportive instrument, "The Travel Chain Enabler", for assessing urban public bus transportation accessibility for persons with functional limitations [11], [12].

4. Measurement model of accessibility

We propose a triangular model in which the accessibility to railway travelling results from interactive processes among three cornerstone constructs (Figure 1, Model A): travellers' functional limitations, barriers in the transport system, and overall travelling behavior. By replacing the invariant construct travellers' functional limitations (Model A) with the variant construct travellers' *abilities* or *disabilities* (Model B), our measurement model of accessibility will involve three reciprocal interactions instead of one. Thus, in Model B, each of the three constructs constitutes complex, multi-variable quantities (or patterns) among which reciprocal interaction outcomes are of paramount importance for measuring the quantity *accessibility*. Thus, it is fruitful to study the abilities of persons with functional limitations, rather than focusing on their disabilities. By reducing the barriers, the ability of travelling will be improved, and consequently allow for more flexible travel behaviors.

Emardson et al. [13] have defined an accessibility measure for persons with functional limitations in which barriers have different weights for the different persons depending on their functional ability and travel behavior. This gives the probability of facing a certain barrier when travelling to a certain destination; that is, a measure of accessibility for the individual. The more weight placed on a certain barrier, the less probable is the particular journey. Moreover, disability and functional limitation should be understood from the perspective of travellers' personal experience and interpretation, which may differ among individuals even if the functional limitations, and additional disabilities, are of the same kind and severity.

As said above, barriers can vary among individuals as well as for the same individual over time. For example, having encountered a barrier during a trip can result in an adverse perception of subsequent barrier. However, for simplicity we will not consider, in the following, that the perception of barriers is affected by previous situations. Using the approach of viewing barriers as constraints in the travel chain, Emardson et al. [13] modeled the accessibility reported by a person i with regard to destination j, A_{ij} as:

$$A_{ij}^{m} = \prod_{b}(1 - p_{jb}\psi_{i}(d_{b})) \quad (1)$$

where d_b is barrier number b and p_{jb} is the probability of facing barrier b when going to destination j and

$$\psi_{i}(d_{b}) = \phi_{i}(d_{b}) + \nu_{ib} \quad (2)$$

Here, we introduce an accessibility indicator, ϕ, which is a function operating on the set of barriers, and it is defined on a scale from 0 to 1 [14]. These scale values denote the perceived effort if facing a certain barrier, and it can be interpreted in terms of the probability of cancelling a journey. The function value 1 indicates 100% probability of cancelling the journey, and the value of 0 indicates a barrier causing no problem to the traveler. The symbol ν represents the measurement noise and is a random variable, which is normally distributed with a mean value u and a standard deviation, σ_{ν}. For a specific person and destination, the presented model is separable on the barriers. It is not additive however.

The measured accessibility for a group of travelers can then be modeled as:

$$A_{j}^{m} = \frac{1}{I}\sum_{i}\widehat{p}_{ij}\prod_{b}(1 - p_{jb}\psi_{i}(d_{b})) \quad (3)$$

Where I is the number of persons and $\widehat{p}_{ij}$ the probability that person i is aiming for destination j. The measured accessibility will hence be a number between 0 and 1. Simulations [10] indicate a measurement uncertainty of the accessibility around 0.03 given a situation with 20 barriers. In difference to the accessibility model for a specific person, this model is not separable on the barriers.

In our study, we will use a set of persons to grade identified barriers. Emardson *et al.* [13] showed, using Monte Carlo simulations, that in order to have a standard deviation of the measured accessibility of a size typically a tenth of the measured value; we need a group of about 40 persons grading the barriers.

In order to guarantee the validity of the accessibility measures, it is important that the barriers are scaled by a representative group of persons. To find these groups of persons, we conducted a questionnaire study on perceived accessibility in the railway system. We chose older persons as the target group because functional limitations are more common with older age. Moreover, the growing number of older persons in society poses challenges to the transport system as regards services to meet the special needs of this group. From the questionnaire data, we constructed a reference group of representative *'typical older persons'* (65-85 years). They are identified by a procedure involving two stages: Stage 1, persons are removed who are very dissimilar from the typical 'older train traveller' on the basis of binary coding of a large number of criterion variables. Stage 2, participants are selected on the basis of fuzzy resemblance of the typical 'older person' on the most important criteria (see further [15]).

With the reference group constituted and in place, individuals are recruited for an interview study. The research goal is to explore the range of barriers encountered in the railway system and how they are perceived. The respondents are selected based on inclusion criteria for the reference group (age, gender, travel frequency, travel mode, etc.). Thus, persons would be included who have various functional limitations and represent a wide range of travel behavior. We will explore the range of barriers encountered by them. The different kinds of barriers would be measured according to Eq. 1, which delivers *individual accessibility measures*. This is accomplished by putting a perceived-effort weight on each barrier for each individual. These weights will be obtained in forthcoming research on the perception of a set of various scenarios representing barriers.

Acknowledgments
This work is supported by the Swedish Transport Administration.

References

[1] Gould P 1969 *Spatial Diffusion.* Association of American Geographers Resource, Paper No. 17, (Washington DC, USA: Commission on College Geography)
[2] TSG–Transport Studies Group 2005 *TSG Research: Measuring Accessibility as Experienced by Different Socially Disadvantaged Groups* (London, UK: University of Westminster) http://home.wmin.ac.uk/transport/projects/samp.htm
[3] Envall P 2007 *Accessibility Planning: A Chimera?* (Leeds, UK: University of Leeds, Doctoral Thesis)
[4] Makrí M-C and Folkesson C 1999 *Accessibility Measures for Analyses of Land Use and Travelling with Geographical Information System* (Lund: Lund Institute of Technology and Lund University and University of Karlskrona/Ronneby)
[5] WHO 2013 World Health Organization. http://www.who.int/topics/disabilities/en/
[6] SIKA 2007 *Res 2005-2006: The National Travel Survey* (Stockholm: SIKA Statistics)
[7] Linder P 2007 *Äldre människors res- och aktivitetsmönster – en litteraturstudie* Arbetsrapport (Umeå: Transportforskningsenheten, Umeå universitet)
[8] Alsnih R and Hensher D A 2003 The mobility and accessibility expectations of seniors in an aging population *Transportation Research Part A* **37**(10) 903-916.
[9] Transek 2005 Äldre personers resvanor och aktiviteter (Stockholm: Transek AB)
[10] Rimmer J H, Riley B, Wang E and Rauworth A 2004 Development and validation of AIMFREE: Accessibility instruments measuring fitness and recreation environments *Disability and Rehabilitation* **26**(18) 1087-1095
[11] Iwarsson, S, Jensen G and Ståhl A 2000 Travel chain enabler: Development of a pilot instrument for assessment of urban public bus transport accessibility *Technology and Disability* **12** 3-12
[12] Jensen S, Iwarsson S, and Ståhl A 2002 Theoretical understanding and methodological challenges in accessibility assessments, focusing the environmental component: an example from travel chains in urban public bus transport *Disability and Rehabilitation* **24**(5) 231-24
[13] Emardson R, Jarlemark P, Pendrill L, Sundling C, Nilsson M E and Berglund B 2012 Measurements of accessibility to rail transport systems *Advanced Mathematical and Computational Tools in Metrology and Testing IX* ed F. Pavese, M Bär, J-R Filtz, A B Forbes, L Pendrill and K Shirono (Singapore: World Scientific) pp 136-142
[14] Weibull J W 1980 On the numerical measurement of accessibility *Environment and Planning A*, **1**, 53-67
[15] Botteldooren D Berglund B and Lercher P 2005 *Perceived Noise Annoyance Caused by Maglev Trains* (Gent, Belgium: Scientific Report from the Department of Information Technology, Ghent University)

Journal of Physics: Conference Series **459** (2013) 012022 doi:10.1088/1742-6596/459/1/012022

Measuring Study Habits in Higher Education: The Way Forward?

E D Fitkov-Norris and A Yeghiazarian
Kingston University, Kingston-upon-Thames KT2 7LB, UK

E-mail: e.fitkov-norris@kingston.ac.uk

Abstract. This article reviews existing study habit measurement instruments and discusses their drawbacks, in the light of new evidence from neuroscience on the workings of the brain. It is suggested that in addition to traditional frequency based past behavioural measures, the predictive accuracy of study habit measurement instruments could be improved by including measures of habit strength that take into account behaviour automaticity and efficacy, such as the Self-Report Habit Index (SRHI) developed by [1]. The SRHI has shown high reliability and internal validity in a wide range of contexts and its applicability and validity in the context of learning and higher education as an enhancement to study habit measurement instruments is as yet to be tested.

1. Introduction

The growth of the knowledge based economy has highlighted the need for developing a country's human capital and increasing the support for lifelong learning in the workplace [2] and several government and EU lead initiatives supporting lifelong learning have been launched [3]. A number of factors have been identified as facilitators of lifelong learning, such as learner's motivation, skills and ability to assess their own educational needs [4,5], while academic specific anxiety and low self-esteem were found to be an important negative predictor of performance and engagement in learning [6,7]. Recent meta analysis has also highlighted that in addition to motivation and anxiety, study habits are the third most significant predictor of academic performance [8].

A number of instruments have been designed to measure the effectiveness of students' study habits and strategies [8-10] and have been adapted and applied to provide educators with insight into the way students approach their studies and the impact that has on student learning and academic achievement [10]. However, advances in neuroscience are beginning to reshape the ideas held about the ways in which information is perceived, processed and managed by the brain. Theories suggesting that information is processed using two separate pathways are confirmed using brain scanning [11]. Brain scanning suggests that incoming information is "matched" to existing knowledge and if similarity between new and existing knowledge is found, the information is processed by the reflexive system (X-system). The reflexive system uses past experiences and information to make decisions and is automated, parallel and very fast. This system is very efficient and it is believed to be the default processing system used by the brain for decision making and information processing. When the brain encounters a significant dissimilarity between the incoming and existing knowledge, it wakes up its reflective system (C-system) to processes the information and make a decision. The reflective system uses propositional logic to draw conclusions and in comparison the reflexive system it is effortful, slow and sequential [11].

Researchers have started to take into account the implications suggested from the two part model and new theories and implicit behavioural measures are emerging in social science and organisational behaviour [12,13]. The use of the model in education and learning has been very limited, but the two-part model has been used to explain teacher bias towards students' ethnic background [14], and to model the way in which experienced teachers transfer knowledge and experience to students [15], although the implications of the two-part model in the actual learning process has not been discussed. Study habits are one aspect of the learning process that utilises the two-part decision making process as habituated behaviour is by default processed by the automated reflexive system. Thus inventories measuring the strength of students' study habits ought to recognise and reflect the dual nature of the student study habit behaviour. The purpose of this positioning article is two fold. Firstly it reviews existing study habit instruments and examines the components of study habits and the scales used to measure them. Secondly, the article discusses the implications of the two-process model decision making and information processing model [11]. The article starts with a brief introduction to the field of learning and definitions of key terms. This is followed by an overview of most commonly used study habit measurement instruments, identifying their main components and scales. The implications of the two-part decision making and information processing theoretical model on study habit measurement inventories are discussed next and finally recommendations for study habit inventory improvement and future work are made.

2. Learning Styles, Preferences and Habituation

Many factors determine the way in which different people learn, and a large number of models have been developed to describe peoples learning preferences, styles and habits [8,16]. There is no agreed taxonomy for classifying learning styles and preferences and a number of researchers have attempted to identify appropriate dimensions for learning style classification (see [16] for an overview). The difficulty is in part due to the lack of a consistent naming convention for terms associated with learning and the learning process and the overlap in their functionality and dimentions. Five terms relating to learning are often used interchangeably by researchers: namely, learning preference, learning strategy, learning style, cognitive strategy and cognitive style. Their definitions according to [17] are given below:

- Learning preferences: these relate to the learner's preference for one method of teaching over another [17] and reflect the way in which a learner prefers to receive information. There are four ways for conveying and absorbing information and learning: Visual (via pictures and diagrams), Auditory (by hearing), Read/write (by text) and Kinaesthetic (by doing). The preferences of students for using combinations of information input pathways are measured using VARK type questionnaires [16,18].
- Learning strategy is defined by [17] as: "adopting a plan of action in the acquisition of knowledge, skills or attitudes". Study strategies relate to the students ability to manage resources such as available time, regularly reviewing lecture notes, under-lining passages in textbooks, preparing several drafts of essays, and the use of other students for peer support [7,19]. In addition, some instruments measure the propensity of students for deep, surface or strategic learning which are associated with different learning approaches such as use of evidence to link ideas or simply rote learning [20].
- Cognitive strategies form a subset of learning strategies and are concerned with the process of organising and processing information. They lead directly to learning results in terms of knowledge, understanding, and skill. Examples of cognitive strategies are: looking for relations among parts of the subject matter (relating), distinguishing main and minor points (selecting), thinking of examples (concretizing) and looking for applications (applying) [21].
- Cognitive style is defined as: "a systematic and habitual mode of organising and processing information" [17,22]. Several cognitive style indices and measures have been developed (see [16] for details of some of the most influential models).

- Learning style is the most all encompassing of the five terms and refers to "adopting a habitual and distinct mode of acquiring knowledge" [17,23]. A systematic review of learning styles was carried out by [16] and identified five different families of learning style depending on their degree of flexibility and propensity to change and evolve over time.

The definitions above show that habituation is an essential component of two of the terms associated with learning: learning styles and cognitive styles. Surprisingly, study habits are not explicitly mentioned in the definition above although, as mentioned previously, it is one of the predictors of academic performance [8]. A number of instruments for measuring the strength of students study habits have been developed over the past 60 years and the following section gives a flavour of their evolution over time.

3. Overview of Study Habits Inventories

One of the first study habits inventories was developed by [24] and it assessed study habits and attitudes which high and low scholarship groups possess to different degrees even though they have similar scores on aptitude tests. The inventory evaluates the effectiveness of the students' reading and note taking techniques, concentration, the amount of continuous time spend studying, both by oneself and as part of a group, the extent to which a students suffers from exam stress and the degree to which they plan their answers. The inventory consists of 28 questions measuring the frequency with which student engaged with a particular behaviour on a 3-point Likert scale ("rarely or never", "sometimes", "often or always").

The Survey of Study Habits and Attitudes (SSHA) was developed by [10] and consists of 75 items. These items assess learners'

study methods such as neatness of study space, effective use of study time, delay avoidance and planning of learning, as well as attitudes towards teachers, teacher approval and beliefs about teacher bias towards the student. In addition, the SSHA evaluated the students' motivation for studying, goal orientation and related attitudes towards importance of scholastic success. Although used extensively in research and for policy setting [25,26] it was not possible to verify the measurement scale of the SSHA instrument.

SSHA was adapted to British students by [9] as the Students Attitudes Inventory (SAI) which is an instrument with 47 true/false items considering four scales: student motivation, study methods, examination techniques and lack of distractions.

Another Learning and Study Strategy Inventory (LASSI) was developed by [27]. It is a 10 scale, 80-item inventory measuring the extent to which students adopt strategic learning as a function of their "study skills", "will" and "self-regulation". "Study skills" consist of approaches to information processing, summarising main ideas and test strategies, the scale for "will" assesses the extent of students' anxiety, attitudes and motivation, while the "self-regulation" component tests student concentration, time management and self-testing skills and use of study aids. Students self report the degree to which they consider a statement typical of them on a 5 point scale rated: a = "most atypical", and e = "very much typical".

One of the most recent developments is a 10 item, 3 scale, study habits questionnaire by [28] measuring three scales: access to good quality notes, scheduling and time management, and ability to concentrate. Students are asked to report the frequency of a particular behaviour on a 5 point Likert scale ("never", "almost never", "sometimes", "fairly often", "very often").

4. Discussion

The overview of study habit inventories outlined above show that all of the study habits inventory measures were designed to measure the effectiveness of the students' learning strategies, rather than their learning or cognitive style habits. As study strategies involve active planning and management of available resources it is difficult to tally this conscious effort on the part of the students with the expected automaticity and unconscious effort demanded by habitual behaviour. This contradiction could be explained by looking at the interactions between the reflective and reflexive information processing systems [11]. Past behaviour consists of some routine or habitual behaviour processed by the automated reflexive X-system and some intentional, norm or situation governed behaviour managed by the logic driven reflective C-system and those two systems constantly interact [11,29]. Students who consistently use the same study skills and strategies benefit in the long run by gradually strengthening their decision pathways and shifting from using their reflective to using their more efficient and less effortful reflexive system [11]. In addition, the repeated actions lead to situational familiarity which is in turn more likely to trigger the same habitual behavioural response [29], further reinforcing the habituation of that particular behaviour. This reinforcing feedback loop may explain how students could be trapped in using inefficient study habits and strategies, even when they make a conscious effort to study hard or work efficiently [19,20]. This suggests that assessing the strength of habituation of a students study habits may improve the predictive power of the study habit construct. More accurate understanding of the strength of students study habits would enable educators to offer adequate means of support for individual students. For example, a student with weak bad study habits may find it easier to change it compared to a student with a stronger habit. Similarly, a student with weak good study habits may be at a higher risk of losing focus than a student with strong habits during periods of transition such as for example, when transferring from studying from school to university.

The main measure used to determine the strength a student's study habits is self reported past behavioural frequency, with the students who engage with the behaviour more frequently having stronger habits [10,28]. Although frequency of past behaviour is a very good indicator of future behaviour [30], the measure of past frequency introduces bias and error, as by definition, habitual behaviour is automatic and so it may be difficult recall accurately the frequency of occurrence of the behaviour in the past [31]. One approach for improving the measurement accuracy of habituated behaviour is to take into account the degree of automaticity and lack of consciousness present within that behaviour. It has been shown that the strength of habitual behaviour measured using an instrument taking into account both process past frequency of occurrence and automaticity had an additional discriminant validity over a measure of past frequency occurrence on its own, and this applied in the case of both overt habits (unhealthy food consumption) and mental habits (thinking negative thoughts) [30]. Similar results emerged when [30] carried out controlled experiments on information processing. These findings suggest that measuring the degree of habituation present in a particular behavioural process, in addition to the past behavioural frequency, could improve the predictive power of a construct. In light of this, the question arises on what is the best approach for measuring the strength of students study habits and whether the study habit constructs could benefit from the introduction of an instrument for measuring habituation as well as past behavioural frequency.

Theoretical development and empirical studies from other disciplines hint at an affirmative answer. As our understanding of the way in which the brain processes new information grows, it is becoming apparent that the complex relationship and interactions between the reflective and the reflexive information processing systems offer a challenging and exciting avenues for research. The impact of the automated reflexive system is of particular interest in studies in areas involving the formation of habits, such as for example study and learning habits. Appropriate instruments to measure the strength of study habits are emerging and, for example, [1] propose a very flexible instrument for measuring the degree of habituation of a particular behaviour, the Self-Report Habit Index (SRHI). The SRHI is a 12-item self report instrument of habit strength that measures five aspects of habits: history of repetition of behaviour, the difficulty of controlling that behaviour, the lack of awareness of the behaviour, efficiency, and how closely the behaviour is associated with the self. The construct has the advantage of being applicable to any type of behaviour as subjects are asked to agree or disagree with the extent to which a particular behaviour is performed automatically, i.e. without thinking or belongs to their (daily, weekly or monthly) routine. The construct also tries to ascertain the degree to which behaviour is associated with the self. It is worth noting that one of the study habit measures used a scale that reflected the degree to which a behaviour was typical of the student [27], indicating a possible convergence in past frequency behaviour measures and habit strength measurement in the context of study habits. Therefore, including a measure of habit strength in existing study habit instruments may offer a way forward for improving the accuracy of study habit measurement instruments. The SRHI instrument has been tested and shown high validity and reliability in a wide range of everyday contexts such as reading newspapers, watching TV, eating junk food, thinking negative thoughts [1,30,32]. However, further empirical studies and work is required to test the practical suitability, reliability and validity of SRHI as an enhancement to study habit measurement instruments, in the context of higher education.

5. Conclusions

This article carried out an overview of existing study habit measurement instruments and discussed the implications of new evidence from neuroscience suggesting that information is processed by the brain using two distinct pathways: the fast, automatic reflexive pathway and the slower, reflective pathway. As strong habituation of inefficient study strategies or poor study skills may have a significant impact on student performance, accurate assessment the strength of a student's study habits could enable educators to put in place adequate academic support mechanisms. The study found that the majority of study habit instruments use the past frequency behaviour of students to assess the strength of their study habits and it is suggested that the study habit constructs could be improved by adding measures for process automaticity and habituation as suggested by [1]. The Self-Report Habit Index (SRHI) construct was identified as a potential candidate for improving the measurement power of study habit constructs, as it has been shown to provide valid and reliable measurement oh habit strength in many different contexts. Despite these promising results further work is required to test the practical suitability, reliability and validity of SRHI as an enhancement to study habit measurement instruments in the context of learning and higher education.

Acknowledgements

The authors would like to thank Dr Richard van den Berg for his encouragement and generous help in shaping their ideas.

References

[1] Verplanken B and Orbell S 2003 Reflections on Past Behavior: A Self-Report Index of Habit Strength *J. Appl. Soc. Psychol.* **33**(6) 1313–30
[2] Wilton N 2011 Do employability skills really matter in the UK graduate labour market? The case of business and management graduates *Work Employ. Soc.* **25**(1) 85–100
[3] Thomas L 2012 Building student engagement and belonging in Higher Education at a time of change. Paul Hamlyn Foundation p.100
[4] Nicol DJ and Macfarlane DD 2006 Formative assessment and self-regulated learning: a model and seven principles of good feedback practice *Stud. High. Educ.* **31**(2) 199–218
[5] Rushton A 2005 Formative assessment: a key to deep learning? *Med. Teach.* **27**(6) 509–13
[6] Zakaria E and Nordin N 2008 The effects of mathematics anxiety on matriculation students as related to motivation and achievement *Eurasia*. **4**(1) 27–30
[7] Bennett R 2003 Determinants of undergraduate student drop out rates in a university business studies department *JFHE* **27**(2) 123–41
[8] Credé M and Kuncel NR 2008 Study Habits, Skills, and Attitudes: The Third Pillar Supporting Collegiate Academic Performance *Perspect. Psychol. Sci.* **3**(6) 425–53
[9] Entwistle N, Thompson J and Wilson J 1974 Motivation and study habits *High. Educ.* **3**(4) 379–96.
[10] Brown W and Holtzman W 1955 A study-attitudes questionnaire for predicting academic success *J. Educ Psychol.* **46**(2) 75-84
[11] Lieberman MD, Gaunt R, Gilbert DT and Trope Y 2002 Reflexion and reflection: A social cognitive neuroscience approach to attributional inference *Adv. Exp. Soc. Psychol.* **34** 199–249
[12] Bohl V and van den Bos W 2012 Toward an integrative account of social cognition: marrying theory of mind and interactionism to study the interplay of Type 1 and Type 2 processes *Front. Hum. Neurosci.* **6** 1–15

[13] Uhlmann EL, Leavitt K, Menges JI, Koopman J, Howe M and Johnson RE 2012 Getting Explicit About the Implicit: A Taxonomy of Implicit Measures and Guide for Their Use in Organizational Research *Organ. Res. Methods* **15**(4) 553–601
[14] Glock S and Krolak-Schwerdt S 2012 Does nationality matter? The impact of stereotypical expectations on student teachers' judgments *Soc. Psychol. Educ. 16*(1) 111–27
[15] Stolpe K and Björklund L 2012 Seeing the Wood for the Trees: Applying the dual-memory system model to investigate expert teachers' observational skills in natural ecological learning environments *Int. J. Sci. Educ.* **34**(1) 101–25
[16] Coffield F, Moseley D, Hall E and Ecclestone K 2004 Learning Styles and Pedagogy in Post-16 Learning. Learning and Skills Research Centre p. 173
[17] McLoughlin C 1999 The implications of the research literature on learning styles for the design of instructional material Aust. J. Educ. Tech. 15(3) 222–41
[18] Fleming N 2001 VARK: a Guide to Learning Styles. The VARK Questionnaire
[19] Stegers-Jager KM, Cohen-Schotanus J, Themmen APN 2012 Motivation, learning strategies, participation and medical school performance. Med. Educ. **46**(7) 678–88
[20] Diseth A, Pallesen S, Brunborg GS and Larsen S 2010 Academic achievement among first semester undergraduate psychology students: the role of course experience, effort, motives and learning strategies *High. Educ.* **59**(3) 335–52
[21] Vermunt JD 1996 Metacognitive, cognitive and affective aspects of learning styles and strategies: A phenomenographic analysis *High. Educ.* **31**(1) 25–50
[22] Lee C, Cheng Y, Rai S and Depickere A 2005 What affect student cognitive style in the development of hypermedia learning system? *Comput. Educ.* **45**(1) 1–19
[23] Bartram B and Bailey C 2010 Assessment preferences: a comparison of UK/international students at an English university *Res. Post-Compulsory Educ.* **15**(2) 177–87
[24] Wrenn CG and Humber JW 1941 Study habits associated with high and low scholarship *J. Educ. Psychol.* **32**(8) 611–6
[25] Credé M, Roch SG and Kieszczynka UM 2010 Class Attendance in College: A Meta-Analytic Review of the Relationship of Class Attendance with Grades and Student Characteristics *Rev. Educ. Res.* **80**(2) 272–95.
[26] Magno C 2010 Predicting Grades in Mathematics and English Through Study Habits *TIJRR* **5** 40–50
[27] Weinstein C and Palmer D 2002 LASSI H&H Publishing Company, Inc.
[28] Nonis SA and Hudson GI 2010 Performance of College Students: Impact of Study Time and Study Habits *J. Educ. Bus.* **85**(4) 229–38
[29] Klockner C and Matthies E 2004 How habits interfere with norm-directed behaviour: A normative decision-making model for travel mode choice *J. Environ. Psychol.* **24**(3) 319–27
[30] Verplanken B 2006 Beyond frequency: Habit as mental construct. *Brit. J. Soc. Psychol.* **45**(3) 639–56
[31] Verplanken B and Faes S 1999 Good intentions, bad habits, and effects of forming implementation intentions on healthy eating *Eur. J. Soc. Psychol.* **29**(5-6) 591–604
[32] Verplanken B, Friborg O, Wang C, Trafimow D and Woolf K 2007 Mental habits: Metacognitive reflection on negative self-thinking *J. Pers. Soc. Psychol.* **92**(3) 526–41

The measurement of statistical reasoning in verbal-numerical and graphical forms: a pilot study

M Agus [1], M Peró-Cebollero[2], J Guàrdia-Olmos[2], M P Penna[1]

[1]Department of Pedagogy, Psychology, Philosophy, Faculty of Humanistic Studies, University of Cagliari, Via Is Mirrionis, 1, 09123 Cagliari, Italy

[2]Department of Methodology of the Behavioural Sciences, Faculty of Psychology, University of Barcelona, Pg. Vall d'Hebron, 171, 08035 Barcelona, Spain

E-mail: mirian.agus@unica.it

Abstract. Numerous subjects have trouble in understanding various conceptions connected to statistical problems. Research reports how students' ability to solve problems (including statistical problems) can be influenced by exhibiting proofs. In this work we aim to contrive an original and easy instrument able to assess statistical reasoning on uncertainty and on association, regarding two different forms of proof presentation: pictorial–graphical and verbal–numerical. We have conceived eleven pairs of simple problems in the verbal–numerical and pictorial–graphical form and we have presented the proofs to 47 undergraduate students. The purpose of our work was to evaluate the goodness and reliability of these problems in the assessment of statistical reasoning. Each subject solved each pair of proofs in the verbal-numerical and in the pictorial–graphical form, in different problem presentation orders. Data analyses have highlighted that six out of the eleven pairs of problems appear to be useful and adequate to estimate statistical reasoning on uncertainty and that there is no effect due to the order of presentation in the verbal–numerical and pictorial–graphical form.

1. Introduction

In recent research we observe a strong interest in statistical reasoning and in judgment making in uncertainty situations [1,2]. Generally the concept of statistical reasoning is mentioned as the manner in which individuals reflect on statistical concepts and form conclusions on statistical data [3]. Garfield [2] suggests that six kinds of reasoning (on data, on representations of data, on association, on statistical measures, on samples, and on uncertainty) can be recognised in statistical reasoning. Each type of reasoning can be elicited by specific activities related to data comprehension, analysis, and communication. Specifically, the reasoning of association is important in statistics because this process is used to identify statistical relationships between two or more variables [2]. Reasoning on uncertainty is defined as the ability to apprehend and apply concepts of probability, casualness and chance, such as the capability to make conclusions with uncertainty, and with an awareness that not all results are obtained in the same way [3,1]. But many studies show that individuals make mistakes when they are reasoning on association and on probabilities [4]. Subjects appear not to be familiar with statistical association and probabilistic information [5,6]. For example, when individuals try to use uncertainty information, they tend to incorrectly estimate the probability of events; this occurs for both expert and non expert individuals [7]. Referring to these aspects, the use of several visual aids could appear as an important method for clarifying and improving reasoning on uncertainty and on association [8,9]. Green [10] states that graphs could elicit an in-depth understanding of data complexity. On the other hand, Knauff and Johnson-Laird [11] affirm that visual aids could impede probabilistic reasoning in some problems, because they could overload the cognitive system with additional information.

2. Aim and method

This work represents the middle stage in our research which investigates the use of imagery in statistical reasoning, in order to detect the effect of graphical facilitation [12]. In this phase of the research we aim to explore statistical reasoning on uncertainty in association with appropriate proofs, in relation to two different forms of proof presentation: verbal–numerical and pictorial–graphical. We appoint a measure instrument of these articulated dimensions, which presents several pairs of the analogous proof; in this way we can compare, in the same individual, the statistical reasoning developed in comparable problems presented in two forms. We intend to analyse the reasoning in subjects without statistical skills, in order to evaluate the real contribution of the presentation method in the development of statistical reasoning. Then, we investigate reasoning on uncertainty and on association, with reference to basic mathematical abilities learnt at high school. Therefore we decided to exclude from our assessment those types of statistical reasoning which require a specific statistical knowledge.

In this pilot study we apply specific statistical analyses in order to detect the adequacy of proofs in this investigation. Moreover, in our work we aim at evaluating the potential effect of the order of presentation between proofs in verbal–numerical and pictorial–graphical form.

2.1 Participants, instrument and procedure

Participants consisted of 47 undergraduates at the University of Cagliari (Italy), attending the first year of the degree course in Psychology. The average age of the participants was 22.2 years (SD = 5.7). The group consisted of 32 females (68.1%) and 15 males (31.9%). Data collection took place in December 2012. The sample involved students who voluntarily participated in the research and the sampling was non probabilistic. We submitted for resolution eleven pairs of analogous items in paper format, to a large group in a classroom.

Each item was given to the same student to be solved using two modalities: verbal–numerical (form A) and pictorial–graphical (form B). Each problem necessitated an open response, where the subjects were required to explain the reasoning applied in the resolution. We analysed the students' answers as given to each problem and then in each pair of problems and we coded the correct and the incorrect answers (1/0). In the Appendix we show some exemplificative pairs of items in two dimensions.

Double forms of problems were presented in two separate sessions (one week apart), in different orders of subgroup students. We analysed the students' open answers and then we coded the responses as correct or incorrect. We computed the descriptive statistics the Bivariate Chi Square, the McNemar test, the Discrimination index, the Difficulty index, and the Alpha Reliability.

3. Results

In Figure 1 we show the percentages of correct responses in the two forms in each pair of items of two dimensions: uncertainty (pairs 1,2,3,4,5,6,7) and association (pairs 8,9,10,11). By this figure we can observe that our participants frequently had a better performance in the resolution of problems in the verbal–numerical form of each pair of proofs.

Next we applied a Bivariate Chi Square for each problem, in order to detect the effect of the order of presentation (verbal–numerical / pictorial–graphical; pictorial–graphical / verbal–numerical) in the correct/incorrect answers of each item (see Table 1). The sequence of the order of proof presentation did not have an effect on the number of correct responses; indeed we did not find significant effects by the application of these inferential statistics.

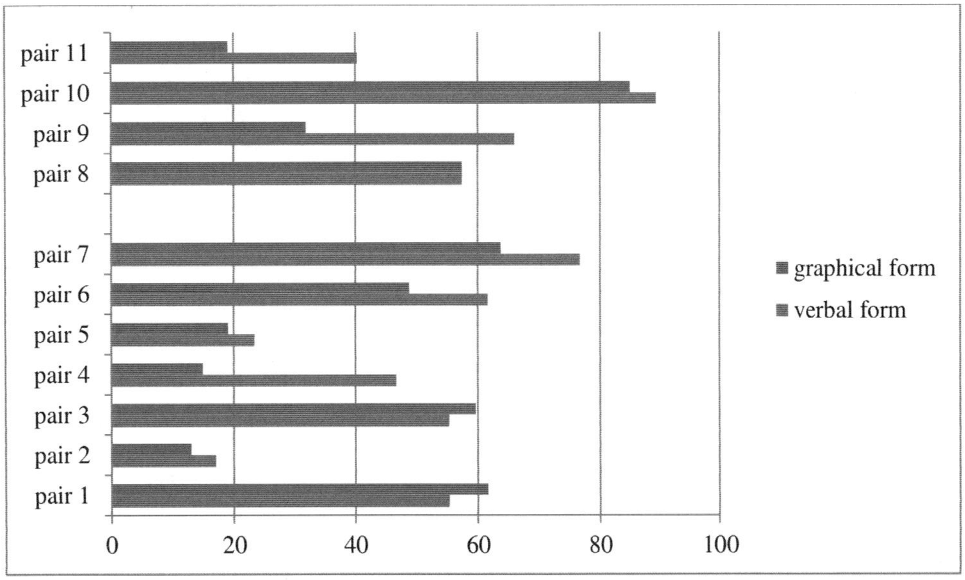

Figure 1. Percentages of correct answers for each pair of proofs

Table 1. Bivariate Chi Square: responses correct/incorrect by order of problem presentation

	Proof Verbal form	Pearson Chi Sqr. (*df*=1)	*p*	Proof Graphical form	Pearson Chi Sqr. (*df*=1)	*p*
Reasoning on uncertainty	**FA 1**	1.561	0.212	**FB 1**	0.774	0.379
	FA 2	2.623	0.105	**FB 2**	0.000	1.000
	FA 3	0.000	1.000	**FB 3**	0.000	0.999
	FA 4	0.038	0.846	**FB 4**	0.094	0.759
	FA 5	0.083	0.774	**FB 5**	0.127	0.721
	FA 6	0.000	1.000	**FB 6**	0.000	1.000
	FA 7	0.000	1.000	**FB 7**	0.003	0.953
Reasoning on association	**FA 8**	0.112	0.738	**FB 8**	0.000	1.000
	FA 9	0.047	0.828	**FB 9**	1.281	0.258
	FA 10	0.064	0.800	**FB 10**	0.094	0.759
	FA 11	3.194	0.074	**FB 11**	0.127	0.721

In assessing the dimension of reasoning on uncertainty for the verbal–numerical form we obtained an Alpha Reliability =0.556 (St. Alpha =0.553) and for the pictorial–graphical form an Alpha Reliability =0.656 (St. Alpha=0.666). Then, in evaluating the dimension of reasoning on association for the verbal–numerical form we had an Alpha Reliability = 0.357 (St. Alpha = 0.320) and for the pictorial–graphical form an Alpha Reliability = 0.416 (St. Alpha= 0.432). We also calculated the indices of Difficulty and Discrimination in order to evaluate the goodness of each proof (see Table 2 and Table 3). The reliability of the two dimensions was low; particularly in the dimension of reasoning by association some statistics appear inept in order to pursue our aim to construct an evaluation instrument of statistical reasoning.

Table 2. Descriptive statistics and indices on the verbal–numerical form

	Proof	Number of correct responses	Difficulty index	Discrimination index	Alpha deleting each item in turn	Std. Alpha deleting each item in turn	r (item, total)
Reasoning on uncertainty	**FA 1**	26	0.553	0.866	0.454	0.453	0.424
	FA 2	8	0.170	0.333	0.545	0.547	0.197
	FA 3	26	0.553	0.666	0.527	0.527	0.259
	FA 4	22	0.468	0.733	0.449	0.451	0.435
	FA 5	11	0.234	0.533	0.503	0.501	0.322
	FA 6	29	0.617	0.400	0.600	0.589	0.066
	FA 7	36	0.766	0.333	0.516	0.512	0.284
Reasoning on association	**FA 8**	27	0.570	0.733	0.179	0.161	0.277
	FA 9	31	0.660	0.600	0.268	0.218	0.211
	FA 10	42	0.894	0.200	0.431	0.430	0.013
	FA 11	19	0.404	0.733	0.226	0.195	0.243

Table 3. Descriptive statistics and indices on the pictorial–graphical form

	Proof	Number of correct responses	Difficulty index	Discrimination index	Alpha deleting each item in turn	Std. Alpha deleting each item in turn	r (item, total)
Reasoning on uncertainty	**FB 1**	29	0.617	0.866	0.579	0.593	0.490
	FB 2	6	0.128	0.400	0.628	0.640	0.352
	FB 3	28	0.596	0.400	0.668	0.671	0.220
	FB 4	7	0.149	0.266	0.627	0.637	0.350
	FB 5	9	0.191	0.533	0.581	0.581	0.512
	FB 6	23	0.489	0.933	0.554	0.578	0.555
	FB 7	30	0.638	0.466	0.684	0.692	0.161
Reasoning on association	**FB 8**	27	0.574	0.800	0.429	0.408	0.166
	FB 9	15	0.319	0.600	0.379	0.379	0.203
	FB 10	40	0.851	0.400	0.345	0.382	0.240
	FB 11	9	0.191	0.533	0.240	0.276	0.342

4. Discussion

This research presents a pilot study in order to develop a measurement instrument to assess undergraduate students' statistical reasoning on uncertainty and on association, in relation to methods of proof presentation. By the construction of paired items in two forms, we could compare the reasoning applied to problem resolution, with regard to the specific problem structure.

With evaluation of the statistical analysis, we could presume that the dimension of reasoning on association did not have good indices; therefore these pairs of items could be discarded by our questionnaire [13]. In relation to the reasoning on uncertainty, we could assume that one item (FB 4) did not have a good Discrimination Index, therefore it is not appropriate for this measurement [13].

We could also assume that the pair corresponding to this item could be discarded by the instrument, using the remaining six couples for the assessment of reasoning on uncertainty.

Moreover, in order to improve the agility and the promptness of this instrument, we could prepare a new version characterized by closed response options for these six pairs of items; then each problem could be followed by an open question, inquiring as to the reasoning implemented in the proof solution. In this way we could reduce the solution time of the proofs and simultaneously investigate the cognitive resolution process applied.

Furthermore we have highlighted that the proof disposition could not affect the problem resolution. Nevertheless, in order to avoid and control this contingent effect, it could be useful to balance the two forms of proofs by a casual disposition of problems in the random sub-samples of students.

The results of this work could be useful regarding the intention to develop statistical teaching and to project appropriate learning experiences in order to improve statistical reasoning.

5. Appendix

Some exemplificative pairs of items in verbal–numerical format and pictorial–graphical format

5.1. Reasoning on uncertainty - VERBAL-NUMERICAL FORMAT

A factory produces electronic games, but not all work well. Of every 100 game products: 20 may have an electrical problem; 80 can function correctly. The company has developed control systems to identify faulty games; however, these systems do not work properly. In reality, half of the games with electrical problems continue in the production line where they are considered as functional.

If you randomly extract a game that has been addressed to the shops and evaluated free of defects, what is the probability that it is defective?

5.2. Reasoning on uncertainty - PICTORIAL–GRAPHICAL FORMAT

A factory that produces personal computers has problems in the production process. Some of the computers are defective (problems with the video card). Such problems are not always identified by the quality control and consequently some defective computers are sent forward in the production line. The graphic below shows this process. What is the probability that a computer addressed to the shops, evaluated free of defects, is defective?

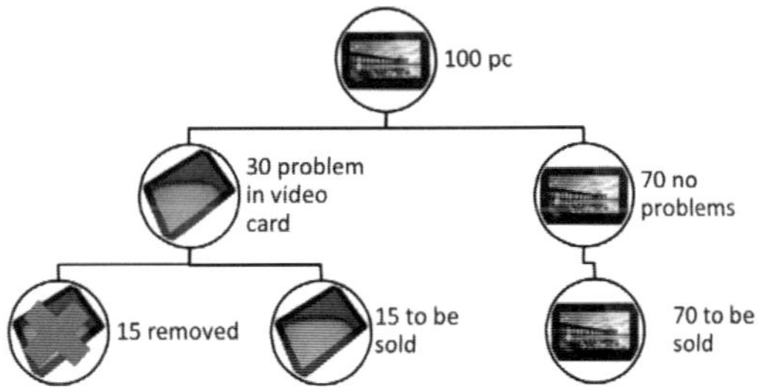

Figure 2. Pictorial graphical format – Reasoning on uncertainty

5.3. Reasoning on association - VERBAL–NUMERICAL FORMAT

X and Y are two variables in the relationship between them. The values of these variables are listed below:

X	1	2	3	4
Y	1	8	27	64

Taking into account this information, define the mathematical expression that describes the relationship between the two variables.

5.4. Reasoning on association - PICTORIAL–GRAPHICAL FORMAT

A and B are two variables in the relationship between them. On the basis of the graphical representation below, can you identify the type of mathematical relationship between the two variables?

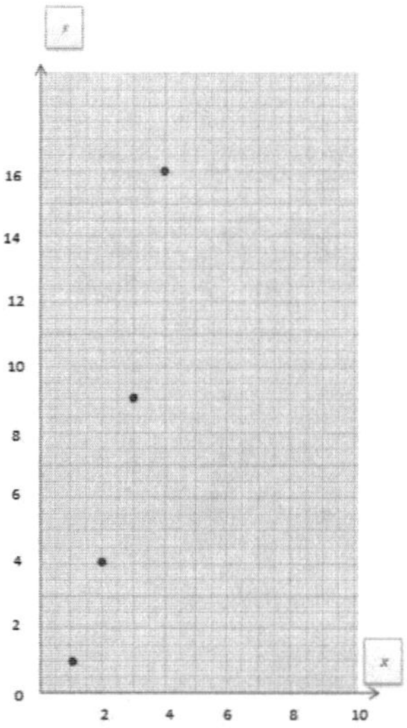

Figure 3. Pictorial graphical format – Reasoning on association

References

[1] Garfield J. Assessing statistical reasoning *Stat. Educ. Res. J.* 2003; **2** 22-38
[2] Lavigne NC, Lajoie SP. Statistical reasoning of middle school children engaged in survey inquiry *Contemp. Educ. Psychol.* 2007; **32** 630-66
[3] Garfield J, Chance B. Assessment in statistics education: Issues and challenges *Math. Think. Learn.* 2000; **2** 99-125
[4] Ibrekk H, Morgan MG. Graphical Communication of Uncertain Quantities to Nontechnical People *Risk Anal.* 1987; **7** 519-29
[5] Hoffrage H, Gigerenzer G. Using natural frequencies to improve diagnostic inferences *Acad. Med.* 1998; **73** 538-48
[6] Van Dijk E, Zeelenberg M. The discounting of ambiguous information in economic decision making *J. Behav. Decis. Making* 2003; **16** 341-52
[7] Shafir E, Tversky A. Thinking through uncertainty: Non consequential reasoning and choice *Cognit. Psychol.* 1992; **24** 449-74
[8] Moro R, Bodanza GA. El debate acerca del efecto facilitador en problemas de probabilidad condicional: ¿Un caso de experimentación crucial? *Interdiscipl.* 2010; **27** 163-74
[9] Brase GL. Pictorial representations in statistical reasoning *Appl. Cogn. Psychol.* 2009; **23** 369-81
[10] Green H. Grapharti: A new visual summary of data, *J. Stat. Ed.* 2007; **15** 3
[11] Knauff M, Johnson-Laird PN. Visual imagery can impede reasoning *Mem. Cognition.* 2002; **30** 363–71
[12] Penna MP, Agus M, Peró-Cebollero M, Guàrdia-Olmos J, Pessa E. The use of imagery in statistical reasoning by university undergraduate students: a preliminary study *Qual. Quant.* 2012; 1-15, doi:10.1007/s11135-012-9757-5
[13] Reynolds CR, Livingston RB, and Willson V. T V *Measurement and Assessment in Education* Boston: Pearson; 2009

Journal of Physics: Conference Series **459** (2013) 012024 doi:10.1088/1742-6596/459/1/012024

Overcoming the Invisibility of Metrology: A Reading Measurement Network for Education and the Social Sciences

William P Fisher Jr
Research Associate, BEAR Center, Graduate School of Education, University of California, Berkeley, CA (USA) & Principal, LivingCapitalMetrics Consulting
E-mail: william@livingcapitalmetrics.com

A Jackson Stenner
CEO, MetaMetrics, Inc.
Durham, North Carolina, USA
E-mail: jstenner@lexile.com

Abstract. The public and researchers in psychology and the social sciences are largely unaware of the huge resources invested in metrology and standards in science and commerce, for understandable reasons, but with unfortunate consequences. Measurement quality varies widely in fields lacking uniform standards, making it impossible to coordinate local behaviours and decisions in tune with individually observed instrument readings. However, recent developments in reading measurement have effectively instituted metrological traceability methods within elementary and secondary English and Spanish language reading education in the U.S., Canada, Mexico, and Australia. Given established patterns in the history of science, it may be reasonable to expect that widespread routine reproduction of controlled effects expressed in uniform units in the social sciences may lead to significant developments in theory and practice.

1. Introduction

Metrology is the means by which a common frame of reference is provided for measurement applications in widely separated industrial, scientific, and practical tasks. Large volumes of resources—comprising significant fractions of many nations' economic productivity—are invested in ensuring traceability to standards for various units of measurement. Ironically and paradoxically, the value of the return on these investments is almost entirely dependent on their transparency and the invisibility of the processes by which they were made transparent.

1.1. Transparent measures, invisible processes

Metrology is doing its job best when no one knows it is there. Experimental and theoretical scientists often take little interest in instrumentation beyond the satisfaction of their immediate needs for precision measures. It is not surprising, then, to also find that the intensive work of establishing traceability is largely invisible to the general public and to researchers in psychology and the social sciences, with significant consequences [1-4].

The uniformity of the various phenomena described by natural laws allows scientists themselves, moreover, the convenient efficiency of not needing to specify scale units in statements of laws. This conveniently supports a division of labour in science that separates theoretical work from the calibration of instruments and both of these from the use of theory and instruments in experiments [5].

But the separation of theoretical, experimental and instrument-focused communities has its drawbacks, too. The lack of metrological awareness among theoretical and experimental scientists is especially problematic. Not knowing when or how reference standard units are established reinforces historically long standing assumptions—such as the idea that the universe or nature is inherently and innately numerical, quantitative, or mathematical—that rarely become explicit objects of attention.

The effect of these presuppositions is significant. Huge social, industrial, and economic efficiencies are gained by universal consensus on the facts of complex phenomena like electricity, temperature, and time. Making quantities seem natural is a cultural achievement of the highest order.

But when the undeniably historic and historical quantitative status of scientific objects becomes reified as unquestioned and unquestionable, then the advancement of science must inevitably be put at risk. The two questions few are asking are: (1) how did the natural sciences succeed in making some things seem so thoroughly natural [1-2,6-8], and (2) how might the social sciences learn from those successes? It will be shown that recent advances in reading measurement embody important lessons in this regard for the social sciences.

1.2. Shortsightedly focusing attention on the local measurement outcome

Historians and philosophers of science have been known to systematically cut the technical infrastructure of measurement out of the picture, sometimes literally, as when a woodcut of a laboratory scene printed in its entirety in one place is trimmed in a later publication to exclude the means by which the effect was produced [9]. Transparency in measurement is a two-edged sword. Widely accessible transparent meaningfulness is achieved only to the extent that technical complexities can be ignored. This point was emphasized by Whitehead [10], who observed that "Civilization advances by extending the number of important operations which we can perform without thinking about them" (p. 61). But what happens when those who make those advances do not record—or do not themselves fully understand—*how* they extended the number of important operations that can be performed by persons unversed in their technicalities?

This general and very difficult problem was noted by Husserl [11]. In his philosophical consideration of the geometric assumptions informing Galileo, Husserl was sensitive to the ways in which an unexamined hidden agenda was allowed to set priorities. Thus we find ourselves in a situation, in accord with the philosophical problems attending the foundations of measurement, in general, where

> Metrology has not often been granted much historical significance. ... Intellectualist condescension distracts our attention from these everyday practices, from their technical staff, and from the work which makes results count outside laboratory walls [3].

Researchers in the natural sciences make use of commercially available precision tools calibrated to universally uniform reference standards, standards capitalizing on the value of invariant laws. The transparency of the substantive qualitative meanings shared in an elaborated metrological network renders the effects of metrology's lack of historical significance relatively harmless. There is enough general awareness of the role metrology plays in science and commerce that standards engineering institutions and traditions with the express purposes of enabling standards are in place and can be accessed when they are needed.

2. Metrological traceability for reading measurement

2.1. Consequences for psychology and the social sciences

But in the social sciences, the lack of metrological institutions, methods, and traditions has been catastrophic. As social scientists have long recognized for themselves [12-13], mainstream research methods and statistical models employ scale-dependent ordinal data in a search for a kind of significance that is often irrelevant to and even antithetical to the production of new knowledge. Even when regularities akin to natural laws are sought and found in psychological and social phenomena [14-15], the results are typically assessed in the language and methods of statistics rather than of measurement and metrology. The human, social, economic, and scientific consequences of this can be profound.

The lack of institutions and traditions concerning metrological traceability and standards in psychology and the social sciences may have more to do with broad and deep cultural presuppositions than with an actual lack of a basis for them in evidence. After all, what systematic program of experimental evaluation has ever irrefutably established that uniform metrics based in lawful regularities are impossible in psychology and the social sciences? Evidence that has been sought indicates that provisional possibilities exist in some circumstances [14-15].

3. Metrological traceability for reading measurement

Educators have long understood the need to provide students with reading challenges appropriate to their reading abilities. The problem of matching individual readers to texts is usually approached in terms of general curricular structures, and teacher training and experience. Theory has not been a focus of primary interest [16-17]. Rasch's development of a new class of measurement models in the 1950s was an important step forward in improving the quantification of reading ability [18]. An unsought advantage of this research was improved precision in the matching of readers to text.

When Rasch's concept of specific objectivity obtained in local measures was combined with a general predictive theory of English text complexity in the 1980s, following the work of Stenner and colleagues [19-21], the stage was set for the efficient creation of a network of reading measurement instruments calibrated in a common unit. By the late 1990s, all of the major English reading tests had been brought into the system. These are today complemented by the hundreds of thousands of books, tens of millions of short articles and hundreds of millions of readers that have been brought into the system.

Reader abilities and text complexities are measured in the same unit. The scale ranges from below 200 for beginning readers to over 1600 for very high level readers and texts. Knowing the text measure of a book and the reader's measure predicts the degree to which the book will be comprehensible to the student. When the two measures match, a 75% comprehension level is expected. In other words, as has been repeatedly verified in experimental tests, a student with a measure of 500 would be expected to answer correctly 75% of the questions on an assessment made from that text.

More than 30 million measures annually are reported in the U.S. from state and commercial assessments, and from classroom reading program assessments, in a common unit of measurement [21]. The 21 state departments of education that have formally adopted this unit for use statewide are shown in the map in Figure 1.

Traceability to the common unit is determined via both empirical, data-based equating studies and theory-based text analyses [21].

Additional features of the system include electronic tools integrating instruction and assessment for mass customized diagnostics [22], and others charting growth in reading ability relative to college and career readiness [23-24]. Establishing this network of comparable assessments required formal relationships with book and test publishers, teachers, schools and school districts, state departments of education, and psychometric researchers. Furthermore, a new array of material practices was needed to give all the parties involved ongoing and verifiable confidence in the theory. Though great efficiencies stood to be gained, credibility demanded a cautious approach to their implementation. Formal documentation of the birth of the traceability system would be a valuable contribution to the sociotechnical qualities of education.

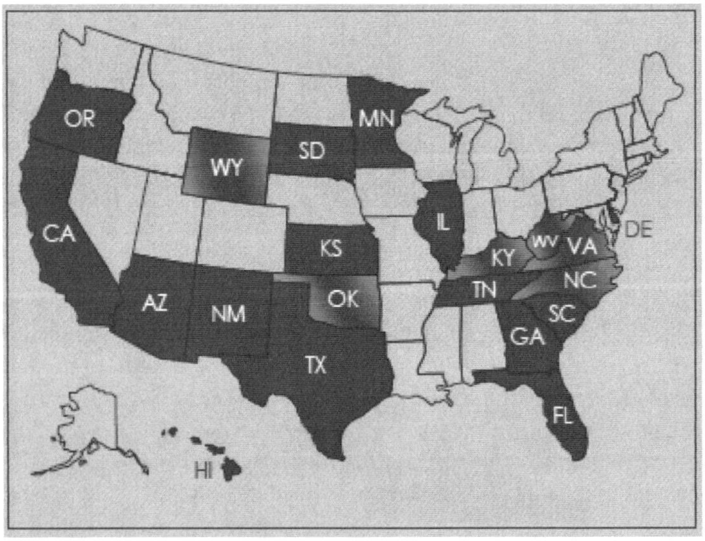

Figure 1. Map of U.S. states employing a common unit of reading measurement (two tones indicate common units in use for both reading and mathematics)

4. Implications for psychology and the social sciences

In 1965, the National Academy of Sciences published a report articulating common assumptions as to the sequence of events supposed to take place in the development of new instrumentation [25]. Four stages were identified:

(1) discovery of suitable means of observing some phenomenon,
(2) exploration of this phenomenon with special, homemade instruments or commercial prototypes,
(3) widespread use of commercial instruments,
(4) routine applications of the instrument to control industrial production as well as research."

Textbook assumptions and presentations of this sequence have indoctrinated researchers in the human sciences to believe, mistakenly, that this is the normal sequence of events, and we have indeed largely succeeded in imposing this sequence on ourselves, against our own interests, and against the grain of actual events in the natural sciences. Rabkin [25] points out that

> this scheme seems to be at variance with much of the evidence in the history of science. It has been shown that the integration of instruments has been rarely due to the demand on the part of the researcher. Rather it occurs through vigorous supply of advanced instruments on the part of the industry. The company that proposes these four stages in the report has itself had experience when stages 3 and 4 occur in the reverse order and, moreover, stage 4 is by far the most decisive factor in the development of new instrumentation.

The "vigorous supply of advanced instruments", and not demand, also characterizes the introduction of popular electronic appliances. Just as Rabkin points out has been the case in research, there was little or no clamour among the public for telephones, televisions, faxes, the internet, blenders, or cell phones before they were developed and introduced.

Scientists and the public both tend to think of instrumentation only as tools employed in the service of the individuals who use them. This perspective is at odds with the historical evidence as well as with philosophers' observations, such as, for instance, Thoreau's realization that humanity has become the tool of its tools [26] and Nietzsche's insight that the victory of science is better cast as a victory of method over science [27].

This alternative perspective is important because, in the history of science, theory follows from extensive experience with instruments more often than instruments are designed and built from theoretical projections. Standardized and commercially available instrumentation make possible the predictable and routine reproduction of scientific effects essential to the conduct of controlled experiments—and so also to the development of precise and accurate theoretical predictions. As stated by Price,

> Historically, we have almost no examples of an increase in understanding being applied to make new advances in technical competence, but we have many cases of advances in technology being puzzled out by theoreticians and resulting in the advancement of knowledge. It is not just a clever historical aphorism, but a general truth, that 'thermodynamics owes much more to the steam engine than ever the steam engine owed to thermodynamics. ...historically the arrow of causality is largely from the technology to the science [28].

In the context of reading measurement, the repeated reproduction of consistent results following the work of Rasch and others led to the Anchor Test Study in the 1970s [29]. This study equated seven major reading tests in the U.S. and involved over 350,000 students in all 50 states. But the purely empirical basis of the calibration and the lack of predictive theory meant that the value of the common unit of measurement was lost as soon as new items were added to the tests, which was immediately.

A plain feature of the equated test results, however, was the similarity of the items from different tests that calibrated in the same locations. More common words and shorter sentences characterized the texts associated with easier questions calibrating at the bottom of the scale, in contrast with the rarer words and longer sentences in the texts with the harder questions at the top of the scale. The stability of this phenomenon will come as no surprise to anyone able to read, but its practical application in a predictive theory relating text complexity, comprehension rates, and reading ability was difficult to achieve [30].

5. Conclusion

Historians of science have repeatedly documented the roles in theory development played by researchers with hands-on experience with instrumentation, as when Kuhn [31] notes that seven of the nine pioneers in quantifying energy conversion process were either trained as engineers or were working with engines when they made their contributions. Indeed, this attitude that an instrument can make a science was taken from physics into economics by both Stanley Jevons and Irving Fisher in their uses of the balance scale as a model of market equilibrium [32-33].

Perhaps new instances of this phenomenon will be created by widespread use of Rasch's models to calibrate multiple instruments measuring the same things. Consistent behaviour on the part of these instruments may lead in due course to clearer understandings of the constructs involved, and so also to developments in predictive theory and more efficient implementations. These positive outcomes may be made more likely by recent work emphasizing the qualitative progression of learning defined by item hierarchies in educational assessments generally [34].

6. References

[1] Latour B 1987 *Science in action: How to follow scientists and engineers through society* (New York: Cambridge University Press)
[2] Latour B 2005 *Reassembling the social: An introduction to Actor-Network-Theory* (Oxford, England: Oxford University Press)
[3] Schaffer S 1992 In R Bud & S E Cozzens (Eds.), *Invisible connections: Instruments, institutions, and science* pp. 23-56 (Bellingham, Washington: SPIE Optical Engineering Press)

[4] Wise M N 1995 *The values of precision* (Princeton, New Jersey: Princeton University Press)
[5] Galison P 1999 In M. Biagioli (Ed.) *The science studies reader* pp. 137-160 (New York: Routledge)
[6] Dear P 2012 *Osiris* **27** 37-55
[7] O'Connell J 1993 *Social Studies of Science* **23** 129-173
[8] Sundberg M 2011 *Social Studies of Science* **41** 107-125
[9] Shapin S 1989 *American Scientist* **77** 554-563
[10] Whitehead A N 1911 *An introduction to mathematics* (New York: Henry Holt and Co.)
[11] Husserl E 1970 *The crisis of European sciences and transcendental phenomenology* (Evanston, Illinois: Northwestern University Press)
[12] Berkson J 1938 *American Statistical Association Journal* **33** 526-536
[13] Salsburg D S 1985 *The American Statistician* **39** 220-223
[14] Burdick D S, Stone M H, Stenner A J 2006 *Rasch Measurement Transactions* **20** 1059-1060
[15] Fisher W P 2009 *Measurement* **42** 1278-1287
[16] Sadoski M, Paivio A 2007 *Scientific Studies of Reading* **11** 337-356
[17] Engelhard G 2001 *J Appl Meas* **2** 1-26
[18] Rasch G 1960 *Probabilistic models for some intelligence and attainment tests* (Reprint, with Foreword and Afterword by B. D. Wright, Chicago: University of Chicago Press, 1980 (Copenhagen, Denmark: Danmarks Paedogogiske Institut)
[19] Stenner A J, Smith M 1982 *Perceptual and Motor Skills* **55** 415-426
[20] Stenner A J, Smith M, Burdick D S 1983. *J Ed Meas* **20** 305-316
[21] Stenner A J, Burdick H, Sanford E E, Burdick D S 2006 *J Appl Meas* **7** 307-322
[22] Rivero V 2011, June 10 *Edtech Digest.* Retrieved 23 May 2012, from http://edtechdigest.wordpress.com/2011/06/10/interview-jack-stenner-takes-education-beyond-the-metrics/
[23] Williamson G L 2008 *J Advanced Academics* **19** 602-632
[24] Williamson G L, Fitzgerald J, Stenner A J 2013 *Educational Researcher* **42** 59-69
[25] Rabkin Y M 1992 In R Bud, S E Cozzens (Eds) *Invisible connections: Instruments, institutions, and science* pp. 57-82 (Bellingham, Washington: SPIE Optical Engineering Press)
[26] Thoreau H D 1854 *Walden, or life in the woods* (Boston: Ticknor and Fields)
[27] Nietzsche F 1967 *The will to power* (New York: Vintage)
[28] Price D J 1986 In *Little science, big science--and beyond* pp. 237-253 (New York: Columbia University Press)
[29] Rentz R R, Bashaw W L 1977 *J Ed Meas* **14** 161-179
[30] Stenner A J, Stone, M 2010 *J App Meas* **11** 244-252
[31] Kuhn T S 1977 *The essential tension: Selected studies in scientific tradition and change* (Chicago, Illinois: University of Chicago Press)
[32] Boumans M 2001 In M S Morgan & J. Klein (Eds.) *The age of economic measurement* pp. 313-344 (Durham, North Carolina: Duke University Press)
[33] Maas H 2001 In M S Morgan & J Klein (Eds.), *The age of economic measurement* pp. 277-302 (Durham, North Carolina: Duke University Press)
[34] Wilson M R 2009 *Journal of Research in Science Teaching* **46** 716-730

Journal of Physics: Conference Series **459** (2013) 012025 doi:10.1088/1742-6596/459/1/012025

Metrological Traceability in the Social Sciences: A Model from Reading Measurement

A Jackson Stenner
CEO, MetaMetrics, Inc.
Durham, North Carolina, USA
E-mail: jstenner@lexile.com

William P Fisher Jr
Research Associate, BEAR Center, Graduate School of Education, University of California, Berkeley, CA (USA) & Principal, LivingCapitalMetrics Consulting
E-mail: william@livingcapitalmetrics.com

Abstract. The central importance of reading ability in learning makes it the natural place to start in formative and summative assessments in education. The Lexile Framework for Reading constitutes a commercial metrological traceability network linking books, test results, instructional materials, and students in elementary and secondary English and Spanish language reading education in the U.S., Canada, Mexico, and Australia.

1. Introductory history

The ability to read is fundamental to education, and it is accordingly tested and measured more often than any other subject area. The index to the eighteenth edition of the Buros Mental Measurements Yearbook [1] includes over 140 tests with the word "reading" in their titles. This count does not include tests focused on vocabulary or word meaning, which are also numerous.

Though the issues are complex, literacy remains essential to productivity in the global economy [2]. The need for effective and efficient reading education will only intensify as communication, teamwork, information, and information management are increasingly demanded as basic skills [3].

2. Theory for reading measurement

Despite the longstanding fundamental importance of reading as the tool most essential to learning, reading research remained atheoretical until 1953 [4], and interest in a unified theory of reading is a relatively new phenomenon [5]. Further, in the years since 1953, available reading theories have not been used to inform the design or interpretation of assessments of reading ability[4].

Though it may seem counterintuitive, this failure to apply theory in the course of empirical measurement research is not unusual, nor is it restricted to reading research. On the contrary, measurement technologies in the natural sciences have historically been developed through socially-contextualized trial-and-error solutions to practical engineering problems, such as consistent, stable results, and not directly from theoretical principles [6,7]. Theory generally comes later, after researchers have had the opportunity to employ standardized technologies in the routine and repeated reproduction of a controlled phenomenon. Only then do applicable general principles emerge as useful insights that can be fed back into technical refinements.

1.1 Syntactic and semantic elements

Science inevitably involves reducing complex and rich phenomena to simpler models. The truth of the models is less an issue than their usefulness [8,9]. Simplification is usually achieved only in a context that respects necessary constraints and accepts the realization of a limited goal. The efficiency and power obtained when limited goals can be realized in the form of a useful tool, however, can sometimes confer great value on a simplified process.

In the 1950s, Rasch's formulation of a parameter separability theorem, the concept of specific objectivity, and models useful in practical measurement applications combined into an important step forward in educational measurement [10]. These developments were followed by Wright's introduction of improved estimation algorithms, model fit tests, and software in the 1960s, along with his vigorous championing of Rasch's ideas [11]. By the 1970s, enough data from reading tests had been successfully fit to Rasch models in the U.S. to support the viability of the Anchor Test Study, an equating of seven major reading tests involving over 350,000 students in all 50 U.S. states [12].

Lacking a theory capable of uniting different tests into a common framework, however, the uniform scale resulting from the equating study was made obsolete as soon as the tests were changed, which was, of course, immediately. Successful research predicting item difficulties on the Peabody Vocabulary Test and the Knox Cube Test (a measure of short term memory and attention span) [13,14], however, led to a new effort focused on reading.

Reading theories build on the fact that all symbol systems share two features: a semantic component and a syntactic component. In language, the semantic units are words. Words are organized according to rules of syntax into sentences [15]. Semantic units vary in familiarity and the syntactic structures vary in complexity. The readability of a text passage is dominated by the familiarity of the semantic units and by the complexity of the syntactic structures used in constructing the message. Many readability equations therefore use a two-variable equation to forecast text difficulty. The word-frequency and sentence-length measures combine to produce a regression equation, known as a construct specification equation [13,14]. This equation provides a theoretical model that is evaluated in terms of the proportion of the variance of reading comprehension task difficulties (or, more recently, the means of specification-equivalent ensembles of item difficulties [16]) that can be explained.

2.1 The specification equation

One approach to such a specification equation first employs the mean of the logarithm of the frequencies with which words in a text appear in a 550-million word corpus of K-16 texts. More specifically, the log frequency of the word family, which is more highly correlated with word difficulty, comprises one term in the equation. Word families include the stimulus word, all plurals, adverbial forms, comparatives, superlatives, verb forms, past participles, and adjectival forms. The frequencies of all words in the family are summed and the log of that sum is used in the specification equation.

The second term of the specification equation is the logarithm of the text's mean sentence length. This parameter is operationalized simply by counting and averaging the number of words in each sentence. The relationship of word frequency and sentence length to text readability was investigated in research that extended a previous study on semantic units [14]. This analysis involved calculation of the mean word frequency and the log of the mean sentence length for each of the 66 reading comprehension passages on the Peabody Individual Achievement Test. The observed difficulty of each passage was the mean difficulty of the items associated with the passage (provided by the publisher) converted to the logit scale.

A regression analysis based on the word-frequency and sentence-length measures produced a regression equation that explained much of the variance found in the set of reading comprehension tasks. The resulting correlation between the observed logit difficulties and the theoretical calibrations was 0.97 after correction for range restriction and measurement error [14].

The regression equation was further refined based on its use in predicting the observed difficulty of the reading comprehension passages on eight other standardized tests (see Table 1). Repeated and

Journal of Physics: Conference Series **459** (2013) 012025 doi:10.1088/1742-6596/459/1/012025

ongoing comparisons of theoretically expected calibrations with data-based estimates produced from test data analysis provide continually updated validity evidence.

The regression equation links the syntactic and semantic features of text to the empirically determined difficulty of text. That link, in turn, is reproduced across thousands of test items and millions of examinees in applications. The consistent display of the link over time provides a basis for using the equation to perform theory-based calibrations of test items and texts, thus rendering empirical calibrations necessary only as checks on the system.

Table 1 Correlations of theory-based calibrations produced by the specification equation and data-based item difficulties.

Test	# of Questions	# of Passages	$r_{(OT)}$ [a]	$R_{(OT)}$ [b]	$R^*_{(OT)}$ [c]
SRA	235	46	.95	.97	1.00
CAT-E	418	74	.91	.95	.97
Lexile	262	262	.93	.95	.97
PIAT	66	66	.93	.94	.97
CAT-C	253	43	.83	.93	.96
CTBS-U	246	50	.74	.92	.95
NAEP	189	70	.65	.92	.94
Battery	26	26	.88	.84	.87
Mastery	85	85	.74	.75	.77
Totals	1780	722			
Means			.84	.91	.93

[a] $r_{(OT)}$ = raw correlation between observed difficulties (O) and theory-based calibrations (T)
[b] $R_{(OT)}$ = correlation between observed difficulties (O) and theory-based calibrations (T) corrected for range restriction
[c] $R^*_{(OT)}$ = correlation between observed difficulties (O) and theory-based calibrations (T) corrected for range restriction and measurement error.

The theory-based calibration process begins from an analysis of a text in which the numbers of words per sentence are counted, and the commonality of the individual words is estimated. The theoretical logit is then a function of sentence length and word frequencies in the language stated in the specification equation:

$$(9.82247*LMSL)-(2.14634*MLWF)-constant \tag{1}$$

where LMSL is the log of the mean sentence length and MLWF is the mean of the log of the word frequencies. LMSL and MLWF are used as proxies for syntactic complexity and semantic demand [17]. These logits are then scaled as follows:

$$(logit + 3.3)*180 + 200 \tag{2}$$

The uncertainty (standard error) of the individual measures [18] is

$$SE = X * [L/r(L-r)]^{\wedge}1/2 \tag{3}$$

which is the square root of the test length L divided by the count correct r times the L-r count incorrect, times an expansion factor X that depends on test width. This logit is then converted to the standard unit. The original study found success on items at about -3.3 logits as indicating the earliest reading ability, and set that level at 200. A practical top to the scale was at 2.3 logits, and this was set 1000 units higher,

Journal of Physics: Conference Series **459** (2013) 012025 doi:10.1088/1742-6596/459/1/012025

to 1200. A standard unit uncertainty for a well targeted 36-item test measuring with an uncertainty of about .40 logits is the original logit range of 2.3 - (-3.3) = 5.6 divided into the 1000 L range, times .40, which comes to about 71.

3. Practical implementation issues

Projected comprehension rates should not be the only factor influencing text selection. To make the quantified measure the sole determinant of a curricular decision would be analogous to reducing a table to its physical dimensions when its sentimental or historical value might also be relevant.

Initial efforts at deploying the unit of measurement quickly encountered a chicken and egg question from book publishers: why should they adopt the unit as a means of indicating the text complexity of their books and articles if there were no schools or students prepared to take advantage of that information? Conversely, state departments of education and school districts asked, why should they be interested in a universally uniform measure of reading ability if there were no books or articles to match with students' ability measures?

The solution arose when one publisher incorporated the unit in their own system, which involved both a reading curriculum and a reading assessment system. This bootstrapping of one instance of coordinated reader-text matching made the link to the unit more attractive to testing agencies, who could now point to an additional use for their results; to book publishers, who now were assured of a population of students with measures to match with their books; and to state departments of education and school districts, who could now effectively put the matching system to work.

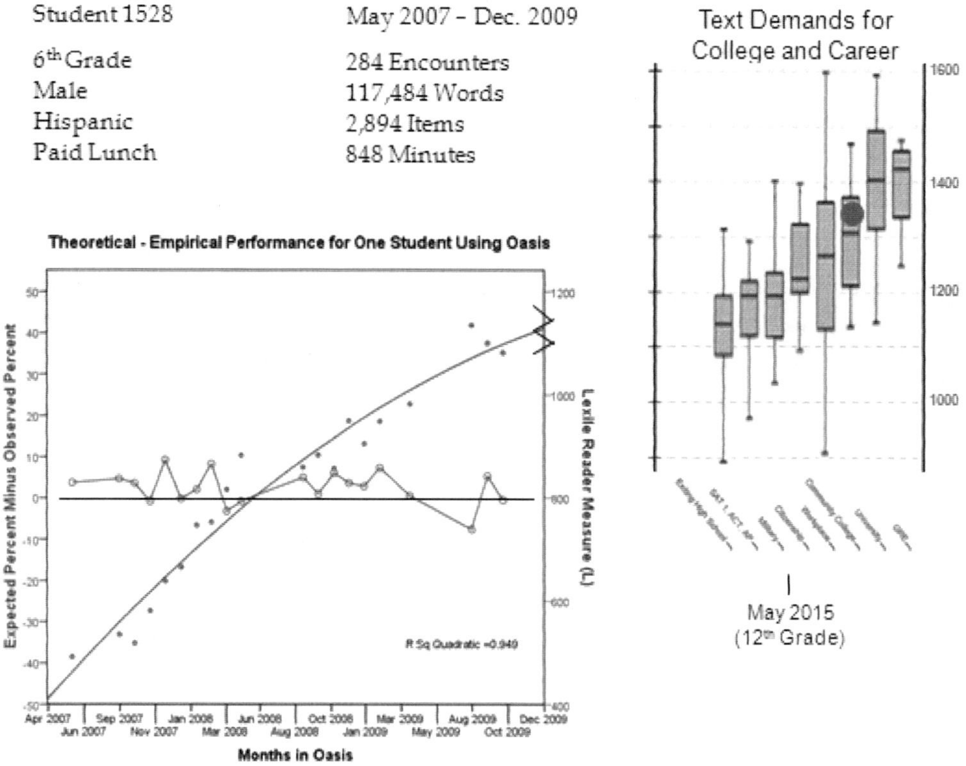

Figure 1. Individual student online reading measurement tracking system report

4. Conclusion

Over 28 million U.S. students, over half of all children in school, currently receive at least one reading ability measure per school year in the standard unit. The text complexity of tens of millions of magazine

Journal of Physics: Conference Series **459** (2013) 012025 doi:10.1088/1742-6596/459/1/012025

articles and hundreds of thousands of books are expressed in the standard unit, and all major reading assessments report measures in it.

The English-based system is in use in the U.S., Canada and Australia (with applications emerging in New Zealand, South Africa, and England), and in ESL applications in Korea, Japan, Malaysia, Hong Kong, and elsewhere in Asia. A Spanish system for matching readers and texts in the same unit is in use in Mexico and the Philippines. Researchers in various parts of the world are exploring possibilities for expanding the reader-text matching system to Mandarin, French, and other languages.

Educational textbook and curriculum publishers have developed online software applications for tracking individual student growth in reading ability. A report from one such system is shown in Figure 1. The value of repeated measures of a student over time and across texts is evident in the growth trajectory and the expected convergence of the student's ability with the reading demands of adult life.

Figure 2 shows the relationship between expected and observed text complexity measures in the online system. This plot illustrates the power of theory. Traceability to the standard unit is achieved not only by estimating student reading ability measures from data, but by gauging text complexity from its syntactic and semantic makeup. Given theory-based estimates of item difficulty, items can be adaptively selected for custom-tailored individualized administration, and those students' measures may then be estimated from their comprehension rates relative to the scale values of those items.

The specification equation operationalizes Rasch's notion of a frame of reference in a way that extends the frame beyond the specific objectivity obtained in the context of a particular test or set of equated tests to an indefinitely large collection of actual or virtual instruments. Theory-based instrument calibration eliminates the need to use data to both calibrate instruments and measure persons. The pay-off from using theory instead of data to calibrate instruments is large and immediate. When data fit a Rasch model, *differences* among person measures are free of dependencies on other facets of the measurement context (i.e., the differences are specifically objective). When data fit a causal or theory-enhanced Rasch model, *absolute* person measures are free of the conditions of measurement (items, occasions, etc.) making them objective beyond the limits of a specific frame of reference. In the theory - referenced context, person measures are individually-centred statistics in that no reference to another person(s) figures in their estimation.

One of the most important uses of reading test scores is to predict how a reader will perform on non-test tasks. For example, educators may be interested in the likelihood that a particular reader completes her first year of studies at a four year college. A regression equation could be fit to first year completion outcome data using reading test scores as predictors to estimate an individual first year completion likelihood on the basis of the relationship between test scores and completion found to hold in the population at large.

An alternative strategy is to imagine that first year college textbooks are virtual reading tests with item calibrations provided by the specification equation. Arbitrarily, but usefully, fixing a success rate (say 75%) on the virtual items for each textbook enables solving for the reader measure needed to correctly answer 75% of those virtual items. The individual reader's measure is then interpreted relative to the text complexity measure for each text in the freshman book bag. If the likely success rate in correctly answering the virtual items is high, so is the expectation of completing the first year.

High school graduates' reading measures can thus be compared to college text demands and a theoretically satisfying prediction can be made about the likelihood of first year completion. The efficiencies this system realizes from its use of scientifically validated predictive theory shows special promise as a tool for tracking reading readiness for post-secondary experiences in college, the work place, and the responsibilities of citizenship [19,20].

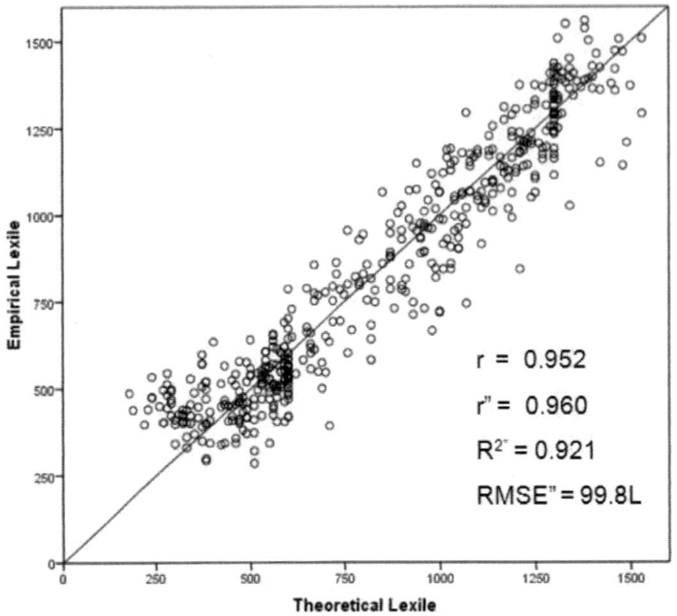

Figure 2. Empirical vs. theoretical Lexile text complexity estimates

5. References

[1] Spies R A, Carlson J F, Geisinger K F (Eds) 2010 *The Eighteenth Mental Measurements Yearbook* (Lincoln, Nebraska: University of Nebraska Press)

[2] Hamilton M, Pitt K 2011 *International Journal of Educational Development* **31** 596-605

[3] Neuman S B, Roskos K 2012 *The Reading Teacher* **66** 207-210

[4] Engelhard G 2001 *J App Meas* **2** 1-26

[5] Sadoski M, Paivio A 2007 *Scientific Studies of Reading* **11** 337-356

[6] Wise M N (Ed) 1995 *The values of precision* (Princeton, New Jersey: Princeton University Press)

[7] Bijker W E, Hughes T P, Pinch T (Eds) 2012 *The social construction of technological systems: New directions in the sociology and history of technology* (Cambridge, Massachusetts: MIT Press)

[8] Rasch G 1973/2011 *Rasch Measurement Transactions* **24** 1309

[9] Box G E P 1979 In R L Launer & G N Wilkinson (Eds.) *Robustness in statistics* pp. 201-235 (New York: Academic Press)

[10] Rasch G 1960 *Probabilistic models for some intelligence and attainment tests* (Reprint, with Foreword and Afterword by B. D. Wright, Chicago: University of Chicago Press, 1980 (Copenhagen, Denmark: Danmarks Paedogogiske Institut)

[11] Rasch G 1972/1988 *Rasch Measurement Transactions* **2** 19

[12] Rentz R R, Bashaw W L 1977 *J Ed Meas* **14** 161-179

[13] Stenner A J, Smith M 1982 *Perceptual and Motor Skills* **55** 415-426

[14] Stenner A J, Smith M, Burdick D S 1983. *J Ed Meas* **20** 305-316

[15] Carver R P 1974. *J Reading Behav* **6** 249-74

[16] Burdick D S, Stone M H, Stenner A J 2006 *Rasch Measurement Transactions* **20** 1059-60

[17] Stenner A J, Burdick D S 1997 The objective measurement of reading comprehension. www.lexile.com (visited 10 March 2013) (Durham, North Carolina: MetaMetrics, Inc.)

[18] Wright B D, Stone M H 1979 *Best test design* (Chicago, Illinois: MESA Press)

[19] Williamson G L 2008 *J Advanced Academics* **19** 602-632

[20] Williamson G L, Fitzgerald J, Stenner A J 2013 *Educational Researcher* **42** 59-69

Journal of Physics: Conference Series **459** (2013) 012026 doi:10.1088/1742-6596/459/1/012026

From Concrete to Abstract in the Measurement of Length

Andreas Stephanou[1]
Senior Research Fellow, Australian Council for Educational Research, 19 Prospect Hill Rd, Camberwell VIC Australia 3124, phone: +61 3 9277 5555
mailto:Stephanou@ACER.edu.au

William P. Fisher, Jr.,
BEAR Center, Graduate School of Education, University of California, Berkeley, CA (USA) & Principal, LivingCapitalMetrics Consulting
mailto:william@livingcapitalmetrics.com

Abstract. The concatenation of units of length is widely viewed as the paradigmatic expression of fundamental measurement. Survey, assessment, and test scores in educational and psychological measurement are often interpreted in ways that assume a concatenation of units to have been established, even though these assumptions are rarely stated or tested. A concatenation model for measurement is shown to be equivalent to a Rasch model: any two units of measurement placed end to end must together be of the same length as either one of them added to itself. This additive principle and a concatenation model of measurement together serve as a heuristic guide for organizing two experimental approaches to calibrating instruments for measuring length. The capacity to reproduce the unit of measurement from theory with no need for repeated empirical calibration experiments, as in the geometrical bisection of the line and the resultant halving of the length measure, is highlighted as essential to demonstrating a thorough understanding of the construct.

1. Introduction

When we measure we compare a property of two objects and express the result in counts of an amount of that property. What do we count? We count repetitions of a unit. Where do we get this unit? It is typically built into an instrument and can be traced to a reference standard. But where did the reference standard come from? It was, of course, instituted at a particular time and place, and was preceded by earlier standard units. History shows us that these earlier units are typically less precise and more variable than later ones. Furthermore, some things became measureable for the first time at specific moments in history [1-3].

How do things become measureable? What was counted before there were units? How do vague comparisons of less and more become focused on differences relative to some unit amount? After all, counts are ambiguous. I may have five rocks, and you may have 15, but I may nonetheless have more rock than you do. How does one overcome the ambiguity of counting numbers? What basis for fair comparisons can be formulated?

These are not just rhetorical questions. Though the natural sciences have long since established reference standard units, psychology and the social sciences continue to grapple with problems of fair comparisons. The natural inclination is, of course, to find something to count, and to use those counts as

[1] To whom any correspondence should be addressed

a basis for comparisons of less and more. But just as a higher count of rocks may ultimately amount to less rock than a lower count, so, too, do we need some way of knowing how to relate counts of correct answers on tests, or of rating scale responses on surveys, to fair measures of amount. After all, a high score on a test or survey can result from a person with very low abilities encountering very easy questions, while a person with much higher abilities could have a much lower score as a result of encountering very difficult questions.

How can this problem be resolved? Early contributions in this area were made by Thorndike [4], who raised the question back in 1904:

> If one attempts to measure even so simple a thing as spelling, one is hampered by the fact that there exist no units in which to measure. One may arbitrarily make up a list of words and observe ability by the number spelled correctly. But if one examines such a list one is struck by the inequality of the units. All results based on the equality of any one word with any other are necessarily inaccurate. (p. 7)

Wright [5] shows how the problem of a fair unit has been resolved by a number of different approaches involving concepts and methods associated with concatenation [6], conjoint additivity [7], statistical sufficiency [8], infinite divisibility [9], and parameter separation [10]. In every approach, counts provide a natural place to start, but, being based in concrete events or things in the real world, inevitably must be replaced by an abstract and ideal unit. Such units are defined in mathematical models specifying the consistency, homogeneity, monotonicity, etc. to which observations of concrete events must conform for measurement to be achieved.

The principles informing this distinction between the concrete and the abstract frame the question as to whether observations of differences in physical extension in space would provide a basis for a estimating a unit linearly equivalent to centimeters?

More specifically, would the calibrations and measures produced by data conforming to a Rasch model [5,10] of differences between physically extended objects add up and plot out in a manner equivalent to the measures of the same objects made with a meterstick? These are the questions explored in this research.

2. Two experiments in measuring length

Wright [5] describes a basis for additivity in measurement in terms of comparing different sized sticks each held up against the same sample of persons or objects. The sticks are calibrated to a standard determined by the consistency of the observed data, and not via reference to a predetermined unit external to the data. The proportions of marks the sticks receive across the persons or objects indicates whether they are longer (or shorter), and the persons or objects are measured according to the proportion of longer to shorter marks they receive across the sticks. The invariant consistency of the comparisons of the sticks across the persons or objects, and of the persons or objects across the sticks, is of paramount importance.

This scheme was simplified in two experiments: first, by comparing the lengths of two collections of different sized sticks, each with the others [11], and second, by placing all the 'sticks' on the same edge of a sheet of paper [12]. In order to imitate the situation of the social sciences, where the precise difficulty of the items measuring is not known in advance and the differences between them impossible to make exactly uniform, the units of measurement employed in both of these experiments were deliberately arranged to vary in size in an uneven fashion. Sometimes the units were clumped together like questions of similar difficulty, and other times they were spread apart with gaps between them.

In both experiments, the resulting logit measures were correlated and plotted against the corresponding millimetre or centimetre measures of the match sticks or objects. By arranging the study of ordinal observations in this way, the manner in which, and the extent to which, length might be taken as the paradigm of measurement in the social sciences can be tested.

2.1 The first experiment

In the first experiment, a pairwise-comparison method for the measurement of length was implemented by breaking 99 matches in two pieces, each time obtaining a headed match and a simple stick. A random sample of 33 sticks was then selected and used as a set of reference lengths. The 99 headed matches and the 33 reference lengths can be displayed along a scale of length, the headed matches on the left and the reference lengths on the right. When a headed match is at the same location as a reference length, the probability for the comparison to result in 1 (headed match longer than reference length) is 0.5; for reference lengths below the headed match the probability is greater than 0.5 and for reference lengths above it is less than 0.5.

The length of each of the 99 headed matches was compared with the length of each of the 33 reference lengths, recording the result of each of the (99x33) comparisons as 1 (headed match longer than reference length) or 0 (headed match shorter than reference length). In the recorded data, a row of 1s and 0s was formed for each headed match, with each reference stick in its own column.

2.2 The second experiment

In the second experiment, a number of small objects such as books, photographs, picture frames, ashtrays, and a television screen were compared to unevenly spaced lines marked along the edge of a sheet of paper (hash marks) and a score read off. All objects were aligned on one end with the bottom edge of the sheet of paper. The hash marks on the side of the paper were each treated as asking, "Does the object extend from the bottom of the paper to here, or is it longer than this?" Each answer of "longer than" was scored 2, each answer of “to here” was scored 1, and each "not up to" was assigned a 0. A row of 2s, 1s and 0s was then formed for each object, keeping each hash mark in its own column.

2.3 The attenuation paradox

Both experiments resulted in situations exhibiting what is known as the attenuation paradox [13]. Perfect scores, obtained when all observations for a question are in the same extreme category, contribute nothing to an analysis beyond indicating that the test is poorly targeted. Such items are too difficult or too easy for the people tested, or a person is too able or unable for the test. It is as though some objects are shorter than the shortest length unit or longer than the longest, and some units are shorter and others longer than all objects. In either case, there is no evidence available that indicates how much longer, shorter, or too easy, difficult, able or unable the item or person is, and estimation is therefore impossible. There are ways around this problem in education and psychology (for instance, by asking questions that are more closely spaced in difficulty, or that reach up or down into the difficulty range where they are needed), but the measurement of length raises a special problem.

The information available in data of this kind is inadequate for calibrating a new instrument. The problem is that the data exhibit a degree of precision producing a deterministic order more suited to a Guttman scale than to Rasch's model [14]. There is no information about the length between two consecutive reference matches in the data from the first experiment because each reference match provides limited information along the scale. In fact the information it provides is only at its actual length. Any shorter headed match will be found to be shorter, and any longer headed match longer, thus failing to discriminate between different lengths of headed matches except at its length, within measurement error. Measurement error in this case is very small compared to the values of length that are measured because of the careful nearby comparison of headed matches with reference lengths.

If the comparisons of the headed matches with the reference lengths were less precise (e.g., the two sticks were held far away from each other when compared) for each headed match there would be some overlapping of 0s for smaller and 1s for longer reference lengths: the results of the comparisons would be governed by some probability law and the boundary between 1s and 0s would not be as sharp. Figure 1 illustrates the two types of comparisons, the nearby comparisons on the left and the far-away comparisons on the right. These probability curves are for a reference length of 25 mm and give the probability for a headed match to be longer than this reference length. The comparison distance determines the discrimination between shorter and longer reference lengths: the closer the distance, the greater the discrimination. The more gradual slopes to the asymptotes in the Rasch ogive provide

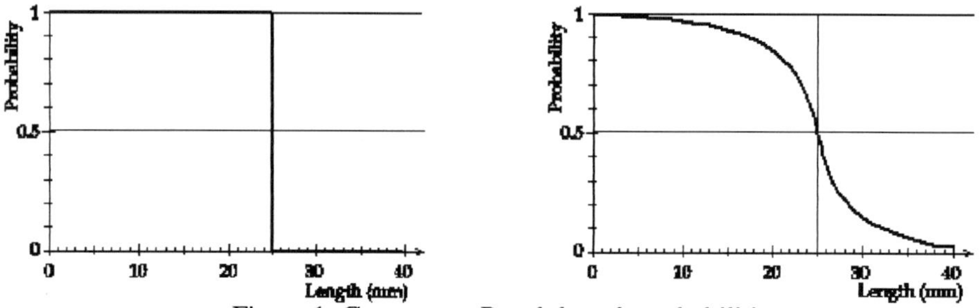

Figure 1. Guttman vs Rasch length probabilities

information on the relative distances between items, making it possible to estimate their locations relative to one another. This information is absent from the Guttman data, meaning that answers to the questions of how much more each item represents cannot be had.

Because the estimation algorithm will not converge in the absence of overlapping probabilities, a small number of individual observations (less than 1%, 3 cases in the second experiment) were changed to provide the information needed.

2.3 Results

The millimeter values for the 99 match heads units have been plotted against the estimated logits in Figure 2, and centimeter values for the 17 ordinal hash marks in the second experiment are plotted against their corresponding logits in Figure 3. There is a steady progression and one to one correspondence between the meter and logit values that approximates an identity line.

In Figure 3, items 2 and 3 calibrated to the same logit value because there was no object measured that landed between them; there is thus no way of discriminating between them on the basis of these data. Where units 12, 13, 14 and 15 are bunched up within 1.5 centimeters of each other, their logit values are nearly within standard errors of each other. Unit 15 calibrates a little higher than the centimeter values indicate it should have; this is because so few of the objects measured reached it. Overall, the Rasch model places the units of measurement in exactly the same order and with very nearly the same spacing as a meterstick does. Can we say that the units add up the way centimeters do?

Given the linearity of the plots in Figures 2 and 3, with some degree of uncertainty, apparently there is a basis in evidence for the claim that a linear and additive unit of measurement has been defined. What does this mean?

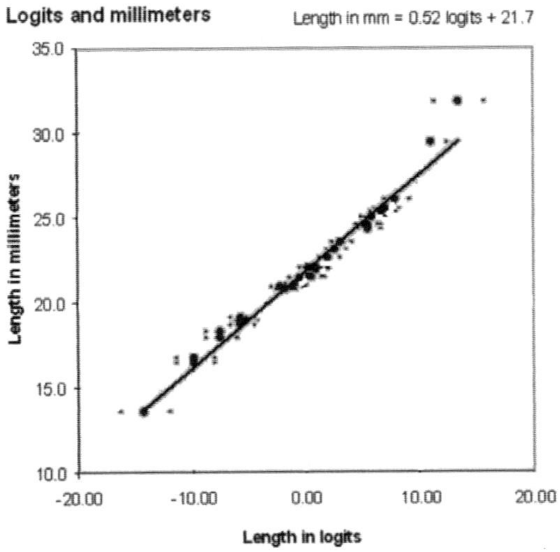

Figure 2. Results from the first experiment

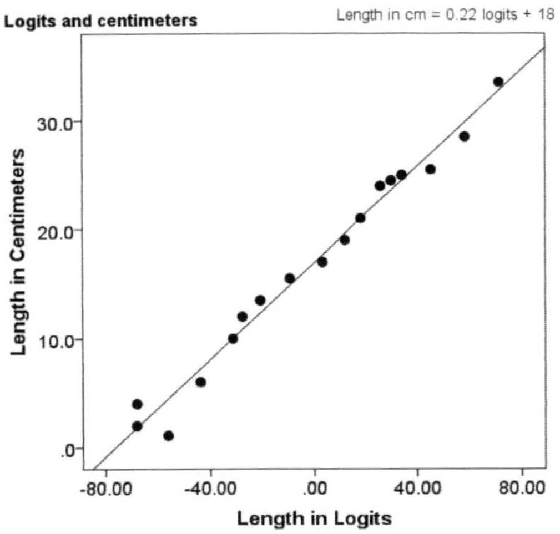

Figure 3. Results from the second experiment

3. Conclusion

Scientific laws assert abstract ideals never observed to hold with perfect precision in real life [15, 16]. Rasch [17] and Box [18] both repeatedly pointed out that scientific and statistical models are not meant to be true, but to be useful.

Structurally invariant additivity exhibited within and across different ways of observing a phenomenon can be scaled, equated, and anchored to a common unit, which is then disseminated for use as the shared language in a field of research and practice. The two different studies reported here reported their results in different log-odds units, but because both conformed to established millimetre and centimetre units, they could easily be equated and anchored to a shared common language.

The results of these two experiments are similar to those seen in other studies of distance, weight, and density [19-21]. In all of these investigations, standard units of physical measurement were reproduced from ordinal observations. If ordinal observations can support the calibration of instruments of physical measurement that demonstrably exhibit conformity to existing unit standards, is there not a strong possibility that ordinal observations might also support the calibration of instruments and the establishment of unit standards for educational, psychological, and social measurement [2]?

References

[1] Heilbron J L 1993 *Weighing imponderables and other quantitative science around 1800* (Berkeley, California: University of California Press)
[2] Roche J 1998 *The mathematics of measurement: A critical history* (London: The Athlone Press)
[3] Zupko R E 1977 *British weights and measures: A history from antiquity to the seventeenth century* (Madison, Wisconsin, USA: University of Wisconsin Press)
[4] Thorndike E L 1904 *An introduction to the theory of mental and social measurements* (New York: Teacher's College)
[5] Wright B D 1999 Fundamental measurement for psychology pp. 65-104 in S E Embretson, S L Hershberger (Eds), *The new rules of measurement: What every educator and psychologist should know* (Hillsdale, New Jersey: Lawrence Erlbaum Associates)

Journal of Physics: Conference Series **459** (2013) 012026 doi:10.1088/1742-6596/459/1/012026

[6] Campbell N R 1920 *Physics, the elements* (Cambridge, England: Cambridge University Press)
[7] Luce R D, Tukey J W 1964 *J Mathematical Psychol* **1** 1-27
[8] Fisher R A 1922 *Philosophical Transactions of the Royal Society of London* **222** 309-368
[9] Kolmogorov A N 1950 *Foundations of the theory of probability* (New York: Chelsea Publishing)
[10] Rasch G 1961 *Proceedings of the Fourth Berkeley Symposium on Mathematical Statistics and Probability* **4** 321-333
[11] Stephanou A 2007 *The measurement of conceptual understanding in physics* PhD thesis, School of Physics, University of Melbourne, Victoria, Australia
[12] Fisher W P 1988 *Truth, method, and measurement: The hermeneutic of instrumentation and the Rasch model* Dissertation Abstracts International (Dept. of Education, Division of the Social Sciences, University of Chicago) **49** 0778A
[13] Engelhard G 1993 *Rasch Measurement Transactions* **6** 257 www.rasch.org/rmt/rmt64.htm
[14] Wilson M 2011 *Joint International IMEKO TC1 + TC7 + TC13 Symposium*, August 31st – September 2nd, Jena, Germany, urn:nbn:de:gbv:ilm1-2011imeko-005:8
[15] Butterfield H 1957 *The origins of modern science* (New York: The Free Press)
[16] Cartwright N 1983 *How the laws of physics lie* (Oxford: Clarendon Press)
[17] Rasch G 1973/2011 *Rasch Measurement Transactions* **24** 1309
[18] Box G E P 1979 In R L Launer & G N Wilkinson (Eds.) *Robustness in statistics* pp. 201-235 (New York: Academic Press)
[19] Choi S E 1997 *Rasch Measurement Transactions* **11** 557 www.rasch.org/rmt/rmt112.htm#Ounces
[20] Pelton T and Bunderson V 2003 *J App Meas* **4** 269-281
[21] Moulton M 1993 *Rasch Measurement Transactions* **7** 268 www.rasch.org/rmt/rmt71b.htm
[22] Fisher W P 2009 *Measurement* **42** 1278-1287

Merging Psychophysical and Psychometric Theory to Estimate Global Visual State Measures from Forced-Choices

Robert W. Massof,[1] Karen M. Schmidt,[2] Daniel M. Laby,[3] David Kirschen,[4] and David Meadows[5]

[1]Johns Hopkins University, [2]University of Virginia, [3]Harvard University, [4]University of California Los Angeles, [5]University of Florida

Email: bmassof@jhmi.edu

Abstract. Visual acuity, a forced-choice psychophysical measure of visual spatial resolution, is the *sine qua non* of clinical visual impairment testing in ophthalmology and optometry patients with visual system disorders ranging from refractive error to retinal, optic nerve, or central visual system pathology. Visual acuity measures are standardized against a norm, but it is well known that visual acuity depends on a variety of stimulus parameters, including contrast and exposure duration. This paper asks if it is possible to estimate a single global visual state measure from visual acuity measures as a function of stimulus parameters that can represent the patient's overall visual health state with a single variable. Psychophysical theory (at the sensory level) and psychometric theory (at the decision level) are merged to identify the conditions that must be satisfied to derive a global visual state measure from parameterised visual acuity measures. A global visual state measurement model is developed and tested with forced-choice visual acuity measures from 116 subjects with no visual impairments and 560 subjects with uncorrected refractive error. The results are in agreement with the expectations of the model.

1. Background

Figure 1. Examples of Landolt rings at different orientations. The observer must identify the orientation of the gap.

Visual resolution (a.k.a. visual acuity) often is measured by determining the smallest size Landolt ring for which the orientation of the gap in the ring can be identified reliably [1]. As shown in Figure 1, on each trial the Landolt ring is presented at one of four randomly chosen orientations and the observer is forced to respond with one of four categories: "up", "down", "left, or "right". The response is scored dichotomously as "correct" (1) or "incorrect" (0). Different size stimuli are presented multiple times and the frequency of correctly identifying the stimulus orientation is estimated as a function of stimulus size.

Journal of Physics: Conference Series **459** (2013) 012027 doi:10.1088/1742-6596/459/1/012027

As illustrated with simulated data in Figure 2, by convention the visual resolution threshold is defined as the stimulus size that corresponds to the probability of correctly identifying the stimulus orientation that is halfway between 100% and chance performance (25% in this example). [2] The curve fit to the simulated data in Figure 2 is a cumulative logistic (approximation to a cumulative log normal distribution) with the lower bound constrained to chance, i.e., $P_c(S|\hat{S}) = 0.25 + \frac{.75\exp(a[S-\hat{S}])}{1+\exp(a[S-\hat{S}])}$, where S is the log stimulus size and $\hat{S}$ is the log stimulus size at threshold.

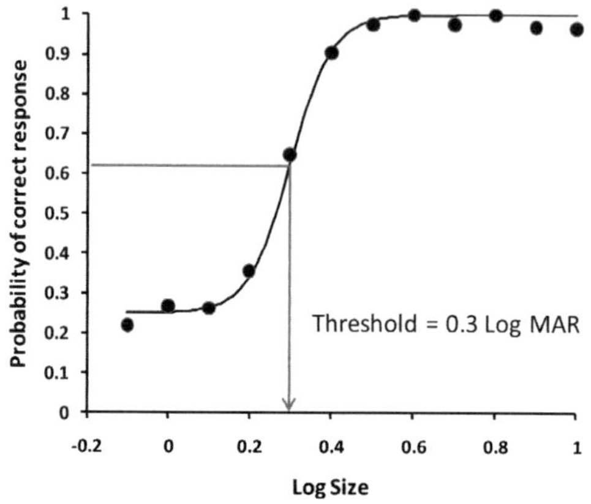

Figure 2. Simulation of typical psychophysical visual acuity data plotted as probability of correctly identifying the orientation of the gap in the Landolt ring as a function of stimulus size, expressed as log minimum angle of resolution (MAR). The solid curve drawn through the data is a cumulative logistic with a lower bound at chance performance. Visual acuity, or the size threshold, is defined as the stimulus size that results in a probability correct half way between chance and 100%.

The visual resolution threshold depends on how far off axis the stimulus is from the line of sight, the average luminance of the stimulus, stimulus contrast, stimulus exposure duration, and other physical parameters of the stimulus. The visual resolution threshold also depends on the observer's light exposure history prior to the stimulus presentation and on the health state of the observer's visual system. Besides size thresholds, one can use the Landolt ring resolution task to measure contrast thresholds, exposure duration thresholds, etc. as a function of the other stimulus parameters.

Figure 3 illustrates log size thresholds as a function of log contrast and exposure duration displayed at different orientations of the axes and Figure 4 similarly shows log contrast thresholds as a function of log size and exposure duration. These

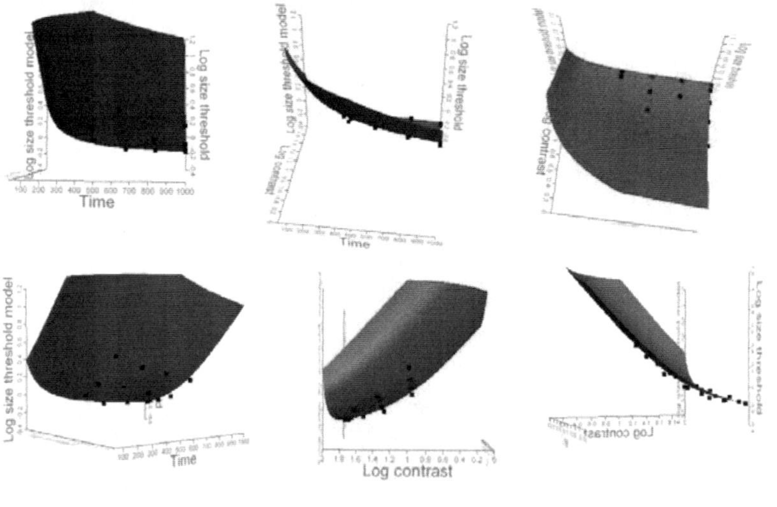

Figure 3. Log size threshold (ordinate), measured as a function of contrast (Log contrast) and stimulus exposure duration (Time). The black points are average thresholds for 116 normally-sighted subjects. The red surface is the fit of a vision psychophysical model based on stimulus reciprocities. The different panels illustrate six different orientations of the threshold surface.

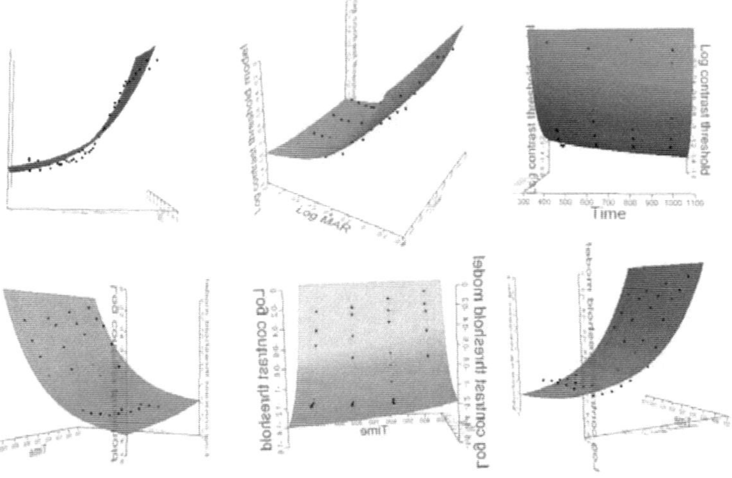

thresholds (points) are averages for 116 observers who have normal vision (points).

Figure 4. Log contrast threshold (ordinate) for detecting the orientation of a Landolt ring was measured as a function of stimulus size (Log MAR) and exposure duration (Time). The black points are average thresholds for 116 normally-sighted subjects. The green surface is the fit of a vision psychophysical model based on stimulus reciprocities.

The threshold surfaces in Figures 3 and 4 are generated by a vision psychophysical model that is based on well known stimulus reciprocity laws (Ricco's law, Bloch's law, Piper's law, and Pieron's law). The log size threshold surface as a function of log contrast (C) and exposure duration (T) is defined as $\hat{S}(C,T) = ln\left(\frac{k_1}{Te^C} + k_2\right)$ [3-4] and the log contrast threshold surface as a function of log size (S) and exposure duration is defined as $\hat{C}(S,T) = k_3 ln\left(\frac{k_4}{T}\right) + \frac{k_5}{e^S}$ [4-5].

The probability of correctly identifying the orientation of a stimulus with log size, log contrast, and exposure duration S,C,T is $P_c(S,C,T) = P_c\left(S,C,T|\hat{S}(C,T)\right) = P_c\left(S,C,T|\hat{C}(S,T)\right)$. This truism gives rise to the possibility that the parameterized size and contrast thresholds represent a single latent trait of the person, θ, relative to a function of S,C, and T, i.e., $P_c\left(S,C,T|\theta,I(S,C,T)\right)$, where $I(S,C,T)$ is a constructed latent variable that represents processing of the multi-parameter stimulus by the average visual system for a defined population. This concept could be extended to an arbitrary number of stimulus parameters, i.e., $P_c\left(S,C,T,\cdots|\theta,I(S,C,T,\cdots)\right)$. Thinking of resolution thresholds in this way opens the door to the possibility of measuring the global state of the visual system, relative to the average visual system, with a single parameter-independent variable, the person measure θ.

2. Theory

Landolt ring *j* is an object on a display screen that has the spatial-temporal dependent spectral radiant energy distribution $\boldsymbol{Q}_j(\boldsymbol{x},\boldsymbol{y},\boldsymbol{t},\boldsymbol{\lambda})$. The environmental optics and the optics of the eye of observer *n* produce an image of the Landolt ring on the observer's retina with spectral radiant energy distribution $\boldsymbol{Q}_{nj}(\boldsymbol{x},\boldsymbol{y},\boldsymbol{t},\boldsymbol{\lambda})$. The retinal image is sampled by a matrix of cone and rod photoreceptors with space-time dependent spectral absorption coefficients specific to person *n*: $\boldsymbol{L}_n(\boldsymbol{x},\boldsymbol{y},\boldsymbol{t},\boldsymbol{\lambda}), \boldsymbol{M}_n(\boldsymbol{x},\boldsymbol{y},\boldsymbol{t},\boldsymbol{\lambda}), \boldsymbol{S}_n(\boldsymbol{x},\boldsymbol{y},\boldsymbol{t},\boldsymbol{\lambda}), \boldsymbol{R}_n(\boldsymbol{x},\boldsymbol{y},\boldsymbol{t},\boldsymbol{\lambda})$ for the long, middle, and short wavelength-sensitive cones and rods, respectively. Thus, the retinal image of Landolt ring *j* for person *n* is transformed to the retinal image light absorption vector field

$\boldsymbol{Q}_{nj}(\boldsymbol{x},\boldsymbol{y},\boldsymbol{t}) = \left[\int_0^\infty \boldsymbol{a}_n(\boldsymbol{x},\boldsymbol{y},\boldsymbol{t},\boldsymbol{\lambda})\, \boldsymbol{Q}_{nj}(\boldsymbol{x},\boldsymbol{y},\boldsymbol{t},\boldsymbol{\lambda}) d\boldsymbol{\lambda}\right]$ for $\boldsymbol{a} = \boldsymbol{L},\boldsymbol{M},\boldsymbol{S},\boldsymbol{R}$. A vector of neurophysiologic space-time operators transform the light absorption vector field to a visual sensation vector field, $\boldsymbol{V}_{nj}(\boldsymbol{x}',\boldsymbol{y}',\boldsymbol{t}') = \boldsymbol{\Phi}_{\mathrm{n}}\{\boldsymbol{Q}_{nj}(\boldsymbol{x},\boldsymbol{y},\boldsymbol{t})\}$, which corresponds to a visual sensory representation of Landolt ring *j* for observer *n*. Visual sensation is a vector field because the visual sensory representation of the retinal image corresponds to spatial and temporal variations in color and brightness, a personal experience of the observer. [6]

Observer *n* knows what he is looking for and has expectations for the appearance of the Landolt ring at each orientation ω, $\widehat{\boldsymbol{V}}_{n\omega}(\boldsymbol{x}',\boldsymbol{y}',\boldsymbol{t}')$. Observer *n* performs a comparator operation on the visual sensation vector field and expected vector field to produce a scalar decision variable for stimulus *j* that is a function of orientation, $\boldsymbol{D}_{nj}(\boldsymbol{\omega}) = \boldsymbol{\Psi}_{\mathrm{n}}\{\boldsymbol{V}_{nj}(\boldsymbol{x}',\boldsymbol{y}',\boldsymbol{t}'), \widehat{\boldsymbol{V}}_{n\omega}(\boldsymbol{x}',\boldsymbol{y}',\boldsymbol{t}')\}$ (e.g., vector cross correlation at each orientation across *x',y',t'*). Observer *n*'s response for stimulus *j* would be the orientation that produces the maximum (or minimum, depending on the comparator operation) value of $\boldsymbol{D}_{nj}(\boldsymbol{\omega})$.

For a given population, the decision variable for stimulus *j* can differ between people because of fixed and stochastic between and within person differences in expectations and decision rules, i.e., differences in $\widehat{\boldsymbol{V}}_{n\omega}(\boldsymbol{x}',\boldsymbol{y}',\boldsymbol{t}')$ and $\boldsymbol{\Psi}_{\mathrm{n}}$, and because of fixed and stochastic differences between and within people in the global state of the visual system, i.e. differences in $\boldsymbol{Q}_{nj}(\boldsymbol{x},\boldsymbol{y},\boldsymbol{t},\boldsymbol{\lambda})$ due to differences in retinal image formation, $\boldsymbol{a}_n(\boldsymbol{x},\boldsymbol{y},\boldsymbol{t})$ due to space-time differences in the spectral absorbance of photopigments, and $\boldsymbol{\Phi}_{\mathrm{n}}$ due to spatial and temporal variations in amplitude, gain, space and time constants, and other neurophysiological processing characteristics. If $\overline{\boldsymbol{D}}_j(\boldsymbol{\omega})$ is the expected value for the population of the decision variable for stimulus j as a function of orientation, then $\boldsymbol{D}_{nj}(\boldsymbol{\omega}) = \overline{\boldsymbol{D}}_j(\boldsymbol{\omega}) + \boldsymbol{\theta}_n + \boldsymbol{e}_n$, where $\boldsymbol{\theta}_n$ is the fixed difference and $\boldsymbol{e}_n$ is the stochastic difference for person *n* from the expected value.

The task of the observer is to identify the correct orientation of the Landolt ring, $\boldsymbol{\omega}_c$. In terms of the theory, the observer chooses the orientation that corresponds to the maximum value of $\boldsymbol{D}_{nj}(\boldsymbol{\omega})$. The probability that observer *n* makes the correct choice for stimulus *j* is the probability that $\boldsymbol{D}_{nj}(\boldsymbol{\omega}_c) > \boldsymbol{D}_{nj}(\boldsymbol{\omega})$ for all $\boldsymbol{\omega} \neq \boldsymbol{\omega}_c$. If the density function for $\boldsymbol{e}_n$ is a logistic that approximates a standard normal distribution, then the probability that $\boldsymbol{D}_{nj}(\boldsymbol{\omega}_c) < D$ is $\boldsymbol{Pr}(\boldsymbol{D}_{nj}(\boldsymbol{\omega}) < D) = \mathbf{1} - \frac{\exp\,(D-\overline{D}_j(\omega)-\theta_n)}{1+\exp\,(D-\overline{D}_j(\omega)-\theta_n)}$ and the probability that $\boldsymbol{D}_{nj}(\boldsymbol{\omega}_c) = \boldsymbol{D}$ is $\boldsymbol{Pr}(\boldsymbol{D}_{nj}(\boldsymbol{\omega}_c) = \boldsymbol{D}) = \frac{\exp\,(D-\overline{D}_j(\omega_c)-\theta_n)}{(1+\exp\,(D-\overline{D}_j(\omega_c)-\theta_n))^2}$. If $\boldsymbol{e}_n$ is stochastically independent with respect to orientation, given all possible values of the decision variable *D*, then the probability of person *n* making the correct choice for stimulus *j* is

$$\boldsymbol{Pr}\left(\boldsymbol{D}_{nj}(\boldsymbol{\omega}_c) > \boldsymbol{D}_{nj}(\boldsymbol{\omega})\right) = \int_{-\infty}^{\infty} \boldsymbol{Pr}(\boldsymbol{D}_{nj}(\boldsymbol{\omega}_c) = \boldsymbol{D}) \prod_{\omega \neq \omega_c} \boldsymbol{Pr}(\boldsymbol{D}_{nj}(\boldsymbol{\omega}) < D)\, \boldsymbol{dD},$$

and the probability of making the incorrect choice of orientation *i* for stimulus *j*, $\boldsymbol{\omega}_i$, is

$$\boldsymbol{Pr}\left(\boldsymbol{D}_{nj}(\boldsymbol{\omega}_i) > \boldsymbol{D}_{nj}(\boldsymbol{\omega})\right) = \int_{-\infty}^{\infty} \boldsymbol{Pr}(\boldsymbol{D}_{nj}(\boldsymbol{\omega}_i) = \boldsymbol{D}) \prod_{\omega \neq \omega_i} \boldsymbol{Pr}(\boldsymbol{D}_{nj}(\boldsymbol{\omega}) < D)\, \boldsymbol{dD}.$$

The probability of choosing the correct orientation out of $\boldsymbol{N_\omega}$ equally likely and mutually exclusive orientations is $\boldsymbol{P_c} = \frac{\boldsymbol{Pr}\left(\boldsymbol{D_{nj}(\omega_c) > D_{nj}(\omega)}\right)}{\boldsymbol{Pr}\left(\boldsymbol{D_{nj}(\omega_c) > D_{nj}(\omega)}\right) + \sum_{i=1}^{N_\omega - 1} \boldsymbol{Pr}\left(\boldsymbol{D_{nj}(\omega_i) > D_{nj}(\omega)}\right)}$.

If $\boldsymbol{Pr}\left(\boldsymbol{D_{nj}(\omega_c) > D_{nj}(\omega)}\right) = \boldsymbol{Pr}\left(\boldsymbol{D_{nj}(\omega_i) > D_{nj}(\omega)}\right)$ for all *i*, then $\boldsymbol{P_c} = \frac{\boldsymbol{1}}{\boldsymbol{N_\omega}}$, which is chance performance. This condition for chance performance is met when $\boldsymbol{\Psi_n\{V_{nj}(x', y', t'), \widehat{V}_{n\omega_i}(x', y', t')\} = \Psi_n\{V_{nj}(x', y', t'), \widehat{V}_{n\omega_c}(x', y', t')\}}$ for all incorrect orientations *i*.

Subjects' choices of Landolt ring orientations for different combinations of stimulus parameters are scored as correct (1) or incorrect (0). From the assumption of local independence of observations, a maximum likelihood estimation routine is used to estimate $\boldsymbol{\theta_n}$ for every person and $\boldsymbol{\overline{D}_j(\omega_c)}$ for every stimulus from the matrix of response scores. [7] For Landolt ring size, contrast, and exposure duration, we define $\boldsymbol{I(S, C, T)_j \equiv \overline{D}_j(\omega_c)}$.

The hypothesis of a global visual system state variable predicts that $\boldsymbol{\theta_n - I(S, C, T) = a_{nS}\left(S - \widehat{S}(C, T)_n\right) + b_{nS}}$ and $\boldsymbol{\theta_n - I(S, C, T) = a_{nC}\left(C - \widehat{C}(S, T)_n\right) + b_{nC}}$ for all combinations of S,C,T. Figure 5 is a scatter plot of the average $\boldsymbol{\theta_n - I(S, C, T)}$ vs the average $\boldsymbol{S - \widehat{S}(C, T)_n}$ of 560 subjects with uncorrected refractive error (points) and Figure 6 is the similar scatter plot of the average $\boldsymbol{\theta_n - I(S, C, T)}$ vs $\boldsymbol{C - \widehat{C}(S, T)_n}$ for the same subjects (each point corresponds to a pair of mean values for a unique combination of S,C,T).

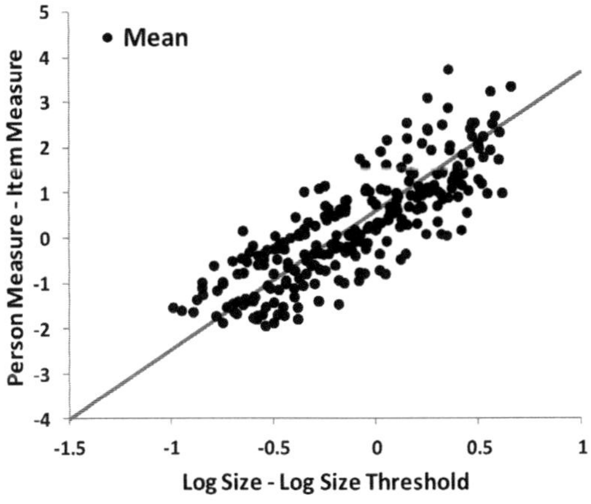

Figure 5. Scatter plot of the difference between the estimated global visual state variable (i.e., person measure, θ_n for person *n*) and the stimulus-dependent expected value of the decision-variable for the population (i.e., item measure, $I(S,C,T)_j$ for stimulus *j*) vs. the difference between log stimulus size (S_j for stimulus *j*) and the log size threshold ($\hat{S}(C, T)_n$ for person *n*). Each point is the average pair of values across 560 subjects for each combination of stimulus contrast (*C*) and exposure duration (*T*). The red line is the predicted relationship.

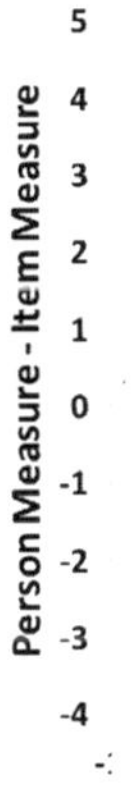

Figure 6. Scatter plot of the difference between the estimated global visual state variable (i.e., person measure, θ_n for person *n*) and the stimulus-dependent expected value of the decision-variable for the population (i.e., item measure, $I(S,C,T)_j$ for stimulus *j*) vs. the difference between log stimulus contrast (C_j for stimulus *j*) and the log contrast threshold ($\hat{C}(S, T)_n$ for person *n*). Each point is the average pair of values across 560 subjects for each combination of stimulus size (*S*) and exposure duration (*T*). The red line is the predicted relationship.

The observed linear relationships in Figures 5 and 6 are consistent with our hypothesis that a single parameter-free variable, $\boldsymbol{\theta_n}$, can be used to measure the global state of the visual system (at least for uncorrected refractive error) and predict multivariable visual resolution thresholds.

3. References

1. International Organization for Standardization. Optics and Optical Insturments. Viusal Acuity Testing: Method of Correlating Optotypes, ISO 8597:1994. Geneva, Switzerland: International Organization for Standardizations; 1994.
2. Alexander KR, Wei X, Derlacki DJ. Visual acuity and contrast sensitivity for individual Sloan letters. Vision Res. 1997;37:813-819.
3. Nachmias J. Effect of exposure duration on visual contrast sensitivity of square-wave gratings. J Opt Soc Am. 1967;57:421-427.
4. Kelly DH. Motion and vision. II. Stabilized spatio-temporal threshold surface. J Opt Soc Am. 1979;69:1340-1349.
5. Alexander KR, McAnany JJ. Determinants of contrast sensitivity for the tumbling E and Landolt C. Optom Vis Sci. 2010;87:28-36.
6. Massof RW, Bird JF. A general zone theory of color and brightness vision. I. Basic formulation. J Opt Soc Am. 1978;68:1465-1471.
7. Royal KD, O'Neill TR. Using the CUTLO procedure to investigate guessing. Rasch Measur Trans 2011;251:1319-1320.

Journal of Physics: Conference Series **459** (2013) 012028 doi:10.1088/1742-6596/459/1/012028

Measurement as cognitive and applied problem

V A Granovskii[1]
General Research Fellow, Concern CSRI Elektropribor, JSC,
30, Malaya Posadskaya St., 197046 St. Petersburg, Russia

E-mail: vgranovsky@eprib.ru / valgr39@mail.ru

Abstract. Measurement is analyzed as a process of task-oriented actions with material and model objects. It is emphasized that model elements of measurement are of crucial importance in the formulation of measurement tasks and implementation of measurement experiments. It is shown that there is a methodological gap between the rows of material and model elements of measurement, which belong to fundamentally incompatible systems. This gap determines an adaptive nature of measurement, during which the attribute of an object being measured is specified more accurately. The need for constructing equivalence systems aimed at overcoming the gap is substantiated. The measurement traceability system fulfils the role of an internal equivalence system for measurement as a cognitive procedure. The application system in the interests of which the measurements are made fulfils the role of an external equivalence system.

1. Introduction

Metrology is an ancient scientific and practical discipline. It originates from Sumerian culture, in which measurements were a significant instrument of social life [1]. It is surprising that despite the conservatism, which is inherent in it by definition, metrology continues developing for six thousand years. This progress can only be stimulated by the presence of a permanent problem. The latter is obvious and it has two sides: the need to meet the required accuracy in measuring known quantities and the need to estimate previously unknown properties of objects, or attributes of previously unknown objects. In turn, the source of the above-mentioned never-ending renewable problem lies in an inherent contradiction: we mean a fundamental contradiction which is the basis for the development of measurements and metrology.

2. Measurement structure

Bearing in mind both sides of the problem, we can consider the establishment of relationship between a set of objects and a set of numbers to be in fact the source for the development of measurements [2]. Indeed, this problem does not have a general unambiguous solution and it resumes again and again. However, for measurement practice, it is preferential to formulate the problem more specifically, in terms of reality and its simulation.

It is argued that the basic contradiction of measurement (and metrology) consists in the fact that the purpose of measurement is formulated and attained in two fundamentally incompatible systems. The purpose of measurement is formulated as quantitative estimation of an actual manifestative

[1] To whom any correspondence should be addressed.

property of a material object. In other words, we need to obtain a result in a system of real objects: material objects, processes, attributes, and properties. The term ‘quantitative’ does not change this representation in principle because the nature of measurement is to compare similar objects (attributes, properties) and express one of them through another. Therefore, it is assumed that procedure of measurement, i.e., a series of practical actions, is carried out in the above-mentioned real system.

Of course, the notion “number” is necessary for evaluation of two objects (of abiocoen) relation. There is no problem because mathematics gives us this notion. But the problem exits and it turns out to be the deeper one. The reason is that not only number but notion “model”, i. e., mathematical category of more general system level, should be used. In other words, starting point that measurement as procedure (set of practical actions) is carried out in reality system (RS) referred above seems to be at least incomplete.

As regards measurement as a process of motion towards a goal, the process includes, besides practical actions, a number of manipulations in the system of models, i.e., a system of conventional and abstract elements. Moreover, it is impossible to start practical measuring procedures, first of all, formulate a practical measurement task without using a ‘model (reflection)’ system (MS). The fact is that the concept of a measurand — the key notion in the formulation of the measurement problem — is introduced on the model of the attribute (property) being measured, which, in turn, serves as an element of a more general model for the measurement object. On top of everything, it is necessary for measurement realization that second and first models are sufficiently detailed. Many practical examples confirm that exceedingly general object model and, as consequence, property model, lead to the result which does not meet the practical aim or even has no physical sense.

Thus, the property to be measured is expressed for the purpose of measurement by a model, i.e., an immaterial (information) object representing the relationship between a specified property and other known, with higher or less certainly, properties of objects, including the object of measurement. In other words, we intend to measure an attribute of a material object, but, instead, we measure an attribute of a model. This being so, we need to interpret (verify) the measurement result (MR). Interpretation is carried out in two stages (see Fig. 1): internal (MR correspondence to the measurement models) and external (correspondence of the result of interaction between the object of measurements and its environment—other objects to our expectations).

3. Measurement structure

Based on the foregoing reasoning, we define an object of measurement as an element of RS with an attribute (property) which is of interest to us. Here, we should draw your attention to the paradox of actions (interaction) with RS: as soon as we name an object, we move to MS, especially it is true for the attribute. Indeed, the name of an object follows from its classification, i.e., the assignment of an object to a group of homogeneous, in a sense, objects. Objects are classified based on a type, or group, model. Hence, denomination is establishing a correspondence between an object and a certain group model. This note is valid for an attribute too. Besides, there is no possibility to pick out the attribute from a set of all attributes of an object without modelling. To do our reasoning simpler, we ignore below this inevitable stage of modelling by name. So, an object and an attribute can be accepted as elements of RS but it not relevant to measurand.

Really, as regards the attribute, it requires a model of the subsequent (more complicated, specific, and detailed) level. The measurand in this series of representations appears as a model possessing certain features (characteristics, parameters) of a model of an attribute, or ‘a model of a model’ (except for object model).

Journal of Physics: Conference Series **459** (2013) 012028 doi:10.1088/1742-6596/459/1/012028

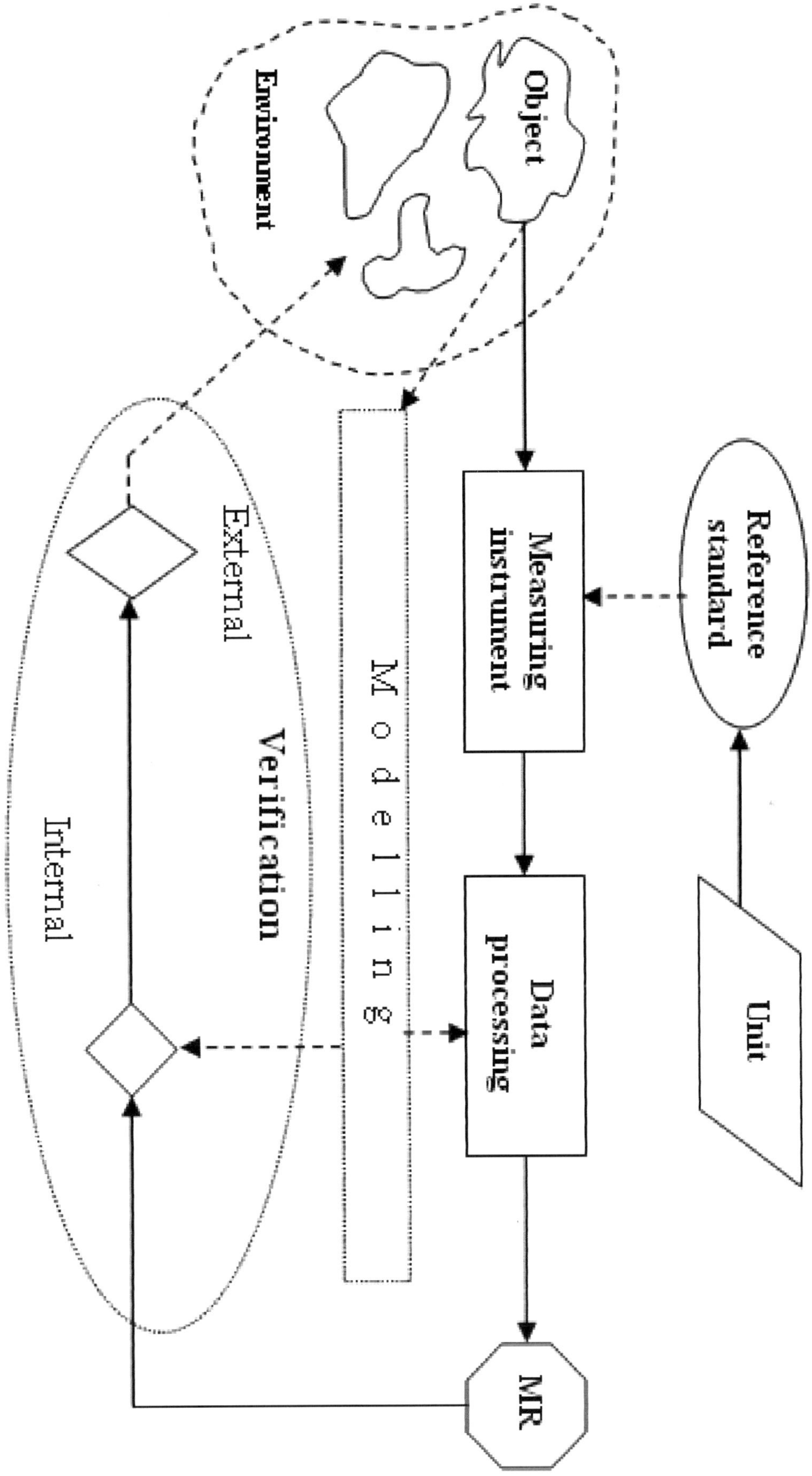

Fig. 1. Measurement and verification of its result.

Thus, object has to be represented in measurement not only as reality object with detailed property (attribute) but as totality of models (for attribute and measurand). Measurand as "second model derivative", on the one hand, and the input physical impact on the MI, on the other hand, are separated by a gap. It is inevitable because the RS and MS belong to different planes in the model of cognition (learning, investigation) of the universe. To overcome the gap it is necessary, after obtaining the output physical effect as a result of physical transformation of an input action, we may return, using this basis, from the RS to the MS, where the measurement result must be compared with the measurand.

The gap is overcome by constructing an equivalence system (ES). It is only within this system that the reflection of 'attribute – number' makes sense. We can speak about general and specific ES's. The general ES applies to all measurements of a specific type, whereas a specific model applies to measurements made within a specific measurement task.

Obviously, the ES should rely on some reference points; it remains to find out which ones exactly.

For a specific ES, the reference points are certification of measurement procedure and MI calibration. More precisely, a reference point is obtained from both of these procedures, which ensures traceability of measurements, or MR. The reason is that going up to the reference standard and, further, to the unit, makes it possible to rely on the well-known model-reality relationship. The primary reference standard is the highest point, in the traceability chain, of model adequacy.

To put it otherwise, to search models which are adequate to reality objects, artificial reality objects are constructed, namely reference standards. They are constructed specially in order to ensure their required adequacy to initial model, i. e., measurement unit (if there is an initial etalon) or more high level etalon model.

For a general ES, the reference point is a measurement unit. It is in the unit that the maximum model adequacy, namely, identity, is attained. Indeed, the unit is a specification of a material object (artifact, physical process), an attribute of which is, by definition, identical to itself.

Therefore in this case too, the model is constructed specially which is adequate completely to the model, namely, to itself. Above-mentioned means that the problem of RS and MS connection, or ES construction, represents essentially the problem of reality objects adequacy to models, in relations object – model, and object – object, and model - model. The problem decision is in creation of real objects and construction of models with assured range of adequacy; examples are measurement unit and reference standard.

Hence, it makes sense to fix the fact that measurement as an object of introspection is related to two incompatible systems (RS and MS), which are interrelated by an ES. The ES itself is constructed by modelling specially selected objects.

Considering measurement as a process, we find that it evolves by cyclic handling with RS and MS. The structure of measurement based on two parallel rows of material and model elements has been introduced by us earlier [3] and we have used it in previous publications. This static structure 'revives' when measurement is considered as a process, i.e., it becomes dynamic [see fig. 2]. As a measurement process evolves, real and model elements are used alternately, which provides a confirmation and validation of the proposition that model and real elements of measurement go hand in hand inseparably. Indeed, it is impossible to proceed with the next actions towards real elements without introducing another model (object, attribute, MI, etc.).

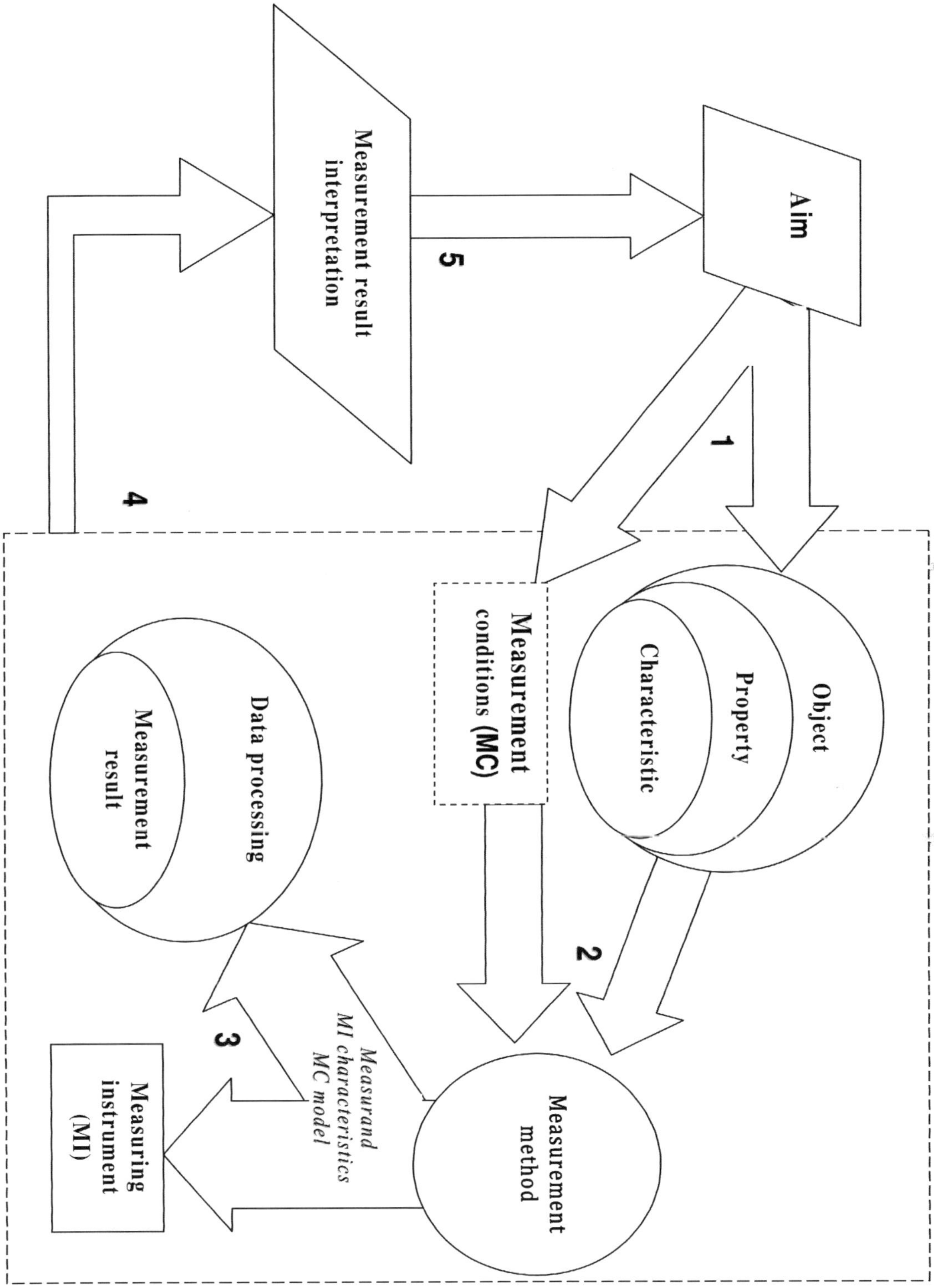

Fig. 2. Logically-temporal measurement structure

Basically, the dynamics of the measurement structure stems from the desire to overcome the gap. Moreover, trying to overcome the gap, we actually try to find out during the measurement process what exactly we are measuring. It is important because a clear a priori separation of measurement into steps of measurand identification and its estimation is an illusion. The latter representation appears when a static structure is automatically assigned a meaning of a process structure.

From the above standpoint, the UNCERTANTY concept is preferential to the ERROR concept. On the other hand, the ERROR concept traditionally involves analysis (structurisation of the gap) and rests on a break point. There was a further advance to be made, but the UNCERTANTY concept makes the problem irrelevant without resolving it.

4. Practical aspect of measurement

In the foregoing, we considered measurement as a cognitive process, which it is on the merits. However, a measurement task is always an applied problem, which has to meet certain requirements of a specific application. Therefore, for the purposes of analysis, we should immerse the above-mentioned systems, in which the measurement is to be performed, in the application system (AS). The immersion in an AS should be carried out step by step: from the local system (e.g., a laboratory or enterprise) to a group system (belonging to the same group, e.g., instrument-making enterprises) and a general system (e.g., national industry), and further, up to the world economy. It is needless to say that in this case the AS levels can be configured in different ways.

An AS should be treated as a supersystem for the MS. The word 'application' should not be understood in a narrow sense as opposed to the words 'scientific' or 'cognitive', especially as the measurement system itself is cognitive.

The AS is an ES, which can be called external, unlike the above-mentioned internal ES (a system of standardization, traceability). Entering the AS, the measurement result agrees (or disagrees) with the RS through the measurement object, which must be in agreement with the set of RS objects.

5. Conclusions

A measurement result as a model system element represents an object (a reality system element) only within a totality of {RS, MS, ES}, dynamically evolving in a process called measurement. Metrology as a discipline aimed at ensuring uniformity of measurements (traceability) dictates the need and allows comparing the results of measurements of the same object property (or a property of similar objects) only in the conditions when totalities $\{RS_1, MS_1, ES_1\}$ and $\{RS_2, MS_2, ES_2\}$ related to the measurements being compared can be considered equivalent. In other words, we should compare the tuples $\{MR_1; RS_1, MS_1, ES_1\}$ and $\{MR_2; RS_2, MS_2, ES_2\}$.

This work is supported by Russian Foundation for Fundamental Researches, grant No. 13-08-00688.

6. References

[1] Kramer S N 1981/3e History Begins at Sumer: Thirty-Nine Firsts in Man's Recorded History (Philadelphia: University of Pennsylvania Press) p 416
[2] Finkelstein L 1994 Measurement 14 3
[3] Granovskii V A, Siraya T N 1984 Measurement data processing (St Petersburg: Energoatomizdat) p 103

Journal of Physics: Conference Series **459** (2013) 012029 doi:10.1088/1742-6596/459/1/012029

PCA-based algorithm for calibration of spectrophotometric analysers of food

Roman Z. Morawski[1] **and Andrzej Miękina**
Warsaw University of Technology, Faculty of Electronics and Information Technology, Institute of Radioelectronics, Warsaw, Poland

E-mail: *r.morawski@ire.pw.edu.pl*

Abstract. Spectrophotometric analysers of food, being instruments for determination of the composition of food products and ingredients, are today of growing importance for food industry, as well as for food distributors and consumers. Their metrological performance significantly depends of the numerical performance of available means for spectrophotometric data processing; in particular – the means for calibration of analysers. In this paper, a new algorithm for this purpose is proposed, *viz.* the algorithm using principal components analysis (PCA). It is almost as efficient as PLS-based algorithms of calibration, but much simpler.

1. Introduction

The term *spectrophotometric analyser of food* is applied to various spectrophotometric sensors, devices, instruments, probes and testers dedicated to measurements of physical and (bio)chemical parameters characterising food products (including beverages). The author's review paper published by the end of 2012 [1] contains a comprehensive overview of basic concepts related to food analysers and of the methodology of measurement data processing in such analysers, as well as a justification of their practical importance – all these illustrated with 250 references (which are not repeated here because of the space limitation). It contains, in particular, the definition, typology and review of the methods for calibration of food analysers. In this paper, a new algorithm of calibration is proposed, the algorithm based on the method of principal components analysis (PCA) which up to now has been used in chemometrics for identification of data structures rather than for parameter estimation. This method is applied here for obtaining an approximate solution of the calibration equation of the form:

$$\dot{\mathbf{X}}\Big|_{N\times M} \cdot \dot{\mathbf{p}}\Big|_{M\times 1} = \dot{\mathbf{y}}\Big|_{N\times 1} \tag{1}$$

with respect to the vector of parameters $\dot{\mathbf{p}}$, on the basis of an error-corrupted data matrix $\tilde{\mathbf{X}}$ representative of $\dot{\mathbf{X}}$ and an-error corrupted data vector $\tilde{\mathbf{y}}$ representative of $\dot{\mathbf{y}}$.

2. Principal components analysis

If the rows of the matrix $\dot{\mathbf{X}}\Big|_{N\times M}$ are realisations of a zero-mean random vector $\underline{\mathbf{x}}$ whose covariance matrix is $\mathbf{\Sigma_x}$, then the covariance matrix of the vector $\underline{\mathbf{z}}\Big|_{M\times 1} \equiv \mathbf{L}\Big|_{M\times M} \cdot \underline{\mathbf{x}}\Big|_{M\times 1}$ is:

$$\mathbf{\Sigma_z} \equiv \mathrm{Cov}[\underline{\mathbf{z}}] = \mathbf{L} \cdot \mathrm{E}\left[\underline{\mathbf{x}} \cdot \underline{\mathbf{x}}^T\right] \cdot \mathbf{L}^T = \mathbf{L} \cdot \mathbf{\Sigma_x} \cdot \mathbf{L}^T \tag{2}$$

[1] To whom any correspondence should be addressed.

Let's construct a row orthonormal matrix $\mathbf{L} \equiv [\mathbf{l}_1 \ldots \mathbf{l}_N]^T$ whose rows $\mathbf{l}_n^T$ maximise the variances of the random variables $\underline{z}_n \equiv \mathbf{l}_n^T \cdot \underline{\mathbf{x}}$. The application of the method of Lagrange multipliers to this optimisation problem leads to the conclusion that the vectors $\mathbf{l}_n$ should satisfy the equations: $\boldsymbol{\Sigma}_\mathbf{x} \cdot \mathbf{l}_n = \lambda_n \cdot \mathbf{l}_n$, which means that they are eigenvector of the covariance matrix $\boldsymbol{\Sigma}_\mathbf{x}$ and λ_n are the corresponding eigenvalues of this matrix. If the matrix $\mathbf{L}$ is composed of the (normalised) eigenvectors of the matrix $\boldsymbol{\Sigma}_\mathbf{x}$, ordered according to the decreasing magnitude of the corresponding eigenvalues λ_n, then the vector $\underline{\mathbf{z}} \equiv [\underline{z}_1 \ldots \underline{z}_N]^T$ is ordered according to the decreasing variance of its components. Since $\underline{\mathbf{z}}^T \equiv \underline{\mathbf{x}}^T \cdot \mathbf{L}^T$, the corresponding transformation of the realisations of the random vectors has the form: $\dot{\mathbf{Z}} \equiv \dot{\mathbf{X}} \cdot \mathbf{L}^T$, where $\dot{\mathbf{X}} \equiv [\dot{\mathbf{x}}_1 \ldots \dot{\mathbf{x}}_N]^T$ and $\dot{\mathbf{Z}} \equiv [\dot{\mathbf{z}}_1 \ldots \dot{\mathbf{z}}_N]^T$. The above transformation may be conveniently implemented using the singular-value decomposition (SVD) of the matrix $\dot{\mathbf{X}}$: $\dot{\mathbf{X}} = \dot{\mathbf{U}} \cdot \dot{\boldsymbol{\Sigma}} \cdot \dot{\mathbf{V}}^T$, where the matrices $\dot{\mathbf{U}}\big|_{N\times N}$ and $\dot{\mathbf{V}}\big|_{M\times M}$ are orthonormal, and the matrix $\dot{\boldsymbol{\Sigma}}$ has the form:

$$\dot{\boldsymbol{\Sigma}}\Big|_{N\times M} \equiv \begin{cases} \left[diag\{\dot{\sigma}_1, \ldots, \dot{\sigma}_M\} \;\; \mathbf{0} \right] & \text{if } N > M \\ \begin{bmatrix} diag\{\dot{\sigma}_1, \ldots, \dot{\sigma}_M\} \\ \mathbf{0} \end{bmatrix} & \text{if } N < M \end{cases} \tag{3}$$

Under an assumption that the singular values $\dot{\sigma}_n$ are ordered ($\dot{\sigma}_1 \geq \dot{\sigma}_2 \geq \dot{\sigma}_3 \geq \ldots$), the variances of the columns of the matrix $\dot{\mathbf{Z}} \equiv \dot{\mathbf{X}} \cdot \dot{\mathbf{V}} = \dot{\mathbf{U}} \cdot \dot{\boldsymbol{\Sigma}}$ are also ordered according to the decreasing magnitude. The mean values of the columns of $\dot{\mathbf{Z}}$ are all equal to zero because:

$$\sum_{n=1}^{N} \dot{z}_{n,m} = \sum_{n=1}^{N} \left(\sum_{\mu=1}^{M} \dot{x}_{n,\mu} \dot{v}_{\mu,m} \right) = \sum_{\mu=1}^{M} \left(\sum_{n=1}^{N} \dot{x}_{n,\mu} \right) \dot{v}_{\mu,m} = 0 \quad \text{for } m = 1, \ldots, M \tag{4}$$

where $\dot{z}_{n,m}$, $\dot{x}_{n,\mu}$ and $\dot{v}_{\mu,m}$ are the elements of the corresponding matrices. The mth column of the matrix $\dot{\mathbf{Z}}$ has the form $\dot{\sigma}_m \dot{\mathbf{u}}_m$, where $\dot{\mathbf{u}}_m$ is the mth column of the matrix $\dot{\mathbf{U}}$; so, its variance is $\dot{\sigma}_m^2 / N$ due to the orthonormality of $\dot{\mathbf{U}}$ implying $\|\dot{\mathbf{u}}_m\|_2^2 = 1$. If the matrix $\dot{\mathbf{X}}$ is not centred, then the columns of the matrix $\dot{\mathbf{Z}}$ should be sorted to order them according to the decreasing variance.

3. Criterion of discrimination

A criterion for rejection of the least informative components is a key element of the PCA-based algorithm for solving Eq.(1). In order to derive such a criterion, let's perform the singular value decomposition of $\tilde{\mathbf{X}}$: $\tilde{\mathbf{X}} = \tilde{\mathbf{U}} \cdot \tilde{\boldsymbol{\Sigma}} \cdot \tilde{\mathbf{V}}^T$, compute the matrix $\tilde{\mathbf{Z}} = \tilde{\mathbf{U}} \cdot \tilde{\boldsymbol{\Sigma}}$, and the variances of its columns $\tilde{v}_{Z,m}$ for $m = 1, \ldots, M$. Now, the corresponding reverse accumulated variances defined as:

$$\tilde{a}_{Z,m} = \sum_{\mu=m}^{M} \tilde{v}_{Z,m} \quad \text{for } m = 1, \ldots, M \tag{5}$$

may be used for elimination of non-informative components which have appeared due to the insufficient number of measurements ($N < M$) or due to the presence of random errors in the x-data. If $N < M$, then the variances $v_{Z,m}$ with indices larger than M are zero; so, the signal-to-noise ratio (SNR_X) seems to be a natural criterion for the rejection of the least informative components due to the errors in the x-data. Under an assumption that those errors are realisations of zero-mean independent random variables with the variances $\sigma_{X,n,m}^2$, SNR_X may be defined and calculated as follows:

$$SNR_X \equiv \|\dot{\mathbf{X}}\|_F^2 \Big/ \mathrm{E}\left[\|\Delta\tilde{\mathbf{X}}\|_F^2 \right] \cong \sum_{n=1}^{N} \sum_{m=1}^{M} \tilde{x}_{n,m}^2 \Big/ \sum_{n=1}^{N} \sum_{m=1}^{M} \sigma_{X,n,m}^2 \tag{6}$$

where $\|\circ\|_F$ is the Frobenius norm, $\tilde{x}_{n,m}$ are elements of the matrix $\tilde{\mathbf{X}}$. One may reasonably assume that the share of rejected components in the accumulated variance of all components should be the same as the share of the power of noise in the power of signal, *i.e.* that a threshold for cutting-off the least informative components should be $a_{th} = \tilde{a}_{Z,1}/SNR_X$, and the indices of rejected components should start from $m_{th} = \inf \arg_m \{ \tilde{a}_{Z,m} < a_{th} \}$.

4. Proposed algorithm of calibration

The method of principal components may be used not only for identification of data structures, but also for estimation of model parameters or for model response prediction. The corresponding estimation method is most frequently called principal component(s) regression. Here, for the sake of nomenclature consistency, the name PCLS will be used for implementation of that method. Its derivation is based on an assumption that the statistical model of the data has the form:

$$\underline{\tilde{\mathbf{y}}} = \dot{\mathbf{X}} \cdot \dot{\mathbf{p}} + \Delta\underline{\tilde{\mathbf{y}}} \quad \text{and} \quad \underline{\tilde{\mathbf{X}}} = \dot{\mathbf{X}} + \Delta\underline{\tilde{\mathbf{X}}} \tag{7}$$

where $\underline{\tilde{\mathbf{y}}}$ and $\underline{\tilde{\mathbf{X}}}$ are column-centred. For the given realisations $\tilde{\mathbf{y}}$ and $\tilde{\mathbf{X}}$ of $\underline{\tilde{\mathbf{y}}}$ and $\underline{\tilde{\mathbf{X}}}$, respectively, the following operations are performed to obtain an estimate $\hat{\mathbf{p}}$ of $\dot{\mathbf{p}}$. First, the matrix $\tilde{\mathbf{Z}} = \tilde{\mathbf{U}} \cdot \tilde{\mathbf{\Sigma}}$ is determined using matrices resulting from singular value decomposition of $\tilde{\mathbf{X}}$. Next, the columns of $\tilde{\mathbf{Z}}$, whose indices are larger than m_{th}, are removed. The matrix $\tilde{\mathbf{Z}}_0$ obtained in this way is then used for solving the following OLS normal equation:

$$\tilde{\mathbf{Z}}_0^T \cdot \tilde{\mathbf{Z}}_0 \cdot \hat{\mathbf{q}} = \tilde{\mathbf{Z}}_0^T \cdot \tilde{\mathbf{y}} \tag{8}$$

with respect to $\hat{\mathbf{q}}$ which is next used for computing $\hat{\mathbf{p}} = \tilde{\mathbf{V}} \cdot \hat{\mathbf{q}}$. The latter equality is implied by the following identity:

$$\dot{\mathbf{X}} \cdot \dot{\mathbf{p}} = \left(\dot{\mathbf{U}} \cdot \dot{\mathbf{\Sigma}}\right) \cdot \left(\dot{\mathbf{V}}^T \cdot \dot{\mathbf{p}}\right) = \dot{\mathbf{Z}} \cdot \dot{\mathbf{q}} \tag{9}$$

where $\dot{\mathbf{q}} \equiv \dot{\mathbf{V}}^T \cdot \dot{\mathbf{p}}$ or $\dot{\mathbf{p}} = \dot{\mathbf{V}} \cdot \dot{\mathbf{q}}$ (because $\dot{\mathbf{V}}$ is orthonormal).

5. Example of spectrophotometric application

5.1. *Problem formulation*

This section is entirely devoted to an example of application of the proposed algorithm for determination of edible oil mixtures on the basis of spectrophotometric data. It is assumed here that an oil mixture to be analysed is composed of three known components – nut oil, corn oil and olive oil – and that the M-dimensional vectors of exact data – $\dot{\mathbf{s}}_1$, $\dot{\mathbf{s}}_2$ and $\dot{\mathbf{s}}_3$ – representative of their absorbance spectra, are available. The study is based on demi-synthetic data generated using the denoised and baseline-corrected real-world data representative of nut oil ($\dot{\mathbf{s}}_1$), corn oil ($\dot{\mathbf{s}}_2$) and olive oil ($\dot{\mathbf{s}}_3$), acquired by means of the FTIR spectrophotometer *Perkin Elmer System 2000*, set to the resolution 1 cm^{-1}.

According to the Lambert-Beer's law, the vector of exact absorbance data $\dot{\mathbf{s}}$, representative of the spectrum of the mixture, satisfies the equation:

$$\dot{\mathbf{s}} = c_1 \cdot \dot{\mathbf{s}}_1 + c_2 \cdot \dot{\mathbf{s}}_2 + c_3 \cdot \dot{\mathbf{s}}_3 \tag{10}$$

where $\mathbf{c} = [c_1 \; c_2 \; c_3]^T$ is the vector of concentrations of all components satisfying the following constraints: $c_1, c_2, c_3 \in [0, 1]$ and $c_1 + c_2 + c_3 = 1$. It is moreover assumed that the real-world absorbance data $\tilde{\mathbf{s}}$, representative of the spectrum of a mixture, are corrupted with random errors $\Delta\tilde{\mathbf{s}}$ resulting both from inaccurate preparation of the mixture and from imperfections of the spectrophotometer. Since $c_3 = 1 - c_1 - c_2$, the spectral data, representative of the mixture, may be modelled by means of the following set of M linear algebraic equations:

$$\tilde{\mathbf{s}} - \dot{\mathbf{s}}_3 = c_1 \cdot \left(\dot{\mathbf{s}}_1 - \dot{\mathbf{s}}_3\right) + c_2 \cdot \left(\dot{\mathbf{s}}_2 - \dot{\mathbf{s}}_3\right) + \Delta\tilde{\mathbf{s}} \tag{11}$$

The problem under study consists in estimation of the concentrations c_1 and c_2. It is numerically ill-conditioned because the spectra of nut oil, corn oil and olive oil are very close to each other.

5.2. Generation of semi-synthetic data

Each vector of spectral data for calibration is synthesised according to the methodology imitating the laboratory procedure used for obtaining the real-world data. First, the reference (exact) values of the concentrations of nut oil ($\dot{c}_1$) and of corn oil ($\dot{c}_2$) are selected, and the value of the concentration of olive oil is calculated: $\dot{c}_3 = 1 - \dot{c}_1 - \dot{c}_2$. Next, the error-corrupted values of all concentrations are determined by emulation of the process of sample preparation which consists in mixing the measured-out volumes of the components: $\tilde{V}_1$, $\tilde{V}_2$ and $\tilde{V}_3$:

$$\tilde{c}_j = \tilde{V}_j / \tilde{V} \quad \text{for} \quad j = 1, 2, 3 \quad \text{with} \quad \tilde{V} = \tilde{V}_1 + \tilde{V}_2 + \tilde{V}_3 \quad \text{and} \quad \tilde{V}_j = \dot{V}_j (1 + \vartheta_j) \tag{12}$$

where $\dot{V}_j$ are error-free values of the volumes, and ϑ_j are realisations of zero-mean random variables $\underline{\vartheta}_j$ with the variances modelling relative errors of measuring out the volumes. The resulting concentrations $\tilde{c}_j = \dot{c}_j (1 + \zeta_j)$ are subject to relative errors:

$$\zeta_j = \frac{1 + \vartheta_j}{\dot{c}_1 (1 + \vartheta_1) + \dot{c}_2 (1 + \vartheta_2) + \dot{c}_3 (1 + \vartheta_3)} - 1 = \frac{1 + \vartheta_j}{1 + \dot{c}_1 \vartheta_1 + \dot{c}_2 \vartheta_2 + \dot{c}_3 \vartheta_3} - 1 \quad \text{for} \quad j = 1, 2, 3 \tag{13}$$

which may be viewed as realisations of random variables $\underline{\zeta}_j$ related in the same way to $\underline{\vartheta}_1$, $\underline{\vartheta}_2$ and $\underline{\vartheta}_3$ as ζ_j are related in to ϑ_1, ϑ_2 and ϑ_3 in the above equation. The moments of those random variables depend on concentrations and on the variance σ_ϑ^2 of $\underline{\vartheta}_j$ in the following way:

$$\mathrm{E}\left[\underline{\zeta}_1\right] \cong \left(\dot{c}_1^2 + \dot{c}_2^2 + \dot{c}_3^2 - \dot{c}_1\right)\sigma_\vartheta^2 \quad \text{and} \quad \mathrm{E}\left[\underline{\zeta}_2\right] \cong \left(\dot{c}_1^2 + \dot{c}_2^2 + \dot{c}_3^2 - \dot{c}_2\right)\sigma_\vartheta^2 \tag{14}$$

$$\mathrm{Var}\left[\underline{\zeta}_1\right] \cong \left((1 - \dot{c}_1)^2 + \dot{c}_2^2 + \dot{c}_3^2\right)\sigma_\vartheta^2 \quad \text{and} \quad \mathrm{Var}\left[\underline{\zeta}_2\right] \cong \left(\dot{c}_1^2 + (1 - \dot{c}_2)^2 + \dot{c}_3^2\right)\sigma_\vartheta^2 \tag{15}$$

$$\mathrm{Cov}\left[\underline{\zeta}_1, \underline{\zeta}_2\right] \cong \left(\dot{c}_1^2 + \dot{c}_2^2 + \dot{c}_3^2 - \dot{c}_1 - \dot{c}_2\right)\sigma_\vartheta^2 \tag{16}$$

The spectral data $\tilde{\mathbf{s}} \equiv [\tilde{s}_1 \; \tilde{s}_2 \; \ldots]^T$, corresponding to a given triplet of concentrations – $\tilde{c}_1$, $\tilde{c}_2$ and $\tilde{c}_3 = 1 - \tilde{c}_1 - \tilde{c}_2$ determined after Eq.(12) – are generated using the formulae:

$$\tilde{\mathbf{s}} = \tilde{c}_1 \dot{\mathbf{s}}_1 + \tilde{c}_2 \dot{\mathbf{s}}_2 + \tilde{c}_3 \dot{\mathbf{s}}_3 + \Delta\mathbf{z} \tag{17}$$

where $\Delta\mathbf{z} \equiv [\Delta z_1 \; \ldots \; \Delta z_M]^T$ is a realisation of a random vector $\underline{\Delta\mathbf{z}}$ modelling additive errors of spectrum measurement. By comparing Eq.(17) with Eq.(11) and randomising the result of comparison, one may get:

$$\underline{\Delta\tilde{\mathbf{s}}} = \mathbf{A} \cdot \underline{\zeta} + \underline{\Delta\mathbf{z}} \quad \text{with} \quad \mathbf{A} \equiv \left[\dot{c}_1 \cdot (\dot{\mathbf{s}}_1 - \dot{\mathbf{s}}_3) \;\; \dot{c}_2 \cdot (\dot{\mathbf{s}}_2 - \dot{\mathbf{s}}_3)\right] \quad \text{and} \quad \underline{\zeta} \equiv \left[\underline{\zeta}_1 \;\; \underline{\zeta}_2\right]^T \tag{18}$$

Under an assumption that $\underline{\Delta\mathbf{z}}$ is a zero-mean random vector whose components are statistically independent of $\underline{\zeta}_1$ and $\underline{\zeta}_2$:

$$\mathrm{Cov}\left[\underline{\Delta\tilde{\mathbf{s}}}\right] = \mathbf{A} \cdot \boldsymbol{\Sigma}_\varsigma \cdot \mathbf{A}^T + \boldsymbol{\Sigma}_\mathbf{z} \quad \text{with} \quad \boldsymbol{\Sigma}_\varsigma \equiv \begin{bmatrix} \mathrm{Var}\left[\underline{\zeta}_1\right] & \mathrm{Cov}\left[\underline{\zeta}_1 \cdot \underline{\zeta}_2\right] \\ \mathrm{Cov}\left[\underline{\zeta}_1 \cdot \underline{\zeta}_2\right] & \mathrm{Var}\left[\underline{\zeta}_2\right] \end{bmatrix} \tag{19}$$

where $\boldsymbol{\Sigma}_\mathbf{z}$ is the covariance matrix of $\underline{\Delta\mathbf{z}}$, and $\boldsymbol{\Sigma}_\varsigma$ is the covariance matrix of $\underline{\zeta}$. Using this equation, one may determine the average power $\mathrm{E}\left[\underline{\Delta\tilde{\mathbf{s}}}^T \cdot \underline{\Delta\tilde{\mathbf{s}}}\right]$ of the noise $\underline{\Delta\tilde{\mathbf{s}}}$:

$$\mathrm{E}\left[\underline{\Delta\tilde{\mathbf{s}}}^T \cdot \underline{\Delta\tilde{\mathbf{s}}}\right] \cong \sum_{m=1}^{M} \mathbf{a}_m^T \cdot \boldsymbol{\Sigma}_\varsigma \cdot \mathbf{a}_m + \sum_{m=1}^{M} \sigma_{z,m}^2 \tag{20}$$

where $\mathbf{a}_m^T$ is the m th row of the matrix $\mathbf{A}$. For generation of the relative errors of the volumes (ϑ_k), uncorrelated pseudorandom numbers, following the 3-std-truncated normal distribution $\mathcal{N}_3(0, \sigma_\vartheta^2)$, have been used; for generation of the errors of spectrum measurements (Δz_n) – uncorrelated pseudorandom numbers following the distribution $\mathcal{N}_3(0, \sigma_z^2)$. The set of data for calibration $\tilde{\mathbb{D}}^{cal}$ has been assumed to contain $N = 121$ pairs of the following values of concentrations:

$$\dot{c}_1^{cal}, \dot{c}_2^{cal} \in \{k \cdot 0.01 \mid k = 0, 1, ..., 10\} \tag{21}$$

and the corresponding sequences of spectral data, each composed of $M = 512$ points. The set of data for validation $\tilde{\mathbb{D}}^{val}$ has been assumed to contain $N' = 441$ pairs of the following values of concentrations:

$$\dot{c}_1^{val}, \dot{c}_2^{val} \in \{k \cdot 0.005 \mid k = 0, 1, ..., 9\} \tag{22}$$

and the corresponding sequences of spectral data, each composed of $M = 512$ points.

According to the inverse-model-based approach, the calibration consists in solving a set of linear algebraic equations defined by Eq.(1) with:

$$\tilde{\mathbf{X}} \equiv \tilde{\mathbf{S}}^{cal} - \dot{\mathbf{S}}_3 \quad \text{and} \quad \tilde{\mathbf{y}} \equiv \dot{\mathbf{c}}_j^{cal} \quad \text{for} \quad j = 1, 2 \tag{23}$$

where $\dot{\mathbf{c}}_1^{cal}$ and $\dot{\mathbf{c}}_2^{cal}$ are the vectors containing the reference concentrations defined by Eq.(21), $\tilde{\mathbf{S}}^{cal}$ is the matrix whose rows contain corresponding spectral data, and $\dot{\mathbf{S}}_3$ is the matrix whose all rows are equal to $\dot{\mathbf{s}}_3^T$. If $\tilde{\mathbf{y}} = \dot{\mathbf{c}}_j^{cal}$, then the procedure of calibration yields an estimate $\hat{\mathbf{p}}_j$ of the vectors of parameters $\dot{\mathbf{p}}$, which may be used for computing an approximate value of $\dot{c}_j$ on the basis of any vector of acquired spectral data $\tilde{\mathbf{s}}$, using the formula: $\hat{c}_j = \hat{\mathbf{p}}_j \cdot (\tilde{\mathbf{s}} - \dot{\mathbf{s}}_3)$ for $j = 1, 2$.

5.3. Methodology and selected results of comparison

The data for calibration $\tilde{\mathbb{D}}^{cal}$ have been synthesised according to the methodology described in Subsection 5.2 using the values $\sigma_\vartheta = 2.2 \cdot 10^{-3}$ and $\sigma_z = 3.0 \cdot 10^{-6}$. The data for validation $\tilde{\mathbb{D}}^{val}$ have been generated using the values $\sigma_\vartheta = 0$ and $\sigma_z = 3.0 \cdot 10^{-6}$. The latter assumption is reflecting the laboratory situation where the concentrations for validation may be measured with a significantly smaller uncertainty than that expected to be reached by the algorithm under study. It has been used to facilitate the interpretation of the results of numerical experimentation since in this way they are freed form that component of error of estimation which is propagated from validation data with the amplification coefficient equal to 1. The performance of the PCLS algorithm – with respect to the uncertainty of the final result of measurement – has been assessed using $R = 100$ versions of the set of data $\tilde{\mathbb{D}}^{cal}$ and $R' = 1000$ versions of the set of data $\tilde{\mathbb{D}}^{val}$, each corresponding to a different realisation of errors. For each version $\tilde{\mathbb{D}}^{cal}(r)$, $r = 1, ..., R$, of the set $\tilde{\mathbb{D}}^{cal}$, the full calibration procedure has been performed, and its results have been validated by means of R' versions $\tilde{\mathbb{D}}^{val}(r')$ of $\tilde{\mathbb{D}}^{val}$, $r' = 1, ..., R'$. For each pair of the values of concentrations $\dot{c}_1^{val}$ and $\dot{c}_2^{val}$, defined by Eq.(22), the estimates of two uncertainty indicators have been calculated, *viz.* of the worst-case error:

$$\hat{e}\left[\Delta\hat{c}_j^{val}\right] \equiv \sup\left\{ \left|\Delta\hat{c}_j^{val}(r, r')\right| \,\middle|\, \begin{matrix} r = 1, ..., R \\ r' = 1, ..., R' \end{matrix} \right\} \tag{24}$$

and of the error standard deviation:

$$\hat{s}\left[\Delta\hat{c}_j^{val}\right] \equiv \sqrt{\frac{1}{RR'-1}\sum_{r=1}^{R}\sum_{r'=1}^{R'}\left(\Delta\hat{c}_j^{val}(r, r') - \hat{m}\left[\Delta\hat{c}_j^{val}\right]\right)^2} \quad \text{with} \quad \hat{m}\left[\Delta\hat{c}_j^{val}\right] \equiv \frac{1}{RR'}\sum_{r=1}^{R}\sum_{r'=1}^{R'}\Delta\hat{c}_j^{val}(r, r') \tag{25}$$

where $\Delta\hat{c}_j^{val}(r, r') \equiv \hat{c}_j^{val}(r, r') - \dot{c}_j^{val}$. Those indicators have been used for comparing the results of calibration obtained by means of the proposed algorithm with the corresponding results obtained by means of the OLS and PLS estimators. The first of them has been implemented according to the formula:

$$\hat{\mathbf{p}} = \arg_{\mathbf{p}} \inf\left\{ \tilde{\mathbf{X}}^T \cdot \tilde{\mathbf{X}} \cdot \mathbf{p} = \tilde{\mathbf{X}}^T \cdot \tilde{\mathbf{y}} \right\} \tag{26}$$

For implementation of the PLS estimator, the procedure ***pls*** from *PLS_Toolbox* [2] – with manual optimisation of the number of latent variables – has been used. The indicators of estimation uncertainty $\hat{s}\left[\Delta\hat{c}_1^{val}\right]$ and $\hat{e}\left[\Delta\hat{c}_j^{val}\right]$, obtained for the PCLS algorithm and two reference algorithms, are compared in Figure 1.

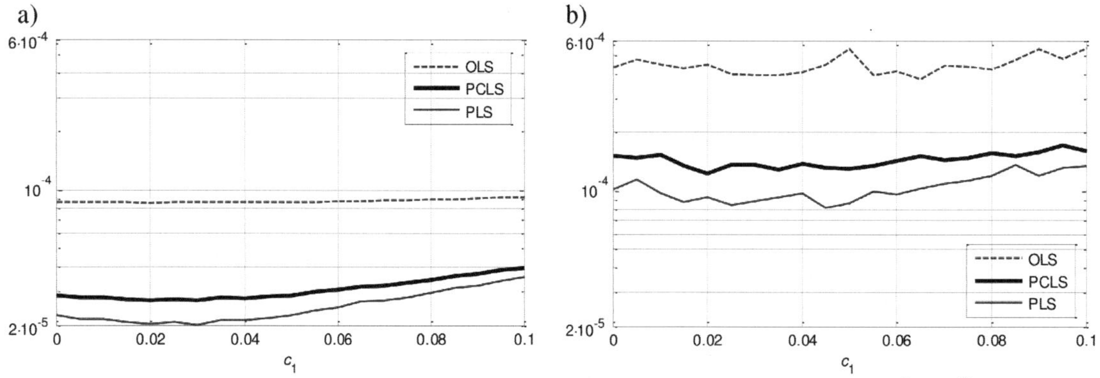

Figure 1. The uncertainty indicators: a) the error standard deviation $\hat{s}\left[\Delta\hat{c}_1^{val}\right]$, b) the worst-case error $\hat{e}\left[\Delta\hat{c}_j^{val}\right]$.

6. Conclusion

A new PCA-based algorithm for calibration of spectrophotometric analysers of food has been proposed – the algorithm including fully automatic selection of the most informative components. It has been compared with two standard tools currently used for this purpose, *viz.* with the OLS estimator (being the simplest, but least accurate tool) and with the PLS estimator (being the most complex and most accurate one). The comparison has been based on a test problem consisting in determination of the composition of trinary oil mixtures. The results of comparison have demonstrated that the proposed algorithm is significantly more accurate than the OLS estimator, and only slightly less accurate than the PLS estimator. Being much simpler than the latter and fully automatic, the proposed algorithm seems to be an attractive alternative for the PLS estimator which is currently a favourite calibration tool in many areas of chemometrics.

Acknowledgements

This work has been supported by the National Science Centre in Poland (grant No. N N505 464832). The authors express their sincere gratitude to Dr. Grażyna Zofia Żukowska from the Faculty of Chemistry, Warsaw University of Technology, for the acquisition of data used for numerical experimentation reported in this paper.

References

[1] Morawski R Z 2012 *Metrol. Meas. Syst.* **19** 623–652
[2] *PLS_Toolbox*, Eigenvector Research, Inc., *http://www.eigenvector.com/software/pls_toolbox.htm* [as of February 10, 2013]

Journal of Physics: Conference Series **459** (2013) 012030 doi:10.1088/1742-6596/459/1/012030

Non-invasive measurement of cholesterol in human blood by impedance technique: an investigation by 2D finite element field modelling

E Aristovich and S H Khan

City University London, School of Engineering and Mathematical Sciences, Northampton Square, London EC1V 0HB, UK

E-mail: Ekaterina.Aristovich.1@city.ac.uk

Abstract. This paper concerns detection of solid particles suspended in conductive media by impedance technique. The technique is based on changes in impedance measured between two electrodes placed across a given volume of conducting medium. It presents a methodology for modelling and investigation of the feasibility of such a technique for particle detection by 2D finite element (FE) field modelling. This is based on modelling and computation of electric field distribution between the above electrodes. It establishes the modelling approach, the complexity involved and justifies the need for modelling in 3D to incorporate some of the effects that cannot be taken into account in 2D models. It reports on the modelling investigation for a specific case of detecting, by impedance technique cholesterol particles suspended in human blood and points to a possible instrument for non-invasive measurement of blood cholesterol level.

1. Introduction

Driven by concerns over public health associated with cardiovascular and other related diseases an easy, non-invasive, reliable and effective method for measuring cholesterol level in human blood is of increasing importance. There are at present a large number of direct and indirect methods that are available in clinical and research laboratories for the quantitative measurement of cholesterol in blood [1-3]. These measurements are being done increasingly frequently and often for prophylaxis and awareness against cardiovascular diseases. All these methods require a sample of the blood obtained from the body by an invasive procedure. The procedure may carry a risk of bruise, infection and/or hematoma. There are, however some indirect methods by which the level of total cholesterol can be estimated using bioelectrical impedance measurements carried out elsewhere in the body [4]. As these methods are indirect and based on inference of cholesterol level from statistical data obtained by multiple impedance measurements of substances other than blood, they are complicated, time consuming and often lack accuracy and reproducibility.

Thus, there is a growing need to develop a simple, non-invasive method for measuring total blood cholesterol. It is proposed to do this by using electrical impedance technique in which the conductivity distribution of objects (e.g. cholesterol particles carried by lipoproteins) within a volume (e.g. blood plasma) is measured in the form of impedance (Z) between a pair of electrodes. The impedance technique has been widely used in many areas (industry, chemical production, pharmacy, medicine, biosciences) and it has been proven to be reliable for imaging (EIT, ERT) [5-7] and quantitative analyses (BMI, fuel purity, impedance spectroscopy) [8-11].

The impedance technique has been used successfully for many biological measurements of human body, especially for measuring characteristics of blood. These include measurement of body fluid volume [12], fat [13, 14], glucose [15], blood coagulation [16], hematocrit (erythrocyte volume fraction, EVF) [17], white blood cell count and other blood measurements [18]. However the blood measurement methods used involve an invasive procedure for obtaining blood from the body.

As mentioned earlier, the proposed impedance technique for cholesterol measurement is non-invasive and, in this paper the feasibility of this technique has been established by 2D finite element modelling. This essentially comprises modelling of particles suspended in conductive media.

2. 2D mathematical modelling of particles suspended in conductive medium and its finite element realisation

2.1. Mathematical model and impedance calculation

Although the mathematical physics of conduction is well understood, the mathematics of describing the current and potential fields between and around impedance electrodes, particularly those attached to organic tissue, is more challenging.

For given potentials φ assigned to driving electrodes the electric field distribution in the 2D problem domain Ω (x, y) between the electrodes is given by the following Laplace's equation [19]:

$$\nabla^2\varphi = 0 \text{ in } \Omega \tag{1}$$

Under appropriate boundary conditions the above equation is solved by finite element method (FEM) [20] in terms of unknown electric potential φ for given conductivity (σ) and permittivity (ε) distributions in the problem Ω. It is assumed that the material properties are linear, piece-wise homogeneous and isotropic. Following the solution of equation (1), field intensity and flux density vectors $\mathbf{E}=-\text{grad}V$ (V is the potential difference between the driving electrodes) and $\mathbf{D}=\varepsilon\mathbf{E}$ can be calculated. From these the current $I=\mathbf{J}A$ (A is the cross-sectional area) is calculated using current density $\mathbf{J}=\sigma\mathbf{E}$. Finally, the impedance Z is calculated from the simple relationship between current (I) and voltage (V) between the electrodes:

$$Z = \frac{V}{I} \tag{2}$$

For the modelling studies reported in this paper the above field equation was solved using the commercial software package COMSOL [21].

2.2. Assumptions

The modelling and computation of electric field distribution in the problem domain is based on a number of assumptions some of which are either justified by modelling experiments or discarded by introducing further complexities in the subsequent FE models.

The main assumption taken in the studies reported here is that the electric field distribution between the driving electrodes mentioned above is two-dimensional and, it has been shown by modelling experiments that this is appropriate as a first approximation.

The human blood is a complex medium, which is composed of cells of different sizes and shapes (red blood cells, white blood cells, platelets, etc.) distributed in a conductive aqueous solution (blood plasma). About 55% of the blood fluid is composed of blood plasma (mostly water – 91.5%) and the remaining 45% consists of various blood components (various blood cells and other inclusions) [22]. This means from the electric field and conductivity points of view blood is essentially a non-homogenous and anisotropic medium. Specifically, this also means that the electrical properties of the whole blood (e.g. electrical conductivity, σ and the dielectric permittivity, ε) are different from those of blood plasma, which contains no cells [23, 24].

For all modelling purposes it was assumed that blood was linear, piecewise homogeneous and isotropic. The piecewise homogeneity can account for non-homogeneous properties of blood and its isotropy is justified by the fact that in the 2D problem domain the main components of electric field

(**E**, **D**) and current density vectors (**J**) are mainly confined in the *x-y* plane. Furthermore, it should be noted that for frequencies lower than 100 kHz most living tissues are assumed to be electrolytic conductors. At higher frequencies the dielectric properties of bio-tissue may dominate. The higher the frequency the closer the tissue properties come to those of water.

There are no eddy current effects for the range of frequencies (50 Hz to 100 MHz) which is normally used for measuring impedance of living tissues. At higher frequencies electromagnetic properties of tissue become very close to the properties of water, so no useful information can be obtained. The lowest skin depth in this context is 30 mm (for blood plasma at 100 MHz), which is significantly higher than the characteristic size of features and problem domain involved. Hence there are no eddy-current effects.

2.3. *Finite element models*

Figure 1 shows the basic 2D FE model of the problem domain Ω (x, y) in which Equation (1) was solved under the boundary conditions shown in Figure 2. The active problem domain in between the driving electrodes is surrounded by air regions ($\sigma = \sigma_1$) to account for any fringe and leakage fields. The blood-plasma region ($\sigma = \sigma_2$) in between the electrodes is also extended in the *y*-direction to account for fringe field effects, which need to be quantified. As shown in Figure 1 the cholesterol particles ($\sigma = \sigma_3$) are confined in the blood plasma between the electrodes.

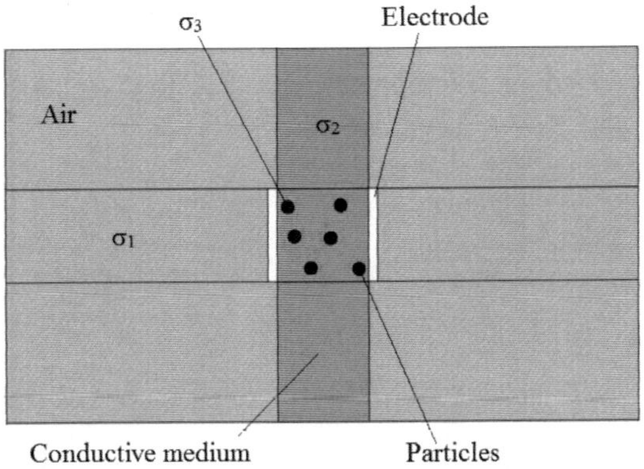

Figure 1. Basic 2D finite element (FE) model of particles in a conductive medium.

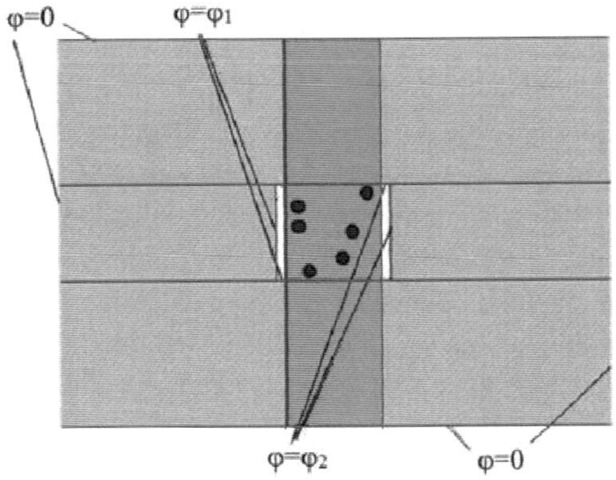

Figure 2. 2D finite element model showing boundary conditions.

The above FE models were used to investigate the effects of various parameters on the distribution of alternating currents and electric potential (hence voltage drops) in the problem domain. The parameters included various physiological (e.g. level and distribution of lipoproteins/cholesterol, other blood components), material (e.g. dielectric permittivities (ε) and conductivities (σ) of blood components including lipoproteins carrying cholesterol particles). This can also be used to investigate the effects of electrical variables (e.g. measurement circuitry, excitation frequency, electrode geometry and their placement). Since the unknown potential distribution is a function of many of the physiological components in blood and tissue, the primary modelling aim here was to, by simulation establish an adequate and identifiable functional relationship between this distribution and the level of cholesterol in the blood. This would then provide a unique 'impedance fingerprint' corresponding to a given cholesterol level.

2.4. Inclusion of particles in the problem domain

As mentioned before the human blood is a complex conductive medium. Fatty substances (lipids) are present in the plasma in suspension and in solution. Cholesterol, the particles used in this simulation study is almost insoluble in water (0.095 mg/l at 30 °C) and it is transported in the blood stream by lipoproteins (made up of proteins and lipids) that are water-dispersible and carry cholesterol and triglycerides internally. Based on the size, lipoproteins vary from chylomicrons (100-1000 nm in diameter), very low density lipoprotein (VLDL, 30-80 nm), intermediate density lipoprotein (IDL, 25-50 nm), low density lipoprotein (LDL, 18-28 nm) and high density lipoprotein (HDL, 5-15 nm). LDL-bound cholesterol is referred to as 'bad cholesterol' because high LDL level (in concentration and smaller particle size) is linked to cardiovascular diseases. On the other hand, HDL-bound cholesterol is often termed 'good cholesterol' because HDL can remove cholesterol from cells and atheroma and transport it back to the liver. Triglycerides, which form major components of VLDL and chylomicrons are also linked to cardiovascular diseases. Hence the Total cholesterol in the blood is defined as the sum of HDL, LDL, and VLDL. Given the wide variations in size and density of lipoproteins and the concentration of cholesterol they contain, it should be possible to measure by impedance technique not only the Total cholesterol but also its constituent components. Furthermore, these particles provide high contrast ratio in terms of electrical properties compared to other components in the blood. For example, because of different water content blood has a resistivity of about 1.6 Ωm, whereas fat has a resistivity of about 25 Ωm – 15 times higher than blood. Tissues other than those of heart and lungs have an approximate resistivity of 10 Ωm.

From the modelling point of view it is important to note that total cholesterol of 5 mmol/l corresponds to about 10^{12} lipoprotein particles (5-30 nm in size) in each mm^3 of blood. This clearly poses a major challenge for taking into account every individual particle for modelling purposes. This would not only introduce enormous complexity in building FE models but also dramatically increase modelling times to an unacceptable level.

This problem was tackled by using a particle aggregation (coagulation) strategy in which smaller number of larger particles, N_C was used keeping the total volume of cholesterol particles, V_C, corresponding to a given blood cholesterol level constant. Knowing the total volume of cholesterol in 1 mm^3 of blood plasma, the total volume of cholesterol in the blood volume, V_b being modelled can be calculated. These parameters were then adjusted for FE models in two dimensions. However, this simplifying modelling strategy needed to be quantitatively justified by modelling experiments. This was to establish the maximum number of cholesterol particles above which no significant change in the modelling results could be obtained by further increasing the number of particles. Furthermore, the effects of randomness in the distribution (placement) of these N_c particles on modelling results also needed to be quantified. As it is evidenced from some of the modelling results given below in Section 3 this strategy for particle aggregation (coagulation) simplifies FE models (Figure 3) without significantly affecting the results of FE modelling. Figure 4 shows typical FE discretisation used at the individual particle level. For simulation runs, the random distribution of particles in the active volume was achieved at the pre-processing stage by using appropriate Matlab scripts interfacing the FE software package COMSOL. This allowed automatic random generation and placement of particles in the problem domain for subsequent FE discretisation.

Figure 3. 2D FE models showing variable number of particles, N_C between the driving electrodes.

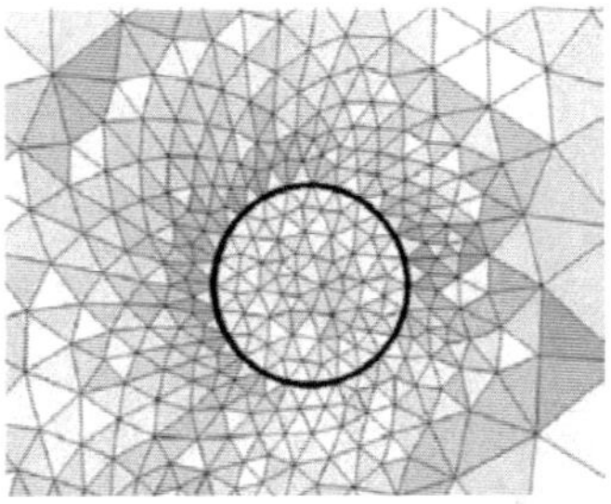

Figure 4. Details of 2D finite element discretisation in and around a single particle.

3. Some of the modelling results and discussions

3.1. Comparison with simple analytical solutions

To test the 2D FE models, a number of simple cases were tested for which accurate analytical solutions exist. Figure 5 shows the two cases in which 3 impedances are placed in series and in parallel occupying fully the problem domain between the electrodes. More realistic cases for this are

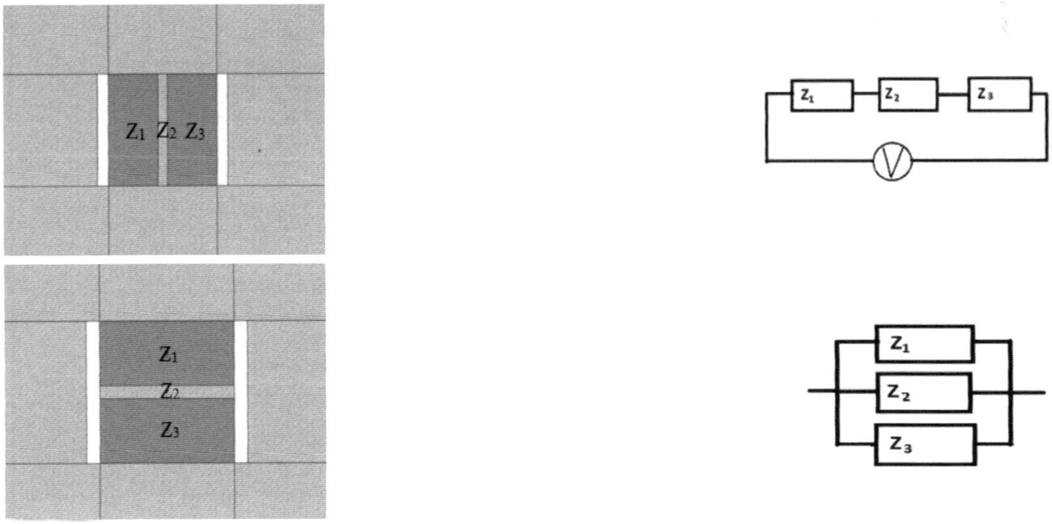

Figure 5. Simple series (above) and parallel (below) placement of impedances and their FE models.

shown in Figure 6 in which spherical particles are placed in series and in parallel with the driving electrodes. FE modelling results for these cases gave good agreement with those obtained from analytical calculations.

Figure 6. Series (left) and parallel (right) arrangements of particles in FE models for comparison with simple analytical solutions.

3.2. Effects of the number of particles, N_c and their distribution

For total volume of cholesterol particles, V_c corresponding to a given cholesterol level (3 mmol/l) the effects of particle numbers, N_c are shown in Figure 7. It shows that above a given threshold ($N_c \approx 200$-300, further refined to 500 by 3D modelling) no appreciable change in the modelling results can be obtained. This justified the particle aggregation strategy discussed in Section 2.

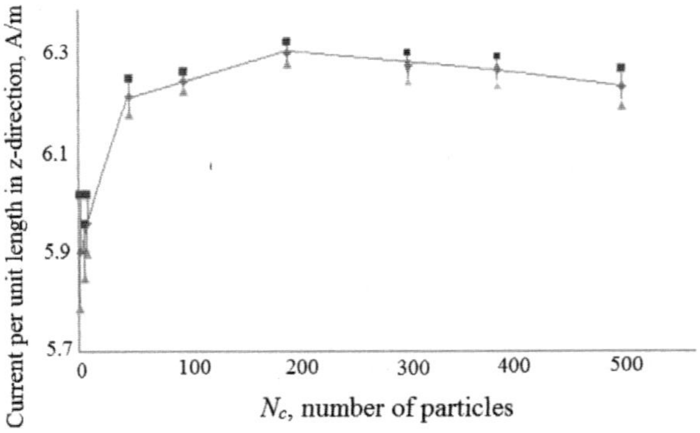

Figure 7. Variation of modelling results with number of particles, N_c used in 2D FE modelling (total volume of particles, V_c=const.).

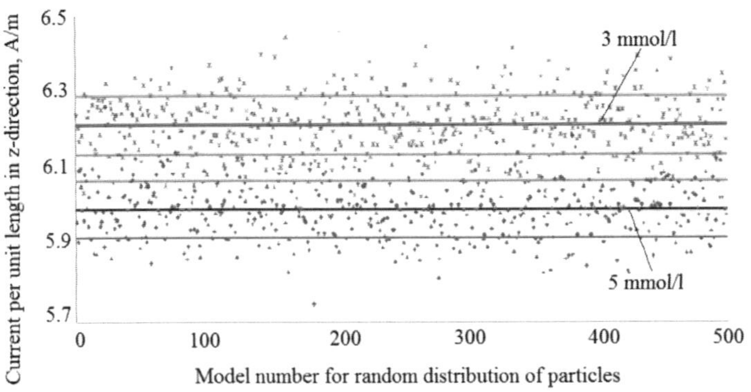

Figure 8. Effects of the distribution (placement) of particles on modelling results for 500 FE models containing randomly distributed particles (total number of particles, N_c=200=const.).

Figure 8 presents the effects of particle placement on modelling results which essentially shows that for a threshold number of particles, N_c the effects of placements of particles on modelling results average out to a clearly identifiable value corresponding to a given cholesterol level.

The most important consequence of the above two graphs is that despite the enormous number of cholesterol particles in a given volume of blood there exists a much smaller threshold number of particles which is enough to simulate the effects of these larger number of particles. Furthermore, for this threshold number of particles the effects of their random distribution averages out.

3.3. Fringe field effects

Figure 9 shows the fringe field effects at the edges of driving electrodes which justify the extension of problem domain beyond that bounded by the electrodes. This points to a need for 3D modelling since such effects are also present in the z-direction. The fringe field effects can be minimised in practice by effective screening which is well practiced in other areas.

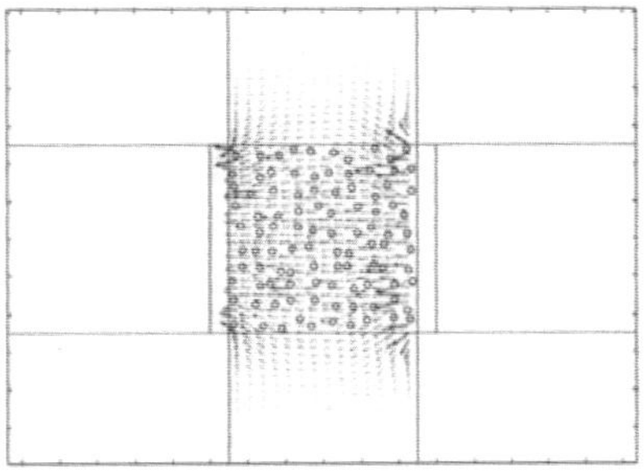

Figure 9. Fringe field effects at the edges of driving electrodes for a given distribution of particles (the arrows represent current density vectors **J**).

3.4. Effects of cholesterol level

The 2D FE modelling results presented in Figure 10 show that the impedance technique can be used to differentiate between different levels of cholesterol in the blood. However, they also point the need for accurate 3D FE modelling to refine these results and to take into account some of the important '3D effects'.

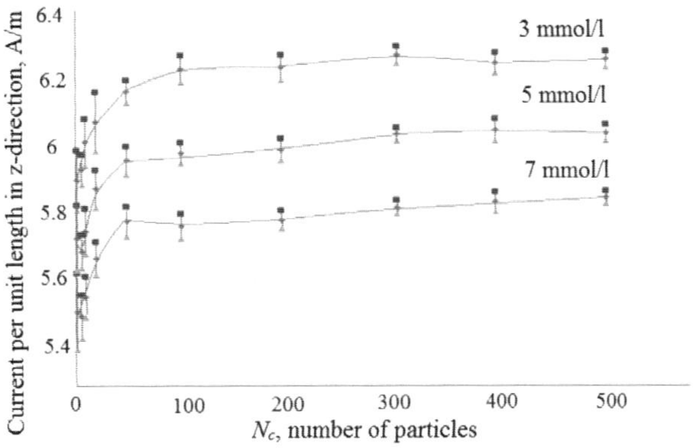

Figure 10. Electrical 'fingerprints' corresponding to different cholesterol levels (total volume of particles, V_c=const. for a given cholesterol level)

4. Conclusions

From the 2D modelling results presented this paper the following main conclusions can be made: (a) noninvasive impedance measurements can be used to differentiate between levels of Total cholesterol

Journal of Physics: Conference Series **459** (2013) 012030 doi:10.1088/1742-6596/459/1/012030

in blood, (b) this is based on distinct electrical 'fingerprints' mentioned above, (c) detail 3D FE modelling is needed to take into account some of the electrical effects which cannot be seen in 2D.

This work is on-going and the focus of the current research is to incorporate in the accurate 3D FE models major blood components and quantify their effects.

References

[1] Bairaktari E, Seferiadis K and Elisaf M *Journal of Cardiovascular Pharmacology and Therapeutics* 2005 **10** 45-54

[2] Demacker P, Hessels M, Toenhake-Dijkstra H and Baadenhuijsen H *Clinical Chemistry* 1997 **43** 663-668

[3] Baudhuina L, Hartmana S, O'Briena J, Meissnerb I, Galenc R, Warda J, Hogena S, Branuma E and McConnell J *Clinical Biochemistry* 2004, **37** 481-488

[4] Mohktar M, Ibrahim F and Ismail N, *Expert system for non-invasive classification of total cholesterol level using bioelectrical impedance* in Ibrahim F, Abu Osman N, Usman J and Kadri N (Eds.): Biomed 06, *IFMBE Proceedings* **15** 2007 63-66

[5] Holder D 2005 *Electrical impedance tomography: methods, history and applications* (Bristol: Institute of Physics)

[6] Yungal S 1996 *Electrical methods in geophysical exploration of deep sedimentary basins* (London: Chapman & Hall)

[7] Scharfetter H, Brunner P, Mayer M, Brandstätter B, and Hinghofer-Szalkay H *IEEE Transactions on Biomedical Engineering* (New York) **52** 975-82

[8] Grimnes S and Martinsen ØG 2008 *Bioimpedance and Bioelectricity Basics* (London: Academic)

[9] Tura A, Sbrignadello S, Barison S, Conti S and Pacini G 2007 *Biophys Chem* **129** 235-41

[10] Shiga T, Oshima Y, Kanai H, Hirata M, Hosoda K and Nakao K 2007 *13th International Conference on Electrical Bioimpedance and the 8th Conference on Electrical Impedance Tomography : ICEBI 2007, August 29th-September 2nd 2007, Graz, Austria* (Berlin; New York: Springer) **17** 687-690

[11] Grossi M, Lanzoni M, Pompei A, Lazzarini R, Matteuzzi D and Ricco B 2008 *Biosensors and Bioelectronics* **23** 1616-23

[12] Tender B 1985 *In vivo fluid-volume monitoring with impedance technique* PhD thesis, Department of Medical Engineering, Karolinska Institute, Stockholm, Sweden

[13] Hermann S, Brunner P, Mayer, M Brandstätter B, and Hinghofer-Szalkay H, *IEEE Transactions on Biomedical Engineering* 2005 **52** 975-982

[14] Nagai M, Komiya H, Yutaka M, Ohta T, Kasahara Y, and Ikeda Y, *Tohoku Journal of Experimental Medicine* 2008 **214** 105-112

[15] Talary M, Dewarrat F, Huber D, Falco-Jonasson L and Caduff *A Non-invasive impedance based continuous glucose monitoring system* in H. Scharfetter and R. Merv (Eds.): ICEBI 2007, *IFMBE Proceedings* 2007 **17**, 636–639

[16] Von Kaulla K, Ostendor P and Von Kaulla E *Journal of Medicine* 1975 **6** 73-87

[17] Yamakoshi K, Shimazu H, Togawa T, Fukuoka M and Ito H *IEEE Transactions on Biomedical Engineering* 1980 **27** 156-161

[18] Mannucci L, Redaeili R and Tremoli E *Thrombosis Research* 1988 **52** 143- 151

[19] Cheng D K 1989 *Field and wave electromagnetics* (Addison-Wesley)

[20] Rao S S 2005 *The finite element method in engineering* (Oxford: Butterworth-Heinemann)

[21] Online: www.comsol.com 2013

[22] Tortora G J and Derrickson B 2009 *Principles of anatomy and physiology.* (Hoboken, N.J.: Wiley) **2** 691

[23] Hirsch F G, Texter C E, Wood L A, Ballard W C, Horan F E and Wright I S *American Society of Hematology. Blood* (Philadelphia, PA) **5** 1017-1035

[24] Texter C E, Hirsch F G, Wood L A, Ballard W C, Horan F E, Wright I S, Starr D and Balandin B *American Society of Hematology. Blood* (Philadelphia, PA) **5** 1036-1048

Journal of Physics: Conference Series **459** (2013) 012031 doi:10.1088/1742-6596/459/1/012031

Implementation of perceptual aspects in a face recognition algorithm

F. Crenna[1]*, E. Zappa°, L. Bovio*, R. Testa°, M. Gasparetto° e G.B. Rossi*
* Measurement Lab, Dept. of Mechanic DIME-mec,
University of Genova, Via all'Opera Pia 15A, 16145 GENOVA, Italy.
° Dept. of Mechanical Engineering,
Politecnico di Milano, Via G. La Masa 1, 20156 MILANO, Italy.

E-mail: francesco.crenna@unige.it - emanuele.zappa@polimi.it

Abstract. Automatic face recognition is a biometric technique particularly appreciated in security applications. In fact face recognition presents the opportunity to operate at a low invasive level without the collaboration of the subjects under tests, with face images gathered either from surveillance systems or from specific cameras located in strategic points. The automatic recognition algorithms perform a measurement, on the face images, of a set of specific characteristics of the subject and provide a recognition decision based on the measurement results. Unfortunately several quantities may influence the measurement of the face geometry such as its orientation, the lighting conditions, the expression and so on, affecting the recognition rate. On the other hand human recognition of face is a very robust process far less influenced by the surrounding conditions. For this reason it may be interesting to insert perceptual aspects in an automatic facial-based recognition algorithm to improve its robustness. This paper presents a first study in this direction investigating the correlation between the results of a perception experiment and the facial geometry, estimated by means of the position of a set of repere points.

1. Introduction

Biometric techniques are more and more diffused in security applications. Among them face recognition is particularly appreciated since it does not require subject collaboration and the recognition process can take place unbeknown to the subject [1].

Applications regards both authentication and recognition. The former is just a confirmation of a declared identity of a subject, typical applications regards access control where, after and identity declaration, the subject undergo a biometric verification of the identity. In this case the task is a one to one verification. Recognition is a rather more complex task, since there is no a priori knowledge of the expected identity. In fact the goal is to verify if a subject is or not classified in a reference data base, so in this case is a one to many task. Of course in identification task, subject collaboration is generally required and several biometrics techniques find their application, such as voice recognition, retina imaging and so on, while in recognition problems the subject under test is often just moving in the field of view of a camera at the station or at the desk counter, so no collaboration is possible and low

[1] To whom any correspondence should be addressed.

invasivity is required. Such cases are the peculiar application of face recognition, so in the following we will refer to recognition problems, where there will a test image of the subject's face and a reference data base of face images of a set of known subjects.
Face images in the reference data base are captured under controlled conditions, i.e. good illumination, controlled distance and orientation to the camera, and with the subject collaboration, i.e. frontal face images with neutral expression. Face image captured in the field suffers for several unfavorable conditions due to: the imaging device (i.e. poor quality cameras, low resolution, noise), the surrounding environment (i.e. illumination, shadows, camera to subject distance), subject collaboration (i.e. face to camera orientation, non-neutral expression, camouflage).
These are just some example of the possible quantities influencing the face image captured in the field, so generally the situation is rather complex since the task requires the comparison of a reference face with a test captured under non controllable conditions.
This is the major cause of failure for automatic recognition algorithms, as well known in literature [2, 6]. In order to evaluate the recognition robustness ad hoc face images data bases are available and contests are performed to identify the most performing algorithm.
On the other hand human face recognition is really robust, since humans are able to recognize a subject even when just a little portion of face is available or under poor illumination [8-10].
In such a scenario two possible strategies to improve automatic face recognition performances emerge. One considers the possibility to improve the face measurement process at the base of the recognition decision, the other focuses on human perception of similarities among faces, to find out the robust perception mechanisms that could be implemented to improve the automatic face recognition algorithms.
This paper deals with this opportunity and, on the base of a measurement of perceptual similarity of modified faces, focuses on their consequences on a recognition method based on specific facial landmark points.

In particular the goal of this work, although to be considered as a feasibility study, is to identify which geometric entities (distances between features, areas of face regions, angles, ratios between distances…) are more relevant in human perception of similarity/dissimilarity in order to assign to these entities a higher importance in automatic recognition techniques. Since human capability to identify people relying on the facial image is very high, it is expected that trying to mimic some aspects of this recognition process might improve the reliability of automatic recognition systems.

The paper will briefly introduce the basics of the perceptual similarity measurement, some details regarding the algorithm used to measure the facial geometric characteristics, then it will present some experimental results with a set of artificially modified faces and it will consider the relationship between the position of repere points and their effect on perceived similarity identifying the most significant repere points. Final results will be discussed with a look to future possible applications.

2. Perceptual similarity

The direct measurement of perceptual quantities is a rather new field with a set of possible applications that is getting wider, consider for example social science, customer satisfaction, quality perception. After a long period of indifference due to the conviction that such kind of quantities could not be treated at the same scientific level than the traditional physical quantities, the theory has consolidated new, and well established approaches, new experimental approaches have been developed and their general consideration increased obtaining the recognition of the scientific community [3].
In this specific application the interest is the perceived similarity among a set of faces and the measurement has been designed to obtain reliable robust results with a metrological validity.

Journal of Physics: Conference Series **459** (2013) 012031 doi:10.1088/1742-6596/459/1/012031

The measurement procedure consists of two independent tests one focused on an interval scale the other on a ratio scale. The complete detail of the approach is presented in [4, 5], here we will briefly provide some details that we consider important in this specific context.

2.1. *Creating the set of faces*

The set of faces to be evaluated is the central point in the design of the perceptual similarity tests. Here a set of slightly different faces of some subjects is needed and the face should undergo a controllable face modification process, in order to be able to control the modification for each subject. The controlled modification of a face image is rather complex in two dimensions, while it is possible on a three dimensional model of the head. So three dimensional head models were realized starting from the real face images of four subjects, then they underwent a controlled modification process, by acting on specific face characteristics such as mouth to chin distance or forehead height. Five modification for each model were introduced, and from them we obtained a set of 6 frontal images, as presented in figure 1.

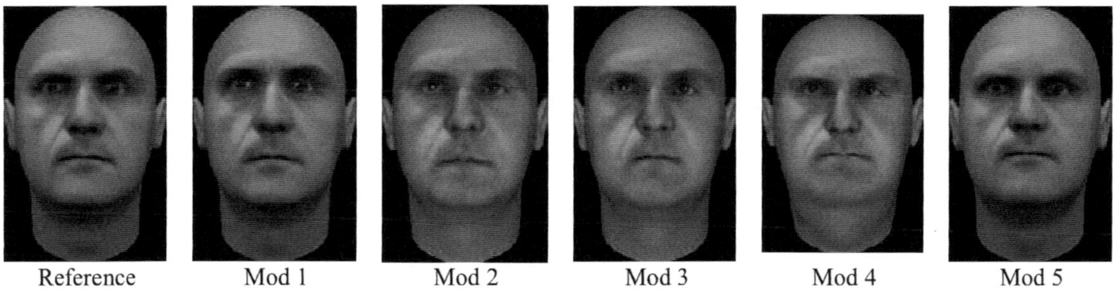

Figure 1. Reference model and related face modifications

2.2. *Perception tests results*

The set of faces including 6 images for 4 different subjects was proposed to a jury of 12 users, to evaluate the perceived similarity among the modifications of each subject. Results from the two test methods were compared to obtain a validated measure as presented in figure 2. For two face modifications (nr 3 and 5) the validation procedure was not successful (incoherent results from the two tests) and they were excluded from further evaluation. So we concluded that modification type 4 deeply affected perceived similarity while both number 1 and 2 had a slight reduction effect. Correlation coefficient between the results of the two tests, for these 3 face modification, is above 0.95, confirming the validation procedure.

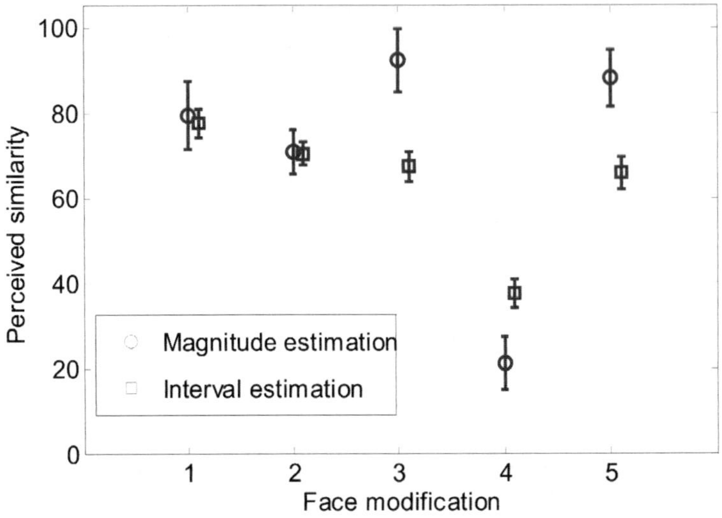

Figure 2. Perceived similarity estimated with "Magnitude estimation" and "Interval estimation.

3. Repere points identification

Several face recognition algorithms are based on the measurement of biometric features and subject identification is undertaken by evaluating geometric discrepancies between facial masks [14]. The purpose of this work is to analyze facial geometric information and to compare them with the results of perception tests with the goal to identify which facial characteristics are more influents/relevant for the perception of similarity/dissimilarity in human recognition. The final goal is to introduce a weighting of the facial geometric details in recognition algorithm [11] in order to try to emulate the human recognition mechanism in machine-based recognition.

Repere points marked on facial images are used to describe face biometrics. In particular, according to [12, 13], 58 points has been used to represent facial geometry.

These 'repere' points, also known as landmarks, are detected on face images through features extraction algorithms, which permit to select most relevant biometric information. In literature it is possible to find many different codes that permit to automatically extract position of relevant landmarks [15]. One of the most popular algorithm is the Active Appearance Model (AAM) [17], a model based code [16], which permits to create a deformable model that allows to detect features from any image of the same type of those used to create the deformable model. The AAM technique uses a statistical approach based on the processing of a training set of images where biometric features are manually outlined to build the model that will be used to automatically detect the features in the following pictures. A principal component analysis of the training images collection generates a set of parameters which represents geometry and texture variation. The resultant model is a compact representation of the training images variability. Once the model is created, it is iteratively matched as well as possible to the new analyzing image and sought features are detected.

The use of the AAM imply three steps:

1) The model creation: the 58 points could be manually labeled on various facial images and this information is analyzed with a statistical approach and the AAM model is created.
2) The fitting of the model to a new image permit to automatically detect the 58 set of landmark.
3) The extraction of coordinates of the points of interest from the matched model.

A training set consisting of 20 face images from 6 different subjects was used in this work to create the model for the AAM technique. Then the model was applied to the same set of 24 faces that was previously evaluated by perception tests (these images are different from the ones used in the AAM training). In this way a quantitative and user-independent estimation of the features position is obtained; this geometric information can be used as a comparison with the results of perception tests with the goal to identify which geometric characteristics of the face are relevant for the perception of similarity/dissimilarity in human beings.

4. Perceptual similarity and face geometry: results and discussion

In this section the relationship between the modifications of the repere points positions in the test faces, and their perceived similarity is analysed. The perceptual results shown in section 2 will be used here as a reference, while the facial geometric information analysed with the technique discussed in section 3 will be considered in order to detect which geometric details modification correspond to the most relevant perceptual variation.

4.1. Quantification of facial features geometric modifications

As explained in section 3, a set of 58 repere points is used to represent the facial geometry, as represented in figure 3.
Analysis of the modifications effect on facial geometry has been based on the observation of 11 geometric indexes, listed in Table 1. In Figure 3 the geometric entities described in Table 1 and their related Center of Gravity (CoG) are shown.

Table 1. List of the biometric indexes.

Index #	**Biometric quantity**	**Reference in figure 3**
1	Eyes distance	A
2	Nose-mouth distance	B
3	Mouth heigth	C
4	Mouth width	D
5	Eyes-mouth distance	E
6	Eyes area (sum of the 2 eyes)	
7	Mouth area	
8	Eyebrows-Face CoG distance	F
9	Average jaw width	G
10	Jaw height	H
11	Mouth-chin vertical distance	I

All the indexes listed in table 1 were calculated for each of the 4 persons considered in the perception test and for each of the 6 images belonging to every person (reference image plus 5 modifications).
The difference between the index evaluated on the reference image and the same index evaluated on each modification is then computed, obtaining the index discrepancies. Finally the mean and standard deviation of the discrepancies on the four persons is computed for each modification. The mean discrepancies will be used in the following of the work as indexes of the effect of face modifications on the geometric changes of the facial biometrics.

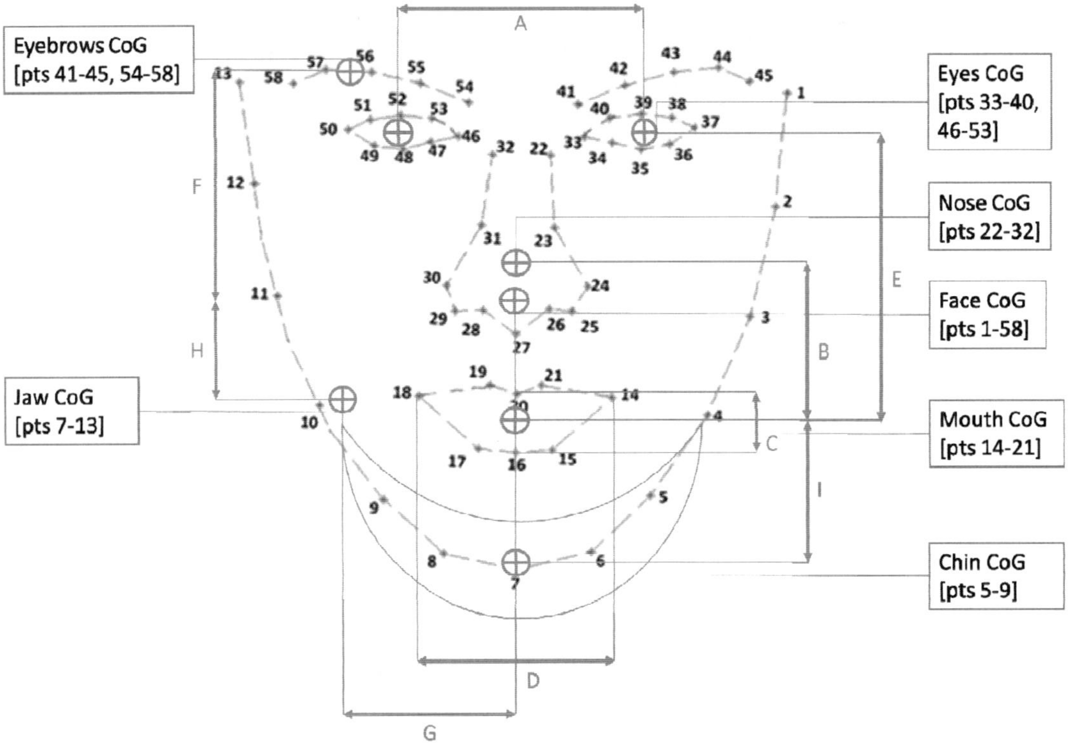

Figure 3. The 58 facial landmarks used in the analysis and the references to the 11 indexes described in table 1.

The obtained geometric indexes show a variability among the four faces with a standard deviation of the means ranging between 0.15 and 2.5 pixel with a conversion factor of 0.8 mm/pixel.
A first analysis can be carried out on the variation of the biometric indexes. Among all the 11 calculated biometric quantities (listed in table 1), the most interesting results, observed for the 4th modification in terms of appreciable mean discrepancy, can be found in the analysis of indexes number 9, 10 and 11 (average jaw width, jaw height and mouth-chin distance respectively). In fact, biometrics indexes 9, 10 and 11 are highly influenced by 4th modification in all 4 individuals while they are not influenced in an appreciable way by other modifications.
Results of this analysis are shown in figure 4, for the three parameters 9, 10 and 11 respectively.
In particular, observing the variation of index no.9, the 4th modification induces an evident widening of the jaw producing an average enhancement of 8.4% of the distance between reciprocal jaw landmarks (i.e. pairs of points 1-13, 2-12, 3-11 … of Figure 3) whereas, observing the mean variation induced by other modifications, it is possible to state that index 9 is affected only by the 4th modification. Very similar results can be found analyzing the index no. 11, which represents the mouth-chin vertical distance: the 4th modification strongly impacts on it, with a change of 8.2% of this parameter, while the effect of the other face modifications is negligible. The 4th modification also affects jaw height variations (index no. 10), although with a less evident effect.
We can assess that the 4th modification produces a significant geometric alteration only of the lower part of the face and correspondingly produces a relevant reduction in the perceived similarity results (see Figure 2). This demonstrates that the geometric changes of the lower part of the face play a role in the perceptual similarity and therefore should be carefully considered in automatic face-based recognition systems.

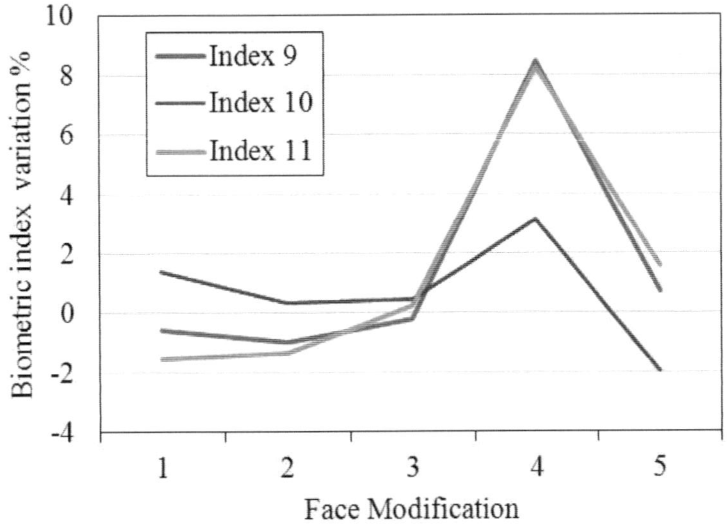

Figure 4. Mean values for % variation of biometric indexes 9-10-11

4.2. Comments

A preliminary analysis of the results shows that not all the considered biometric variations have the same effect on perception. In particular a high effect of face widening on the human perception was found.

However, these results are based on a limited number of facial modifications performed on few individuals; it seems therefore reasonable to expect better and more detailed information by studying a larger individual sample and a wider set of face modifications. The goal of this work was mainly to develop a feasibility study on the possibility to insert perceptual aspects in an automatic facial-based recognition algorithm to improve its robustness and the obtained results, although preliminary, are encouraging.

The outcomes of this work will be useful to further develop feature-based automatic algorithms for facial identification (see for example [12]). In this type of algorithms the recognition decision is based on the relative position of a set of facial feature points and in the evaluation of the similarity among faces, the most stable feature points are considered with a larger weight than the others. Since human capability to identify people relying on the facial image is very high, introducing a sort of perception weight in the recognition process it is expected to increase the recognition performance and robustness against negative influence factors such as expression variations.

5. Concluding remarks

Human recognition of face is a very robust process, for this reason in this work the possibility to insert perceptual aspects in an automatic facial-based recognition algorithm was studied, with the goal to improve the robustness of the automatic facial recognition techniques. The correlation between the results of a perception experiment and the facial geometry modifications was analysed and discussed, showing the presence of correlation between modification of some facial biometrics and human perceived dissimilarity. An extended analysis based on a larger statistical sample would permit to increase the results goodness and to identify the effect of a larger number of biometric modifications on the human recognition.

6. References

[1] Stan Z. Li, Anil K. Jain eds *Handbook of Face Recognition*, Springer-Verlag, 2004

[2] F. Crenna, G. B. Rossi, L. Bovio, *Probabilistic characterization of face measurement*, in F. Pavese, M. Bar et al eds, Advanced mathematical and Computational Tools in Metrology and Testing IX- Series on Advances in Mathematics for Applied science vol. 84, 2012, pp 90-101, ISBN 978-981-4397-94-0

[3] Berglund B, Rossi G.B., Townsend J.T., Pendrill L.R., eds. *Measurement with persons*, Psychology Press - Taylor and Francis, 2012

[4] F. Crenna, G.B. Rossi, and L. Bovio, *Measurement of the perceived similarity in face recognition*, in proc. of the XX IMEKO World Congress - Metrology for Green Growth, September 9–14, 2012, Busan, Republic of Korea, ISBN 978-89-950000-5-2-95400

[5] F. Crenna, G. B. Rossi, L. Bovio , *Loudness measurement by robust magnitude estimation*, proc. of the 14th IMEKO TC1-TC7-TC13 joint Symposium, "Intelligent Quality Measurement - Theory, Education and Training ", JENA, Germany, 2011

[6] M. J. Wenger, & J. T. Townsend, (Eds.). *Computational, geometric, and process perspectives on facial cognition: Contexts and challenges*. Scientific psychology series, Erlbaum 2001

[7] Townsend, J. T., Aisbett, J., Assadi, A., & Busemeyer, J. *General recognition theory and methodology for dimensional independence on simple cognitive manifolds*. In H. Colonius & E. N. Dzhafarov (Eds.), Measurement and Representation of Sensations: Recent Progress in Psychophysical Theory. (2006).

[8] Alice J. O'Toole, *Psychological and Neural Perspectives on Human Face Recognition*, in Li, Jain eds 'Handbook of Face Recognition', pp 349-370, Springer-Verlag, 2004

[9] P. Sinha, B.J. Balas, Y. Ostrovsky, *Face recognition by humans*, in Zhao, Chelappa, Face Processing, pp 257-292, Elsevier, 2006

[10] A. O'Toole, F. Jiang, D. Roark, H. Abdi, *Predicting human performance for face recognition*, in Zhao, Chelappa, Face Processing, pp 293-320, Elsevier, 2006

[11] Zappa, Emanuele, and Paolo Mazzoleni. *Reliability of personal identification base on optical 3D measurement of a few facial landmarks,* Procedia Computer Science 1.1 (2010): 2769-2777.

[12] Zappa, Emanuele, Paolo Mazzoleni, and Yumei Hai. *Stereoscopy based 3D face recognition system,* Procedia Computer Science 1.1 (2010): 2521-2528.

[13] Betta, G., et al. *Face-based recognition techniques: proposals for the metrological characterization of global and feature-based approaches*, Measurement Science and Technology 22.12 (2011): 124005.

[14] Sala, Remo, Emauele Zappa, and Alfredo Cigada. *Personal identification through 3D biometric measurements based on stereoscopic image pairs*, Measurement Systems for Homeland Security, Contraband Detection and Personal Safety, Proceedings of the 2006 IEEE International Workshop on. IEEE, 2006.

[15] Jafri, R. and Arabnia, H. R., 2009. *A Survey of Face Recognition Techniques*. Journal of Information Processing Systems. 5, 2, 41--68.

[16] Heisele, B., Ho, P., Wu, J., Poggio, T.: *Face recognition: component-based versus global approach*, Computer vision and image understanding, vol. 91, no. 1–2, pp. 6–21, (2003)

[17] Cootes, Timothy F., Gareth J. Edwards, and Christopher J. Taylor. *Active appearance models*, Pattern Analysis and Machine Intelligence, IEEE Transactions on 23.6 (2001): 681-685.

Dynamic Investigation Test-rig on hAptics (DITA)

F Cannella[1], L Scalise[2], E Olivieri[1], M Memeo[2] and D G Caldwell[1]

[1] Department of Advanced Robotics, Istituto Italiano di Tecnologia, via Morego, 30, 16163 Genova, ITALY

[2] Dipartimento di Ingegneria Industriale e Scienze Matematiche, Università Politecnica delle Marche, Via Brecce Bianche, 60131, Ancona, ITALY

E-mail: ferdinando.cannella@iit.it

Abstract. Research on tactile sensitivity has been conducted since the last century and many devices have been proposed to study in detail this sense through experimental tests. The sense of touch is essential in every-day life of human beings, but it can also play a fundamental role for the assessment of some neurological disabilities and pathologies. In fact, the level of tactile perception can provide information on the health state of the nervous system. In this paper, authors propose the design and development of a novel test apparatus, named DITA (Dynamic Investigation Test-rig on hAptics), aiming to provide the measurement of the tactile sensitivity trough the determination of the Just Noticeable Difference (JND) curve of a subject. The paper reports the solution adopted for the system design and the results obtained on the set of experiments carried out on volunteers.

1. Introduction

The importance of the sense of touch for human beings is universally known; everybody knows that all the physical interactions of humans with the environment generate touch sensations which are of utmost importance for the subject interactions. Up to now the evaluation of somatic sensation measurement devices used in screening exam, such as groove, Von Frey filaments or Semmes-Weinstein monofilaments, 2 or 3 points discriminator, tuning forks, are quite simple and are mostly operated by means of light touch, pain, vibration or warmth&cold sensation, directed to the human skin [1]. Even if widely used with positive results [2–6], they present a relevant limit: when used in clinical environment on patients, the variety of nervous system disorders or injuries can cause impaired tactile sensibility. Moreover, the literature shows the dependence of the test protocol on the results obtained, in fact different procedures lead to different conclusions about the prevalence of severe neuropathy, with differences ranging from 3.4% to 29.3% [7].

The aforementioned aspects led to the request of the design and development of a novel test apparatus to have a standard scale on which objectively evaluate the assessment of peripheral neuropathies. The high resolution of cutaneous sensation allows to monitor every small difference between measurements; by analyzing stimuli comparison the novel apparatus should be able to measure the acuity of the patient's sense of touch and to compare it with the standard cutaneous sensations levels.

From the previously reported observations, this paper founds its main scope which is to design and build a novel measurement system, that has been called DITA (Dynamic Investigation Test-Rig on hAptics) and is aiming to the measurement of tactile sensitivity.

The concept is based on known experimental tests widely used in haptic, the task is a well-known psychophysical method for eliciting responses from a person about his or her experiences of a stimulus where the experimental paradigms classify it as simple forced choice (also named as yes-no task) [8]. From these data the subject performance and the Just Noticeable Difference (JND) were determined. The JND, also known as the difference threshold, is the minimum difference in stimulation that a person can detect 50 percent of the times [9]. As a consequence, the objective sensitivity scale obtained from this apparatus will provide quantitative data to help the current peripheral neuropathy screening procedures, [10, 11]; 40 subjects have been tested measuring their tactile sensitivity.

2. Materials and methods

The DITA measurement system is reported in figure 1 and it is possible to observe how the fingers of the subject can be moved on a stimulus with or without array. The tactile peripherals (array and stimuli) are shown in figure 2; their values are listened in table 1 and table 2.

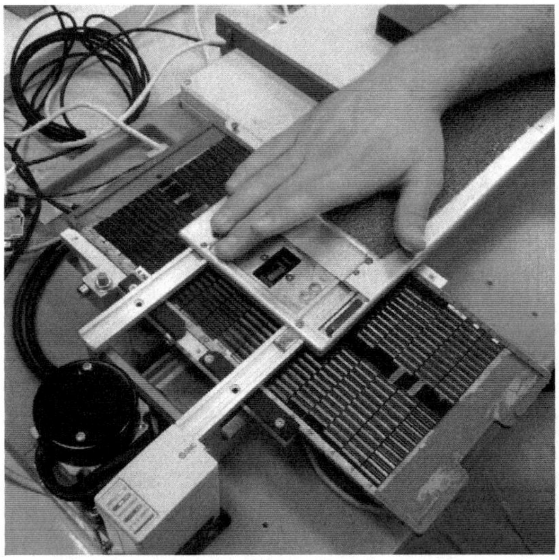

Figure 1. Dynamic Investigation Test-rig on hAptics (DITA).

Subjects' right arm rests on a metallic support where there is a non-slip surface placed (it is the blue part under the forearm); for exploring the gratings one movement along only one direction is allowed. In this work the gratings are the tactile cues and they are 17, as shown in the table 2; each of them has two elements (called stimuli) separated by a flat part, as shown in figure 2. One of the stimuli is called "reference stimuli", its wavelength is 5.09 mm and it is the same for all the gratings, the other one is called "main stimuli" and the wavelength changes depending on the grating, as shown in the table 2. The test procedure is the following: the subject slides the index fingertip along the grating beginning from the 1st stimulus and then on the 2nd stimulus. Finally the subject has to indicate which stimulus is perceived as the one with the higher spatial frequency.

In order to have a scale for determining the level of the peripheral neuropathy, 5 arrays shown in the (figure 2) are used in the experimental tests as an artificial tactile handicap. These arrays have the following pins distribution: 7x7, 9x9, 11x11, 13x13, 17x17 (respectively numbered as array number 1 up to 5) and the bare finger which corresponds to the array number 6, as shown in the table 1. A higher number of pins means that touch sensitivity is similar to the bare fingertip, while a lower number of pins means that the skin will be stimulated in fewer point than the bare fingertip thus the touch will be impaired.

Figure 2. Tested arrays with different spatial resolution, it increases from the top left to the bottom right. On the bottom of the picture a grating is also shown.

Table 1. Array list: first column is the identification number, second is the resolutions (number of pin) and the third is the interaxis distance between pins.

Number	Resolution	Pin distance (mm)
1	7x7	2.00
2	9x9	1.50
3	11x11	1.20
4	13x13	1.00
5	17x13	0.75
6	Bare finger	No pins

3. Results

The experimental test was undertaken on 40 subjects (1 was subsequently discharged), aged 24 ±5 years; all subjects were right handed, in good physical conditions and were recruited among the student population. The testing procedure is described in details in a previous paper [10]. To improve subjects' attention on the tactile sensation, they were visually and acoustically isolated from the laboratory environment.

The analysis has been conducted with two different examinations: the first is focused on subjects' answers depending on each grating (as first or second position) and the second depends on stimulus wavelength according to the different arrays used.

Figure 3 illustrates the correlation between the correct answers and the grating stimuli difference, listed in the table 2: larger is the difference, higher is the guess rate. In fact, the best performance is where the main stimulus is 3.77 mm (−1.32 mm of difference) or 6.41 mm (+1.32 mm of difference) and worst one where the main stimulus is close to the reference (±0.33 mm). This relationship can be associated to the tactile sensitivity of the subjects. The first group of gratings has negative wavelength difference (the second stimulus explored has higher spatial frequency), that seems more perceivable. A particular mention is given for the "fake" grating where the difference is 0 mm: the two stimuli have the same wavelength, as shown in the table 2, because this grating is used for determining the "neutral" value of the subjects. Normally both the answers are given with the same rate: 50%. The non respect of this condition could be due to some tactile issues.

The percentage of correct answers related to the difference between the grating stimuli represents the psychometric function: this function typology describes the relationship between the subjective responses and a parameter of a physical stimulus. In the case of plotted stimulus wavelength, the psychometric curves have a standard profile, as shown in the figure 4. When the stimulus wavelength is near the longest value (6.41 mm) or the shortest one (3.77 mm), the correct answer percentage is near to 100. In between, around 5.09 mm, there is a transition value which represents the just noticeable difference (JND). This value is the minimum level of stimulation that a person can detect 50 percent of the time [12].

Starting from these data, it was set a global threshold referred to the highest detectable percentage (at 80%) and JND for each subject. The results are evaluated depending on two different parameters: the difference between grating stimuli, as shown in the figure 3, and the stimulus wavelength, as shown in the figure 4.

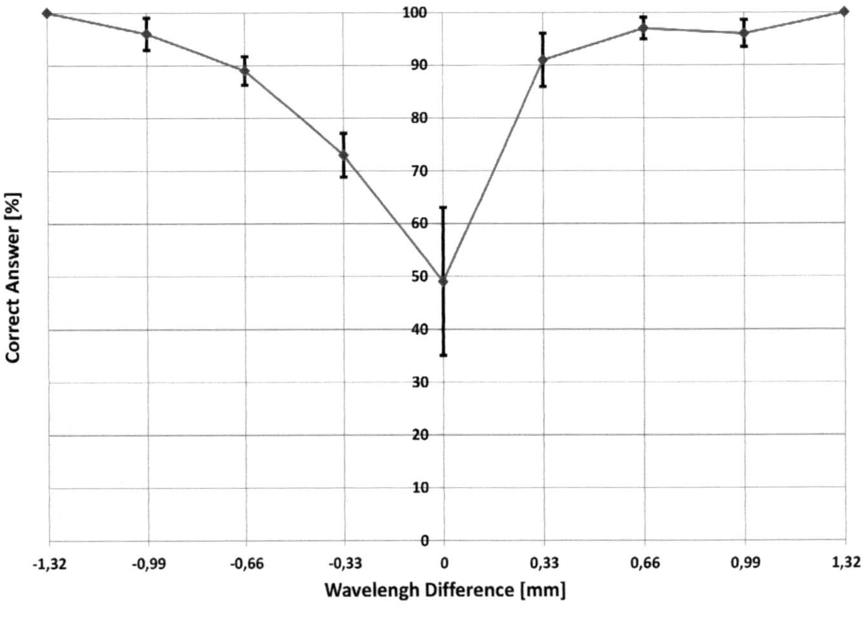

Figure 3. Correct answer depending on grating *perceptilility*, expressed as the difference between the wavelengths of the two stimuli of gratings.

The analysis conducted with the JND psychometric curves of figure 4 is different from the previous one. In this case subject classification is double because they are divided into six groups according to the different arrays and the psychometric curves are drawn for each group depending on stimulus wavelength. In statistics, the method used to investigate differences between groups and the differences variation among and between groups is the ANOVA analysis [13].

In this case the difference among groups is described with the six arrays used instead the difference between groups is dependent from the stimuli wavelength. This analysis uses the F-test with the hypothesis that the means (in this case the JND values) of these six groups are equal [13]. In JND analysis, ANOVA results show that JND values have a statistically significant difference, with accuracy of 5% (*F-test* = 4.87, $\rho = 0.001$). JND values found, for each array

Table 2. Gratings with main and reference stimuli wavelength and their difference.

Stimulus 1 (mm)	Stimulus 2 (mm)	Difference (mm)
6.41	5.09	−1.32
5.09	3.77	−1.32
6.08	5.09	−0.99
5.09	4.10	−0.99
5.75	5.09	−0.66
5.09	4.43	−0.66
5.42	5.09	−0.33
5.09	4.76	−0.33
5.09	5.09	0
5.09	5.42	0.33
4.76	5.09	0.33
5.09	5.75	0.66
4.43	5.09	0.66
5.09	6.08	0.99
4.10	5.09	0.99
5.09	6.41	1.32
3.77	5.09	1.32

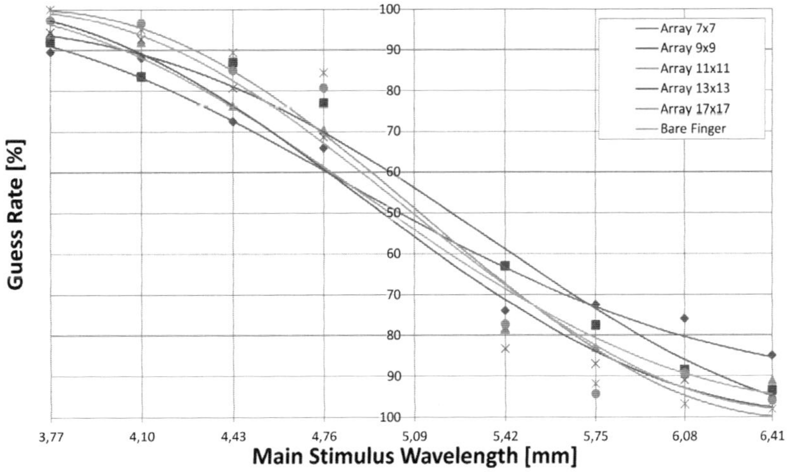

Figure 4. Psychometric curve (JND) for each array. On the x axis there is the main stimulus of the grating, in the ordinate the corresponding percentile guess rate. From 5.42 value (on x-axis), the JND values have been inverted to show the slope of the passage around the 50%.

are: *7x7* 0.49 mm, *9x9* 0.48 mm, *11x11* 0.41 mm, *13x13* 0.38 mm, *17x17* 0.34 mm, *Bare Finger* 0.33 mm.

It is also possible to use a more powerful test, the Fisher Post Hoc, in order to provide specific information on the causes of the JND difference. This test shows that the significant differences are between arrays 4, 5, 6 and 1, 2, 3. As expected, the arrays 1, 2 and 3 have worse performance than the 4, 5 and 6. No significant differences were found among bare finger and arrays 5 and 4. It is important to note that the standard deviation of JND reaches always the minimum for the array 6 (bare finger).

4. Discussion

From the aforementioned results, it is possible to derive how the accuracy of the DITA is adequate to the scope of the system and this is derivable from figure 3 and from figure 4 for the JND values; the arrays and the gratings can be distinguished (according to the analysis ANOVA).

The authors claim that a tactile sensitivity scale is determined and the subject can be compared with it in order to determine his/her own sensitivity; the overall accuracy of DITA is 9.4% (within the range 3.4-29.3% of the current medical instrumentation [7]), the main improvement is that an objective sensitivity scale is determined and then the subject test results can be stored and recorded accurately. Also, such values of accuracy are considered by the authors adequate for the measurement of the subject sensitivity. This is valuable also with reference to the well know of the scarce repeatability of the sensitivity test conducted with other methodologies.

Thus this device should be adopted to support and improve screening in diagnosis concerning the tactile perception (such as carpal tunnel syndrome, Parkinson disease, muscular dystrophy and so on).

References

[1] Brodal A *Neurological anatomy in relation to clinical medicine* Oxford Medicine Publications ISBN 9780195026948

[2] Bell-Krotoski J and Tomancik E 1987 *J Hand Surg Am* **12** 155–161

[3] Lee S, Kim H, Choi S, Park Y, Kim Y and Cho B 2003 *J. Korean Med. Sci.* **18** 103–107

[4] Keizer D, Fael D, Wierda J M and van Wijhe M 2008 *Clin J Pain* **24** 463–466

[5] Pestronk A, Florence J, Levine T, Al-Lozi M T, Lopate G, Miller T, Ramneantu I, Waheed W and Stambuk M 2004 *Neurology* **62** 461–464

[6] Keizer D, van Wijhe M, Post W J and Wierda J M 2007 *Clin J Pain* **23** 85–90

[7] McGill M, Molyneaux L, Spencer R, Heng L F and Yue D K 1999 *Diabetes Care* **22** 598–602

[8] Smith J 1969 *Journal of Verbal Learning and Verbal Behavior* **8** 633 – 636 ISSN 0022-5371 URL `http://www.sciencedirect.com/science/article/pii/S0022537169801155`

[9] Garcia-Hernandez N, Tsagarakis N and Caldwell D 2011 *Haptics, IEEE Transactions on* **4** 100–110 ISSN 1939-1412

[10] Valente M, Cannella F, Scalise L, Memeo M, Liberini P and Caldwell D 2012 *Haptics: Perception, Devices, Mobility, and Communication* (*Lecture Notes in Computer Science* vol 7282) ed Isokoski P and Springare J (Springer Berlin Heidelberg) pp 589–600 ISBN 978-3-642-31400-1 URL `http://dx.doi.org/10.1007/978-3-642-31401-8_52`

[11] Scalise L, Memeo M, Cannella F, Valente M, Caldwell D G and Tomasini E P 2012 *AIP Conference Proceedings* vol 1457 ed Tomasini E P (AIP) pp 287–293

[12] Savage C *The measurement of sensation: a critique of perceptual psychophysics* ISBN 9780520015272

[13] Mahajan B 2002 *Methods in Biostatistics* (Jaypee Brothers, Medical Publishers) ISBN 9788171795208 URL `http://books.google.it/books?id=y-F1fMHP5hgC`

Journal of Physics: Conference Series **459** (2013) 012033 doi:10.1088/1742-6596/459/1/012033

The evaluation of uncertainty in mass calibration: possible approaches in a comparison study

J A Sousa[1], A B Forbes[2], A S Ribeiro[3], P M Harris[2], F Carvalho[4], L Bacelar[4]
[1]LREC, 9000-264 Funchal, Portugal
[2]NPL, Teddington TW11 0LW, UK
[3]LNEC, 1700-066 Lisboa, Portugal
[4]TAP Air Portugal, 1704-801 Lisboa, Portugal

E-mail: jasousa@lrec.pt

Abstract. In this paper we give the results of four methods of calculating uncertainty associated with a mass calibration problem, three based on different implementations – the first and second order law of propagation of uncertainty and the Monte Carlo method – of the general methodology described by the Guide to the Expression of Uncertainty in Measurement, the fourth based on a Bayesian formulation. The nonlinearities present in the model for the calibration problem means that the first order approach can be an unreliable method for evaluating uncertainties, relative to the other three approaches.

1. Introduction

The application of the GUM (Guide to the Expression of Uncertainty in Measurement) [1] uncertainty framework to the evaluation of uncertainty is accepted as adequate in most fields of metrology, including mass metrology. The methodology has, as its starting point, an input-output model describing the measurand as a function of influence quantities and the assignment of probability distributions or probability density functions (PDFs) to the influence quantities. Once these two elements are in place, the PDF associated with the measurand, the output quantity is defined. The issue is how to determine information about this PDF in a computationally convenient way.

The GUM uncertainty framework, as usually implemented, involves two types of approximation, i) linearization, i.e., first order Taylor expansion, of the function relating the output to the inputs, and ii) an appeal to asymptotic results derived from the Central Limit Theorem to associate a Gaussian distribution to the output quantity. (The GUM also provides a methodology for associating a student *t*-distribution to the output quantity, using a scheme to determine effective degrees of freedom of the *t*-distribution. We do not consider this more general scheme in this paper.) For linear models and Gaussian inputs, the GUM uncertainty framework is exact. In all other cases, it will only provide an approximate solution. The quality of the approximation is however difficult to predict, a priori. The GUM also discusses the use of a second order Taylor expansion, which should lead to more accurate approximations. GUM Supplement 1 [2] describes the use of a Monte Carlo method (MCM) to propagate the distributions associated with the input quantities through to that associated with the

[1] To whom any correspondence should be addressed.

output quantity. The MCM does rely on approximating the function nor on asymptotic results associated with the output distribution and can be used to test the validity of the GUM approach or as the main computational tool for evaluating the uncertainty [3]. Moreover, an important advantage of MCM is that it goes beyond the evaluation of the measurement result and its associated standard uncertainty since it propagates the probability density functions instead of just the uncertainties of the input quantities. This leads to an estimate of the PDF of the output quantity, and thus any required statistic, including the measurement result, the associated standard uncertainty and coverage intervals, can be obtained from this distribution, as represented by the Monte Carlo sample. Another important advantage of MCM is its applicability regardless of the nature of the model, e.g., those incorporating strong nonlinearities. As mentioned above, the GUM uncertainty framework does in fact encompass second order methods to deal, at least partially, with nonlinearity.

The three approaches mentioned so far, the GUM first and second order and MCM, can be viewed as computational approaches that attempt to evaluate summary information about the same PDF, that associated with the output quantity, defined in terms of the input-output model and the PDFs assigned to the input quantities [4]. In a Bayesian formulation of uncertainty evaluation, the starting point is somewhat different, in that a prior PDF has also to be assigned to the measurand. However, the MCM method coincides with the Bayesian approach for a particular choice of prior and the sample produced using the MCM approach is a sample for the Bayesian posterior distribution constructed using that particular prior [5,6]. The MCM sample can be used to generate a sample from the posterior distribution constructed using another (perhaps more appropriate) prior for the measurand through the application of an extremely simple Markov chain Monte Carlo (MCMC) algorithm [7]. In this context, the GUM and MCM methods can be viewed as approximate computational approaches to Bayesian uncertainty evaluation.

2. The mass calibration model

See for example [3, section 9.10]. The model concerns the calibration of a weight *W* of mass density ρ_W against a reference weight *R* of mass density ρ_R. It is assumed that the two masses are nominally the same. The calibration is performed using a balance operating in air of density ρ_a and provides an estimate of the mass δm_R of a small weight δR, also of density ρ_R, needed to achieve a balance. Taking into account buoyancy effects, the application of Archimedes' principle leads to the following model equation.

$$m_W(1-\rho_a/\rho_W)=(m_R+\delta m_R)(1-\rho_a/\rho_R) \tag{1}$$

involving the masses m_W and m_R of W and R, respectively. It is usual to work in terms of conventional masses relating to a standard density ρ_0 = 8 000 kg/m^3 for the weight and that of ρ_{a_0} = 1.2 kg/m^3 for air, so that, for example, the conventional mass $m_{W,c}$ corresponding to m_W is given by

$$m_W(1-\rho_{a_0}/\rho_W)=m_{W,c}(1-\rho_{a_0}/\rho_0)$$

Working in terms of conventional masses, (1) becomes

$$m_{W,c}\left(\frac{1-\rho_a/\rho_W}{1-\rho_{a_0}/\rho_W}\right)=(m_{R,c}+\delta m_{R,c})\left(\frac{1-\rho_a/\rho_R}{1-\rho_{a_0}/\rho_R}\right)$$

from which we obtain (using an approximation adequate for most purposes)

$$m_{W,c} = (m_{R,c} + \delta m_{R,c})\left[1 + (\rho_a - \rho_{a_0})\left(\frac{1}{\rho_W} - \frac{1}{\rho_R}\right)\right] \tag{2}$$

The calibration problem is to determine an estimate of $m_{W,c}$, and its associated uncertainty, from knowledge of $m_{R,c}$, $\delta m_{R,c}$, ρ_a, ρ_W and ρ_R. Eq. (2) gives $m_{W,c}$ as a function $m_{W,c} = f(\boldsymbol{\xi})$ of the five parameters $\boldsymbol{\xi} = (m_{R,c}, \delta m_{R,c}, \rho_a, \rho_W, \rho_R)^{\mathrm{T}}$. Thus, Eq. (2) provides the input-output model to enable the GUM and MCM approaches to be executed.

In this paper, we compare the MCM against the first and second order propagation of uncertainties and also compare the MCM method with a Bayesian evaluation of uncertainty.

3. The different approaches used in this study for the evaluation of uncertainty

3.1. Uncertainty evaluation using the first and higher order GUM approach

If $\eta = f(\boldsymbol{\xi})$ is a function of n input parameters ξ_j, the GUM gives a general approach based on the law of propagation of uncertainty (LPU) for propagating uncertainties associated with estimates of the inputs to that associated with an estimate of the output. In its simplest form, the approach states that if each input parameter ξ_j is associated a probability distribution with mean x_j and standard deviation $u(x_j)$, then the output parameter η is associated with a distribution with mean $y = f(\mathbf{x})$, $\mathbf{x} = (x_1, \ldots, x_n)^{\mathrm{T}}$, and standard deviation

$$u^2(y) = \sum_{j=1}^{n} c_j^2 u^2(x_j), \qquad c_j = \frac{\partial f}{\partial \xi_j}(\mathbf{x}), \tag{3}$$

assuming that the ξ_j are independently distributed. The expression for the uncertainty $u^2(y)$ given in Eq. 3 is derived from a first order approximation to the function *f*. A more accurate estimate can be made by using a higher order approximation. The general approach is as follows [3][8]. If we assume that the input quantities are independently distributed, and make the further assumption that the input distributions are symmetric, there will be important simplifications to take advantage of, and the some of expansion terms will be zero. With these assumptions, a higher order estimate of η given by

$$y + \Delta y = f(\mathbf{x}) + \frac{1}{2}\sum_{i=1}^{n}\frac{\partial^2 f}{\partial \xi_i^2}u^2(x_i),$$

and its associated uncertainty is estimated by

$$u^2(y + \Delta y) =$$

$$\sum_{i=1}^{n}\left(\frac{\partial f}{\partial \xi_i}\right)^2 u^2(x_i) + \sum_{i=1}^{n-1}\sum_{j=i+1}^{n}\left(\left(\frac{\partial^2 f}{\partial \xi_i \partial \xi_j}\right)^2 + \frac{\partial f}{\partial \xi_i}\frac{\partial^3 f}{\partial \xi_i \partial \xi_j^2} + \frac{\partial f}{\partial \xi_j}\frac{\partial^3 f}{\partial \xi_j \partial \xi_i^2} + \frac{1}{2}\frac{\partial^2 f}{\partial \xi_i^2}\frac{\partial^2 f}{\partial \xi_j^2}\right)u^2(x_i)u^2(x_j)$$

$$+ \sum_{i=1}^{n}\left(\left(\frac{\partial^2 f}{\partial \xi_i^2}\right) + \frac{1}{3}\frac{\partial f}{\partial \xi_i}\frac{\partial^3 f}{\partial \xi_i^3}\right)E(\Delta \xi_i^4) - \left(\frac{1}{2}\sum_{i=1}^{n}\frac{\partial^2 f}{\partial \xi_i^2}u^2(x_i)\right)^2. \tag{4}$$

If $\Delta\xi_j$ is associated with a normal or rectangular distribution centred at 0, then $E(\Delta\xi_j^4) = 3(u^2(x_i))^2$ and $(9/5)(u^2(x_i))^2$, respectively. The calculation also requires the evaluation of $\frac{\partial f}{\partial \xi_i}$, $\frac{\partial^2 f}{\partial \xi_i \partial \xi_i}$, and $\frac{\partial^3 f}{\partial \xi_i \partial \xi_j^2}$, usually the main deterrent in implementing this approach. For the mass calibration problem with the data as in Table 1, the only nonzero higher order partial derivatives that will be evaluated as nonzero (taking into account that $x_3 = \rho_{a0}$ and $x_4 = x_5$) are

$$\frac{\partial^2 f}{\partial \xi_3 \partial \xi_4} = -(\xi_1 + \xi_2)/\xi_4^2, \qquad \text{and} \qquad \frac{\partial^2 f}{\partial \xi_3 \partial \xi_5} = (\xi_1 + \xi_2)/\xi_5^2.$$

The second order estimate of η is the same as the first order estimate, since $\Delta y = 0$, but its associated uncertainty is given by

$$u^2(y + \Delta y) = u^2(x_1) + u^2(x_2) + \left(\frac{\partial^2 f}{\partial \xi_3 \partial \xi_4}\right)^2 u^2(x_3)u^2(x_4) + \left(\frac{\partial^2 f}{\partial \xi_3 \partial \xi_5}\right)^2 u^2(x_3)u^2(x_5) \quad (5)$$

3.2. Uncertainty evaluation using MCM

The Monte Carlo Method (MCM) [1][2] is a way of evaluated the uncertainty associated with the estimate $m_{W,c}$ derived from Eq. 2 without linearising approximations or, for the calculation of coverage intervals, assumptions of normality. For many problems, the method is straightforward to implement. If the input variable ξ_j is associated with a distribution with probability density function $p(\xi_j)$, then for $q = 1,\ldots.M$, draws $\boldsymbol{\xi}_q = (\xi_{1,q},\ldots,\xi_{n,q})^{\mathrm{T}}$ are made from the input distributions. Then $y_q = f(\boldsymbol{\xi}_q)$ are draws from the distribution associated with $\eta = f(\boldsymbol{\xi})$. Means, standard deviations and coverage intervals can be estimated easily from the corresponding sample statistics derived from $\{y_q, q = 1,\ldots,M\}$.

3.3. Uncertainty evaluation using a Bayesian approach

The Bayesian methodology defines a posterior distribution for $m_{W,c}$ to be determined on the basis of the measurement of $\delta m_{R,c}$ and the prior information on the other parameters. The noninformative prior $p(m_{W,c}) \propto 1$ was chosen for $m_{W,c}$. The posterior distribution is related to the distribution $p(\eta)$ for $\eta = f(\boldsymbol{\xi})$ by a factor that depends on the partial derivative of f with respect to $\delta m_{R,c}$. The MCMC approach follows that described in [7] and starts with the draws $y_q = f(\boldsymbol{\xi}_q)$ but also requires $d_q = \frac{\partial f}{\partial \delta m_{R,c}}(\boldsymbol{\xi}_q)$ to be recorded. The aim of the MCMC algorithm is to modify the MCM sample in a simple sequential accept-reject scheme involving d_q, so that the modified sample represents a sample from the Bayesian posterior distribution for $m_{W,c}$. Means, standard deviations, etc., can be estimated from the sample statistics as in the MCM approach.

4. Discussion of results

As an example calculation, we assume that the information about $m_{R,c}$ and $\delta m_{R,c}$ is taken from calibration certificates and for which it is appropriate to assign Gaussian distributions. We assume that the density information is given in terms of upper and lower limits for which it is appropriate to assign rectangular distributions. This information is summarised in Table 1. The uncertainty associated with the density estimate ρ_W is perhaps unrealistically high but it is useful to be able to illustrate how nonlinearities in the model affect the uncertainty evaluations.

Table 1. Distributions associated with the input quantities with $m_{W,c}$.

ξ_j	Distribution	Mean	Standard uncertainty
$m_{R,c}$	Gaussian	100 000.000 mg	0.050 mg
$\delta m_{R,c}$	Gaussian	1.234 mg	0.020 mg
ρ_a	Rectangular	1.20 kg/m^3	$(1/\sqrt{3})\times 0.10$ kg/m^3
ρ_W	Rectangular	8.0×10^3 kg/m^3	$(1/\sqrt{3})\times 1.0\times 10^3$ kg/m^3
ρ_R	Rectangular	8.00×10^3 kg/m^3	$(1/\sqrt{3})\times 0.05\times 10^3$ kg/m^3

For the data in Table 1, the estimate of the deviation of $m_{W,c} - m_0$ from a nominal mass of 100 g evaluated from Eq. 2 is $\delta\hat{m} = 1.234\,0$ mg.

The fact that the sensitivity coefficients associated with the density quantities are zero means that, to first order, the uncertainties associated with these quantities do not contribute to the uncertainty associated with the estimate of $m_{W,c}$ so that $u(\delta\hat{m}) = (0.050^2 + 0.020^2)^{1/2}$ mg = 0.053 9 mg. However, for the second order approach, with the same example data, the measurement uncertainty yields $u(y+\Delta y) =$ 0.075 mg, suggesting a clear underestimation entailed by the simpler implementation.

Compared to the second order approach (and even the GUM first order approach), the MCM method is straightforward to implement. Fig. 1 shows the normalised histogram of 100,000 MCM samples generated using the input distributions specified in Tab. 1 and the input-output model given in Eq. (2). These samples were used, along with the partial derivatives $d_q = \frac{\partial f}{\partial \delta m_{R,c}}(\xi_q)$, as the input to the MCMC algorithm to produce a sample from the Bayesian posterior distribution for $m_{W,c}$ using a noninformative prior $p(m_{W,c}) \propto 1$. The degree to which the MCMC algorithm modifies the MCM sample depends on the extent that d_q varies. For the mass calibration example, these partial derivatives vary by less than 5 parts in 10^6, which means that the MCMC sample is essentially the same as the MCM sample. In fact, all of the 100,000 of the MCM samples were accepted so that the two samples are identical. For larger MCM samples, some would be rejected but would not introduce any meaningful differences. The good agreement between the two approaches is due to the fact that the dependence of the measurement equation on $\delta m_{R,c}$ is almost linear. For such cases, the MCM approach is an extremely effective approach for sampling from the Bayesian posterior distribution [6,7].

The results of the four methods of evaluating an estimate of $m_{W,c} - m_0$ are given in Table 2. The first three methods all attempt to provide summary information about the same probability distribution, namely the output distribution associated with Eq. 2. The differences are in the level of approximation, in the case of the GUM methods, a first or second order approximation, in the case of MCM, the approximation of a continuous distribution by a discrete sample. As the number of MCM trials increases, the discrete approximation can becomes more exact. Due to nonlinearities in the model, the first order GUM method significantly underestimates the standard deviation of the output distribution. (For a smaller uncertainty associated with the density quantity ρ_W, the 1st order GUM gives a more accurate estimate of the uncertainty.) The GUM 2nd order method gives satisfactory estimates and can be implemented for the mass calibration problem without much complicated calculation. For a general model, there can be a considerable (or prohibitive) amount of work involved.

As indicated above, the MCM and MCMC approaches give identical results. Fig. 1 also shows the Gaussian distributions determined by the GUM 1st and 2nd order methods. It is noted that the GUM 2nd order distribution looks a very good representation of the frequency distribution derived from the Monte Carlo samples. This can be explained by the fact that the dominant influence quantities $m_{R,c}$ and $\delta m_{R,c}$ are associated with Gaussian distributions and these quantities appear linearly in the measurement equation, Eq. 2.

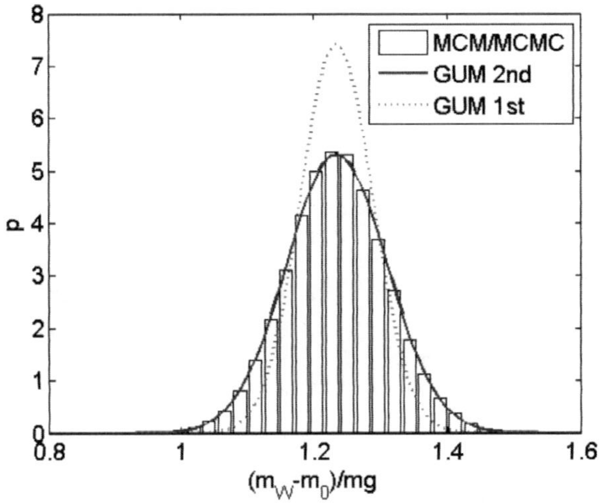

Figure 1. Frequency histogram for MCM compared with the Gaussian distributions determined by the GUM 1st and 2nd order approaches.

Table 2. Parameter estimates and associated uncertainties associated with the mass calibration problem for four methods of evaluation.

Method	$\delta\hat{m}$ /mg	$u(\delta\hat{m})$ /mg
GUM(1st order)	1.234	0.054
MCM	1.234	0.075
GUM(2nd order)	1.234	0.075
Bayes	1.234	0.075

5. Conclusions

Four computational approaches have been applied to evaluating the uncertainty associated with a mass comparison. Three, the GUM 1st order, GUM 2nd order and the Monte Carlo Method (MCM) all involve the same distribution determined by the measurement equation and the assignment of distributions to the influence quantities. The differences between them therefore reflect the degrees of approximation implicit in the computational approaches. The GUM 1st and 2nd order approaches involve truncation approximations and assign Gaussians of the basis of the estimated first and second moments of the distribution. The MCM method is approximate in the sense that only a finite sample is constructed. The MCM method and GUM 2nd order method are in close agreement for the calculated example.

The fourth method involved a Bayesian formulation of the mass comparison measurement and necessarily involves the assignment of a prior distribution for the quantity of interest. If a noninformative prior $p(m_{W,c}) \propto 1$ is chosen, the Bayesian posterior distribution is essentially the same distribution from which the MCM determines a sample. This means that the MCM sample can also be regarded as a sample from the Bayesian posterior corresponding to the assigned prior, as confirmed by the application of the Markov chain Monte Carlo (MCMC) scheme. Thus, three approaches give essentially the same results: GUM 2nd order, MCM and the Bayesian approach. The GUM 1st order approach is not recommended for the mass comparison measurement problem. Of the three approaches, the MCM method is the easiest to implement. The MCMC scheme implementing a Bayesian approach requires a modest extension of the MCM approach.

6. References

[1] JCGM 100:2008 Evaluation of measurement data – Guide to the expression of uncertainty in measurement (GUM 1995 with minor corrections, Joint Committee for Guides in Metrology).

[2] JCGM 101:2008 Evaluation of measurement data – Supplement 1 to the "Guide to the expression of uncertainty in measurement" – Propagation of distributions using a Monte Carlo method (Joint Committee for Guides in Metrology).

[3] Cox M G and Harris P M 2006 Software Support for Metrology Best Practice Guide N° 6, Uncertainty Evaluation. NPL Report DEM-ES-011. National Physical Laboratory, UK.

[4] Forbes A B 2012 Approaches to evaluating measurement uncertainty, *International Journal of Metrology and Quality Engineering*, **3**:71-77.

[5] Elster C, Wöger W and Cox M G 2007 Draft GUM supplement 1 and Bayesian analysis, *Metrologia*, **44**:L31:L32.

[6] Forbes A B and Sousa J A 2011 The GUM, Bayesian inference and the observation and measurement equations, *Measurement*, **44**:1422-1435.

[7] Forbes A B 2012 An MCMC algorithm based on GUM Supplement 1 for uncertainty evaluation, *Measurement*, **45**:1188-1199.

[8] Nielsen, L 2002 Private communication.

Uncertainty in airflow field parameters in a study of shock waves on flat plate in transonic wind tunnel

M L C C Reis[1,3], J B P Falcao Filho[2] and M C C Araújo[3]

[1,2] Institute of Aeronautics and Space
Praça Mal. Eduardo Gomes, 50, CEP 12228-904, São José dos Campos, SP, Brasil
[3] University of Taubate
Av. 9 de Julho, 199, CEP 12020-200, Taubate, SP, Brasil

marialuisamlccr@iae.cta.br

Abstract. When flat plates are submitted to wind tunnel tests and the upstream Mach number is greater than 1.0 this gives rise to shock and expansion waves. This study experimentally evaluates the airflow field parameters ahead of and behind the shock region and compares the data obtained in the tests with the expected theoretical results. For the case of expansion waves only theoretical values are supplied in this preliminary part of the study. The drag and lift forces acting on the model being tested are also estimated from the pressure values measured on the plate surface. Data reduction includes evaluation of the uncertainties associated with the measured quantities, as well as their propagation to the output quantities, employing the Monte Carlo method. Numerical simulation of the phenomena using Computational Fluid Dynamics is also done. Data obtained analytically, experimentally and numerically are compared.

1. Introduction

Shock waves occur in nature when an object moves faster than the speed of sound and when the presence of the object causes an abrupt decrease in the airflow area. If the airflow area increases, a different phenomenon is observed: an expansion fan [1,2]. Shock and expansion waves cause changes in the gas properties and this is the aim of the present study: to experimentally evaluate the conditions associated with the shock and expansion caused by the presence of the flat plate in the airflow. Aerodynamic forces acting on the plate are also calculated. The uncertainties in the estimated flow field parameters are obtained using the Monte Carlo method. The mean value and standard deviation of measured parameters are computed and this information is used to determine the probability density function for output quantities [3]. The study of flat plates is important because they have typical characteristics of all supersonic airfoils. The inclined flat plate is the simplest case of lift airfoil. The tests were carried out in the TTP pilot transonic wind tunnel facility. In the first part of the test campaign, the flat plate was positioned at an angle of attack equal to -4°. The angle between the direction of the incident airflow and the direction of the plate is called angle of attack α.

2. The TTP pilot transonic wind tunnel

The wind tunnel TTP is located at the Aerodynamics Division of the Institute of Aeronautics and Space, IAE/ALA/TTP (figure 1a). The tunnel is continuously driven by an 830 kW main axial compressor and can also be operated intermittently by means of an injection system, which supplies airflow for around 30 seconds. It is a variable-pressure wind tunnel with control capability to independently vary Mach number, stagnation pressure, stagnation temperature and humidity. The test section is 0.25 m high and 0.30 m wide and has longitudinally slotted walls to favor the uniformity of the airflow. At TTP, the Mach number range can vary from 0.20 to 1.30. Some configurations can be changed in order to allow better flow control such as reentry flaps position and the rate of forced mass extraction, determined by the Plenum Evacuation system, PES. A schematic picture of the aerodynamic circuit is presented in figure 1b, highlighting its main parts: the axial compressor, cooler, stilling chamber, plenum chamber, high velocity diffuser and corners (numbered from 1 to 4). Arrows indicate the direction of the airflow. The test section, which receives the test article, is located inside the plenum chamber.

a) The circuit.

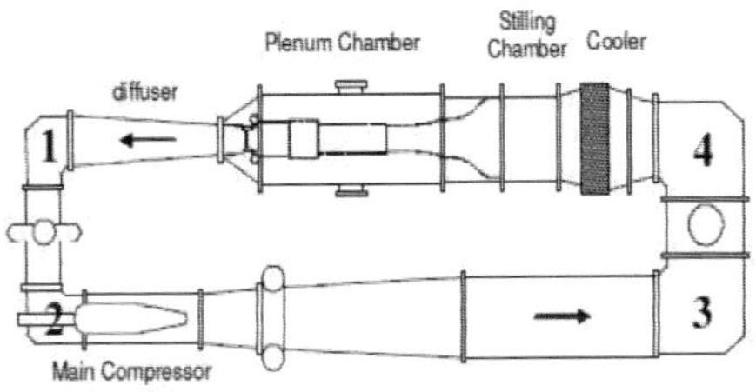

b) The schematic diagram.

Figure 1. The TTP wind tunnel.

3. The flat plate

Airfoils designed for supersonic flows are slender and have a sharp trailing edge. An extreme example of slender airflow is the flat plate. To represent the flat plate a wedge-like model is used (figure 2). The wedge is 0.15 m long, 0.12 m wide and 0.007 m thick. Its superior surface has 15 pressure taps connected to tubes which transmit the static pressure of the airflow to the instrumentation (figure 3). The diameter of the pressure tap holes is 0.5 mm. Only the six anterior central taps are included in the data reduction. The plate is fixed to the wind tunnel test section through a sting support (figure 4).

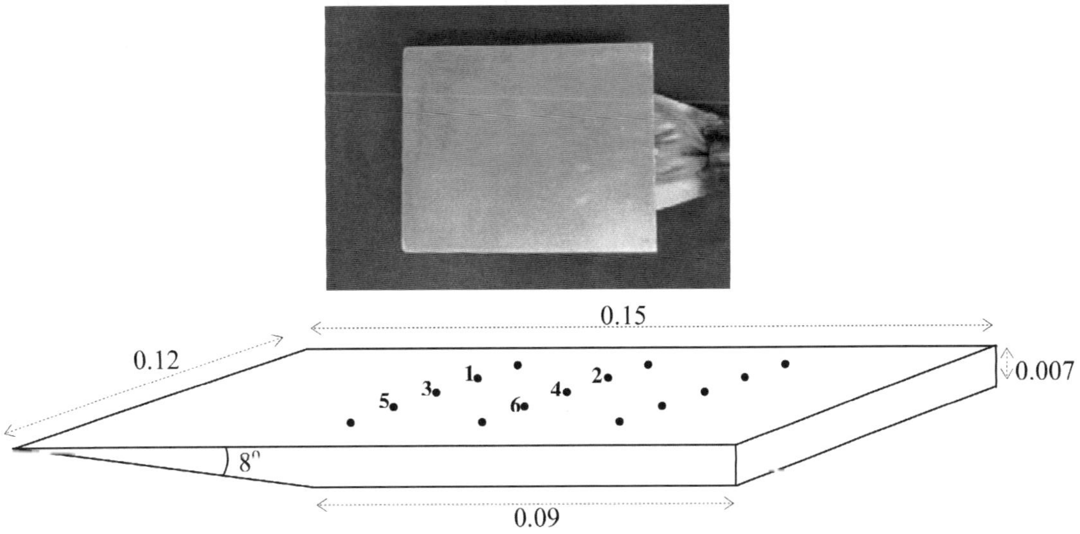

Figure 2. The wedge model picture and schematic. Unit: meter.

a) Pressure taps on the surface.

b) Flat plate being prepared.

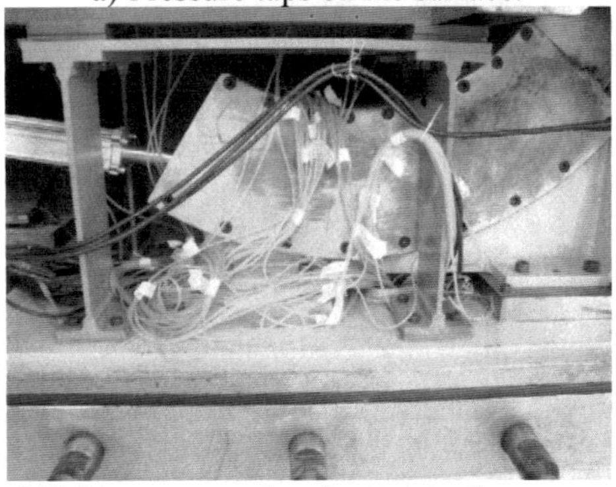

c) Tubing connecting the taps to the sensor.

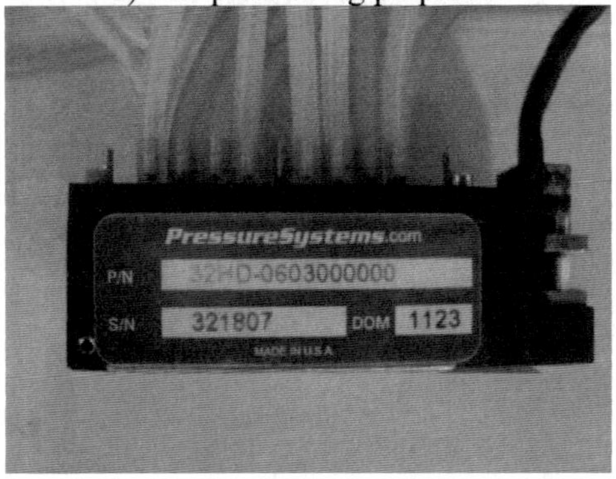

d) Pressure scanner.

Figure 3. The flat plate and instrumentation.

a) Flat plate alignment.

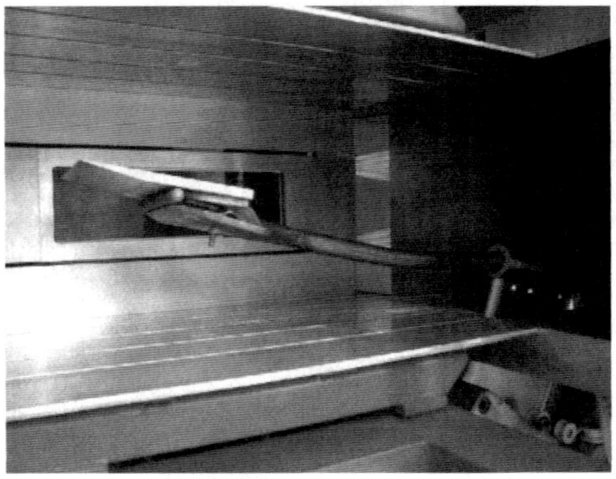

b) Sting support.

Figure 4. The flat plate installed in the TTP test section.

4. Methodology

In order to evaluate the airflow parameters ahead of the shock and expansion waves caused by the presence of the plate, the freestream airflow static pressure, p_1, and total pressure, p_0, were measured. The value of p_1 was obtained at the first pressure tap of the wind tunnel superior wall (figure 5a) and the measurement of p_0 was taken in the plenum chamber (figure 5b).

The mean value and standard deviation of the pressure signals, p_0 and p_1, supplied by the instruments during the runs of the wind tunnel tests are computed. This information is used to determine the probability density function for Mach number using the Monte Carlo method [3].

a) Static pressure tap, p_1.

b) Total pressure tap, p_0.

Figure 5. Pressure taps on the superior wall (a) and in the plenum chamber (b).

The static pressure behind the shock and expansion wave regions is estimated through the average value $\overline{p}_{plate}$ calculated from the six static pressure taps located on the flat plate surface.

The angle of the leading edge of flat plate is equal to +8°. When positioning the plate at an angle of attack α equal to -4°, the airflow encounters the superior surface of the plate at +4°.

In the following sections we will separate the aerodynamic phenomena which take place on the superior and inferior surfaces of the flat plate.

The uncertainties in the estimated parameters are evaluated by the Monte Carlo method. MatLab® codes were elaborated for data reduction. The number of trials employed is 500,000. The distribution assigned to the input quantities is the Gaussian.

4.1. *Superior surface*

In the superior surface of the plate, the airflow encounters a concave corner and the streamlines are deflected upward through the angle θ equal to +4°. An oblique shock occurs in this case (figure 6).

The angle between the shock wave and the upstream flow direction is defined as the wave angle and is denoted by β.

Knowing the flow field parameters in region 1 ahead of the shock, one can estimate the parameters in region 2, behind the shock.

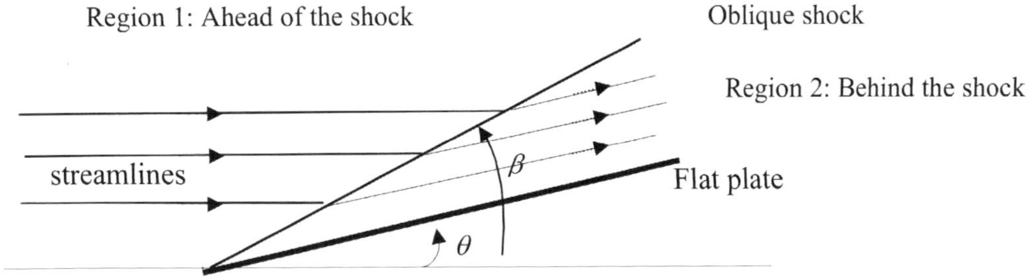

Figure 6. Deflection of a supersonic flow. θ: deflection angle, β: shock wave angle.

The Mach number in region 1 is evaluated by the expression [1]:

$$M_1^2 = \frac{2}{\gamma - 1}\left[\left(\frac{p_0}{p_1}\right)^{\frac{(\gamma-1)}{\gamma}} - 1\right] \tag{1}$$

where:
p_1: freestream static pressure;
p_0: freestream total pressure; and
γ: ratio of specific heats.

The equation relating the deflection angle, θ, the Mach number, M, in region 1 and the shock wave angle, β, is [1]:

$$\tan\theta = 2\cot\beta\frac{M_1^2 sen^2\beta - 1}{M_1^2(\gamma + \cos 2\beta) + 2} \tag{2}$$

where γ is equal to 1.4 for air at normal conditions.

Knowing the deflection angle, θ, and the Mach number M, in region 1, one obtains the shock wave angle, β, using equation (2) or theθ-β-M diagram [4].
The normal component of the of the upstream Mach number, $M_{n,1}$, is:

$$M_{n,1} = M_1 \mathrm{sen}\beta \tag{3}$$

The pressure ratio across the shock wave is:

$$\frac{p_2}{p_1} = 1 + \frac{2\gamma}{\gamma + 1}\left(M_{n,1}^2 - 1\right) \tag{4}$$

The normal component of the Mach number behind the shock, $M_{n,2}$, is:

$$M_{n,2}^2 = \frac{1 + [(\gamma - 1)/2]M_{n,1}^2}{\gamma M_{n,1}^2 - (\gamma - 1)/2} \tag{5}$$

The Mach number behind the shock, M_2, is calculated by the expression:

$$M_2 = \frac{M_{n,2}}{sen(\beta - \theta)} \tag{6}$$

The total pressure in region 2, $p_{0,2}$, is evaluated using the equation:

$$p_{0,2} = p_2\left(1 + \frac{\gamma - 1}{2}M_2^2\right)^{\frac{\gamma}{\gamma-1}} \tag{7}$$

4.2. *Inferior surface*
On the inferior surface of the plate, the airflow encounters a convex corner and an expansion wave is formed. Such wave is denoted as Prandtl-Meyer expansion wave (figure 7).

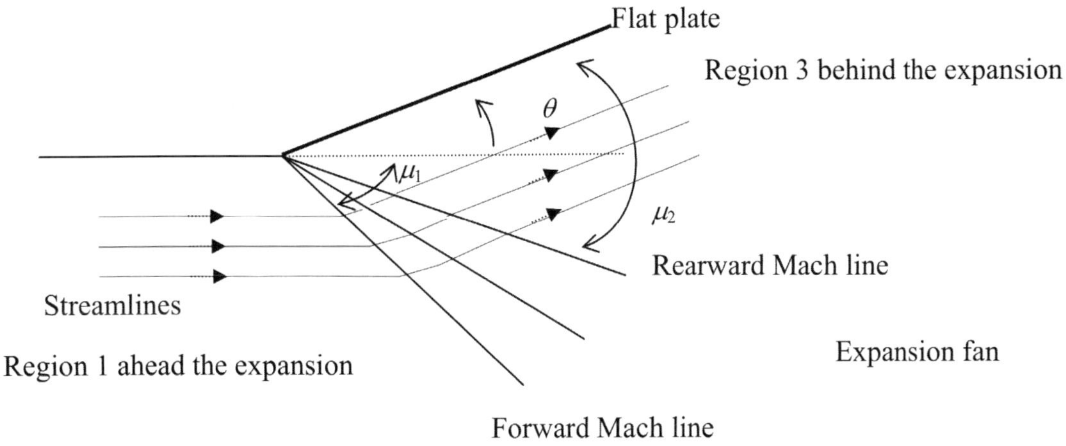

Figure 7. Prandtl-Meyer expansion. θ: deflection angle = +4°. μ_1: angle of the forward Mach line. μ_2: angle of the rearward Mach line.

In order to obtain the properties of the flow in region 3 behind the expansion wave, knowing the values of p_1, M_1 and θ, one first estimates the value of the upstream Prandtl-Meyer function, $\nu(M_1)$, using the equation:

$$\nu(M_1)=\sqrt{\frac{\gamma+1}{\gamma-1}}\tan^{-1}\sqrt{\frac{\gamma-1}{\gamma+1}\left(M_1^2-1\right)}-\tan^{-1}\sqrt{\left(M_1^2-1\right)} \quad (8)$$

or using the table "Prandtl-Meyer Function and Mach Angle", presented in Appendix C of reference [1].

One evaluates the Prandlt-Meyer function in region 3 by employing the following equation:

$$\theta=\nu(M_3)-\nu(M_1) \quad (9)$$

The table of Appendix C [1] supplies the Mach number M_3, from the value ν (M_3) given by equation 9.

To obtain the static pressure value behind the expansion wave, p_3, one uses the expression:

$$p_3=p_1\left(\frac{\left(1+[(\gamma-1)/2]M_1^2\right)}{\left(1+[(\gamma-1)/2]M_3^2\right)}\right)^{\gamma/(\gamma-1)} \quad (10)$$

Expansion is an isentropic phenomenon and therefore the total pressure is constant across the expansion.

Figure 8 summarizes the phenomena which occur on the flat plate at an angle of attack equal to -4° in a supersonic flow, for the freestream Mach number M_1 = 1.21. Flow is left to right. The results were obtained employing Computational Fluid Dynamics, CFD.

At the leading edge, one observes shock waves on the top surface and expansion waves on the bottom surface.

At the trailing edge, the flat plate experiences the opposite behavior. Mach number distribution of the airflow is shown. Only the leading edge region was investigated in this first part of the wind tunnel test campaign.

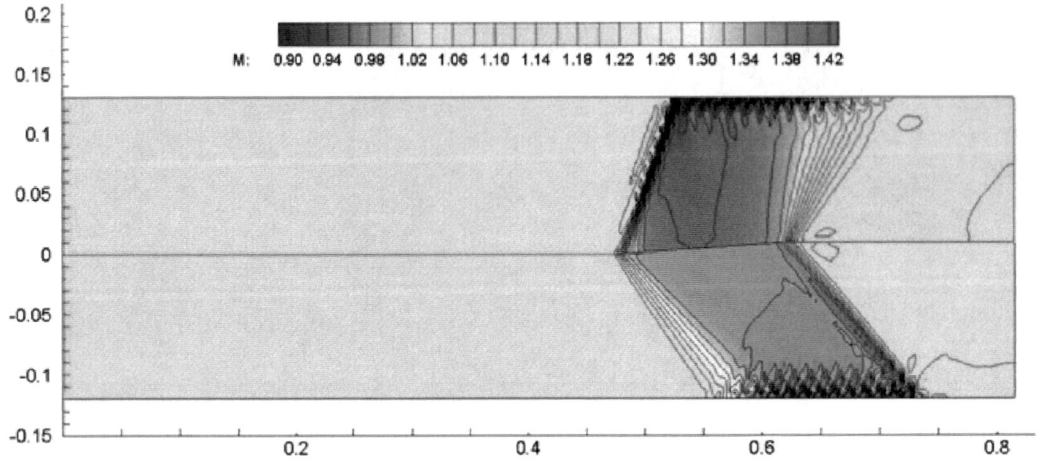

Figure 8. CFD results for the flat plate in a supersonic flow (α = -4°). Axis unit: meter.

The numerical code was based on finite difference and structured mesh with first-order implicit time discretization and second-order spatial central discretization. The code follows the main characteristics of the Alternating Direction Implicit scheme, ADI, and linear artificial viscosity. The mesh is composed by two juxtaposed blocks. Each block has 191 points in the longitudinal direction and 51 points in the vertical direction and the Δy minimum is equal to 0.002 m. Inlet boundary conditions were imposed, outlet boundary conditions were extrapolated by zero-th order, slip condition were imposed on wall surfaces together with adiabatic wall consideration and pressure extrapolation. The test section slotted walls were simulated alternating far field static pressure and solid wall configurations.

4.3. Forces in the flat plate

A flat plate of length *c* at an angle of attack α in a supersonic flow causes oblique shock and expansion waves (figure 9). The plate surfaces are subjected to different pressures, p_3 and p_2, which generates a resultant aerodynamic force *R*.

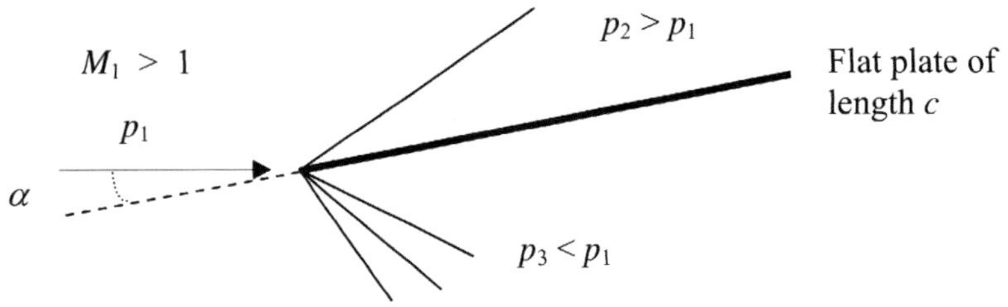

Figure 9. An inclined flat plate in a supersonic flow.

The resultant force for a unit of span, *R*', and its components lift, *L*', and drag, *D*', are expressed by[1]:

$$\begin{aligned} R' &= (p_3 - p_2)c \\ L' &= (p_3 - p_2)c\cos\alpha \\ D' &= (p_3 - p_2)c\,sen\,\alpha \end{aligned} \tag{11}$$

where c is the length of the plate.

The force coefficients, C_F, are calculated using the equation:

$$C_F = \frac{F}{q_\infty S} \tag{12}$$

In equation (13), F represents the lift force, L, or the drag force, D, S is the reference area and q_∞ is the dynamic pressure of the freestream flow, given by:

$$q_\infty = \frac{\gamma}{2} p_1 M_1^2 \tag{13}$$

5. Results

5.1. Superior surface

The mean value of the static pressures, measured by the tap pressures located on the flat plate, is $\bar{p}_{plate}$ = 66491.3 Pa, with a standard deviation equal to 2506.7 Pa.

Deflection angle: θ = 4 degrees.

Through the measurement of the total pressure, p_0, and the static pressure ahead of the shock, p_1, we obtain the Mach number M_1 using equation 1. The value of p_1 was obtained in the first pressure tap on the superior wall of the tunnel and the measurement of p_0 was taken in the plenum chamber. The mean value and standard deviation of the pressure signals, p_0 and p_1, are calculated and used to estimate the M quantity and the standard uncertainty associated with this estimate, u_M. Results are: M_1 = 1,21, u_M = 0.01 for the measured values p_1 = 50853.0 Pa and p_0 = 124348.7 Pa. The standard deviation of p_1 and p_0 are 418.3 Pa and 949.8 Pa, respectively.

The angle of the shock wave with respect to the freestream flow is calculated manipulating equation 2: β= 69.13°, or using the table θ-β-M.

Equation 3 and Monte Carlo supplies $M_{n,1}$ = 1.13 and associated uncertainty $u_{M\ n,1}$ = 0.01. Using equation 1 and propagating distributions of $M_{n,1}$ and the measured p_1, we obtain the static pressure which should occur behind the shock wave: p_2 = 66857.0 Pa, and u_{p2} = 1186.6 Pa. There is a difference equal to 365.7 Pa between the measured static pressure value ($\bar{p}_{plate}$= 66491.3 Pa) and that obtained algebraically (p_2 = 66857.0 Pa).

The normal Mach number estimated using equation 5, $M_{n,2}$ = 0.89, leads to the Mach number M_2 = 0.98, behind the shock wave (equation 6). Both associated uncertainties $u_{M\ n,2}$ and u_{M2} are equal to 0.01.

Using equation 7, the total pressure $p_{0,2}$ is equal 124100.5 Pa and $u_{p0,2}$ = 2425.3 Pa, for p_2 calculated algebraically. This value becomes $p_{0,2}$ = 123319.0 Pa and $u_{p0,2}$ = 4738.6 Pa if the chosen p_2 in equation 7 is the mean value obtained in flat plate pressure taps. The difference of the experimental value in respect to the theoretical value is 781.5 Pa.

5.2. Inferior surface

For M_1 = 1.21, p_1 = 50853.0 Pa and associated uncertainties, and θ = +4°, the upstream Prandtl-Meyer function, calculated using expression (8) is $\nu(M_1)$ – 3.7°, with negligible uncertainty, as obtained through the Monte Carlo method. This results in $\nu(M_3)$ in region 3 equal to 7.7°, obtained through equation 9. Consulting the Appendix C [1], the Mach number behind the expansion is M_3 = 1.35.

Equation 10 supplies the static pressure p_3 = 41905.5 Pa. The uncertainty value is u_{p3} = 583.4 Pa. As the tests for measuring p_3 have not yet been carried out, there is no comparison between experimental and theoretical values in this paper.

The angle between the forward Mach line and the upstream airflow direction is $\mu_1 = 56.0^o$ and the angle between the rearward Mach line and the flat plate direction is $\mu_2 = 47.8^o$. The angle between the rearward Mach line and the upstream airflow direction is $\mu_2 - \theta = 43.8^o$ (figure 7).

Numerical results for Mach number distribution of the flow field using CFD were also obtained and were presented in figure 8.

5.3. Aerodynamic forces

One obtains the aerodynamic forces acting on the flat plate positioned at an angle of attack $\alpha = -4^o$, multiplying the results obtained employing equations 11 by the width of the plate: R = -450,2 N, L = -449.1 N and D = 31.4 N, where the theoretical values of p_2 and p_3 were used. Uncertainties are 24 N, 24 N, and 17 N, for the resultant force, lift and drag, respectively.

This leads to the drag force coefficient, C_D , due to the wave and lift force coefficient, C_L, equal to:

$$C_D = 0,03$$
$$C_L = -0,48$$

estimated using equations 12 and 13.

Applying Monte Carlo supplies u_{CL} = 0.03 and negligible uncertainty in C_D.

These results are compatible with figures 4.20 and equation 4.48b of reference [5], which presents the value 0.17 for the rate C_D/C_L^2. The value obtained in this study is 0.15.

6. Conclusions

This paper presents the results obtained in the tests with the flat plate in the transonic wind tunnel TTP, in the supersonic regime. In this range of airflow speed, shock and expansion waves are expected. The parameters static pressure, p_1, and total pressure, p_0, of the freestream flow field were measured. From the measurement of these parameters, it was possible to evaluate the changes in the airflow properties due to the presence of the plate.

The six pressure taps on the superior flat plate surface supply the experimental data for the comparison between the measured static pressure and that predicted by the shock wave theory. The drag and lift forces acting on the plate were also estimated experimentally. The uncertainties in the estimated flow field parameters were estimated through the Monte Carlo method.

In both cases, static pressure and aerodynamic forces, the experimental results comply with literature.

Measurements of parameters such as temperature of the freestream airflow and total pressure behind the shock, as well as static pressure behind the expansion, will be performed in future tests.

Acknowledgment

The authors would like to express their gratitude to CNPq, The Brazilian National Council of Research and Development, for the partial funding of this research, under Grant 145129/2012-8.

References

[1] Anderson Jr J D 2001 *Fundamentals of Aerodynamics* (Mac Graw Hill).
[2] Anderson Jr J D 2003 *Modern Compressible Flow*, (Mac Graw Hill).
[3] JCGM 101 2008 Evaluation of measurement data – Supplement 1 to the Guide to the expression of uncertainty in measurement – Propagation of distributions using Monte Carlo method.
[4] Ames Aeronautical Laboratory Report 1135: *Equations, Tables and Charts for Compressible Flow* (Moffett Field, Calif).
[5] Schlichting H, Truckenbrodt E and Heinrich J R 1979 *Aerodynamics of the Airplane*, (MacGraw-Hill).

Journal of Physics: Conference Series **459** (2013) 012035 doi:10.1088/1742-6596/459/1/012035

Evaluation of the type A uncertainty in measurements with autocorrelated observations

Z L Warsza

Industrial Research Institute of Automation and Measurement (PIAP)
Al. Jerozolimskie 202, 02-486 Warszawa Poland

E-mail: zlw@op.pl

Abstract. *Described is the proposal of application of the GUM uncertainty type A evaluation to measurements with auto-correlated observations. The first step to it is the identification and cleaning of the raw sample data from the regularly variable components. Then formulas for standard deviation of the sample and of the mean value are expressed with the use correction coefficients or the so-called "effective number" of observations. These quantities depend on number of observations and on the sample autocorrelation function and allow to calculate the expanded uncertainty due to the GUM recommendations. The method of estimation of autocorrelation function for the sample data is also given. Considerations are illustrated by examples.*

1. Introduction

The international guide about evaluation of uncertainty in measurement known under acronym GUM covers only measurement of the variable which data is randomly distributed but no related statistically (i.e. without autocorrelation of their data). Then use of GUM is limited in many types of measurements. In particular, it was not established how to estimate uncertainty in measurements of the processes variable in time or space.

This paper will synthetically discuss the results of some Polish work on the determination of measurement uncertainty of regularly sampled measurand, covered in detail in publications [1] - [6] and their bibliographies. This includes preparing of the raw data sample to further calculations by removal from it regularly variable components (i.e. "cleaning of the raw sample"). Then the estimation of sample proper uncertainty type A for known autocorrelation function ρ is given. Method how to find from the sample data the estimator r of the a priori unknown autocorrelation function is also presented.

2. "Cleaning" the raw data

The The process of collecting measurement observations is now usually automated. The values of the measurement signal or the output readings are discrete as a result of sampling of the input analog signal with properly chosen frequency and A/D conversion.

The dispersion of the "raw" values of measurement observations is caused by reason of both random and determined type, and also as a result of changes of measurand itself, and changes of the internal parameters of the measuring circuit and environmental conditions. It is different for each of the samples taken at different times, measured by different instruments, and even by the same device over its lifetime. Random changes of the observations can be stationary or no stationary. There are short-term noise (called outliers), which prior to the assignment of values and the uncertainty of the measurements should be identified and eliminated from the raw data. Changes in the form of a regular (non periodic) component, i.e. the trend and periodic components affect the shape of the histogram of the sample, the mean value and uncertainty of type A. And there are also undesirable.

Calculations of the result and accuracy of measurement are made usually with the following assumptions:

- time T_n to collect the sample of n observations (width of the measurement window) was chosen properly, i.e. the random component of the sample can be assumed stationary,
- the sampling is uniform with the constant frequency n/T_n, and
- the random component is additive only, i.e. independent from the measurand value, and is described by the normal probability distribution.

Then values $y(i)$ of the discrete output signal obtained by sampling are

$$y(i)= f(i)+f_R(i)+N(0,\sigma) \tag{1}$$

where: $i\in(1,n)$ - the current number of the sample observation $y(i)$ obtained in time t_i, $f(i)$ - values (estimated from measurements) of the known a priori measurand function at times t_i, $f_R(i)$ - the values of other unwanted regular components in time t_i, and N (0, σ) - normal random component with standard deviation σ.

The statistical uncertainty u_A of measurements should be estimated only from values of last random component $N(0,\sigma)$. The values of function $f(i)$ estimating the measurand and of the residual signal contains the unwanted non-known a priori regular component $f_R(i)$ should be subtracted.

Example 1

Figure 1a shows a series of successive n=121 "raw" results of measurement observations (black squares), obtained by regular sampling of the tested process [3]. You can see that the sample contains non-linear decreasing trend, there are not noticeable periodic components, and rather a small laceration of the data graph indicates that this data can be autocorrelated. After removing one outlier, using LSM (least squares) the polynomial was determined:

$$y(i)= (-7{,}65\cdot10^{-7})\,i^2+(6{,}6\cdot10^{-5})i + 0{,}034\,.$$

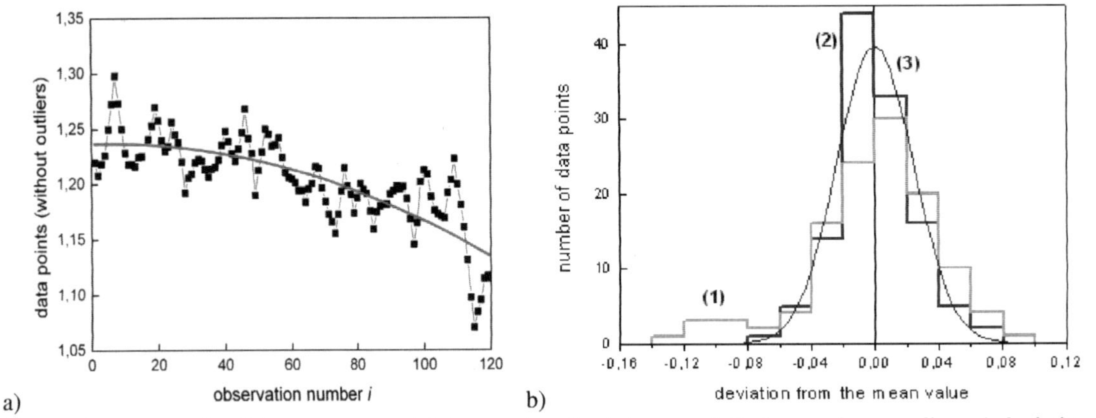

Figure 1. An example of the measurement data sample collected sequentially by regular sampling: a) deviations from the mean value of the raw data and their systematic regularly variable component - drift identified by the least squares method; b) distributions: of the raw data (1), after subtraction of the drift (2) and Gaussian pdf curve fitted to that data (3).

Trend is modeled and removed from the sample data, assuming that it passes through the average value. In Figure 1b a histogram of the raw sample is given - (1) and histogram after removal of the trend - (2) and a matching criterion $\chi2$ its normal (Gaussian) probability density function (pdf) - (3). Trend causing asymmetry and the existence of a large "tail" on the left of histogram (1), which does not meet the criterion of $\chi2$, and with trend uncertainty u_A is about 64% higher than u_A=0.00219 without it. Purification of the raw data significantly reduced u_A. For n=120 the relative standard deviation $s(u_A)/s\approx 8.5\%$.

3. The uncertainty of mean value for correlated observations

Presented is a brief description of the problem. The sequence of measurement data obtained from the sampling process and purified from the deterministic component can be described by a stationary time series. Statistical correlations between realizations X_i, X_{i+k} of such series is characterized by the autocorrelation function

$$\rho_k = \frac{\operatorname{cov}(X_i, X_{i+k})}{\sigma^2} \tag{2}$$

Function ρ_k depends on the frequency spectrum of the test process and is known or its estimate r_k should be found from the measurement data. In measurement of physical quantities the correlation function is positive.

The relationship between standard deviation $\sigma(\bar{x})$ of the mean value and σ of the individual correlated observation x_i results from the variance of the sum of random variables [1, 2, 4]

$$\sigma(\bar{x}) = \frac{\sigma}{\sqrt{n_{eff}}} \tag{3}$$

Where

$$n_{eff} = \frac{n}{1+2\sum_{k=1}^{n-1}(1-k/n)\rho_k} \equiv \frac{n}{1+D_\rho} \tag{3a}$$

For the statistically independent observations $\rho_k \to 0$ (for k $\geq$ 0), consequently $D_\rho = 0$ and formula (2) passes to the commonly known relation

$$\sigma(\bar{x}) = \frac{\sigma}{\sqrt{n}}.$$

In opposite, when the observations are fully correlated (closely linked), i.e. $\rho_k \to 1$, with (3a) results

$$D_\rho \to \frac{2}{n}\sum_{k=1}^{n-1}(n-k)\cdot 1 = n-1 \tag{4}$$

Then the standard deviation of the mean is the same as for a single observation of the sample because in the limit $\rho_k \to 1$ all subsequently repeated observations will be the same.

The value of n_{eff} is needed to proper estimation of the standard deviations $s_a(x_i)$, $s_a(\bar{x})$ of sample of autocorrelated observations [4 -6]. Their relations to the GUM parameters $s(x_i)$, $s(\bar{x}_i)$ are

$$s_a(\bar{x}) = \frac{s_a(x_i)}{n_{eff}} = \left[\frac{1}{n\,(n_{eff}-1)}\sum_{i=1}^{n}(x_i-\bar{x})^2\right]^{1/2} \tag{5}$$

or

$$s_a(\bar{x}) = k_b s(\bar{x}_i) \tag{5a}$$

where:

$$s(\bar{x}_i) = \frac{s(x_i)}{\sqrt{n}}, \qquad k_b = \left(\frac{n-1}{n_{eff}-1}\right)^{1/2} \tag{5b), (5c}$$

Journal of Physics: Conference Series **459** (2013) 012035 doi:10.1088/1742-6596/459/1/012035

$$s^2(x_i) = \frac{1}{n-1}\sum_{i=1}^{n}(x_i - \bar{x})^2 \,, \tag{5d}$$

$$s_a(x_i) = k_a \; s(x_i) \tag{5e}$$

where:

$$k_a = \left[\frac{n_{eff}(n-1)}{n(n_{eff}-1)}\right]^{1/2} \approx 1 \,; \tag{5f}$$

For the autocorrelation data the effective number degrees of freedom ν_{eff} can also be used. Value of ν_{eff} is defined approximately [4-6] as

$$\nu_{eff} \cong \frac{n}{1 + 2\sum_{k=1}^{n-1}\rho_k^2} - 1 \tag{6}$$

But $\nu_{eff} \neq n_{eff} - 1$ [4, 5]. The relative dispersion of the standard deviation is:

$$\frac{u(s_a)}{s_a} = \frac{u(s_a(\bar{x}))}{s_a(\bar{x})} \cong \frac{1}{(2\nu_{eff})^{1/2}} \tag{7}$$

4. Estimator of the autocorrelation function of measurement data sample

The autocorrelation function is usually not known and needs to be estimated from the measurement data. The most commonly used and implemented in computer programs has form

$$r_k = \frac{\sum_{i=1}^{n-k}(x_i - \bar{x})(x_{i+k} - \bar{x})}{s^2(q_i)} \tag{8}$$

Estimate r_k (Figure 2a) has two qualitatively different parts. For small distances k is the falling edge, in which contains real information about the autocorrelation function. The remainder tail is the image of a rather large fluctuations of the correlated noise.

Replacement of the function ρ_k by its estimate r_k in formula (3) gives an estimate of the effective number of observations n_{eff} not of the satisfied properties. The reason is the influence of autocorrelation function of the tail. According Zięba [4, 5] summation in (8) can be reduced to only a few initial estimates of r_k elements, i.e.

$$\hat{n}_{eff} = \frac{n}{1 + 2\sum_{k=1}^{n_c}\left(1 - \frac{k}{n}\right) r_k} \tag{9}$$

The border value n_c is determined by the last non-zero element of the r_k estimate before its first passage through zero (FTZ method – so called from the first transit through zero). For example, this value for the curve on Figure 2a is n_c=3. FTZ method is valid only for positive correlations.

Figure 2b [6], [4] shows two examples of probability distributions of the estimators obtained by Monte Carlo method. The distribution marked as *teor* on figure 2.b is calculated from the formula for uncorrelated observations of the standard deviation $z = s/\sigma$ resulting from the distribution of $\chi 2$, where ν was substituted by the effective number of degrees of freedom ν_{eff}.

The simulation studies using MC [5] show that in this case estimator $\hat{n}_{eff}$ reduce the negative bias of the mean value $\bar{x}$. Obtained value of $\hat{n}_{eff}$ is used for calculations of $s_a(x_i)$ and $s_a(\bar{x})$ by (5e) and (5).

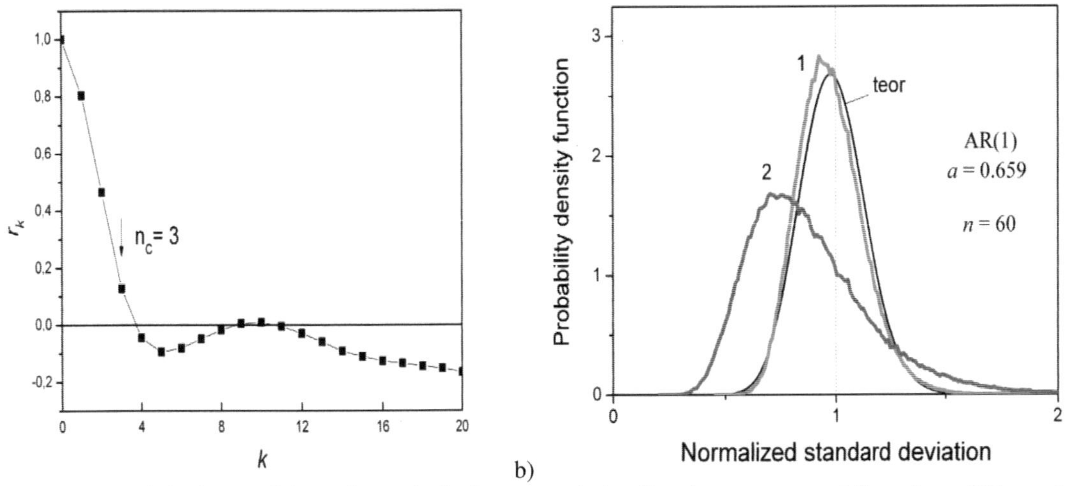

a) b)

Figure 2 a). The initial part of the estimate $\{r_k\}$ of autocorrelation function ρ_k computed from data of Figure 1a (after removing the trend from the raw data); b). Probability density functions for normalized estimators of the standard deviation by the first order autoregressive model AR(1)[1] and the random sample size n=60 [5, 6]. Curves 1 and 2 are derived from MC simulations and relate, respectively, s_a/σ and $s_a(\bar{x})/\sigma(\bar{x})$. Theoretical curve 'teor' has been calculated theoretically using the model (9) for $\nu_{eff} = 22.7$.

Example 2

Let us calculate the standard deviation of single measurement $s_a(q_i)$ and of the mean value $s_a(\bar{q})$ of 120 data of observations q_i from Example 1 after withdrawal trend. The estimator r_k of their autocorrelation function is taken for n_c=3 according to Figure 2a. The large obtained value r_k=0.81 confirms autocorrelation of these data. Standard deviation without considering the autocorrelation is $s(q_i) = 0.0241$. At this value of r_k the formula (9) implies the estimate n_{eff}=32.1 and from (5f) and (5c) the coefficient k_a=1.012 and k_b=1.96. Correlation do not significantly affect the $s_a(q)$=0.0244, and much more the u_A, which increases about 2 times from $s(\bar{q}) = 0.00219$ to $s_a(\bar{q}) = 0.00472$. From (6) and (7): $u(s_a)/s \approx 11\%$.

5. Summary and conclusions

Discussed issues aroused from the purpose of determining the measurement uncertainty for the measurement data obtained by sampling. This paper summarizes the research results of cleaning such raw data from their regular components [3] and the extension of calculating the uncertainty by the method type A of GUM to the autocorrelated data [1, 2] -[4-7]. The developed method is easy to use in the practice of measurement and can provide a basis to adopting it into the GUM upgrading.

Conclusions:

• For a limited time to collect measurement observations, a reduction of measurement uncertainty by increasing the sample size (by increasing the sampling rate) is unreliable, because it leads to the necessity to reflect the impact of autocorrelation function of observation.

• Before calculating the uncertainty $u_A(x)$ appropriate computational methods must first be used to identify and remove from the raw results the systematic (non periodic and periodic) components.

• For such a cleaned values of the observations one needs to know or estimate the autocorrelation function. This function causes a significant increase in uncertainty u_A compared to the calculated according to GUM. It corresponds to the lower effective number of independent measurements to be taken into account in estimating the standard uncertainty. The content specified adjusted formulas.

[1] Model AR (1) - autoregressive time series of the first order, for which $r_k = a^k$ [4]

• Programs for the calculation of uncertainty should be supplemented by algorithms for the identification and elimination from the "raw" data the regular components and to obtain estimators of the autocorrelation function from such cleaned measured data.

• Evaluation of the uncertainty u_A for autocorrelated data discussed here, is valid, as in GUM, for the model of normal distribution. For other distributions, this method requires further investigation.

References

[1] Dorozhovetz M, Warsza Z, 2007 Methods of upgrading the uncertainty …, Part 2 Elimination of the influence of autocorrelation of observations and choosing the adequate distribution, *Proceedings of 15th IMEKO TC4 Symposium*, Sept. 2007 Iasi Romania, part **1** 199-204

[2] Warsza Z L, Dorozhovets M 2008 Uncertainty type A evaluation of autocorrelated measurement observations. *Biuletyn WAT* (*Military Technical Academy*) Warszawa, vol. LVII **2** 143-152

[3] Warsza Z L, Dorozhovets M, Korczynski M J 2007 Methods of upgrading the uncertainty of type A evaluation (1). Elimination the influence of unknown drift and harmonic components. *Proceedings of 15th IMEKO TC4 Symposium*, Sept. 2007 Iasi Romania part **1** 193-198

[4] Zięba A 2010 Effective number of observations and unbiased estimators of variance for autocorrelated data – an overview, *Metrology & Measurement Systems*, **17** (2010) 3-16

[5] Zięba A, Ramza P 2011 Standard deviation of the mean of autocorrelated observations estimated with the use of the autocorrelation function estimated from the data. *Metrology & Measurement Systems*, **18** (2011) 29-25

[6] Warsza Z L, Zięba A 2012 Niepewność typu A pomiaru o obserwacjach samoskorelowanych. *Pomiary Automatyka Kontrola* PAK vol. 58 **2** 157 -161 (in Polish)

Journal of Physics: Conference Series **459** (2013) 012036 doi:10.1088/1742-6596/459/1/012036

Why should correction values be better known than the measurand true value?

Franco Pavese

Torino, Italy

E-mail: frpavese@gmail.com

Abstract. Since the beginning of the history of modern measurement science, the experimenters faced the problem of dealing with systematic effects, as distinct from, and opposed to, random effects. Two main schools of thinking stemmed from the empirical and theoretical exploration of the problem, one dictating that the two species should be kept and reported separately, the other indicating ways to combine the two species into a single numerical value for the total uncertainty (often indicated as 'error'). The second way of thinking was adopted by the GUM, and, generally, adopts the method of assuming that their expected value is null by requiring, for all systematic effects taken into account in the model, that corresponding 'corrections' are applied to the measured values before the uncertainty analysis is performed. On the other hand, about the value of the measurand intended to be the object of measurement, classical statistics calls it 'true value', admitting that a value should exist objectively (e.g. the value of a fundamental constant), and that any experimental operation aims at obtaining an ideally exact measure of it. However, due to the uncertainty affecting every measurement process, this goal can be attained only approximately, in the sense that nobody can ever know exactly how much any measured value differs from the true value. The paper discusses the credibility of the numerical value attributed to an estimated correction, compared with the credibility of the estimate of the location of the true value, concluding that the true value of a correction should be considered as imprecisely evaluable as the true value of any 'input quantity', and of the measurand itself. From this conclusion, one should derive that the distinction between 'input quantities' and 'corrections' is not justified and not useful.

1. Introduction

Since the beginning of the history of modern measurement science, the experimenters faced the problem of dealing with systematic effects, as distinct from, and opposed to, random effects.

Two main schools of thinking stemmed from the empirical and theoretical exploration of the problem, one dictating that the two species should be kept and reported separately, the other indicating ways to combine the two species into a single numerical value for the total uncertainty (often indicated as 'error' in past years). The second way of thinking was adopted by the GUM, and, generally, adopts the method of randomising the systematic effects (i.e. assuming that their expected value is null), by requiring, for all systematic effects taken into account in the model, that corresponding 'corrections' are applied to the measured values before the uncertainty analysis is performed.

On the other hand, about the *value* of the measurand (the quantity intended to be the object of measurement), classical statistics calls it 'true value', admitting that the value should exist objectively (e.g. the value of a fundamental constant), and that any experimental operation aims at obtaining an ideally exact measure of it—GUM omits 'true' as an unnecessary specification. However, due to the uncertainty affecting every measurement, this goal can be attained only approximately, in the sense that a range of values is measured and nobody can ever know exactly how much any measured value differs from the true value. A frequent way to try to overcome the issue is to use a "confidence interval" or a "degree of belief", assumed to allow one to estimate the probability for the true value to fall within the aimed interval.

The paper discusses the credibility of the numerical value attributed to an estimated correction, as compared with the credibility of the estimate of the location of the true value of a quantity.

2. Input quantities and corrections

The *Guide to the expression of uncertainty in measurement* (GUM) [1] introduced since 1995 methods to be used in the field of metrology for evaluating the measurement uncertainty. According to this Guide, "*it is assumed that the results of a measurement have been corrected for all recognised significant systematic effects and that every effort has been made to identify such effects*" (3.2.4), before the issue of evaluating their uncertainty is tackled.

In addition, in the GUM it is also required that a model is used, expressing the functional dependence of the measurand on the identified input quantities X_i: (GUM, 4.1.1) "*In most cases, a measurand Y is not measured directly, but is determined from N other quantities* $X_1, X_2, ..., X_N$ *through a functional relationship f* : $Y = f(X_1, X_2, ..., X_N)$"; (GUM, 4.1.2) "*The* ***input quantities*** $X_1, X_2, ..., X_N$ *upon which the output quantity Y depends may themselves be viewed as measurands and may themselves depend on other quantities, including corrections and correction factors for systematic effects, thereby leading to a complicated functional relationship f that may never be written down explicitly. Further, f may be determined experimentally ... or exists only as an algorithm that must be evaluated numerically. The function f as it appears in this Guide is to be interpreted in this broader context, in particular as that function which contains every quantity,* ***including all corrections and correction factors*** *that can contribute a significant component of uncertainty to the measurement result*" (boldface added here).

The function may remain implicit $f(Y, X_i) = 0$, where the effect of each X_i on Y is obtained through a Type A method of variation of each influence parameter for the determination of the "sensitivity coefficients", or based on Type B information; or, the function can be an explicit mathematical expression, the values of whose parameters are determined experimentally, or via Type B information.

The wished effects of *correction* for systematic effects are, as expressed in the GUM, that: "*it is assumed that, after correction, the expectation or expected value of the error arising from a systematic effect is zero*" (3.2.3); this outcome "*... is based on the concept that there is no inherent difference between an uncertainty component arising from a random effect and one from a correction for a systematic effect.*" (E.1.1), because "*... this Guide treats uncertainty components arising from random effects and from corrections for systematic effects in exactly the same way in the evaluation of the uncertainty of the result of a measurement.*" (E.3).

However, in the GUM, after having admitted in clause (4.1.2) that corrections are included in the model, they are then treated differently from the other 'input quantities' when clause (3.2.4) requires that the results of a measurement must *preliminarily* be "corrected". This induces an ambiguity in the GUM modelling procedure. To show it, let us take a simple explicit model $Y = X_1 \cdot X_2$, where Y is measured indirectly by measuring X_1 and to X_2, with corrections included according to clause (4.1.2):

$$Y = (X_1 + C_1) \cdot (X_2 \cdot C_2) \ , \qquad (1)$$

where X_1 and X_2 are two measured 'input quantities', and C_1 and C_2 are 'corrections',[1] respectively to X_1 and to X_2. The corrections can also be measured quantities, or their values and uncertainties can be obtained from Type B evaluations.

The expression (1) is the GUM model according to clause (4.1.2). However, according to clause

[1] C_1 is an offset correction, typically small in value ($E(C_1) \approx 0$ relative to $E(X_1)$). C_2 is a scale-factor correction, typically with $E(C_2) \approx 1$

Journal of Physics: Conference Series **459** (2013) 012036 doi:10.1088/1742-6596/459/1/012036

(3.2.4), model (1) is only the *initial* model. In fact, after the required corrections are applied, model (1) should be rewritten as

$$Y \approx (X_1^* + C_1^*) \cdot (X_2^* \cdot (1 + C_2^*))^{2} \quad (2)$$

where all quantities on the right of (2) are different from those in (1): $X_1^* = X_1 + \mathrm{E}(C_1)$; $X_2^* = X_2 \cdot \mathrm{E}(C_2)$; $C_1^* = C_1 - \mathrm{E}(C_1)$; $C_2^* = C_2 - \mathrm{E}(C_2)$, where, in general, $\mathrm{E}(C_1) \neq 0$ and $\mathrm{E}(C_2) \neq 0$.[3] According to GUM, the new quantities C_1^* and C_2^* have zero expectation: $\mathrm{E}(C_1^*) = 0$ and $\mathrm{E}(C_2^*) = 0$. Expression (2) is the actual GUM model subject to the uncertainty analysis.

3. Can the value of a correction be known exactly?

Correction for systematic effects may actually be affected by limitations, as also recognised in limited cases in the GUM: "*It is now widely recognized that, when all of the known or suspected components of error have been evaluated and the appropriate corrections have been applied, there still remains an uncertainty about the correctness of the stated result, that is, a doubt about how well the result of the measurement represents the value of the quantity being measured.*" (0.2); "*The result of a measurement after correction for recognized systematic effects is still only an estimate of the [true] value of the measurand because of the uncertainty arising from random effects and from imperfect correction of the result for systematic effects.*" (3.3.1); "*A corrected measurement result is not the value of the measurand—that is, it is in error—because of imperfect measurement of the realized quantity due to random variations of the observations (random effects), inadequate determination of the corrections for systematic effects, and incomplete knowledge of certain physical phenomena (also systematic effects). Neither the value of the realized quantity nor the value of the measurand can ever be known exactly; all that can be known is their estimated values.*" (D.4).

As a consequence, a problem arises about the way to comply with the GUM requirements, and, more in general, to comply with the need of performing corrections for known 'biases', or for 'biases' estimated at various levels of confidence—as occurring also in the testing field. In fact, the above caveats should concern not only the uncertainty, but also the value of the correction, since a model is likely to be imperfect for several reasons, most *not* of random nature:

(i) Non-uniqueness of the measurand (epistemic uncertainty, intrinsic);
(ii) Incomplete list of influence quantities (epistemic uncertainty);
(iii) Imperfect knowledge of the input quantities (re model functional form and the values of its parameter).

In addition, in tackling the problem, one is often confronted with issues arising from practical difficulties, especially when:

(a) Applying a significant recognised correction with large uncertainty;
(b) Handling uncertainty for uncorrected results.

Concerning issue (a), two conditions have to be noted: to be "recognised" (be known) and, to be "significant", also embedded in the GUM: "*In some cases, ... although the uncertainty has been evaluated, it may be ignored if its contribution to the combined standard uncertainty of the measurement result is insignificant. If the value of the correction itself is insignificant relative to the combined standard uncertainty, it too may be ignored*" (3.4.4). However, e.g., CITAC [2] recommends instead correcting also for non-significant effects. As to issue (b), the preference for this choice is widespread.

Actually, the corrections are fundamental good practice in science, because they are to be considered as an integral part of the process of sufficiently understanding and modelling an experiment. They are as essential as the practice of replicating the measurements.

In practice, there are several conditions under which the correction of the measured value is not desirable, and this is not explicitly deprecated in the GUM but allowed in special cases. In the GUM

---

[2] Notice that $\mathrm{E}(1 + C_2^*) \approx 1$, so that the second term is approximately $(X_2^* \cdot C_2)$.

[3] Obviously here we assume that a correction is not a fixed value as in the VIM [3], see Footnote 5.

the difficulties in applying a correction are recognised: "*The error arising from imperfect compensation of a systematic effect cannot be exactly known*" (3.2.3. Note), but the specified way-out is "*The uncertainty of a correction for a known systematic effect may in some cases be obtained by a Type A evaluation while in other cases by a Type B evaluation, as may the uncertainty characterizing a random effect*" and, then, the "*Guide treats uncertainty components arising from random effects and from corrections for systematic effects in exactly the same way in the evaluation of the uncertainty of the result of a measurement*." (E.3).

GUM's recipe is basically Gauß' one: should one recognise a systematic effect, a correction must be applied to measured data before the uncertainty analysis is undertaken. In fact, fixed effects are assumed in the GUM to be fully taken into account by the corrections (zero-expectation after correction), so that only random effects apply to the corrected values.

4. The true value of a correction

The very distinction between 'input quantities' relevant to a measurand and corrections to these quantities is basically artificial, though customary: the fact that a 'correction' may be ignored, or that its value can be zero [4], has possibly promoted this distinction.

However, in principle, 'corrections', when performed, are simply *compensations* for deviations of measured from reference states of specified quantities due to perturbations caused by *other* influence quantities or by off-set conditions. These deviations are customarily called 'bias': in VIM bias is defined "*estimate of a systematic measurement error*" (2.18). Actually, "systematic" should rather mean that bias is relative to a standard condition that is not met (bias value different from zero), or to an unexpected deviation from a standard condition, so that the measurement results are systematically 'off centre' in the variations of their values.

These other quantities are also measured, or the deviations from the reference state are estimated from Type B evaluations, exactly as for the previous 'input quantities': in the vast majority of cases the values of the corrections are non-zero, as are the values of the input quantities. In all respects, as recognised by the GUM, they are part of the input quantities. By definition, once the influence quantities have been identified, no other quantity outside the set of 'input quantities' plus 'corrections' can influence the value of the measurand.

In actuality, due to the fact that *no value of a quantity can be exempt from an uncertainty* about the actual location of the true value, irrespective to the process (either Type A or Type B) bringing to its evaluation, there is not in principle a difference between 'input quantities' X_i and 'biases' B_n—the 'corrections' merely being $C_n = -B_n$:

$$X_i = \mathrm{E}(X_i) + X_i^*, \qquad B_n = \mathrm{E}(B_n) + B_n^*, \tag{3}$$

where, as seen in the expressions (1) and (2) of the initial example, the star quantities have zero mean (the normally non-zero mean value having been 'extracted' from the not-starred quantities) and take into account the variability of the results.

Furthermore, for additive biases, $X_i = X_i^{\mathrm{o}} + B_n$—the symbol of 'standard state' ⊖ is borrowed here from physical chemistry to indicate the reference condition for which $\mathrm{E}(B_n) =: 0$. Thus $X_i^{\mathrm{o}} = X_i - B_n = X_i + C_n$.[4]

The measurement equation for a measurand Y, written in the general way, is:

$$Y = \mu_X + \Delta_X + \varepsilon_\mathrm{r} + \varepsilon_\mathrm{s} \tag{4}$$

where $X = f(X_i)$ with X_i the 'input quantities', μ_X is the mean, Δ_X (sometimes indicated with b or B) is the deviation from the mean, ε_s is the zero-mean random error associated to the systematic effects, and ε_r is the zero-mean random error associated to X (often only $\varepsilon = \varepsilon_\mathrm{r} + \varepsilon_\mathrm{s}$ is indicated). Comparing with

[4] The only functional characteristics of a 'correction' C_n is that it applies to an 'input quantity' X_i, or to another correction.

Journal of Physics: Conference Series **459** (2013) 012036 doi:10.1088/1742-6596/459/1/012036

(3), one can write the following equivalence expressions: μ_X = E(X), $\varepsilon_r = X^*$, $\Delta_X = \Sigma_n B_n$, ε_s = conv(B_n^*)—the convolution of the B_n^*. In addition, $X^o = X - \Delta_X$.

The existence of an uncertainty *always* associated to a correction,[5] should prevent to omit any correction from the initial model, unless, according to the GUM, one wants to label it as non-significant, so neither including its contribution to uncertainty. Thus, as a random variable, any summary statistics of the true value of a correction is only providing a 'best estimate'. *The (true) value of a correction remains as unknowable as that of any 'input variable', and that of the measurand itself.* However, the bias terms should remain explicit also in the model for the *corrected* measurement results [5, 6] because they carry in an uncertainty component.

5. Possible ways out and conclusions

In [4] the main reasons why the most difficult issue, trusting a correction value, cannot be solved easily, or universally, are summarised.

When the decision is to apply a correction, an edge-cutting solution is to consider, and treat, all 'corrections' as they actually are: a type of 'input quantities', to be included in the model—e.g. model (1)—and treated as any other input quantity. From this viewpoint, one does not need anymore to make a distinction between X_i and C_n, and the uncertainty analysis should concern all relevant quantities in a single step.

Therefore, the model (1) should be explicitly modified as

$$Y = f(W_1, W_2, \ldots, W_m, \ldots, W_M),\quad W_m = X_1, \ldots, X_I, C_1, \ldots, C_N, \text{ with } i = 1\ldots I,\ n = 1\ldots N,\ M = I + N \quad (5)$$

where the distinction between the X_i and the C_n is left here only for clarity.

Basically, the practical difference between what is considered a (primary) 'input quantity', X_i, and a correction, C_n, is the fact that the measurement results concerning the former always involve Type A uncertainty components, while the evaluation of a correction may involve only Type B components.

Let us make two examples, one for an 'indirect' measurement, another for a 'direct' measurement.

Gas thermometry. The basic model for the indirect measurement of the intensive quantity T is: $p \cdot V = n \cdot R \cdot T$, with p, pressure; V, volume; n, amount of substance; T, temperature; R, gas constant. In the constant-volume type, V and n are assumed constant during the measurement of p, for the purpose of obtaining measurement results for $T_b = f(p_b)$, where T_b and p_b are the thermodynamic values of the gas *in the thermometer bulb*. In the actual measurements, many "corrections" are to be applied, some due to the gas thermodynamics, some to the technical apparatus, some to the measuring system. Part of them (i) are obtained from measurement of subsidiary quantities, different from p_b, so affected (also) by at least a Type A component of uncertainty; another part (ii) are obtained from calibration certificates, handbook, etc, i.e. by means of Type B components of uncertainty. Among (i), for example, all the corrections for temperature-dependent parameters require the measurement of T in the relevant locations (room, apparatus, equipment, ...), using calibrated thermometers affected by calibration and measurement uncertainty components. Among (ii), one can list, e.g.: (a) as to gas, impurities (chemical and isotopic), non ideality, changes in n (adsorption on walls, diffusion outside the bulb, leaks), amount of gas outside the bulb, difference between p_{meas} and p_b; (b) as to the experimental apparatus, changes in V_b due to thermal expansion or finite stiffness of the bulb material, temperature distribution in the dead-volume and non-uniformities; (c) as to the measuring apparatus, pressure gauge corrections and calibration concerning p_{meas}, calibration of instruments measuring electrical quantities (V, I, R), laboratory standards, and vacuum gauges—also mechanical standards for volume V_b measured dimensionally.

Mass measurement. A model for the direct measurement of the extensive quantity m is $m = \Sigma_i m_{ci}$, with the mass m placed on one balance pan and calibrated masses m_{ci} put on the other pan, and where i

[5] Corrections are variables in the GUM, though the 'bias' intended to be corrected is often considered instead a fixed value—the most notable case is VIM [3], where bias is defined in (2.18)—see above—where the systematic error is a "component of measurement error that in replicate measurements remains constant or varies in a predictable manner" (2.17).

of the latter masses are necessary to obtain the exact balance ($\Delta m \equiv 0$). In the actual measurements, many "corrections" are to be applied, some due to m itself, other to the m_{ci}, other to the technique used. Apart from the possible calibration corrections of each m_{ci} for deviations from the nominal values, all other corrections requires the measurement (or knowledge) of the value of subsidiary quantities. A list of the main, not in a particular order is: for buoyancy, involving air thermodynamic state (p, T and density d) and volume of m and each m_{ci}; for mass magnetism, involving permanent magnetisation and magnetic susceptibility; for mass and balance cleanliness, involving also adsorption; for linearity, sensitivity, drift, hysteresis, eccentricity of the balance.

Incidentally, as shown in these typical examples, in most cases the actual measurement procedure requires a multidimensional capability—in particular temperature and pressure being often involved even in 'direct' measurement, carrying a Type A component—and, traceability of the measurements always requires involving calibration certificates, carrying a Type B component.

Customary, corrections—measured, computed or stipulated—are applied separately. However, an overall model could obviously be constructed by merging all in a single analytical expression whenever an explicit model is feasible. Sometimes the resulting model may look exceedingly complicated. This fact is mitigated today by the use of automatic computing. However, some simple corrections not placing critical problems might remain applied separately.

In many cases, this overall-merging procedure is more straightforward, not needing to separate a first step of application of the corrections of the measurement results from a subsequent distinct step of taking into account its uncertainty. In some cases, it may also overcome the dilemma introduced in [4] and the consequent need for methods of uncertainty-increase evaluation in the absence of a correction. It would also avoid the confusion often observed about the correct handling of 'corrections' [4].

In conclusion, Y should be computed by using the overall model (5), including the explicit functional form of all corrections. This is possible in many cases.

When the model remains instead in an implicit form, or when one likes to avoid embarking in exceedingly complicated analytical computations of uncertainty propagation, a suggestion for a simple way can be to apply the Monte Carlo method. In this case, in addition to the variability distributions of the X_i, one would include, for each 'correction' term, C_i, its probability distribution; E(C_i) is the contingent value attributed to 'bias'—zero, only when justified—i.e. the one that would derive from the GUM clause (3.2.4). In this way, E(Y) directly embeds also the effect of the current 'corrections'.

References

[1] BIPM, IEC, IFCC, ISO, IUPAC, IUPAP, OIML, *Guide to the expression of uncertainty in measurement* (GUM), and its Supplement 1: *Propagation of distributions using a Monte Carlo method*, 2nd edn, Bureau International des Poids et Mesures (BIPM), Sèvres, 1995. Free downloading from http://www.bipm.org/en/publications/guides/gum .

[2] CITAC (2000) *Eurochem Guide 2000*, end edn. http://www.eurachem.org

[3] BIPM, IEC, IFCC, ISO, IUPAC, IUPAP, OIML, *International Vocabulary of Basic and General Terms in Metrology*, BIPM, Sèvres, 2008. Free downloading from http://www.bipm.org/en/publications/guides/vim .

[4] F. Pavese, On the difference of meaning of 'zero correction': zero value versus no correction, and of the associated uncertainties, in *Advanced Mathematical and Computational Tools in Metrology IX* (F. Pavese *et al.*, Eds), vol. 9, World Scientific, Singapore, pp. 297–309, 2012.

[5] F. Pavese, An introduction to data modeling principles in metrology and testing, in *Advances in data modeling for measurements in metrology and testing*, Birkhauser-Springer, Boston, 2009, 1–30.

[6] F. Pavese, *Mathematical and statistical tools in metrological measurements*, in Physical Methods, Instruments and Measurements, Eds. in Encyclopedia of Life Support Systems (EOLSS), Developed under the Auspices of the UNESCO, Eolss Publishers, Oxford, UK, [http://www.eolss.net], pp 68, in publication.

Consideration of the Measurement Uncertainty in the Assembling of Components Characterized by High Dimensional Variation

C Hernández, R Tutsch
Institut für Produktionsmesstechnik, Technische Universität Braunschweig, Schleinitzstrasse 20, D-38106 Braunschweig, Germany.

E-mail: carlos.hernandez@tu-bs.de

Abstract. The Statistical Dynamic Specifications Method (SDSM) relies on in-line measurements to manage dimensional specifications, target and tolerance. SDSM is the cornerstone of an innovative assembling technique based on the Statistical Feed-Forward Control Model (SFFCM) for processes in which the stacked dimensional variation of the assembled components reaches the order of the total allowed tolerance. Since the magnitude of such variation might jeopardize the process capability, it is a matter of interest to study the inclusion of the measurement uncertainty when applying SDSM on a target subprocess. By means of simulating the production of assemblies made of parts having high dimensional variation, a set of experiments were designed to compare the impact of different levels of measurement uncertainty on the capability of a target subprocess. Simulation results showed that depending on the magnitude of the uncertainty the capability index c_p of the target subprocess increases between 2.5% and 34.5% from 1.27 to 1.82 as a direct consequence of adjusting the respective tolerance. Thus, the inclusion of the measurement uncertainty in the proposed SDSM has a significant impact in its practical realization since a decrement in c_p implies an increment of the scrap percentage of the target subprocess.

1. Introduction

The Statistical Feed-Forward Control Model (SFFCM) concentrates the efforts on managing the specifications and tolerances of the inner components of a given assembly by means of applying iteratively the Statistical Dynamic Specifications Method (SDSM) so that the variation of the resulting assemblies' dimension can be kept under control [1].

Assuming the presence of a detectable long-term component or drift in the variation of the dimension of interest, SFFCM divides the component lots in subsets of items produced consecutively in a short-time interval from which some items are inspected to estimate the sample mean and the sample standard deviation of each subset so that necessary adjustments can be determined and triggered.

To carry out the experiments the manufacture of one thousand assemblies made of two components having high dimensional was simulated. The measurement uncertainty was increased stepwise to quantify its influence on the estimation of the adjusted specifications and on the capability of the target subprocess.

2. Statistical Dynamic Specifications Method (SDSM)

SDSM consists of a collection of steps that help managing dimensional specifications and tolerances of inner components of an assembly [1].

Let L_{assy} and t_{assy} be the target and tolerance of an assembly made of two components whose specifications have been set to L_j and t_j and let the variation of the length of the items of Component 1 be the superposition of a random component and a potentially controllable long-term drift.

$$L_{assy} = L_1 + L_2 \quad (1)$$

$$t_{assy} = \sqrt{t_1^2 + t_2^2} \quad (2)$$

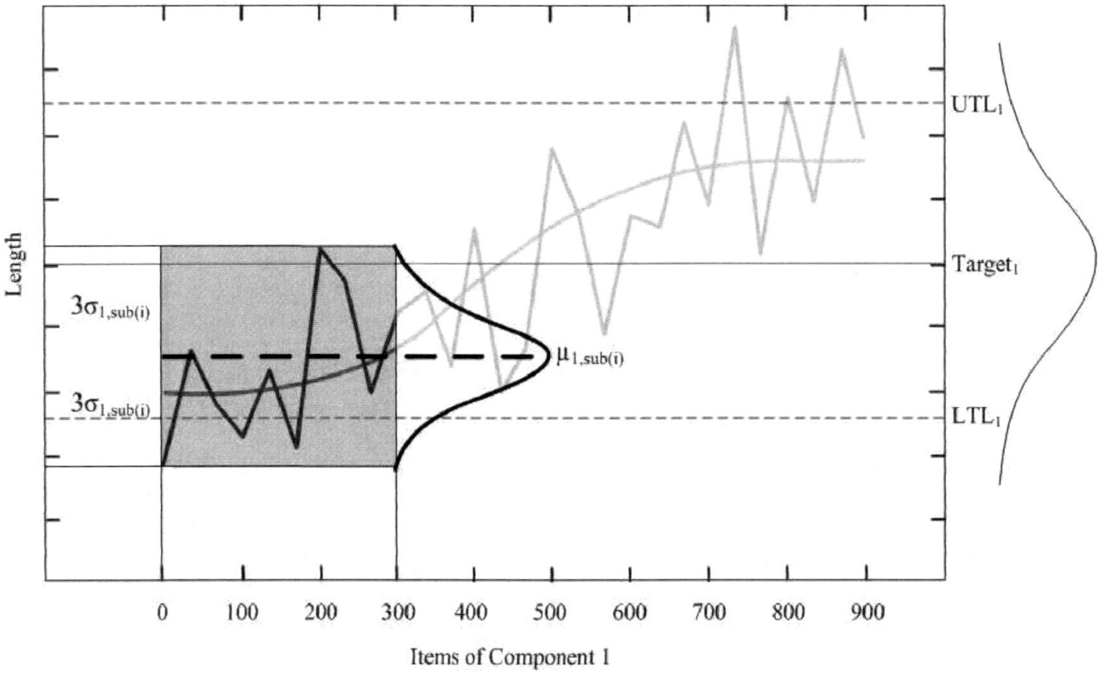

Figure 1. Mean and standard deviation of the subset *i*.

If a small subset *i* of items produced consecutively in a short-time interval were taken from the lot of Component 1, it would be found that 99.73% of the items fall in the band $\mu_{1,sub(i)} \pm 3\sigma_{1,sub(i)}$ (Figure 1). Since the influence of the long-term drift is only partial here, the standard variation of this subset is expected to be smaller than the one of the whole lot [2]. Hence, it is reasonable to think that at least for the subset *i* the nominal tolerance t_1 had not been fully used and that part of it could have been spared to complement the nominal tolerance t_2 of a matching subset *i* of Component 2. In fact, it would have been possible to define another tolerance $t_{2,adj,sub(i)}$ as follows:

$$t_{1,sub(i)} = 3\sigma_{1,sub(i)} \quad (3)$$

$$t_{2,adj,sub(i)} = \sqrt{t_{assy}^2 - t_{1,sub(i)}^2} \quad (4)$$

In the same way, if the mean $\mu_{1,sub(i)}$ of the subset i had been known *a priori* then it would have been possible to define an adjusted target $L_{2,adj,sub(i)}$ for a matching subset i of Component 2 to help meeting the desired L_{assy} as follows:

$$L_{2,adj,sub(i)} = L_{assy} - \mu_{1,sub(i)} \tag{5}$$

3. Statistical Feed-Forward Control Model (SFFCM)

SFFCM separates the system in two, a feeding and a controlled subsystem, to place an intermediate measurement step between. Thus, in the presence of a detectable long-term drift in the output of the feeding Subsystem A, corrective adjustments could be made on the parameters of the controlled Subsystem B by means of applying iteratively SDSM on small subsets of items [3]. The size of the subsets will determine the final number of adjustments needed to assemble the whole lot.

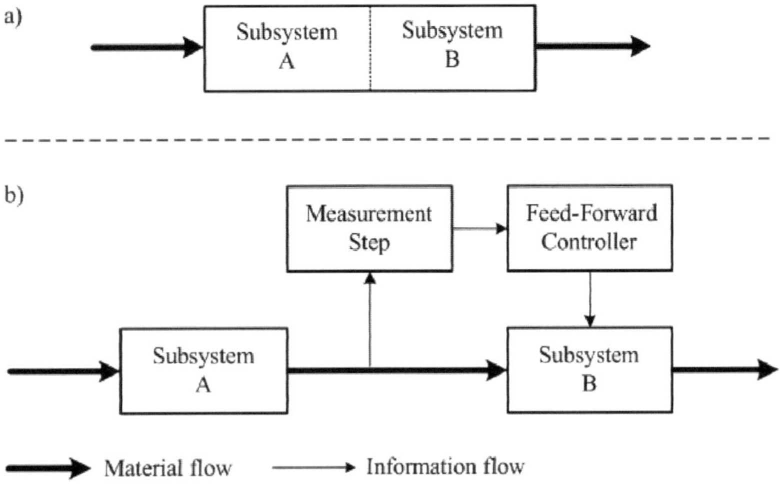

Figure 2. Statistical Feed-Forward Control Model.

4. Measurement Uncertainty

The target of the matching subset i of Component 2 can be computed directly with the following estimator:

$$\hat{L}_{2,adj,sub(i)} = L_{assy} - \bar{x}_{1,sub(i)} \tag{6}$$

However, the computation of the corresponding tolerances is not trivial because the measurement uncertainty has to be considered this time.

$$\hat{t}_{1,sub(i)} = 3s_{1,sub(i)} \tag{7}$$

$$\hat{t}_{1,sub(i),unc} = \hat{t}_{1,sub(i)} + u \tag{8}$$

If $\Delta t_{1,sub(i)}$ defines the difference between the nominal tolerance t_1 and half of the band of $6\,s_{1,sub(i)}$ where 99.73% of the items' lengths are believed to fall and u is the measurement uncertainty (Figure 3), the relation between $\Delta t_{1,sub(i)}$ and u can expressed by means of the following ratio (Equation 10):

$$\Delta t_{1,sub(i)} = t_1 - 3s_{1,sub(i)} \tag{9}$$

$$X\% = \frac{u}{\Delta t_{1,sub(i)}} \tag{10}$$

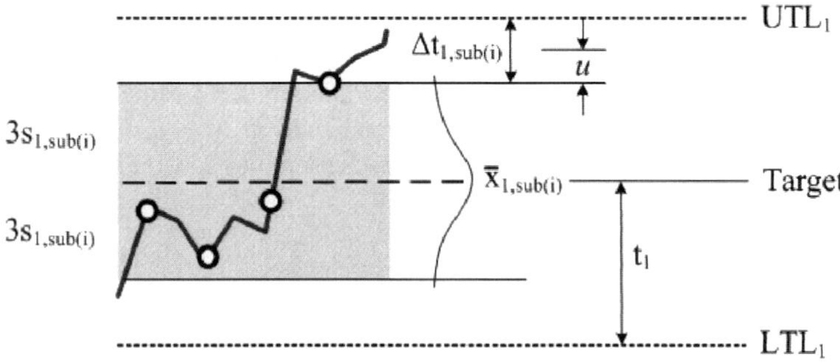

Figure 3. Representation of the measurement uncertainty.

To avoid proceeding with an analysis in terms of a given value u of the measurement uncertainty, Equation 8 was rewritten in terms of the variable ratio $u/\Delta t_{1,sub(i)}$.

$$\hat{t}_{1,sub(i),unc} = \hat{t}_{1,sub(i)} + (X\%)\Delta t_{1,sub(i)} \tag{11}$$

$$\hat{t}_{1,sub(i),unc} = 3s_{1,sub(i)} + (X\%)(t_1 - 3s_{1,sub(i)}) \tag{12}$$

Thus, the adjusted tolerance of a matching subset i of Component 2 can be computed as follows:

$$\hat{t}_{2,adj,sub(i),unc} = \sqrt{t^2_{assy} - \hat{t}^2_{1,sub(i),unc}} \tag{13}$$

5. Simulation

For the purposes of this work the production of a lot of 1,000 assemblies made of two components having high dimensional variation was simulated (Table 2). Each of the experiments mentioned above was replicated 500 times employing new populations generated by a Monte Carlo simulation.

Table 1. Components' nominal specifications [mm] and processes' characteristics

	Target	**Tolerance**	**Mean**	**Std. Deviation**	**c_p**
Assembly	30.00	1.00	29.55	0.29	1.15
Component 1	20.00	0.82	19.60	0.25	1.09
Component 2	10.00	0.58	9.95	0.15	1.29

According to the definition of SFFCM, Component 1 was meant to represent the feeding Subsystem A and Component 2 the controlled Subsystem B.

Simulation results showed that depending on the magnitude of the measurement uncertainty, the adjusted tolerance of the subsets of Component 2 decreased in average from 0.83 to 0.57 (Figure 4). Consequently, the capability of the subprocess corresponding to Component 2 decreased in average from 1.84 to 1.27.

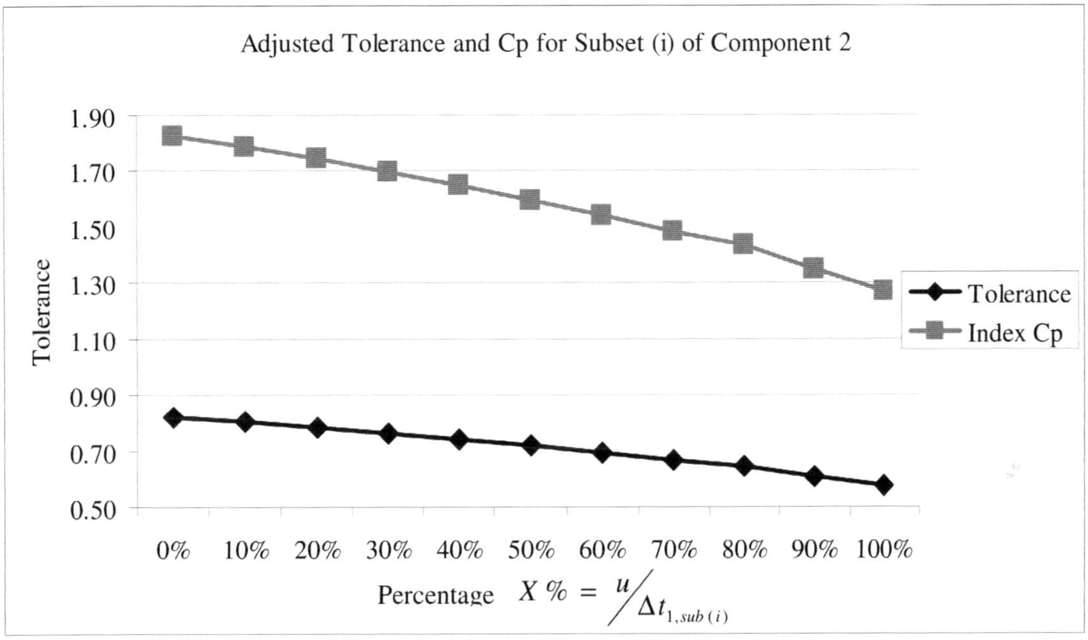

Figure 4. Adjusted tolerance $t_{2,adj,sub(i),unc}$ and $c_{p,2,adj,sub(i)}$ for different ratios $u/\Delta t_{1,sub(i)}$.

In practice, the measurement uncertainty that is added to $t_{1,sub(i)}$ to determine $t_{1,sub(i),unc}$,and thus, to compute the adjusted $t_{2,adj,sub(i),unc}$, will impact directly in the number of items of Component 2 that fall out of their respective tolerance range and that, in consequence, can be considered as defective units or scrap (Figure 5).

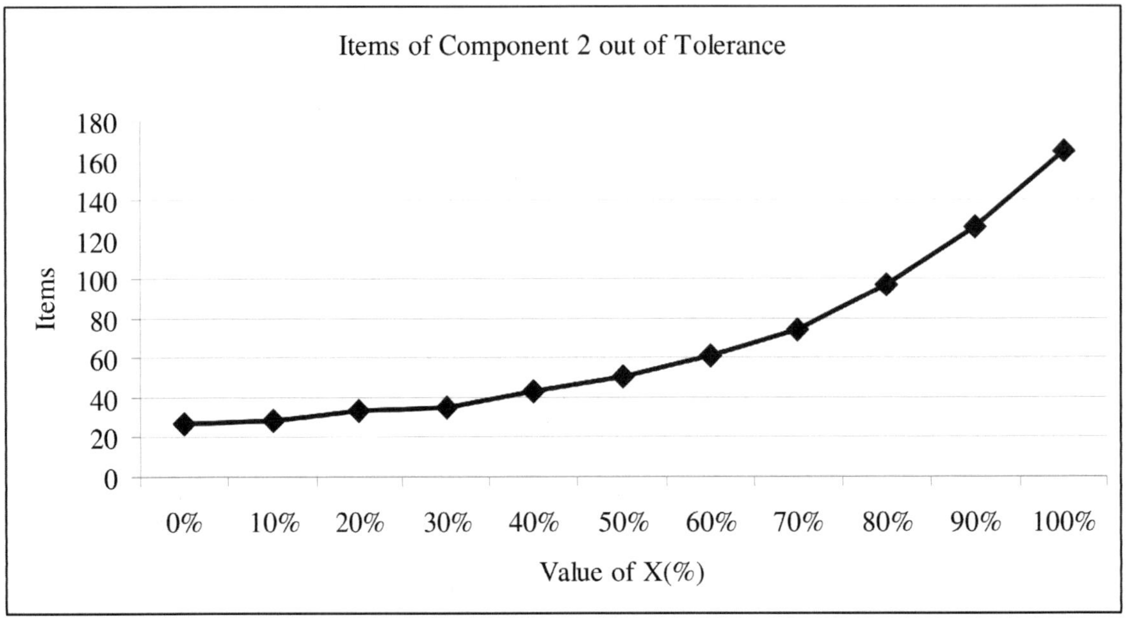

Figure 5. Average number of items of Component 2 out of tolerance.

6. Conclusion

When dealing with assembling processes characterized by high dimensional variation whose cumulative variation reaches the order of the allowed tolerance for the whole assembly, the magnitude of the measurement uncertainty considered to estimate the adjusted specifications using SDSM has a significant impact on its practical realization because any loss of capability (fall of the index c_p) might produce an increment of the scrap percentage of the target subprocess.

7. References

[1] Hernández C and Tutsch R 2012 *Statistical Dynamic Specification Method for Allocating Tolerances: 12th CIRP Conference on Computer Aided Tolerancing (Huddersfield, UK, 18-19 April 2012)*

[2] Burr I 1979 *Elementary Statistical Quality Control* Ch.8 p.192. (Marcel Dekker Inc.).

[3] Hernández C and Tutsch R 2012 *Sampling Strategies and Long Term Variation Modelling for a Statistical Feed-Forward Controller: XX IMEKO World Congress Metrology for Green Growth (Busan, Republic of Korea, 9-14 September 2012)*

Acknowledgments

Carlos Hernández gratefully appreciates the funding of his research stay at TU Braunschweig by CONICYT.

Journal of Physics: Conference Series **459** (2013) 012038 doi:10.1088/1742-6596/459/1/012038

Observations on the worst case uncertainty

Laura Fabbiano[1], Nicola Giaquinto[2], Mario Savino[2], Gaetano Vacca[1]

(1) Dipartimento di Meccanica, Matematica e Management (DMMM)
(2) Dipartimento di Ingegneria Elettrica e dell'Informazione (DEI)
Politecnico di Bari

[giaquinto, savino]@misure.poliba.it; [fabbiano, gvacca]@imedado.poliba.it

Abstract. The paper discuss the computation of the worst case uncertainty (WCU) in common measurement problems. The usefulness of computing the WCU besides the standard uncertainty is illustrated. A set of equations to compute the WCU in almost all practical situations is presented. The application of the equations to real-world cases is shown.

1. Introduction

The Guide to the Expression of Uncertainty in Measurements (GUM) [1] does not consider explicitly the "worst case uncertainty" (WCU), that is, the uncertainty with a 100% coverage probability. The GUM Uncertainty Framework (GUF) is focused on standard uncertainties, and on the computation of expanded uncertainties under the hypothesis of Gaussian distribution. Therefore, while in the GUF an expanded uncertainty with coverage probability of 95% or 99% is easily computed, the expanded uncertainty with 100% coverage probability is not. Indeed, either it is infinite (in the Gaussian approximation), or it needs separate propagation rules [2].

In practice, WCUs are widely used and, in general, worst-case analysis is a well-known and regarded tool in engineering [3], [4]. First of all, accuracy characteristics of almost *all* commercial instruments are given in terms of WCUs [5]. As a consequence, in common engineering and laboratory practice, simple functions of few measurements are assessed using the WCU, using simple and well-known propagation rules [6]. (For example, the absolute WCU of $x_1 - x_2$ is the sum of the absolute WCUs of x_1 and x_2; the relative WCU of x_1 / x_2 is the sum of the relative WCUs; and so on.) WCUs, in general, are the most convenient way of defining a credible interval whenever the involved distributions are bounded and without long tails, which is very common, e.g. when the measurement model involves few input quantities with uniform distribution. Therefore, the lack of an adequate theoretical recognition and sistematization of the WCU concept is not a desirable situation for the measurement science community.

In this paper, the propagation of the WCU is discussed. Firstly, it is pointed out that the information sufficient to compute the standard uncertainty (SU) is, in general, not sufficient to compute the WCU, but, for the same reason, computing *only* the SU may give insufficient information to state correctly the expanded uncertainty. Afterwards, the problem of computing the WCU without a full Monte Carlo analysis is considered. It is demonstrated that simple propagation formulae can be given in a very common situation, i.e. when measurement errors are a linear function of a set of statistically independent measurement errors. For this case, also the propagation of the SU is computed, avoiding explicit computation of correlation coefficients. The propagation formulae are shown to be congruent with accuracy characteristics of instrument manufacturers, characteristics that cannot be written nor understood in the GUF.

Throughout the paper, the concept of measurement error is used. It is the difference $E = \hat{X} - X$, where $\hat{X}$ is the measurement and X is the unknown measurand. According to the Bayesian approach of Supplement 1 of GUM [7], $\hat{X}$ is deterministic, X is a random variable, and therefore E is a random variable. Computing the propagation using E instead of X does not change the final result (GUM, E.5) and has some advantages; in particular, it is necessary to clarify the accuracy characteristics of many instruments, which have an underlying "error model". The WCU is the maximum absolute value that the random variable E can assume: in the following, it is denoted by $U = \max|E|$.

2. Worst-case uncertainty for a general measurement model and general input quantities

2.1. *Computation of WCU using joint distributions*

Let $y = g(x_1,...,x_N)$ be the measurement model, E_i the measurement errors on the input quantities, $U(x_i) = \max|E_i| = U_i$ the WCUs on the input quantities ($i = 1,...,N$), E_y the error on the output quantity, and $U(y) = \max|E_y| = U_y$ the WCU on the output quantity.

Computing the WCU on the output quantity requires, in the general case, complete knowledge of the multivariate joint probability density function (pdf) of the errors $E_1,...,E_N$, and a cumbersome computation of the entire distribution of the error E. In particular, unlike SU, *correlation coefficients are useless* when computing the WCU. This is illustrated by the following example.

Let us consider two measurements x_1, x_2, affected by uncorrelated errors E_1, E_2 with identical (marginal) distributions $f_1(e_1) = f_2(e_2)$, supposed to be symmetric triangular in $[-1,1]$ (and therefore with WCU $U(x_1) = U(x_2) = U_x = 1$). The information on E_1, E_2 is sufficient to evaluate the SU, but not the WCU, of the sum $y = x_1 + x_2$. The inspection of the joint pdf in two different cases of uncorrelated errors clarifies the issue.

Fig. **1** shows the joint distribution of statistically independent triangular E_1, E_2, while Fig. **2** shows the distribution in a case when they are uncorrelated but not independent. The resulting distribution $f(e)$ of $E = E_1 + E_2$ in the two cases is depicted in Figs. **3** and **4**. Uncorrelated errors with the same distribution yield completely different output distributions and WCU ($U(y) = 2$ in the first case, $U(y) = 1$ in the second one).

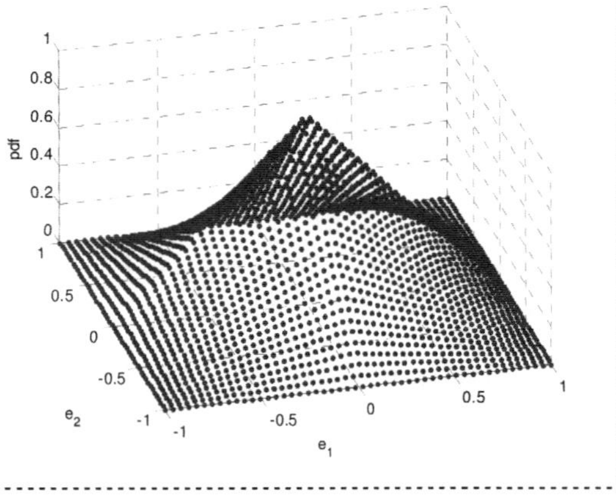

Figure 1. Joint pdf of independent errors with identical marginal pdf (triangular with $U_x = 1$).

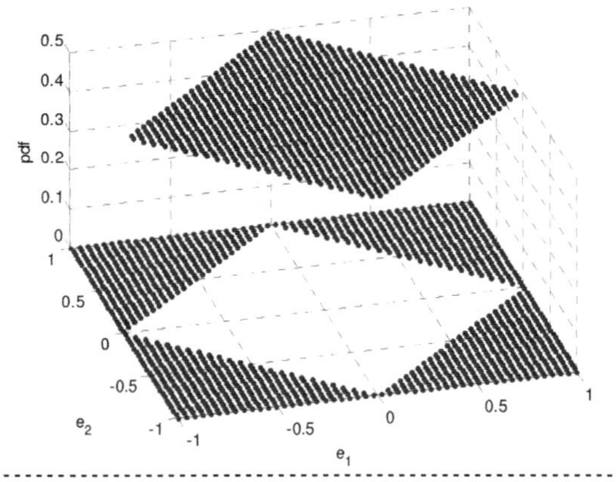

Figure 2. Joint pdf of uncorrelated, but not independent, errors, with the same marginal pdf.

For a generic joint pdf of the input errors E_i, computing the distribution of the output error E can be accomplished only via Monte Carlo simulations, in the same way described in GUM Supplement 1.

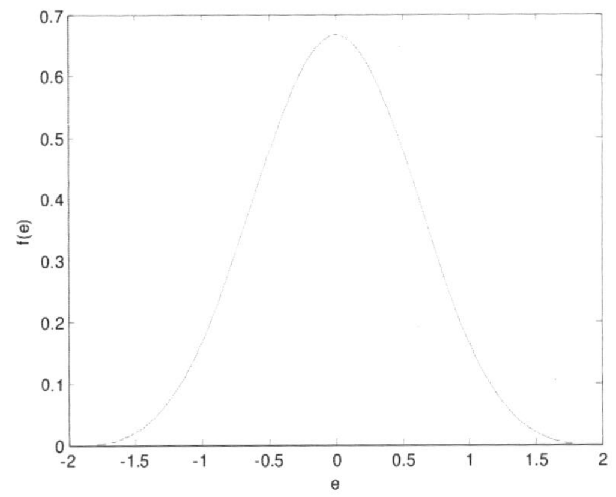

Figure 3. Distribution of $E = E_1 + E_2$ for the joint distribution depicted in Fig. **1**.

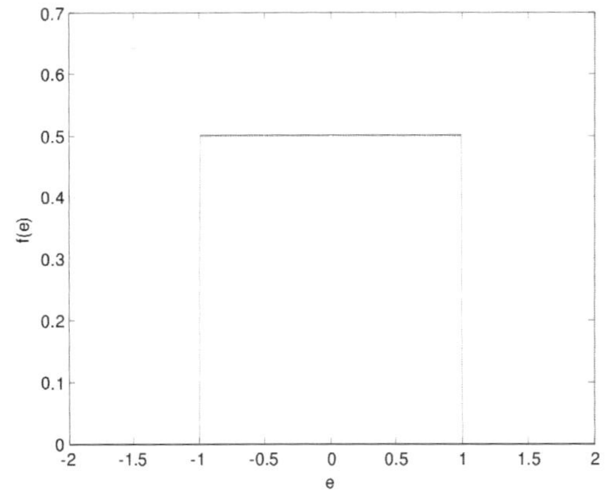

Figure 4. Distribution of $E = E_1 + E_2$ for the joint distribution depicted in Fig. **2**.

2.2. Reasons for computing the WCU

In the two cases above, let us assume knowledge of the WCU *only*, i.e. $U(y) = 2$ in the first case, $U(y) = 1$ in the second. The SU, due to the hypothesis of uncorrelated errors, is $u(y) = \sqrt{u^2(x_1) + u^2(x_2)} = U_x / \sqrt{3} = 0.577$ in both cases. Therefore, the ratio (WCU/SU) is $U(y)/u(y) = 3.464$ in the first case, and $U(y)/u(y) = \sqrt{3} = 1.732$ in the second case.

The ratio U/u is the maximum sensible coverage factor and therefore indicates, even without knowing the actual distributions $f(e)$, the applicability of the normal approximation. In the first case the ratio $U/u = 3.5$ indicates that coverage factors like $k = 2$ or $k = 2.58$ (for probabilities of 95% or 99%, respectively) are applicable and accurate, while, in the second case, the ratio $U/u = 1.732$ demonstrates that these coverage factors do not make sense. In the first case, for a coverage factor $k = 2$, the actual coverage probability p is $p = 95.75\%$, with a negligible difference from the Gaussian value 95.45%. In the second case, instead, the ratio $U/u = 1.732$ immediately suggests that the error must have a uniform distribution, and the measurement uncertainty is effectively quantified by the WCU. Summing up, a simple procedure to compute the WCU, *without* computing the entire distribution, is desirable. The reasons are essentially the following:

- the ratio WCU/SU shows the applicability of the normal approximation (i.e., of the GUF);
- the WCU is the maximum expanded uncertainty that may be attached to the measurement.

3. Practical computation of worst-case uncertainty

3.1. Worst-case uncertainty propagation for statistically independent errors

Under the hypothesis of statistically independent errors, computing the WCU is very simple and even more straightforward then computing the SU. This well-known fact is recalled here to introduce concepts and notation used later. In particular, a compact matrix notation is used. We introduce the symbols $\mathbf{x} = [x_1, ..., x_N]^T$, $y = g(\mathbf{x})$, $\mathbf{E}_x = [E_1, ... E_N]^T$, $\mathbf{U}_x = [U(x_1), ..., U(x_N)]^T$. We denote with $\mathbf{C} = \partial y / \partial \mathbf{x} = [\partial y / \partial x_1, ..., \partial y / \partial x_N]$ the Jacobian matrix of the measurement model (for a scalar output it is a row vector). This is the vector of the sensitivity coefficients. The well-known first-order approximation of the error propagation law is

$$E_y \cong \mathbf{C}\cdot\mathbf{E}_x . \tag{1}$$

If the input errors are statistically independent, the maximum absolute value of the output error, $\max|E_y| = \max|\mathbf{C}\cdot\mathbf{E}_x|$, is simply factorized in $|\mathbf{C}|\cdot\max|\mathbf{E}_x|$, where $|\mathbf{C}|$ denotes the matrix of the absolute values of the elements of **C**. Therefore the WCU is given by:

$$U_y \cong |\mathbf{C}|\cdot\mathbf{U}_x . \tag{2}$$

With the same notation, the SU is simply expressed by

$$u_y^2 = \mathbf{C}\cdot\mathbf{\Sigma}_x\cdot\mathbf{C}^T \tag{3}$$

where $\mathbf{\Sigma}_x$, the covariance matrix, in this case is diagonal $\mathbf{\Sigma}_x = \mathrm{diag}(u^2(x_1),\ldots u^2(x_N))$.

3.2. *Extending the propagation formulae to non-independent errors*

The simple formulae above can be extended to the case of non-independent errors in a case that is very common in practice:

- the errors, although not independent, can be expressed as a linear function of statistically independent errors.

A typical example of this kind of situation is the static error introduced by an A/D converter, which is usually expressed in the form [8]

$$e(x) = \Delta G\cdot x + O + inl(x) + e_q(x) \tag{4}$$

where ΔG is the gain error, O is the offset error, $inl(x)$ is the integral nonlinearity error, and $e_q(x)$ is the quantization error. The experimental results about the variability of the systematic errors in ADC-based instrument are reported in [9], [10].

If many measurements $x_1,\ldots,x_N$ are taken by the A/D converter in repeatability conditions (but with different measurand), gain and offset errors are essentially the same in all the measurements, while the integral nonlinearity and the quantization errors are not. Therefore, the N measurement errors are a function of $2N+2$ independent errors: ΔG, O, $inl(x_1)$, $e_q(x_1)$, ..., $inl(x_N)$, $e_q(x_N)$. Many other examples of this situation are possible, including the common case of "cascaded" measurement models.

In all such cases, the propagation formulae for the WCU are a simple extension of those for independent errors. Let $\mathbf{E}' = [E'_1,\ldots,E'_M]^T$ be the vector of the statistically independent errors. The input errors $\mathbf{E}_x = [E_1,\ldots E_N]^T$ are given by

$$\mathbf{E}_x = \mathbf{T}\cdot\mathbf{E}' \tag{5}$$

where **T** is a transformation matrix. Equations (1), (2), (3) becomes

$$E_y = \mathbf{C}\cdot\mathbf{T}\cdot\mathbf{E}',\ U_y = |\mathbf{C}\cdot\mathbf{T}|\cdot\mathbf{U}',\ u_y^2 = \mathbf{C}\cdot\mathbf{T}\cdot\mathbf{\Sigma}_x\cdot\mathbf{T}^T\cdot\mathbf{C}^T \tag{6)-(7)-(8}$$

Equation (7) is the propagation law for the WCU. An example of its application is easily done, considering errors of the kind (4).

For two measurements x_1, x_2 taken in repeatability conditions, affected by errors (4), the independent errors are the 6×1 (unknown) vector

$$\mathbf{E}' = [\Delta G, O, inl(x_1), inl(x_2), e_q(x_1), e_q(x_2)]^T . \tag{9}$$

with associated (known) WCUs given by

$$\mathbf{U}' = [U_G, U_O, U_{inl}, U_q, U_{inl}, U_q]^T . \tag{10}$$

On the basis of the error model (4), the transformation matrix **T** is:

$$\mathbf{T} = \begin{bmatrix} x_1 & 1 & 1 & 0 & 1 & 0 \\ x_2 & 1 & 0 & 1 & 0 & 1 \end{bmatrix} . \tag{11}$$

It is now immediate to compute the worst-case uncertainty for any given measurement model $g(x_1, x_2)$. For example, if $y = x_1 - x_2$, the sensitivity coefficients are $\mathbf{C} = [1, -1]$, and therefore

$$\mathbf{C} \cdot \mathbf{T} = [x_1 - x_2, 0, 1, 1, 1, 1] . \tag{12}$$

The WCU is:

$$U_y = |\mathbf{C} \cdot \mathbf{T}| \cdot \mathbf{U}' = |x_1 - x_2| U_G + 2U_{inl} + 2U_q . \tag{13}$$

The elements of the vector $\mathbf{C} \cdot \mathbf{T}$ are modified sensitivity coefficients: they are sensitivity coefficients of the independent errors. In the example, $x_1 - x_2$ is the coefficient of the gain error, 0 is the coefficient of the offset error, etc.

4. Application to real-world instruments and to the GUM

The derived propagation formulae applies, of course, to real-world measurements. An example are the accuracy characteristics provided by Agilent Technologies [11], [12]. This manufacturer gives accuracy specification in terms of WCUs, like almost all instrument manufacturers, but usually gives also separate formulae for the uncertainty of the single measurement x, and for the difference of two measurements in repeatability conditions ("dual cursor measurements"), $y = x_2 - x_1$. Therefore, it is very simple to write down the vector $\mathbf{U}'$ and the matrix **T** for a pair of measurements (x_1, x_2).

Below, we report $\mathbf{U}'$ and **T** for different instruments, together with the formulae for the WCU of the single measurement x_1 and the difference $y = x_1 - x_2$. The formulae are computed according to the developed mathematics, and of course coincide with those reported in the technical sheets. In the formulae, V_{FS} is the selected full-scale range, and V_P is the selected "vertical position".

Agilent 54600B:

$$\mathbf{U}' = [1.9\% \;\; 1\% V_{FS} \;\; 0.5\% |V_P| \;\; 0.2\% \cdot V_{FS} \;\; 0.2\% V_{FS}]$$

$$\mathbf{T} = \begin{bmatrix} x_1 & 1 & 1 & 1 & 0 \\ x_2 & 1 & 1 & 0 & 1 \end{bmatrix}$$

$$U(x_1) = 1.9\% |x_1| + 1.2\% V_{FS} + 0.5\% |V_P|$$

$$U(x_1 - x_2) = 1.9\% |x_1 - x_2| + 0.4\% V_{FS}$$

Agilent U1610A:

$$\mathbf{U}' = [4\% \;\; 0.1 \text{ div} \;\; 2 \text{ mV} \;\; 1.6\% |V_P| \;\; 0.2\% \cdot V_{FS} \;\; 0.2\% V_{FS}]$$

$$\mathbf{T} = \begin{bmatrix} x_1 & 1 & 1 & 1 & 1 & 0 \\ x_2 & 1 & 1 & 1 & 0 & 1 \end{bmatrix}$$

$$U(x_1) = 4\% |x_1| + 0.1 \text{ div} + 2 \text{ mV} + 1.6\% V_P + 0.2\% V_{FS}$$

$$U(x_1 - x_2) = U(x_1) = 4\% |x_1 - x_2| + 0.4\% V_{FS}$$

For the U1610A, in the final declared characteristics the manufacturer substitutes $4\% \mid x_1 \mid$ and $4\% \mid x_1 - x_2 \mid$ with a common upper bound, $4\% V_{FS}$.

5. Conclusions

Worst case analysis is a precious and commonly used tool in engineering, but it is not included in the GUM, mainly because the WCU does not fit in the propagation equations for the SU. The paper shows briefly that WCU can be included in the GUM framework with a set of separate equations, which are, however, very simple and formally analogous to those for SU. WCU can be computed simply and automatically when measurement errors are independent, and in almost all the practical cases of non-independent errors. The WCU/SU ratio indicates when the Gaussian approximation of the GUF is applicable, and when not.

Acknowledgements

This work has been partially funded by Project PON R&C 2007-2013 "RES NOVAE".

References

[1] International Organization for Standardization, ISO/IEC Guide 98-3:2008 - Uncertainty of measurement – Part 3: Guide to the expression of uncertainty in measurement (GUM:1995). ISO, 2008.
[2] W. Bich, "How to revise the GUM?," Accreditation and Quality Assurance, vol. 13, no. 4–5, pp. 271–275, 2008.
[3] R. J. N. B. da Silva, J. R. Santos, and M. F. G. F. C. Camões, "Worst case uncertainty estimates for routine instrumental analysis," Analyst, vol. 127, no. 7, pp. 957–963, Jun. 2002.
[4] X. Gu, J. Renaud, S. Batill, R. Brach, and A. Budhiraja, "Worst case propagated uncertainty of multidisciplinary systems in robust design optimization," Structural and Multidisciplinary Optimization, vol. 20, no. 3, pp. 190–213, Nov. 2000.
[5] R. L. Brown, "Using a Manufacturer's Specification as a Type B Error Contribution." Agilent Technologies, 03-Dec-2012.
[6] M. Grabe, "Estimation of measurement uncertainties - an alternative to the ISO Guide," Metrologia, vol. 38, no. 2, Apr. 2001.
[7] International Organization for Standardization, ISO/IEC Guide 98-3:2008/Suppl 1:2008 - Uncertainty of measurement – Propagation of distributions using a Monte Carlo method. ISO, 2008.
[8] F. Attivissimo, N. Giaquinto, and M. Savino, "Worst-case uncertainty measurement in ADC-based instruments," Computer Standards and Interfaces, vol. 29, no. 1, pp. 5–10, 2007.
[9] F. Attivissimo, A. Cataldo, L. Fabbiano, and N. Giaquinto, "Systematic errors and measurement uncertainty: An experimental approach," Measurement, vol. 44, no. 9, pp. 1781–1789, Nov. 2011.
[10] F. Attivissimo, A. Cataldo, N. Giaquinto, and M. Savino, "Assessment of the uncertainty associated with systematic errors in digital instruments: an experimental study on offset errors," Meas. Sci. Technol., vol. 23, no. 3, Mar. 2012.
[11] Agilent Technologies, "54600B 2 Channel 100 MHz Oscilloscope Data Sheet." [Online]. Available: http://www.home.agilent.com.
[12] Agilent Technologies, "U1610A Handheld Digital Oscilloscope Data Sheet." [Online]. Available: http://www.home.agilent.com.

IMEKO 2013 TC1 + TC7 + TC13 IOP Publishing
Journal of Physics: Conference Series **459** (2013) 012039 doi:10.1088/1742-6596/459/1/012039

Fit for purpose models for metrology: a model selection methodology

L Wright, T J Esward

Mathematics and Modelling Group, National Physical Laboratory, Hampton Road, Teddington TW11 0LW, UK

E-mail: louise.wright@npl.co.uk

Abstract. We describe a model selection methodology for partial differential equation (PDE) based models and show the results of applying it to a test problem derived from a model of the laser flash thermal diffusivity measurement technique. A methodology for comparing and choosing simplified models is of benefit to the model development process in metrology. It is assumed that the computational aim is not only to solve the model to obtain the results that the metrologist requires, but to ensure that the model is no more complex than necessary to achieve this and that results can be obtained in a reasonable time using the available computing resources. The advantage of the proposed method is that it avoids the need to solve directly the underlying complex model. We present the results of comparisons of four models of the laser flash problem and identify the further work needed to extend the approach to a wider range of problems and to identify suitable measures for comparing residuals.

1. Introduction

This paper describes a model selection methodology for PDE based models, and shows the results of applying it to a simple test problem, derived from a model of the laser flash thermal diffusivity measurement technique. The majority of PDE models are not amenable to analytical solution, and must be solved using software implementations of numerical approximation techniques. For the purposes of this paper it is necessary to discuss briefly the following terms:

- model complexity: the more complex a model is, the more difficult it is to solve. Nonlinear models are more complex than linear models. Increasing the dimension of a model increases its complexity. Adding more unknown quantities to a model increases the model complexity.
- computational expense: the longer a model takes to solve using a given piece of software running on a given computer system, the more computationally expensive that model is. Increased model complexity leads to increased computational expense.
- computational aim: the computational aim of the model is a specification of the information, preferably including a required level of uncertainty, that the model needs to supply.
- fit for purpose: a fit for purpose model is a model that generates the required results to the necessary level of accuracy within a manageable amount of time, i.e., it achieves its computational aim at a reasonable computational expense.

One approach to model development that balances the need for a detailed description of the physical system and the need for low computational cost is to start from a simple model and progressively add complexity until the model can be considered as validated and hence fit for purpose. Common assumptions used to simplify models include uniformity and constancy of material properties, linearity of boundary conditions and loads, neglect of thermal and electrical effects, and reduction of the dimensionality of the problem by assuming axisymmetry or planarity.

One difficulty with this approach to model development is identifying what change to the model will make the most difference to the results. This problem is particularly common when models have been simplified via assumption of constancy or linearity in their material properties, loads, and boundary conditions and it is not clear which assumption is most likely to be invalid.

Another problem is that it is difficult to know in advance that a given change to a model will have a significant effect on the results, so it is possible to waste a large amount of effort and time implementing and running a complex model, only to find that the new complexity has made no difference to the results. The development of multiphysics models from simpler models, transient models from steady-state models, and the extension of a two-dimensional model into a three-dimensional model are examples of labour-intensive model changes with no guarantee of improvement.

These challenges demonstrate that a methodology for comparing complex model changes that does not involve solution of complex models would be of benefit to the model development process.

2. Model selection

Model selection for PDE models is usually strongly application-driven and the choices made are often motivated by expert knowledge, so as a result there is little work in the literature on generic techniques for comparing different models.

Papers by authors such as Bär et al [3] and Guo and Billings [4] have investigated selecting a set of governing equations for dynamical systems from a parameterised family of nonlinear equations. Applications of this work include development of models of distributed parameter systems and of reaction-diffusion equations. The authors select the models by minimising the difference between model predictions and measured values. Other work [1] has evaluated best-fit values of a spatially varying diffusion coefficient at the nodes of a finite element (FE) mesh by matching model results to known values of the solution at the nodes.

The measured values used by these techniques typically need to be distributed throughout the spatio-temporal domain for a good fit to be possible. This requirement restricts the range of problems to which the approaches can be applied. Many measurement techniques take surface measurements or measurements of bulk behaviour, but cannot measure sub-surface values. Bamabach et al[2] proposed a methodology for choosing the best-fit model for a given experiment from a finite group of governing equations. The methodology does not require measurement results to be distributed throughout the domain, and so is more applicable to metrology problems.

The methodology proposed by Bamabach et al[2] employs optimisation to identify, for every possible pair of models in the group, a set of model input values and model results that create the biggest difference in the chosen results between the two models. It is then necessary to carry out an experiment for each set of input parameters and make measurements equivalent to the model results. The optimal model choice can then be obtained from the measured values and the model results. This approach is a good objective method for choosing amongst a group of models, but the optimisation process to identify the input values that cause most difference between the results of a pair of models will require repeated solution of each model, which will be computationally expensive. The methodology we propose avoids direct solution of complex models, making a significant saving in computational effort.

Journal of Physics: Conference Series **459** (2013) 012039 doi:10.1088/1742-6596/459/1/012039

3. Proposed methodology

The methodology has been developed to compare and assess changes to a model that solves a PDE model using a numerical approximation technique. Throughout the following it is assumed that the same discretisation has been used for all models. At its current stage of development the methodology only addresses steady-state models, so the numerical approximation technique will result in a matrix of equations of the form

$$A(\mathbf{u})\mathbf{u} = \mathbf{b}, \tag{1}$$

where $\mathbf{u} = (u_1, u_2, \ldots, u_N)^\top$ is a vector of unknown solution parameters, $\mathbf{b}$ is a known vector that implements the loading and boundary conditions and is independent of $\mathbf{u}$, and $A(\mathbf{u})$ is a matrix that will be dependent on $\mathbf{u}$ if the model is nonlinear. It is assumed that all matrices that define models are non-singular, so all models are solvable, and that a given model has a unique solution. It is further assumed that all of the entries in the matrix A, the functions $a_{ij}(\mathbf{u})$, are infinitely differentiable functions of the u_k. The majority of physical models lead to functions such that the $a_{ij}(\mathbf{u})$ are either polynomials or can be expanded as polynomials within the range of solution values likely to occur within a given model, so this assumption is unlikely to restrict the validity of the proposed methodology.

Suppose that the solution $\mathbf{u}^0$ to a simple model $A^0(\mathbf{u}^0)\mathbf{u}^0 = \mathbf{b}^0$ is known, and that we wish to assess how different from $\mathbf{u}^0$ the solution to a more complex model $A^1(\mathbf{u}^1)\mathbf{u}^1 = \mathbf{b}^1$ is. Then the ith component of the vector of residuals $\mathbf{r}$ obtained from putting the known solution $\mathbf{u}^0$ into the new model is given by

$$\begin{aligned} r_i &= [A^1(\mathbf{u}^0)\mathbf{u}^0 - \mathbf{b}^1]_i = [A^1(\mathbf{u}^0)\mathbf{u}^0 - A^1(\mathbf{u}^1)\mathbf{u}^1]_i & (2)\\ &= \sum_{j=1}^{N} a_{ij}^1(\mathbf{u}^0)u_j^0 - \sum_{j=1}^{N} a_{ij}^1(\mathbf{u}^1)u_j^1. & (3)\end{aligned}$$

The assumption of the uniqueness of the solution to each model means that these residuals can only be uniformly zero if $\mathbf{u}^0$ and $\mathbf{u}^1$ are identical. We can expand the value of $a_{ij}^1(\mathbf{u})$ as a Taylor series about $\mathbf{u}^0$:

$$a_{ij}^1(\mathbf{u}^1) = a_{ij}^1(\mathbf{u}^0) + \sum_{k=1}^{N}(u_k^1 - u_k^0)\left\{ \left.\frac{\partial a_{ij}^1}{\partial u_k}\right|_{\mathbf{u}_0} + \sum_{l=1}^{N} \frac{(u_l^1 - u_l^0)}{2} \left.\frac{\partial^2 a_{ij}^1}{\partial u_k \partial u_l}\right|_{\mathbf{u}_0} + O((u_m^0 - u_m^1)^3) \right\} \tag{4}$$

so that, abbreviating terms above the linear, the ith component of the residual vector becomes

$$\begin{aligned} r_i &= \sum_{j=1}^{N} \left(a_{ij}^1(\mathbf{u}^0)u_j^0 - a_{ij}^1(\mathbf{u}^0)u_j^1 - u_j^1 \sum_{k=1}^{N}(u_k^1 - u_k^0) \left\{ \left.\frac{\partial a_{ij}^1}{\partial u_k}\right|_{\mathbf{u}_0} + O((u_m^0 - u_m^1)^2) \right\} \right) & (5)\\ &= \sum_{j=1}^{N} \left[a_{ij}^1(\mathbf{u}^0) + \sum_{k=1}^{N} u_k^1 \left\{ \left.\frac{\partial a_{ik}^1}{\partial u_j}\right|_{\mathbf{u}_0} + O((u_m^0 - u_m^1)) \right\} \right] (u_j^0 - u_j^1), & (6)\end{aligned}$$

which is a weighted sum of the components of $\mathbf{u}_0 - \mathbf{u}_1$. The ℓ^2 norm and the ℓ^∞ norm can be used to compare residuals from different models.

Many numerical approximation methods produce matrices and vectors such that the residuals are strongly dependent on the local mesh size. This dependence makes comparison of values within the same vector misleading unless the residuals are weighted by mesh size in some way. All results in this paper have been normalised by multiplying by V_i/V where V_i is the integral of the ith basis function over the domain, and V is the total volume of the domain of the model.

Note that individual residuals are not a direct measure of the difference between the results in a given region. For instance, for models created using finite element analysis, all pairs of models that differ only in boundary conditions have zero residuals for points that do not lie on the boundary.

4. Test problem

A simple steady-state test problem was developed to demonstrate the proposed methodology. The problem was designed such that increases in complexity could be introduced in its boundary conditions and material properties, and potentially an increase in dimension could be introduced. The problem was a two-dimensional steady-state thermal simulation of a cylinder with thermal conductivity $\lambda(T)$, of radius R and thickness H, being heated by a specified flux Q on one flat face, and losing heat via a heat transfer coeffficient $h(T)$ to an atmosphere at fixed temperature T_a from its other faces. The problem was created by considering a steady-state version of a model of the laser flash thermal diffusivity experiment, a problem studied before [5].

The full specification of the problem is:

$$\begin{aligned}
\frac{\partial}{\partial r}\left(\lambda(T)\frac{\partial T}{\partial r}\right)+\frac{\partial}{\partial z}\left(\lambda(T)\frac{\partial T}{\partial z}\right) &= 0, \quad 0 \le r \le R, 0 \le z \le H, && (7)\\
\lambda(T)\partial T/\partial z &= -Q(r), \quad 0 \le r \le R, z = 0, && (8)\\
\lambda(T)\partial T/\partial z &= h(T)(T_a - T), \quad 0 \le r \le R, z = H, && (9)\\
\lambda(T)\partial T/\partial r &= h(T)(T_a - T,) \quad 0 \le z \le H, r = R. && (10)
\end{aligned}$$

The residuals of the discretised form of the equation have units of watts (W).

The initial model fixed λ to be constant, with $\lambda = 55.5$ W/(m K), and fixed Q and h to be constants related to the ambient temperature T_a, the emissivity ϵ (set to 0.8), and the Stefan-Boltzmann constant σ via $Q = 12\epsilon\sigma T_a^3(628 - T_a)$ W/m^2 and $h = 4\epsilon\sigma T_a^3$ W/(m^2 K). The definition of h was derived from a linearisation of the nonlinear boundary conditions for radiative heat loss. The expression for heat flux was developed by using a one-dimensional model that took surface area into account to estimate a heat flux that would give a temperature of 628 K on $z = H$. Two different ambient temperatures were used: $T_a = 528$ K and $T_a = 618$ K.

Two different changes to the initial model were identified, each of which would lead to additional complexity if the model were solved directly. The first change introduced radiation boundary conditions on $r = R$ and on $z = H$ so that $h(T) = \epsilon\sigma(T_a^3 + T_a^2 T + T_a T^2 + T^3)$. The second change used temperature-dependence of the thermal conductivity, specified as $\lambda(T) = 55 + 0.1(T - \bar{T})^2$, where $\bar{T}$ is the average temperature of the simple model for a given ambient temperature. For the model with $T_a = 528$ K, $\bar{T} = 630.55$ K, and for the model with $T_a = 618$ K, $\bar{T} = 628.41$ K

The simple models were solved as a finite element problem using a uniform mesh of 16 linear quadrilateral elements in each direction. The results of the simple model with $T_a = 528$ K are shown in figure 1. The models with added complexity were also solved using the same finite element mesh so that differences in results could be calculated directly.

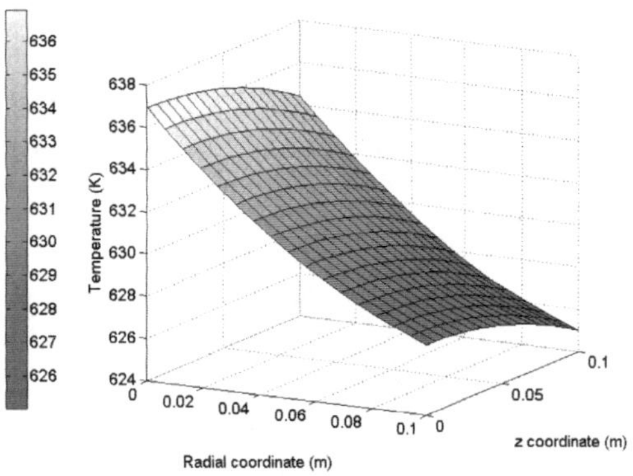

Figure 1. Temperature results for $T_a = 528$ K. The temperature results for $T_a = 618$ K show a similar spatial distribution, but vary between 627.6 K and 629.4 K.

5. Test problem results

The values of the residuals associated with the basis function associated with each node are plotted in figures 2 to 4. Note that these figures plot the results in different orientations so that the interesting features can be seen more clearly: the axes show the orientation of each plot.

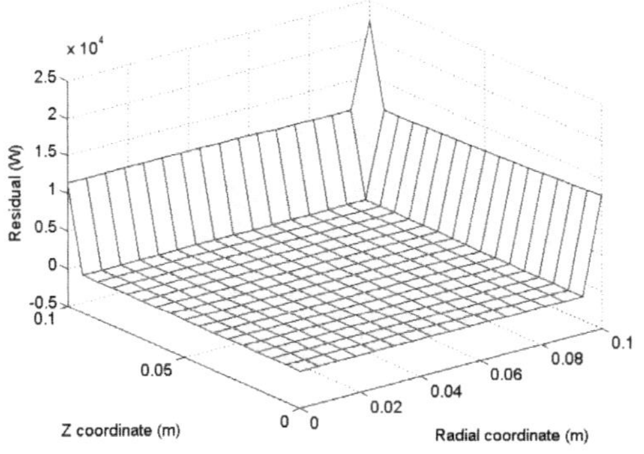

Figure 2. Residuals from the radiation boundary condition model for $T_a = 528$ K. The residuals for $T_a = 618$ K have a similar spatial distribution, but are an order of magnitude smaller.

Figure 2 shows the residuals obtained by putting the solution to the simple model into a model with radiation boundary condition with $T_a = 528$ K. The residuals are zero away from the boundary because the boundary condition change has not altered the equations linking nodes away from the boundary. The residuals are almost uniform where they are non-zero. There is a sharp spike in value at the corner $r = 0.1, z = 0.1$. This spike probably occurs because that corner is affected by the change in boundary condition in two directions. The spike suggests that a different residual scaling strategy is required for nodes affected by boundary conditions.

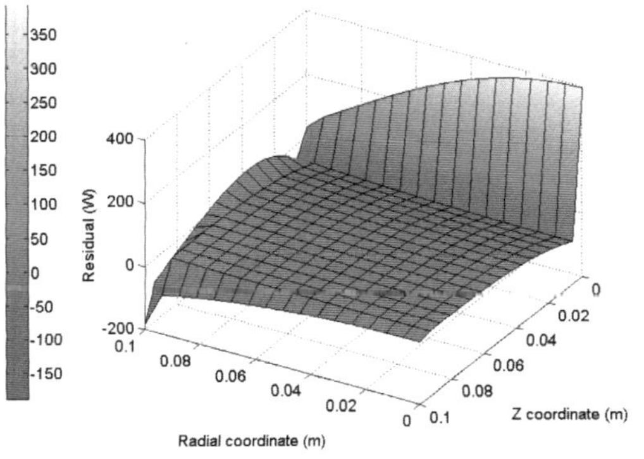

Figure 3. Residuals from the temperature dependent thermal conductivity model for $T_a = 528$ K.

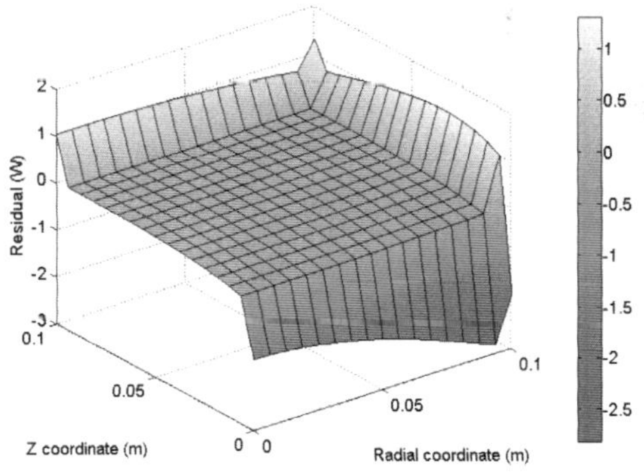

Figure 4. Residuals from the temperature dependent thermal conductivity model for $T_a = 618$ K.

Figures 3 and 4 show the residuals obtained by putting the solution to the simple model into a model with temperature-dependent thermal conductivity. Both plots show that the largest residuals are close to the face $z = 0$. A comparison of the temperature differences from the finite element solutions show that this is the region of largest temperature difference. Figure 5 shows the temperature difference for $T_a = 528$ K as an example.

Table 1 shows the root mean square (RMS) and maximum absolute values of residuals and of difference in nodal temperature ΔT for all four models with added complexity. The numbers

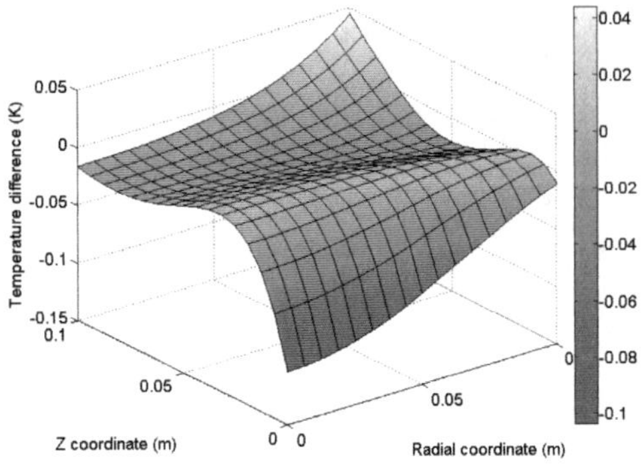

Figure 5. Temperature difference between the results of the simple model and the model with temperature-dependent thermal conductivity and $T_a = 528$ K.

in each row of the table decrease from left to right, showing that the ordering of the models from "most changed" to "least changed" given by the residuals is the same as that given by the temperature results.

Table 1. Maximum absolute values and root mean square averages of changes in temperature caused by a model change, and of the residuals obtained by putting the solution of the original model into the changed model.

	$T_a = 528$ K Radiation b.c.	$T_a = 618$ K Radiation b.c.	$T_a = 528$ K $\lambda(T)$	$T_a = 618$ K $\lambda(T)$
Max. abs. residual	5650	649	97.3	0.704
RMS residual	1020	118	17.9	0.150
Max. abs. ΔT	20.1	0.236	0.104	0.0116
RMS ΔT	20.1	0.234	0.0244	0.00525

6. Conclusions

A methodology has been proposed for choosing between different changes to a model. The proposed methodology does not require solution of the changed model, enabling model changes to be assessed before solution. The methodology has been demonstrated on a test problem, and has been shown to rank possible model changes in an order consistent with the changes in model results.

More work is needed to identify a correct scaling to remove area and volume effects, to investigate the most suitable measure for comparison of residuals, and to demonstrate the methodology on a more complicated problem.

References

[1] Al-Jamal M F and Gockenbach M S *Inverse Problems* **28** 095006
[2] Bambach M, Heinkenschloss M and Herty M 2013 *Inverse Problems* **29** 025009
[3] Bär M, Hegger R and Kantz H 1999 *Phys. Rev. E* **59**(1) 337
[4] Guo L and Billings S A 2006 *IEEE Trans. Circ. Sys. Express Briefs* **53**(8) 657
[5] Wright L, Yang X-S, Matthews C, Chapman L and Roberts S 2011 in *Computational Optimization and Applications in Engineering and Industry*, Springer, 205

Journal of Physics: Conference Series **459** (2013) 012040 doi:10.1088/1742-6596/459/1/012040

A structured strategy of concept definition in measurement: the case of 'sensitivity'

Luca Mari[1#], Arianna Mencattini[2]

[1] School of Engineering, Università Cattaneo – LIUC, Castellanza (VA), Italy
[2] University of Rome Tor Vergata, Italy

E-mail: lmari@liuc.it

Abstract. The paper emphasizes the importance that fundamental concepts in measurement science are defined according to a structured strategy, which provides both a general, qualitative characterization and a specific, type-related, quantitative definition. As a significant case, the concept 'sensitivity' is discussed and a definition for it proposed.

1. Introduction

The scope of measurement is broadening and its importance increasing: "an estimated 80 % [of the world trade] is affected by standards and regulations" and indeed according to various studies "the cost to producers and service providers of complying with "standards" can be 10 % of production costs" [1]. This generates new challenges not only towards the development of innovative (e.g., "smart") instrumentation, but also in the methodologies of data analysis, and at an even more fundamental level in the concept system by which measurement itself is interpreted and its results communicated. Measurement science has then become a moving target, as the *International vocabulary of metrology* (VIM3) [2] excellently witnesses: traditionally consolidated terms are changing their meaning (e.g., "measurement result") and new pivotal concepts are emerging (e.g., 'measurement model'). The VIM3 aim of being a well-structured and consistent concept system is critical and at the same time engaging. Indeed, the VIM3 is a guidance document "meant to be a common reference" for several different professional subjects and several different purposes [2: Scope]. With such an ambitious target, a significant issue is of properly balancing the strategy of concept definition so that definitions are at the same time encompassing and specific. This is particularly significant for those concepts which admit – or possibly require – a general (or "qualitative") characterization but also one or more ("quantitative") specifications. The way the three general features of measuring systems – measurement precision, measurement trueness, and measurement accuracy (whose definitions, by the way, in the VIM3 and in the standards ISO 3534 [3] and ISO 5725 [4] are distinctively different [5]) – are dealt with in the VIM3 is symptomatic on this matter: "measurement precision is usually expressed numerically by measures of imprecision..." [2: 2.15 N1]; "measurement trueness is not a quantity and thus cannot be expressed numerically, but measures for closeness of agreement are given..." [2: 2.14 N1]; "the concept 'measurement accuracy' is not a quantity and is not given a numerical quantity value." [2: 2.13 N1]. Hence, the situation is of

[#] To whom any correspondence should be addressed.
One of the authors is a member of the Joint Committee on Guides in Metrology (JCGM) Working Group 2 (VIM). The opinion expressed in this paper does not necessarily represent the view of this Working Group.

three general/"qualitative" concepts each of them with a claimed different treatment as for its specific/"quantitative" counterpart(s), without stated reasons justifying this position. The need of better constructing this general-specific derivation structure is acknowledged also by the *Guide to the expression of uncertainty in measurement* (GUM), when it states that "because of the lack of different words for this general concept of uncertainty and the specific quantities that provide quantitative measures of the concept, for example, the standard deviation, it is necessary to use the word "uncertainty" in these two different senses." [6: 2.2.1].

A possibly important case of multiple specific concepts under the same general concept has been triggered by the inclusion in the VIM3 of ordinal quantities [2: 1.26] and nominal properties [2: 1.30], complementing the traditional quantities based on measurement units. Even though not so well formalized yet as in the Stevens' theory of scale types [7], then extended in the so called representational theories of measurement [8], the systematic treatment of multiple property types [9] would be a significant step forward. We envisage a structure where a single general concept, say 'measurement uncertainty' can be coupled with a set of specific concepts, one for each type (and therefore 'uncertainty for quantities with unit', 'uncertainty for ordinal quantities', and so on). This would generate a very flexible framework, in which the concept system can be browsed according to several different viewpoints, such as "display the general concepts only" or "display everything which applies to ordinal quantities only".

It is in this perspective that the present work has been developed.

2. On sensitivity

An interesting case study of this possible framework is about the concept 'sensitivity', which is differently defined in different contexts and with different specifications (see, e.g., [10]: today – May 2013 – Electropedia lists 52 different lemmas containing the term "sensitivity"...). The VIM3 itself gives a definition – "quotient of the change in an indication of a measuring system and the corresponding change in a value of a quantity being measured" [2: 4.12] – whose general applicability, in particular in the case of ordinal quantities, is not explicitly discussed.

With the goal of making the many existing definitions comparable, and thus highlighting the possible set of meanings and their compatibility, let us assume:

Hypothesis 1 – Sensitivity is a property of an entity X modeled as a black box and under the only condition that the output $y(t_2)$ of X at the time t_2 causally depends on its input $u(t_1)$, $t_1 \leq t_2$.

X could be a measuring system but also a test, an algorithm, ... If X realizes an experimental process – the case of interest in measurement, where X is a measuring instrument – u and y are properties of the entity that interacts with X (or the environment itself) and X respectively. While the input-output behavior is then an empirical feature of X, related to the involved properties, its quantitative characterization is performed on the corresponding property values. For example, while the input of a spring is a force and its output is a length, the spring sensitivity is evaluated by comparing length values and force values. This distinction between properties and property values (and thus more specifically between quantities and quantity values: note that we are adopting the VIM3 position of considering quantities as specific properties) is crucial for appropriately modeling a measurement process but can be safely omitted here: we are discussing of the metrological characterization of X, not of a measurement by means of X. Hence for simplicity of notation in the following u and y will denote property values.

As the quoted VIM3 definition shows, Hypothesis 1 is immediately applicable to measuring instruments, by just assuming that y = value of the indication and u = value of the quantity being measured.

Hypothesis 1 admits that the analytical form of the input-output relation, i.e., the entity behavior, is not known, and does not impose any constraints on the types of the properties u and y.

Journal of Physics: Conference Series **459** (2013) 012040 doi:10.1088/1742-6596/459/1/012040

As stated above, we are looking for both a general, "qualitative" concept and several type-related, "quantitative" concepts of sensitivity, so that there can be a nominal sensitivity for entities X whose input is a nominal property, an ordinal sensitivity for entities X whose input is an ordinal quantity, and so on, and under the condition that the type-related concepts are specifications of the general one.

Our second basic hypothesis is:

Hypothesis 2 – The sensitivity of X is a feature related to the ability of X to detect variations of its input.

Accordingly, our strategy is to assume that, independently of the type the input property u, X is characterized by its sensitivity by comparing the variation of the generated output y with the variation of the generating input u.

Hence, the concept of variation has to be revised and generalized so to make it applicable to all property evaluation types.

3. A type-aware characterization of 'variation'

In the case of quantities evaluated in interval or ratio scales the concept of variation for their values is intrinsically part of the scale itself, and therefore it can be exploited as is, being represented numerically by the difference of distinct values v_i in the scale:

$$\mathrm{var}(v_i, v_j) = |v_i - v_j|$$

or even the differential dv whenever continuously varying quantities are assumed.

Not so univocal is the concept for ordinal quantities, where the only information meaningfully available on the quantities is ranking and therefore their difference is undefined. On the other hand, variations in a totally ordered scale can be simply evaluated, e.g., by counting the number of values in the scale within the values corresponding to the quantities under consideration. If it is supposed that the scale index i runs monotonically with the scale order, i.e., the scale values are $v_1 < v_2 < v_3 <$..., then trivially:

$$\mathrm{var}(v_i, v_j) = |i - j|$$

It can be easily shown that the so defined function var(v_i, v_j) is a metric.

In the case of nominal properties not even order can be exploited: two properties are either in the same class or in distinct classes, and therefore they are either associated to the same value or to distinct values, but neither distance nor order are meaningfully defined among such values. This suggests a concept of variation such as:

$$\mathrm{var}(v_i, v_j) = 1 - \delta_{i,j}$$

where the function:

$$\delta_{i,j} = 1 \text{ if } i = j \text{ and } = 0 \text{ otherwise}$$

is the Kronecker delta. It can be easily shown that also in this case var(v_i, v_j) is a metric.

Hence, for each property evaluation type at least one variation function var(v_i, v_j) is available that is applied to property values and represents information on the empirical variation of the corresponding properties. The fact that, independently of the type, var(v_i, v_j) is a metric, i.e., a function ranging in R, allows us to exploit it to define a concept of sensitivity which is at the same time qualitatively encompassing and quantitatively specifiable to each type.

4. A type-aware characterization of 'sensitivity'

Let us emphasize again the strategic importance of introducing both a general definition of 'sensitivity' and, on this basis, some type-related specific quantitative definitions. This provides two levels of access to information and reading of documentation:

– the general definition, at the qualitative level, offers to heterogeneous scientific groups, having to do more or less directly with measurement, a wide accessibility and easy interpretation of the concept;

– the specific definitions, at the quantitative level, are oriented to focused targets and can be customized according to specific theoretical assumptions or application needs.

Journal of Physics: Conference Series **459** (2013) 012040 doi:10.1088/1742-6596/459/1/012040

General definition ("qualitative") – Sensitivity is a property of an entity modeled as a black box that characterizes its ability to produce a variation of its output in response to a variation of the received input.

NOTE – The entity could be, e.g., a measuring instrument, a process, a test, an algorithm.
NOTE – Since the entity has generally multiple input properties, a sensitivity can be defined for each input property.
NOTE – Sensitivity can depend on the input value: a constant sensitivity over the whole range of input values is generally a simplification hypothesis.
NOTE – Sensitivity can depend on time delays in the relation between input and output, and on transitory effects: a sensitivity independent of time is generally a simplification hypothesis.
NOTE – To assess the sensitivity of an entity two distinct inputs must be applied to it and the corresponding outputs must be recorded and compared.
NOTE – This definition is based on the generic concept 'variation', which admits different definitions in reference to different property evaluation types. Hence it does not provide any specific way to assign a value to sensitivity.

In all the following definitions, u and y are input and output property values respectively (for simplicity the time dependence is omitted).

General definition ("quantitative") – Sensitivity at the input value u is:

$$S(u) = \frac{\text{var}(y(u), y(u'))}{\text{var}(u, u')}$$

where var is a generic metric function and u' is an input value such that var(u, u') is the minimum positive value allowed for var(u, x), i.e., $u' = x$ where $\min_x \text{var}(u, x) > 0$.

NOTE – A property u' that satisfies the given condition is not necessarily unique.

Specific definition 1 – Sensitivity at the input value u for an interval or ratio evaluation is:

$$S(u) = \frac{dy(u)}{du} \quad \text{or} \quad = \frac{\Delta y(u)}{\Delta u} = \frac{y(u+\Delta u) - y(u)}{\Delta u}$$

for a continuous and a discrete quantity respectively.

Specific definition 2 – Sensitivity at the input value u_i for an ordinal evaluation is:

$$S(u_i) = \frac{\text{var}(y(u_i), y(u_{i+1}))}{\text{var}(u_i, u_{i+1})} = |k_0 - k_1|$$

where k_0 and k_1 are the ordinal indexes of $y(u_i)$ and $y(u_{i+1})$ respectively.

NOTE – In this case var(u_i, u_{i+1}) = 1.
NOTE – S(u_i) counts the number of ordinal positions between $y(u_i)$ and $y(u_{i+1})$.

Specific definition 3 – Sensitivity at the input value u_i and in reference to the input value u_j, $u_i \neq u_j$, for a nominal evaluation is:

$$S_{u_j}(u_i) = \frac{\text{var}(y(u_i), y(u_j))}{\text{var}(u_i, u_j)} = 1 - \delta_{k_0, k_1}$$

where δ is the Kronecker delta and k_0 and k_1 are the classificatory indexes of $y(u_i)$ and $y(u_j)$ respectively.

NOTE – In this case var(u_i, u_j) = 1.

NOTE – Because of the lack of structure on the sets of values, in this case the function S is actually biargumental.

NOTE – The possible values of $S_{u_j}(u_i)$ are either 0 or 1.

5. An example: computing sensitivity in a diagnostic test

The Breast Imaging-Reporting and Data System (BI-RADS) [11] is a quality assurance tool originally designed for use with mammography, published and trademarked by the American College of Radiology (ACR). The system is aimed at standardizing reporting, and is used by medical professionals to communicate a patient's risk of developing breast cancer. In particular, BI-RADS focuses on breast density, assumed as a factor positively correlated to risk of breast cancer, and defines a reference set to be used to evaluate Breast Tissue Composition (BTC). The set includes four elements, D1: the breast is almost entirely fat; D2: scattered fibroglandular densities (25-50%); D3: heterogeneously dense breast tissue (51-75%); D4: extremely dense (> 75% glandular), each element thus indicating the approximate percentage of dense tissue over the entire breast region. On this basis radiologists try to develop objective protocols to evaluate BTC in screening programs, where the protocol sensitivity has to be assessed for scientific purposes and specifically for comparing different protocols. BTC can be at first considered a nominal property by assuming that the four categories identified by the reference set D1–D4 only define a classification. Under this hypothesis the sensitivity of a protocol to evaluate BTC can be assessed, where the input is a mammographic image corresponding to a BTC in a given category u_i = Di and the output is a value $y_j = u(\mathrm{D}i) = \mathrm{D}j$ in the same set as obtained by the protocol (the evaluation procedure could be based on the independent examination of the same image by different expert radiologists, each of them being requested to identify one of the four categories, so that the output is the value corresponding to the most frequently chosen category).

Let us consider the nominal sensitivity of the protocol at the input value u_1 = D1 (fatty tissue), hence by testing a BTC $u_i = \mathrm{D}i$, $i \neq 1$. Then, by applying results of Definition 3:

$$S_{\mathrm{D}i}(\mathrm{D}1)=\mathrm{var}(y(\mathrm{D}1), y(\mathrm{D}i))=1-\delta_{1,i}$$

The sensitivity is equal to 1 if the protocol distinguishes the applied BTC from that of fatty tissue.

Let us now assume that BTC is an ordinal quantity, i.e., such that the relations D1 < D2 < D3 < D4 are empirically meaningful. The sensitivity at the input value D1, S(D1), is assessed by applying the protocol in the cases u_1 = D1 and u_2 = D2, so that var(u_1, u_2) = 1. The possible outputs are such that S(D1) ranges from 0, if y(D1) = y(D2), i.e., the inputs are not distinguished with each other, to 3, in the case y(D1) = D1 and y(D2) = D4 (or vice versa y(D1) = D4 and y(D2) = D1).

6. Some conclusions

The sensitivity of an entity has been defined in the present paper in such a way that it is both qualitatively characterized and then specified quantitatively, in reference to each property evaluation type. The proposed definition is entirely based on the black box modeling of the entity: it assumes that two suitably chosen inputs are applied to the entity and the corresponding outputs are recorded and compared. An important consequence is that such a definition is calibration-independent: were it accepted, all definitions of 'sensitivity' that are instead formulated in terms of reference (or possibly true) values should be revised, because capturing some different concept. On the other hand, as characterized here 'sensitivity' is a so encompassing concept that it can be exploited also to define as specific cases, e.g., selectivity (as lack of sensitivity to properties other than the one intended to be evaluated) and repeatability (as lack of sensitivity to time).

The poor algebraic structure of nominal and ordinal evaluations leads to a characterization in terms of sensitivity that conveys a small amount of information, in particular a binary value in the nominal case. In this view, a possible generalization of the proposed concept builds upon the idea to perform a statistical analysis on the outputs obtained from a sample of inputs in repeatability conditions. Sensitivity would be then a function of a suitable statistic of the variations in such samples.

We believe that the structural strategy used for 'sensitivity' may be generally adopted to produce an accessible, readable, and consistent concept system of measurement science fundamentals.

References

[1] BIPM 2007: Bureau International des Poids et Mesures, Evolving needs for metrology in trade, industry and society and the role of the BIPM [http://www.bipm.org/utils/en/pdf/Kaarls2007.pdf].
[2] JGCM 200:2012, International vocabulary of metrology – Basic and general concepts and associated terms (VIM), Joint Committee for Guides in Metrology, 2012 (2008 edition with minor corrections) [http://www.bipm.org/en/publications/guides/vim.html].
[3] ISO 3534, Statistics - Vocabulary and symbols, International Organization for Standardization, different publication dates for the various parts.
[4] ISO 5725, Accuracy (trueness and precision) of measurement methods and results, International Organization for Standardization, different publication dates for the various parts.
[5] R. Buccianti, M. Cibien, L. Mari, B. Rebaglia, Accuracy, trueness, and precision: considerations based on the International Vocabulary of Metrology (VIM, 3rd ed.) and related standards, proc. XIX IMEKO World Congress, Lisbon, Portugal, 2009.
[6] JGCM 100:2008, Evaluation of measurement data – Guide to the expression of uncertainty in measurement, Joint Committee for Guides in Metrology, 2008 (GUM, originally published in 1993) [http://www.bipm.org/en/publications/guides/gum.html].
[7] S. S. Stevens, On the theory of scales of measurement, Science, 103, 2684, 677–680, 1946.
[8] D. H. Krantz, R. D. Luce, P. Suppes, A. Tversky, Foundations of measurement, vols. 1–3, Academic Press, New York, 1971, 1989, 1990.
[9] A. Giordani, L. Mari, Property evaluation types, Measurement, 45, 437–452, 2012.
[10] IEC 60050, International Electrotechnical Vocabulary (IEV), International Electrotechnical Commission, different publication dates for the various parts [http://www.electropedia.org].
[11] American College of Radiology (ACR) Breast Imaging Reporting and Data System Atlas (BI-RADS® Atlas). Reston, Va: © American College of Radiology; 2003.

Journal of Physics: Conference Series **459** (2013) 012041 doi:10.1088/1742-6596/459/1/012041

The problem of false discoveries from a metrological point of view

Giampaolo E D'Errico

Istituto Nazionale di Ricerca Metrologica (INRIM), Torino (I)

E-mail: g.derrico@inrim.it

Abstract. Valid inference drawn from analysis of experimental results needs scientific grounds, whereas conclusions based on statistical significance tests or hypothesis testing may be problematic, especially when dealing with a multiplicity of tested hypotheses, as in experiments performed on bio-molecules. The problem of false discovery rate is focused in the present paper, aiming at eliciting application of sound criteria for rejection/acceptance of hypotheses and related methods of uncertainty characterisation.

1. Introduction

The qualifier 'statistically significant' attributed to an experimental result is not the same as 'scientifically meaningful'. It is well known that 'significant' is one of the most disputed definitions in scientific literature (see e.g. [1]), and the controversy is such that wondering why – not whether – most published findings are false [2] is an urgent question, for example in the biomedical research field, where harmful consequence to health may arise. This question is formulated in terms of so called false discoveries in the context of simultaneous testing of hypotheses in experimental research.

In this paper Fisher's and Neyman–Pearson's theories developed in classical statistics are focused from a logical point of view; related problems are presented and discussed, and the inference approach based on Bayesian statistics is introduced. The state of the art of multiple hypothesis tests is reviewed with special attention to the problem of estimation and control of false discovery rate (FDR) with applicability to bio-metrology.

2. Frequentist and Bayesian rationals of statistical inference

2.1. Classical approaches

In the so-called frequentist area of interpretation of probability, Fisher [3] advocated the use of p-values in significance testing of so-called null hypothesis (usually denoted by H_0) against the Neyman and Pearson (N–P) theory [4] that introduced the concept of an alternative hypothesis to the null one, and the probabilistic notions of type I/type II (false positive/negative) errors. Both Fisher's and N–P's methods disregarded the interplay of prior and conditional probabilities: this point was taken into account by von Mises in his approach [5] developed in terms of success or error chance of the test (von Mises' use of the term 'chance' rather than probability was intended to point out that no randomization was required in his approach) starting from the assumption of the need of a prior distribution function. Fisherian and N–P theories of testing hypotheses (and related theories of point or interval estimation) were set by Wald in the overall context of the problem of statistical inference [6].

2.2. Role of posterior probability

An early Bayesian approach to this problem was developed by Jeffreys [8]. The conflict between frequentist interpretation and Bayesian interpretation of statistical testing was modelled by Lindley as a paradox showing how the posterior probability of an hypothesis conditional on experimental result *e*, $p(e \mid H)$, is affected by prior probability $p(H)$: for quite small values of $p(H)$, $p(e \mid H)$ may become as high as 0.95, while a significance test may state that result *e* is significant for *H* at the $\alpha = 0.05$ level (in terms of *p*-value, values $p(e \mid H) \leq 0.05$ are compatible with $p(H \mid e) = 0.95$, being dependent on assigned values of prior $p(H)$ as per inverse probability calculation) according to Bayes' formula:

$$p(H \mid +) = p(H)p(+ \mid H)[p(H)p(+ \mid H) + p(\neg H)p(+ \mid \neg H)]^{-1} = 1 - p(\neg H \mid +) \qquad (1)$$

For example, the characteristics of two diagnostic tests are shown in table 1, where the hypothesis tested is $H \doteq$ 'disease is present' and the sign + (–, respectively) stands for 'test result is positive (negative, respectively)'. The significance level is α=0.05 for test1 (0.04 for test2) and the power is $1 - \beta = p(- \mid \neg H) = 0.98$ (for test2: 0.99). Power may be translated into test specificity: thus $p(- \mid \neg H)$ is an index of accuracy in making the negative diagnosis (i.e., correctly reporting absence of disease); $p(+ \mid H) = 1 - \alpha$ is an index of the accuracy in making the positive diagnosis, called the test sensitivity: with data in table 1, for test1 $p(+ \mid H) = 0.95$ (for test2: 0.96).

Table 1. Tests for the presence of a disease (*H*).

	Test results			
	Positive, +		Negative, –	
	test1	test2	test1	test2
H	0.95	0.96	0.05	0.04
$\neg H$	0.02	0.01	0.98	0.99

Table 2. Test2: numerical example, *p*(*H*)=0.005.

	+	–	total
H	48	2	50
$\neg H$	99	9851	9950
total	147	9853	10000

A rationale diagnostic criterion is to require $p(H \mid +) > p(\neg H \mid +)$. Both tests exhibit reliable performance. For test2, comparatively more sensitive and specific, baseline condition to obtain $p(H \mid +) > 1/2$ is $p(H) > 0.01$, i.e. a disease prevalence greater than 1%. Note that rare diseases do not satisfy such a condition (Poisson distribution is the probability paradigm of rarely occurring events). More generally, experiments designed to detect any rare events, for which baseline percentage is under the threshold of 1–2%, should take into account dependence of inverse probability on prior probability – other factors in Eq. (1) being equal. For example, using test2 with $p(H) = 0.005$, Eq. (1) yields $p(\neg H \mid +) \cong 0.675$: in this case, although test2 is quite sensitive (type I error probability is 0.04) and very specific (type II error probability is 0.01), the probability of an incorrect diagnose is almost twice the probability of a correct one. This is known as the false-positive paradox. Moreover, the probability of missed positives (false negatives) $p(H \mid -)$ results slightly larger than 0.0002. Table 2 reports an approximate experimental realization of test2, where the relative frequency of correct diagnosis is $48/147 \approx 0.33 \approx 1 - p(\neg H \mid +)$ and the relative frequency of false negatives is $2/9853 \approx 0.0002 \approx p(H \mid -)$. Data in table 2 show accuracy values in accordance with declared characteristics of test2: namely 0.96 of sensitivity and a specificity of approximately 0.99.

2.3. Controversy: from theory to practice

However this is not a paradox, as it follows from Bayes' formula; the conflict arises in acceptance of an inference criterion or another: is the significance of *H* best expressed by $p(e \mid H)$ or by $p(H \mid e)$?

Journal of Physics: Conference Series **459** (2013) 012041 doi:10.1088/1742-6596/459/1/012041

According to Bayesian scholars, a strict logical line of reasoning leads to the probability of the hypothesis conditional on experimental result, i.e. $p(H|e)$ – the other one being flawed in principle. On the other hand, frequentist scholars insist on p-value $p(e|H)$; this position is stated in famous Fisher's statement: "The force with which such a conclusion is supported [by a significantly small $p(e|H_0)$] is logically that of a simple disjunction: *Either* an exceptionally rare chance has occurred [in our symbols: e] *or* the theory of random distribution [H_0] is not true" [3: p. 42]. (Crediting the first alternative, H_0 may show what can be miscalled its trueness, after the occurrence of an unlikely event.) Ever since, the controversy among schools of thought has been continuously fed from diverse theoretical perspectives (see e.g. [9–18]) and in view of a variety of applied statistics fields, including signal detection techniques (false vs. missed alarms), social sciences (e.g., criminal-court cases), therapeutic industry (e.g., new treatment efficacy vs. placebo), and metrology (outliers in measurements). A thorough review of uses and misuses of null hypothesis significance testing can be found in [19], proposed by "the editor of an empirical journal [...] in attempting to develop a policy that would help ensure the journal did not publish egregious misuses" [19: pp. 241–242]. More recently, Fisher's, N–P's and Jeffrey's approaches were compared and contrasted in [20].

3. False discoveries in multiple hypothesis tests

3.1. Rise and consequence of the FDR problem

Problems with statistical significance become more complex when a multiplicity of hypotheses is under test. Tukey's account of the philosophy of multiple comparison is introduced by a very severe criticism [21: p. 100]: "Statisticians classically asked the wrong question and were willing to ask with a lie [...]". Multiple hypotheses testing is of current interest to applied researchers involved in almost all of scientific disciplines [22]; in biology, e.g. with application to microarray experiments and bioinformatics [23–27]. An issue that gained attention in terms of false discovery can be described after the formulation by Sorić [28]. A provisional rejection of a null hypothesis is called a discovery, false discoveries are related to type I errors; let a large number n of independent experiments be contextually performed at a significance level α each, with an unknown large number N_0 of "true" nulls, and let $\rho=\alpha N_0 r$ denote the quotient of false discoveries among r declared ones: if almost all of those r discoveries are reported in scientific publications, it can be inferred that a non negligible – unless ρ is very small – part of research work may have been misled by false discoveries. Sorić in the above cited paper of 1989 [28] credits Edwards, Lindman, and Savage [29] for having drawn attention to the probability of false discoveries, although diversely alluded to in their Bayesian work published in 1963. However, according to Seeger [30], Eklund in three unpublished reports (in Swedish) archived in 1961–1963 at Uppsala University Institute of Statistics under the title "*Massignifikansproblemet*" already took into account and proposed solution to a problem relevant to large exploratory investigations: to keep low the proportion of predicted number of false significances to the observed number of significances – that was called by Eklund the "mass-significance" problem. However, pioneering work on this matter can be traced back towards the middle of the nineteenth century (references can be found to Cournot [31]: see e.g. [32]). A classical method – proposed by Bonferroni in 1935 [33] (as cited e.g. in [34]) – is known with the name of Bonferroni's correction.

3.2. FDR: estimation and control methods

The main problem in testing a family of hypotheses at a time, is that the overall significance level increases with the number of hypotheses. Suppose the significance level for a single hypothesis is α, and that n such hypotheses are under test: if these statistical hypotheses are mutually independent, the overall significance level is $1-(1-\alpha)^n$, that for small α is approximately $n\alpha$. Bonferroni's correction leads to downscale the individual α by a factor $1/n$ (thus n times the p-value computed for each hypothesis, the so called adjusted p-value, is checked not to exceed α), so that the overall significance level is closely reset to value α. Improving on proposed solutions, the problem to bound the overall type I error has been approached in diverse terms, mainly based on estimating and controlling its probability, or the expected value of its relative frequency or its rate (see Shaffer [32] for a thorough review), till in 1995 Benjamini and Hochberg [35] introduced the formulation of a false discovery rate (FDR) and developed a methodics to control FDR in multiple testing. More historical notes are reported by Benjamini and Hochberg [36]. Another historical perspective is given by Tukey [21]. A unified (frequentist/Bayesian) approach to FDR estimation can be found in [34].

Table 3 summarizes results of multiple test of n hypotheses, of which N_0 are "true" null. R are declared discoveries. N_0, R, U: random variables (RVs), n_0, r, u respective realizations; n is known.

Table 3. U/N_0 is the "false positive" fraction, U/R the fraction of "false discoveries".

	Hypotheses supported by test		
	null	alternative	total
"true" null	N_0-U	U	N_0
"false" null	$n-R-(N_0-U)$	$R-U$	$n-N_0$
total	$n-R$	R	n

The FDR can be expressed in the following form, where the symbol $\vee$ in the denominator signifies that in case R=0 the expected value, E[·], is set to zero (to overcome the problem of the division 0/0).

$$\mathrm{FDR} = \mathrm{E}\left[U/\{R \vee 1\}\right] \tag{2}$$

Conditioning on R>0, a positive FDR is obtained [37]:

$$\mathrm{pFDR} = \mathrm{E}\left[\tfrac{U}{R} \middle| R > 0\right] \tag{3}$$

Diverse control methods (single- or multi-stage) have been proposed to guarantee a FDR under a preset overall significance level, say q. It should however be stressed that FDR control is involved with an expected rate, thus some combinations of observed values of $U=u$ and $R=r>0$ may result into $u/r>q$, even if pFDR$\leq q$. Benjamini and Hochberg [35, 36] introduced a method based on reordering the p-values observed in n simultaneous tests; to distinguish them from the p_i before reordering (when the subscript individuates the i-th test), indices of p-values after monotonic reordering are bracketed:

$$p_{(1)} \leq \cdots p_{(j)} \cdots \leq p_{(n)}, \quad 1 \leq j \leq n \tag{4}$$

This gives rise to a step-up procedure, for each $p_{(j)}$ is matched against the quantity jq/n; eventually, it is defined the index $J = \max\left(j : p_{(j)} \leq jq/n\right)$ that must be estimated, say by $\hat{J}$. After that, all the hypotheses pointed to by the indices $j \leq \hat{J}$ are rejected. Finally, the so-called Benjamini–Hochberg (BH) rule for the correction of the original p_i-values, results into the BH–corrected values, where index(p_i) denotes the position (j) of the observed p_i in the succession of Eq. (4):

$$p_i^{\mathrm{BH}} = np_i/\mathrm{index}(p_i) \tag{5}$$

A twofold insight into the possible usage of FDR can be articulated. On one hand, a threshold on can be pre-set and a multiple test be designed so that the attainable FDR level does not exceed the threshold: this approach is aimed at performing what is appropriately called FDR control [32, 35].

On the other hand, after having fixed the acceptance p-values threshold, a point-wise estimator of the FDR is constructed so that the expectation of this estimator is not less than the FDR value calculated at that threshold: this is the estimation approach. A point estimation approach can be summarized starting from Eq. (2). Putting $U(\alpha)=\#\{\text{"true" null hypotheses}: p_i \leq \alpha\}$ and $R(\alpha)=\#\{\text{null hypotheses}: p_i \leq \alpha\}$, where the operator $\#\{\cdot\}$ returns the cardinality of the set defined inside brackets: $F_{DR}(\alpha)=\mathrm{E}[U(\alpha)/\{R(\alpha)\vee 1\}]$. According to Storey [37], an estimator of $\mathrm{FDR}(\alpha)$ for a given α is $\hat{F}_{DR}(\alpha)=\alpha\hat{N}_0(\tau)/\{R(\alpha)\vee 1\}$, where τ is a tuning parameter and $\hat{N}_0(\tau)$ is the estimate of total number N_0 of "true" null hypotheses:

$$\hat{N}_0(\tau)=(n-R(\tau))/1-\tau \tag{6}$$

The estimator of Eq. (6) is justified by approximate equalities $\alpha\hat{N}_0(\tau)\cong U(\alpha)$, $R(\alpha)\vee 1\cong R(\alpha)$: it was shown [37] that conditions can be stated in order to grant that the inequality $\mathrm{E}[\hat{F}_{DR}(\alpha)]\geq F_{DR}(\alpha)$ holds: based on this estimator $\hat{F}_{DR}(\alpha)$, a unified approach to FDR control and estimation is presented in [38]. More on unified approaches can be found in [34].

A Bayesian approach is illustrated in terms of a so-called two-class model. It assumes that a set of (null) hypotheses $\{H_{0i}:1\leq i\leq n\}$ under test can be represented by identically distributed RVs – supposed sharing a Bernoullian distribution, say H_0– associated to the same test statistics ϑ with significance region ω. By assumption, the set can be bi-partitioned: each H_{0i} is assigned to one or the other partition according to its truth state – either "true" or "false". Let associate H_0 to the binary indicator $\ddot{H}_0$, such that $\ddot{H}_0=0$ with probability $\Pr(\ddot{H}_0=0)=\pi_0$ if H_0 is "true", otherwise $\ddot{H}_0=1$, $\Pr(\ddot{H}_0=1)=1-\pi_0$ (the frequency n_0/n can be an empirical estimate of prior π_0). Storey [39] proves a theorem, equating a "posterior Bayesian type I error probability function" to pFDR (Eq. (3)):

$$\Pr(\ddot{H}_0=0\mid\vartheta\in\omega)=\frac{\pi_0\Pr(\vartheta\in\omega\mid\ddot{H}_0=0)}{\pi_0\Pr(\vartheta\in\omega\mid\ddot{H}_0=0)+(1-\pi_0)\Pr(\vartheta\in\omega\mid\ddot{H}_0=1)}=\mathrm{E}\left[\frac{U(\vartheta)}{R(\vartheta)}\middle|R(\vartheta)>0\right] \tag{7}$$

4. Conclusion

The problem of hypothesis testing about the (unknown) magnitude of a quantity x being estimated (in measurement science term: the measurand) is the reverse of the coin of interval estimation of x along with its measurement uncertainty. However there is no (neither in Bayesian nor in frequentist statistical inference) *experimentum crucis* to discriminate "true" hypotheses from "false" ones. This is a complex problem that rises even extra-difficulties when a multiplicity of hypotheses are tested for significance at a time, as it happens in a variety of bio-scientific experiments.

Techniques for estimation and control of the probability of incurring in false discoveries (i.e., test results erroneously qualified as significant) are available to reduce to manageability – on the average – such a complexity. This paper was focused on main points selected from foundational issues and state of the art developments (a rich bibliography can be found in [40]).

Coverage probabilities of multiple confidence intervals will be the matter of next efforts addressed to evaluation and expression of uncertainties of a multiplicity of quantities under interval estimation.

References

[1] Brumfiel G 2008 Significant *Nature* **455** 1027
[2] Ioannidis J P A 2005 Why Most Published Research Findings Are False *PLoS Medicine* **2** 696
[3] Fisher R A 1973 *Statistical Methods and Scientific Inference* 3rd ed. (London: Macmillan)
[4] Neyman J and Pearson E S 1933 On the Problem of the Most Efficient Tests of Statistical Hypotheses *Philos. Trans. Roy. Soc. London Ser. A* **231** 289
[5] Mises R v 1943 On the Problem of Testing Hypotheses, *Ann. Math. Stat.* **14** 236
[6] Wald A 1942 *On the Principle of Statistical Inference* (Notre Dame, Ind.: Uni. of Notre Dame)
[7] Lindley D V 1957 A Statistical Paradox *Biometrika* **44** 187
[8] Jeffreys H 1961 *Theory of Probability* 3rd ed. (Oxford: Clarendon Press), chaps. 5–6
[9] Gumbel E J 1943 On the Reliability of the Classical Chi-Square Test *Ann. Math. Stat.* **14** 253
[10] Rosenkrantz R D 1973 The Significance Test Controversy *Synthese* **26** 304

[11] Birnbaum A 1977 The Neyman-Pearson Theory as Decision Theory and as Inference Theory; with a Criticism of the Lindley-Savage Argument for Bayesian Theory *Synthese* **36** 19
[12] Cox D R 1977 The Role of Significance Tests *Scand. J. Statist.* **4** 49
[13] Cohen Y 1994 The Earth Is Round (*p*<.05) *American Psychologist* **49** 997
[14] Royall R M 1997 *Statistical Evidence: A Likelihood Paradigm* (London: Chapman & Hall)
[15] Breaugh J A 2003 Effect Size Estimation: Factors to Consider and Mistakes to Avoid *J. of Management* **29** 79
[16] Nakagawa S and Cuthill I C 2007 Effect Size, Confidence Interval and Statistical Significance: A Practical Guide for Biologists *Biol. Rev.* **82** 591
[17] Li X R and Li X-B 2008 Common Fallacies in Applying Hypothesis Testing *Proc. 11th IEEE Conf. on Information Fusion* (Cologne, Germany: IEEE)
[18] D'Errico G E 2009 Testing for Outliers Based on Bayes Rule *Proc. IMEKO XIX World Congress–Fundamental and Applied Metrology* (Lisbon, Portugal: IMEKO) pp. 2368–2370
[19] Nickerson R S 2000 Null Hypothesis Significance Testing: A Review of an Old and Continuing Controversy *Psychological methods* **5** 241
[20] Berger J O 2003 Could Fisher, Jeffreys and Neyman Have Agreed on Testing? (with comments and rejoinder) *Statistical Science* **18** 1
[21] Tukey J W 1991 The Philosophy of Multiple Comparisons *Statistical Science* **8** 100
[22] Gelman A, Hill J and Yajima M 2012 Why We (Usually) Don't Have to Worry About Multiple Comparisons *J. Research on Educational Effectiveness* **5** 89
[23] Dudoit S, Shaffer J P and Boldrick J C 2003 Multiple Hypothesis Testing in Microarray Experiments *Statistical Science* **18** 71
[24] Storey J D and Tibshirani R 2003 Statistical Significance for Genomewide Studies *Proc. Natl. Acad. Sci. USA* **100** 9440
[25] Efron B 2007 Size, Power and False Discovery Rates *Ann. Statist.* **35** 1351
[26] Efron B 2008 Microarrays, Empirical Bayes and the Two-Groups Model *Ann. Statist.* **23** 1
[27] Langaas M, Lindqvist B H and Ferkingstad E 2005 Estimating the Proportion of True Null Hypotheses, with Application to DNA Microarray Data *J. R. Statist. Soc.* B **67** 555
[28] Sorić B 1989 Statistical 'Discoveries' and Effect-size Estimation *J. Am. Statist. Ass.* **84** 608
[29] Edwards W, Lindman H and Savage L J 1963 Bayesian Statistical Inference for Psychological Research *Psychological Review* **7** 193
[30] Seeger P 1968 A Note on a Method for the Analysis of Significances en Masse *Technometrics* **10** 586
[31] Cournot A A 1843 *Exposition de la Théorie des Chances et des Probabilités* (Paris: Hachette)
[32] Shaffer J P 1995 Multiple Hypothesis Testing *Annu. Rev. Psychol* **46** 561
[33] Bonferroni C E 1935 Il calcolo delle assicurazioni su gruppi di teste *Studi in Onore del Professore Salvatore Ortu Carboni* (Roma, I) pp. 13–60
[34] Strimmer K 2008 A Unified Approach to False Discovery Rate Estimation *BMC Bioinformatics* **9** (14 pages, online, doi:10.1186/1471-2105-9-303)
[35] Benjamini Y and Hochberg Y 1955 Controlling the False Discovery Rate: A Practical and Powerful Approach to Multiple Testing *J. R. Statist. Soc.* B **57** 289
[36] Benjamini Y and Hochberg Y 2000 On the Adaptive Control of the False Discovery Rate in Multiple Testing with Independent Statistics *J. Educational and Behavioral Statistics* **25** 60
[37] Storey J D 2002 A Direct Approach to False Discovery Rates *J. R. Statist. Soc.* B **64** 479
[38] Storey J D, Taylor J E and Siegmund D 2004 Strong Control, Conservative Point Estimation, and Simultaneous Conservative Consistency of False Discovery Rates: A Unified Approach *J. R. Statist. Soc.* B **66** 187
[39] Storey J D 2003 The Positive False Discovery Rate: A Bayesian Interpretation and the q-Value *Ann. Statist.* **31** 2013
[40] Rao C V and Swarupchand U 2009 Multiple Comparison Procedures–a Note and a Bibliography *J. of Statistics* **16** 66

Problems in implementing some of the new measurement unit definitions of the SI

Franco Pavese

Torino, Italy

E-mail: frpavese@gmail.com

Abstract. The paper illustrates and discusses some problems that should be taken into account, should the proposed use of fundamental constants in the definition of measurement units of the SI be implemented: (a) more base units being multi-dimensional, instead of fixing the present problems in this respect; (b) the multidimensionality in the definitions; (c) the use of CODATA adjusted values of the constants for this specific purpose; (d) formal issues in stipulating algebraic expressions of the definitions, and in respect to the rounding or truncation of the numerical values in their transformation from uncertain to exact values; (e) formal issues with the use of the integer number N_A; (f) limitations that can arise from the stipulation of the values of several constants for the CODATA Task Group to continue performing in future meaningful least squares adjustments of the fundamental constants taking into account future data.

1. Introduction

The use of some 'fundamental constants' (c_0, h, e, k_B, N_A) with stipulated numerical values has been proposed for the re-definition of some of the present *base* units of the International System of Units, the SI [1]. This proposal was submitted at its 2011 meeting to the CGPM, which decided in its Resolution 1 to "*take note of the intention of the CIPM*" for "*the possible future revision of the International System of Units*" [2].

Problems can be found in the literature over the past five years concerning the proposal on the floor related to four main areas: (a) how meaningful a progress can be achieved by using fundamental constants in the definition of the measurement units; (b) which should the constants used and how the definitions should use them; (c) how the definitions should precisely be formulated; (d) how to deal with future data.

In this paper, some of the issues arising from the above problems, which should be taken into account in the implementation of the proposal, are discussed:

(1) having more base units multi-dimensional instead of fixing the present problems in this respect;

(2) how to take into account the multi-dimensionality in the definitions;

(3) using or not using CODATA 'adjusted values' of the constants for this specific purpose, being the main purpose of the adjustments rather to obtain the best measure of the *consistency* of a much wider set of constants;

(4) formal issues in stipulating algebraic expressions in the definitions, and in respect to the rounding or truncation of the numerical values in their transformation from uncertain to exact values;

(5) formal issues when using the integer number N_A;

(6) handling limitations that can arise from the stipulation of the values of several constants, namely for the CODATA Task Group to continue performing in future meaningful least squares adjustments of the fundamental constants, and taking into account future data.

It is *not* the aim of this paper to discuss the authority of the CODATA Task Group in this respect, nor the general validity of the Least Squares Adjustment (LSA) method [3].

2. Problems of principle

The aim of the International System of Units (SI) is to provide a system of units, one for each of a specified (and limited) list of quantities necessary to express measured properties in nature.

The above items (1) to (3) fall under this heading.

2.1. Meaning of 'base unit' of the SI

A long-lasting debate led to seven quantities that are called fundamental and whose units are called "base" [4]. Each of these units was originally aimed at being defined without having to resort to the unit of any other quantity. They are assumed to be dimensionally *independent* [5]. All the remaining are conversely called "derived units", since one has to express each of them as an algebraic expression of the seven base units.

There are some properties of unit systems that are desirable, and there are some that are essential. A property that is not essential but is desirable is that the system of units is *coherency*. For detail on this see [6].

A property that is mandatory for any system of units, and consequently often assumed achieved also by the SI, is internal *consistency*. For base units defined independently from each other this is necessarily the case.

However, this principle is *presently not* implemented for several base units of the SI (see section 2.2). In this situation, consistency cannot be assumed *a priori*. For non-independent definitions, formal or logical consistency, in the sense of a syntactic property relating unit definitions, can be quite problematic to prove. There is another, more specifically metrological, meaning of the term consistency that is often used. It is defined in [7] as a kind of "metrological compatibility", a property of a set of experimental determinations affected by uncertainty. For details see [14].

With the new proposed definitions, the above property would become greatly complicated to prove. In fact, the constants c_0, h, e, k_B, N_A (and $\nu(^{133}Cs)$, not a "fundamental constant") are *not* each the direct expression of a base SI unit: one has to express the relevant units as algebraic expressions of the constants (where the use of the stipulated values for the constants,[1] indicated with an asterisk, is mandatory—for comments on their numerical values see later):

$$
\begin{aligned}
&[c_0/\nu(^{133}\mathrm{Cs})] = [\mathrm{metre}],\ c_0^*/\nu(^{133}\mathrm{Cs})^* = 3.261\,225\,561\,106\,45\ldots \cdot 10^{-2}\ \mathrm{m} \\
&[h\nu(^{133}\mathrm{Cs})/c_0^2] = [\mathrm{kilogram}],\ h^*\nu(^{133}\mathrm{Cs})^*/c_0^{*2} = 6.777\,265\,181\,006\,83\ldots \cdot 10^{-41}\ \mathrm{kg} \\
&[h\nu(^{133}\mathrm{Cs})/k_B] = [\mathrm{kelvin}],\ h^*\nu(^{133}\mathrm{Cs})^*/k_B^* = 4.411\,764\,153\,914\,89\ldots \cdot 10^{-1}\ \mathrm{K} \\
&[e\nu(^{133}\mathrm{Cs})] = [\mathrm{ampere}],\ e^*\nu(^{133}\mathrm{Cs})^* = 1.472\,821\,956\,237\,59\ldots \cdot 10^{-9}\ \mathrm{A} \\
&[1/N_A] = [\mathrm{mole}],\ 1/N_A^* = 1.660\,539\,000\,996\,49\ldots \cdot 10^{-24}\ \mathrm{mol}
\end{aligned}
\qquad (1)
$$

2.2. Definition of a base unit making use of other base or derived units

As indicated before, already at present some base units are not dimensionally independent: the unit of length involves the second; the unit of amount of substance involves the kilogram; the unit of electric current involves the newton and the metre; the unit of luminous intensity involves the hertz, the watt and the steradian; the unit of amount of substance involves the kilogram. As shown in the expression (1) this fact is now aggravated for the proposed new units. In principle, the text of each definition should also make explicit reference to all other units used. For an alternate possibility for fixing this issue see [6].

---

[1] The present numerical values originate from the set of measurement units that are presently defined—and realised.

Journal of Physics: Conference Series **459** (2013) 012042 doi:10.1088/1742-6596/459/1/012042

2.3. Use of CODATA adjusted values of fundamental constants in the definition of a unit

Since 1969 a CODATA Task Group [8] performs a valuable check of consistency of a large set of 'fundamental constants' by means of the so-called Least Squares Adjustment (LSA) algorithm. At regular intervals, these evaluations produce a set of "adjusted values" and of the corresponding associated uncertainties (see in [8] for the full list of references). The aim of the CODATA Task Group, as expressed in their reports and publications is twofold: (i) "*a least squares adjustment (LSA) is one of the few ways in which the over-all consistency of physical theory can be systematically investigated. Moreover, it provides a consistent set of constants at a particular epoch which can be used by all workers requiring them*" [10]; (ii) to get "*particular numerical values obtained for a set of 'best' or 'recommended' set of constants …*" [9].

For aim (ii), the sentence reported above from [9] ends by saying that this feature "*… is only of secondary importance*", because the main importance is the "*information gained during the course of the critical review which necessarily accompanies* [precedes] *the adjustments*" [9]. Actually, extensive warnings in respect to the use of the LSA method form the whole Section C in [10] (see [8]). Another critical feature requiring attention is that pointed out in a "Warning!" in [9]: "*Because of the intimate relationship which exist among least-squares adjusted values of the fundamental constants, a significant shift in the numerical value of one will generally cause significant shifts in others. Consequently, for any critical application of these numbers, the user is urged to refer to the original article*[s] *as well as to the current literature …*". Their use in the definition of measurement units is certainly a very critical application.

It is a fact that consistency is a most important requirement for a set of constants. However, its check is performed with tools suitable for that purpose, like the LSA, which are *different* from the tools used to obtain the numerical values of the chosen summary statistics directly evaluated from the—critically-selected—experimental data available for each constant. Thus, confounding the results of the LSA with the several choices possible for the latter (e.g., mean, weighted mean, median, …) should be considered incorrect because:

(a) The value of (at least) one member of the set—at arbitrary choice—is to be set fixed in the LSA, so that all the adjusted numerical values are relative to this choice (*relatively*-adjusted values)—while any pair difference is invariant irrespective of the choice of the fixed member;

(b) The original values of the adjustable member(s) of the set depend on the chosen *inter*-subjective criteria, and are *altered* according to the LSA optimisation algorithm. The latter operation sets an "intimate relationship" [9] between *all* members of the set, which may have no physical meaning for some of them in the specific case of the fundamental constants. The issue here is not a possible resulting 'discrepancy' between data in the set, but the fact that the obtained values only optimise the *consistency*, i.e. primarily the value of the statistical parameter used for evaluating the *uncertainty* of the set. For this there is a cost for the values: divorcing by a certain amount from the original physics world. This cost is generally irrelevant in other applications of the CODATA analysis. Also, it is not a characteristic of the LSA only, but of all methods having the same purpose, e.g. the methods with fixed effects. However, in the specific case of the measurement units, the CIPM may want (or need) to take advantage of the best accuracy allowed by the current experimental data, and this is a critical condition making incorrect the use of the CODATA relatively-adjusted values, and that may even result in missing the aimed goal [15]: see next section.

(c) The above difficulties are generally also involving the *adjusted uncertainties* typically lower than the experimental ones, associated with the relatively-adjusted values: it may happen that the uncertainty aimed level is satisfied by some adjusted constants, but not yet achieved in actual experimental determinations.

3. Formal problems

The items (4) to (5) in the Introduction fall under this heading.

3.1. Stipulation of numerical values, namely using algebraic expressions of them

The current proposal requires stipulating the numerical values of c_0, h, e, k_B and N_A (presently only c_0 is). In the general case of stipulation, in principle one could decide to use any arbitrary numerical value. However, this would lead to metrological undesirable consequences. Should a discontinuity in the size of the new units be avoided, as is considered to be preferable in metrology, the stipulated value cannot be chosen anymore arbitrarily. It should coincide with the 'best' value in a statistical sense of the set of experimental data available at the time of the stipulation of that constant, not being the CODATA one.

Another critical issue arising from the suppression of the explicit information on uncertainty in stipulated numerical values is the rounding or truncating of the stipulated numbers. For an uncertain number, this operation follows precise rules: rounding is performed according to the uncertainty level and limited to two uncertain digits (e.g., in the CODATA values), sometimes to one. However, since after stipulation the uncertainty information is lost in the definition, the fact is questionable that the first uncertain digit could be retained in stipulation, a number exact by definition. It seems possible to consistently use only the digits unaffected by uncertainty.

Another issue deserving attention in stipulation is the (frequent) case where a constant is expressed (indirectly) as an algebraic combination of other constants. The algebraic expressions of stipulated values, for example $R = k_B \cdot N_A$ or $F = N_A e$ or $K_J = 2e/h$ or $R_K = h/e^2$, or the expressions (1) above, are often assumed to have in turn stipulated exact numerical values. This opinion is not a direct consequence of the concept of stipulation. In the expressions (1), the numerical values are shown limited to the 15 digits allowed by the use of MS Excel for computation. It obviously happens that the resulting number is sometimes rational, in general irrational, with no obvious rounding or truncation rules applicable. That is a general consequence of algebraic combinations.

The stipulation of the constants forming the algebraic expression does not automatically generate a stipulated numerical value in the definition of the units: a *distinct decision* should be taken in the latter respect, and a *separate specific* stipulation (number rounding or truncation). In particular, the use of the inverse of a constant in the new definitions places the same problem. For more details see [11].

3.2. Representation, use and stipulation of incompletely-known integer numbers

An example of an intrinsic integer number in measurement is the result of a count. An example of a count is the Avogadro number, whose known value is usually expressed as $6.022\ 141\ 29 \cdot 10^{23}$.

Is it correct? This representation does not convey at all the information that the number is an integer, of which, contrarily to usual cases, only some of the *most significant digits are known*. For more details see [6, 11].

4. Problems in perspective for science and metrology

The item (6) falls under this heading.

4.1. Future possibility of considering new determinations of the fundamental constants, and of the use of the Least Squares Analysis by CODATA

A question arises: is it useful (or correct, depending on the viewpoints) that CODATA analysis and outcomes always incorporate constraints arising from a basically regulatory field like metrology, for decisions concerning the measurement units? It is true that the numerical values of the constants completely depend on the size of the units on which the measurements are based. However, may general science, which is basically interested on consistency of the physical theories, be tied to LSA evaluations of consistency that are actually *biased* by metrological issues?

If the CGPM will eventually adopt the CIPM proposal, more constants will have their numerical values stipulated. Whence, many more will also get fixed numerical values because of the inter-relationship between constants placed by the LSA method, considerably reducing the number of the freely-adjustable ones according to the present philosophy. The number of the remaining constants might be so small that the whole CODATA task could become irrelevant or even terminated.

Journal of Physics: Conference Series **459** (2013) 012042 doi:10.1088/1742-6596/459/1/012042

This situation would also place delicate question marks on the handling of future experimental determinations of the constants, namely of the stipulated ones. The traceability checks in the case of constant-based inter-related unit definitions are likely, at least, to be quite more indirect and complex.

Further, how will new determinations be taken into account in future? Possibly, it will not be in the interest of metrology to adjourn the stipulated value of a specific constant. On the other hand, it would certainly be advantageous for general science to have this new information taken in due account, even by the CODATA Task Group, according to the normal scientific method.

5. Possible ways out

Summarising, in this paper issue (1) (section 2.1) concerned point (a) of the Introduction; point (b) was not treated in this paper; the issues (2) to (5) (sections 2.2, 2.3, 3.1 and 3.2) concerned point (c); point (d) was treated in issue (6) (section 4.1).

Concerning point (a), there is an intrinsic contradiction between the SI principle of using 'base' quantities as distinct from 'derived' quantities and using fundamental constants in the base unit definitions, except possibly for the mole—apart from formal problems arising from the use of the inverse of N_A. In fact, there is no possible remedy when relaxing the basic principle of (dimensional) independence of the SI base units, like it would happen by using *multi-dimensional* constants in the definitions, unless the set of base quantities is changed—e.g., by replacing length with velocity. A fact already for some of the base units in the present set will be aggravated by the new proposal.

Furthermore, the new definitions will make necessary to have *practical* realisations of these units, called "mise en pratique": the traceability chain will therefore start from the relationship between the unit definition and its *mise en pratique*. In all instances, the new proposed units will cause the metrological need—note, not simply the scientific will—for future measurements of those constants, in order to provide facts supporting evidence to the traceability chain. National traceability might only demand, within each NMI, (i) to realise an apparatus for the measurement of a constant, (ii) to assign the outcome of the measurements the stipulated value, and, (iii) to establish a relationship between these results and the *mise en pratique* of the relevant unit(s) of the NMI. In the case of constant-based multi-dimensional unit definitions, these checks of the resulting degree of internal traceability are likely to be quite more indirect and complex—and costly—than the present ones. For International *inter-pares* traceability, one *additionally* needs to resort, as usual, to between-laboratories information, the *inter*-comparisons, where the actual differences between realisations will show up [12, 13].

As to point (c), the *values* of the constants, the CODATA *relatively*-adjusted values, computed to test the consistency of the data set, *should not be conceptually confused* with the values obtained, directly or through computations, from the experimental determination(s) of the constants' values, nor with the statistically-evaluated representative values of them—like the mean, weighted mean, ... As to the *uncertainty* associated with the value of a constant, the LSA method reduces, sometimes dramatically, the standard uncertainties associated with the adjusted value. This is a direct benefit of the method arising from statistical features when a larger set of values is involved, even without new direct determinations and beyond current experimental capability. The *uncertainty* levels associated with the adjusted values of the constants *should not be conceptually confused* with the uncertainties actually achieved experimentally.

The way out is to use the values and related uncertainties determined, at the time of the stipulation and for each individual constant, only by selection criteria of the 'best' value and uncertainty, in a strict statistical sense, of the set of available experimental data. Most statistical methods look appropriate—except fixed-effect methods, for the same reason why the LSA is not.

Concerning the expression of the above numerical values in the definitions of measurement units, some issues need attention in the step bringing from the initial uncertain to the stipulated values. The latter being exact by definition, they, with proper rounding, should not involve uncertain digits of the initial numbers, possibly except when the uncertainty affects the last digit by only $\pm \approx 1$ like in the case of c_0. Algebraic expressions of stipulated values should be avoided in the definitions, including the inverse of a constant. In fact, a definition should be in itself comprehensive and self-consistent, and is not supposed to allow arbitrary interpretations by the users. The allowed number of digits of the result of the numerical operations should appear within the definition: it can be set, before stipulation, only from the original uncertain number, information not available to all users. This is to say that only the result of the operations should appear in the definition as a numerical value stipulated, and with the correct number of digits.

Concerning the effects of the new definitions on the future CODATA task and on handling future experimental determinations of the stipulated constants, point (d), a serious concern is expressed for a possible divorcing of the metrological regulatory procedures from the outcomes of interest—and needs—for 'general science'. In order to ensure the future possibility to take into account the fact that new experimental determinations might possibly lead to different 'best' values according to normal scientific practice, a possibility would be that the CODATA Task Group modifies its procedure. In this perspective, the analysed set of constants should concern *exclusively* the check for their best *consistency*, limiting to one, μ_0, the non-adjusted constant. The adjustments should be recalculated back from 1983 with c_0 and ε_0 adjustable.

References

[1] CCU (2010). Draft Chapter 2 of the SI Brochure, BIPM, http://www.bipm.org/en/si/new_si/.
[2] CGPM (2011). On the possible future revision of the International System of Units, the SI, Comptes Rendus des Séances de la XXIV Conférence Générale, Resolution 1, 2011, Bureau International des Poids et Mesures, Sèvres, http://www.bipm.org/en/convention/cgpm/resolutions.html
[3] E.R. Cohen, K.M. Crowe and W.M. Dumond, *The Fundamental Constants of Physics*, G. & J. Chesters, Tamworth, Staff, United Kingdom (1957)
[4] BIPM (2006). *Le Système International d'Unités (SI)*, 8th Edition, Sèvres
[5] B.N. Taylor and A. Thompson, *The International System of Units (SI)*, 2008 ed., NIST Spec. Pub. 330 U.S. Government Printing Office, Washington, DC, 2008
[6] F. Pavese, Some reflections on the proposed redefinition of the unit for amount of substance and of other SI units, ACQUAL 16 (2011) 161-165
[7] BIPM (2008). *International Vocabulary of Metrology—Basic and General Concepts and Associated Terms* (VIM) 3rd edn BIPM/ISO.
[8] CODATA http://physics.nist.gov/cuu/Constants/index.html, http://www.bipm.org/extra/codata/.
[9] NIST (1970). *Precision Measurement and Fundamental Constants*, D.N. Langenberg and B.N. Taylor, eds, Spec. Pub. 343
[10] B.N. Taylor, W.H. Parker and D.N. Langenberg, Determination of *e/h* using macroscopic quantum phase coherence in superconductors: implications for quantum electrodynamics and fundamental physical constants, Rev. Modern Phys. 41 (1969) 375–496
[11] F. Pavese, Rounding and notation, namely when using stipulations in the definition of measurement units, Measurement 46 (2013). http://dx.doi.org/10.1016/j.measurement.2013.04.010.
[12] F. Pavese, An introduction to data modelling principles in metrology and testing, Chapter 1 in *Data Modeling for Metrology and Testing in Measurement Science*, (F. Pavese and A.B. Forbes, eds.), Birkhäuser, Boston (2009) pp. 1–30. ISBN: 978-0-8176-4592-2 (print), 978-0-8176-4804-6 (online).
[13] F. Pavese, *Mathematical and statistical tools in metrological measurement*, Chapter 6.35.78, in Encyclopedia of Life Support Systems (EOLSS), Developed under the Auspices of the UNESCO, Eolss Publishers, Oxford , UK, [http://www.eolss.net] 2013, pp 68, in publication
[14] P. De Bièvre, What is the meaning of the concept 'consistency' in (chemical) measurement?, ACQUAL 18 (2013) 157-158.
[15] CGPM (2007). Comptes Rendus des Séances de la XXIV Conférence Générale, Resolution 12, BIPM, Sèvres, http://www.bipm.org/en/convention/cgpm/resolutions.html.

Journal of Physics: Conference Series **459** (2013) 012043 doi:10.1088/1742-6596/459/1/012043

Are fluctuating measurement results *instable*, *drifting* or *non-stationary*? – A survey on related terms in Metrology.

Karl H. Ruhm

Institute of Machine Tools and Manufacturing (IWF)
Swiss Federal Institute of Technology (ETH), Zurich, Switzerland

Email: ruhm@ethz.ch

Abstract. We describe properties of measurement quantities and measurement processes verbally by using accepted terms, some of which are sanctioned by international standards and guides. However, some of these terms are used inconsistently, as compared to definitions in base sciences like for example Mathematics, Signal and System Theory, Estimation and Optimisation Theory, Stochastics and Statistics. This is the case for different terms in Metrology, like *stability*, *drift* and *stationarity*. In this paper we show systematic relations between these and similar terms and discuss the adequacy of their application in practice. We have to admit that we incorrectly implement the terms *stable* / *instable* unconsciously, when we describe properties of measurement results, due to the use of common colloquial habits.

1. Introduction

Measurement instrumentation and measurement results fulfil defined requirements. We certify them by qualitative verbal descriptions and / or by quantitative logical and mathematical relationships, according to accepted rules.

For example, we say in colloquial language that certain situations and procedures are *stable* or *instable*. Sometimes data are reported to be *drifting*. And many instances appear *nonstationary*. Are such descriptions identical, synonymous, similar or related? Of course, they may serve as hints for the time being. But what shall we do, if we have to deliver quantitative information? We suppose that there must be different meanings behind these terms, which are all used regularly in practice.

Apparently, there must be a concept, which is able to deliver coherent tools and which is applicable to all fields of science and supports top-down considerations. Indeed, there are facilities, provided by Signal and System Theory and by Stochastics and Statistics [1; 2; 3].

In the following sections we will focus on the classification and definition of the terms of interest mentioned above. We will see that some of them refer to *signals* (models of real-world physical or other *quantities*) and some of them refer to *systems* (models of real-world physical or other *processes*) [9]. This means that all terms are well defined properties of signals or systems, which we may describe mathematically, if necessary.

Interestingly enough, no or just a few helpful definitions concerning our field of consideration can be found in the basic Guides of Metrology like the GUM [4] and the VIM [5]. Even the Standards of Statistics refrain from offering significant and precise support, when dealing with these terms [6, 8].

The following investigation will justify, which term has to be applied in which situation. In particular, it will become obvious that the terms *stable* and *unstable* are often used incorrectly for quantitative descriptions (models).

Journal of Physics: Conference Series **459** (2013) 012043 doi:10.1088/1742-6596/459/1/012043

2. Signals and Systems

Terms like stability, drift and stationarity seem to describe properties of signals and of systems as well. However, properties of systems show up in their behaviour first and do so in their output signals afterwards. So, we have to distinguish carefully between definitions for signals and definitions for systems. This will be done in the following sections.

Properties of Signals and Systems

Quantitatively we describe *properties* of a signal by *structures* of equations on the one hand and by the vector of *parameters* of these equations on the other hand. The same is true in the first instance for systems: Quantitatively we describe *properties* of a process by *structures* of equations on the one hand and by the vector of *parameters* of these equations on the other hand. But, for systems the equations have a special meaning: They describe the relations between cause and effect, between input and output, relations, which we will call *behaviour* for defined circumstances.

Behaviour of Systems

Generally speaking, every signal comes from some source, which we call system. Signals depend on systems; they are manifestations of systems. May we then state that signals don´t have their own behaviour and that we can only describe their momentary properties?

On the other side, this is not true for systems; these reveal behaviour indeed. The overall-behaviour links internal signals **x**(t) and output signals **y**(t) as responses to arbitrary excitations (stimulation, impact, action, effect, drive, load, question, disturbance) by the input signals **u**(t) (Figure 1).

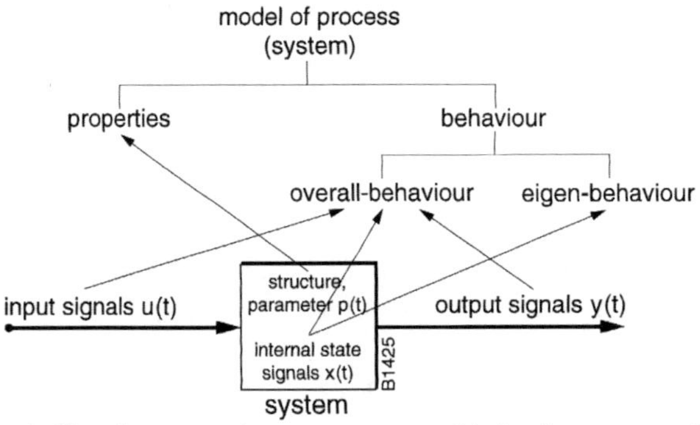

Figure 1. Signals, properties, parameters and behaviour concerning a system

Exactly, this is the immediate interpretation of the mathematical solution of differential equations, which describe systems. The *eigen-solutions*, according to the values of the initial conditions **x**(0), together with the *forced solutions*, according to the input excitations **u**(t) deliver the *total solution* in form of the output signals **y**(t). In this context *solution means behaviour*. Or: Behaviour describes, how a system reacts according to *external* (input) signals and to *internal* (initial) signals. This type of description is called *system transfer response description* [2]. We will use it in order to discuss the term *stability*.

Up to now we have assumed constant parameters **p** within the system. However, according to the cause and effect principle we additionally have to consider a second description of a system as soon as the parameters **p** vary over time and become independent variables. Then we describe the relations between process parameters **p**(t) and output quantities **y**(t). This type of description is called *system sensitivity description* [7]. We will use it in order to discuss the term *drift*.

3. The Term Stability

We frequently say that a procedure is instable if a signal drifts away. Does this mean that *instability* and *drift* are synonymous terms? Not at all! The causes or the mechanisms of instability and drift in systems are completely different (see section 5). There are *instable systems* and there are *drifting systems*. A system can be stable and drifts nevertheless. Actually, this is even the normal situation.

Output signals, emerging from instable systems, cannot be annotated as instable too. Instead, they are qualified as varying, fluctuating, oscillating, limit cycling, drifting, shifting, and as transient, non-constant, nonstationary, unsteady, unbounded, and so on. Again: Those types of signals may come from stable systems alike, due to dedicated input signals, which are able to generate these types of output signals just as well. Quantitatively, this is true for corresponding mathematical descriptions of output signals of course. There exists no definition or mathematical formalism for an "instable signal"!

Therefore we concentrate on *systems* and define *stability of a system* qualitatively as its ability (behaviour) to keep its output signals *bounded*, provided that all input signals are bounded too [2].

Definitions

Qualitative: Once a system is instable, at least one output signal will run away unbounded
- even if all *input signals* remain bounded

and
- even if all *system parameters* remain bounded.

Quantitative: The definition of stability is simple for linear dynamic systems, but cumbersome for nonlinear systems. System Theory normally describes a linear time-independent dynamic system (LTI) by a set of ordinary, linear differential equations (ODE) with constant coefficients. When is it stable?

The stability of a linear system depends on both system properties, on *structures* and *parameters*:

1. System (model) *structure*

There are systems, which cannot get instable for structural reasons:
- linear nondynamic systems
- linear dynamic systems of less than third order

All other systems *may* become instable.

2. System (model) *parameters*

In addition, the stability of a linear dynamic system of higher than second order depends on the *magnitudes* of the parameters, which are elements of the so called characteristic equations of the system:
- In the *time domain*: A linear dynamic system is asymptotically stable, if all poles (eigenvalues) of the set of *characteristic equations* of the *differential equations* have negative real parts and are therefore located in the left half of the complex plane.
- In the *frequency domain*: A linear dynamic system is asymptotically stable if all poles (eigenvalues) of the *denominator polynomial* of the *spectral transfer response function* g(s) have negative real parts and are therefore located in the left half of the complex plane.

There are more stability criteria seen from different points of views. The simplest are the two mentioned above. Mathematically they are equal and the statements are identical. Considering the magnitudes of the parameters, stability criteria deliver stability conditions, stability margins, stability bounds and stability areas, useful for system synthesis and system engineering.

Instability is a binary property (stable / instable) and is based on system properties and not on signal properties, even though it is noticed by apparent output trajectories. Therefore, we also have to avoid the frequently used term *long-term stability*, which concerns *drift* and not stability issues (see section 5).

Besides: In Metrology we seldom encounter instable systems, in particular there are no instable sensors. However, we sometimes have instable processes with implemented stable sensors. Very seldom observer, filter and adaptation structures may raise stability questions.

Journal of Physics: Conference Series **459** (2013) 012043 doi:10.1088/1742-6596/459/1/012043

Note: An *instable* dynamic system does not behave *chaotic*, since its behaviour is completely deterministic and can be predicted, provided the model of the system is known. Chaotic systems follow different rules.

4. The Term Stationarity

The term *stationarity* refers to signals (time series data) and parameters, but not to systems. There are several classes of stationarity. We only mention *weak stationarity*, which shows the main characteristics of interest already. Signals are (weakly) stationary concerning time and space, if characteristic *values* (like mean values, standard deviations values, covariance values and so on) and characteristic *functions* (like probability density functions, correlation functions, spectral power density functions and so on) are independent concerning time and space, or frequency and wavelength.

The simplest example refers to a stationary signal or parameter x(t), for which the following equation (description) holds [10]:

$$\mathcal{AVG}\{x(t)\} = \mu_x = \lim_{t_{obs}\to\infty} \frac{1}{\Delta t_{obs}} \int_{t=-t_{obs}}^{+t_{obs}} x(t)\,dt = \mathcal{AVG}\{x(t)+\Delta\tau\}$$

with

$\mathcal{AVG}\{...\}$	averaging (expectation) operator
μ_x	arithmetic mean value
t_{obs}	observation time value
$\Delta\tau$	arbitrary time shift intervall

This means that for a stationary signal or parameter the arithmetic mean value is independent of time t. However, this seemingly simple definition asks for an infinite range of the independent variable, here of time t, which is never available. So, we normally define *short-time stationarity* within finite observation (averaging) intervals (windows) Δt_{obs} (Figure 2).

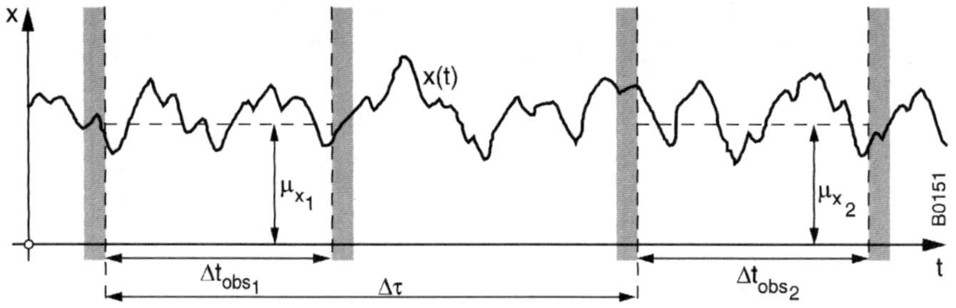

Figure 2. Definition of short-time stationarity

Note: Normally, stationarity will be superposed by other properties of the signal, drift for example. Sometimes, a separation of different signal content can be achieved by dedicated filtering.

Remark: The following source from the National Institute of Standards and Technology (NIST) serves as an example how delicate the use of common language terms can be, if it is not based on scientifically agreed conventions: "A manufacturing process cannot be released to production until it has been proven to be stable. Also, we cannot begin to talk about process capability until we have demonstrated stability in our process. A process is said to be stable when all of the response parameters that we use to measure the process have both constant means and constant variances over time, and also have a constant distribution." [8]. Here stationarity is meant and not stability.

Journal of Physics: Conference Series **459** (2013) 012043 doi:10.1088/1742-6596/459/1/012043

5. The Term Drift

A rough characterization in technical and natural sciences concerning *drift* means that an observed signal changes on average in *time* versus a certain direction. When measuring or collecting data from a process, we may occasionally say: "Our data drift away". Besides, there is the related term *shift* too. It refers to a signal, which changes on average in *space* into a certain direction.

Features

1. Within a dynamic system in steady state condition ($\mathbf{u}(t)$ = constant or stationary) drift appears as a temporal deviation of at least one system parameter from the nominal value in a particular direction.
2. Drift is perceived as a slow movement of behaviour compared to all other temporal patterns in the dynamic system.
3. Regular input-output results of a system and its drift features superpose. Several drift features may interfere with each other.

Definitions

1. Drift, perceived in signals (data), is the special result of *temporal parameter variations* $\mathbf{p}(t)$ of a system and excludes drift as a result of varying input quantities $\mathbf{u}(t)$ of this system.
2. Drifting deviations of system properties or behaviour from explicitly stated nominal properties or behaviour are erroneous properties or behaviour. They give rise to *systematic drift errors* $\mathbf{e}_{drift}(t)$, accompanied by *drift uncertainties* $\boldsymbol{u}_{drift}(t)$ (Figure 3).

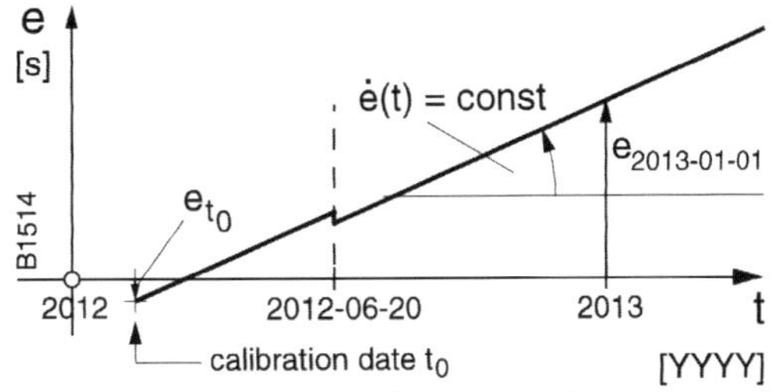

Figure 3. Example of the systematic drift error e of an output signal of a drifting watch.

Since we assume constant or stationary input signals u for the consideration of drift, we concentrate on the *system sensitivity description* $p(t) \rightarrow y(t)$, the second type of system description mentioned above [7].

At least one parameter, the drift parameter $p_{drift}(t)$, changes in one direction. The parameter deviation (error) $\Delta p_{drift}(t)$ makes the dynamic system a *time variant dynamic system.* We usually assume that the parameter drift proceeds linearly with time t: $\Delta p_{drift}(t) = v_{p_{drift}} t$ [{p}]. The constant (average) drift velocity (drift rate) of the parameter p is $v_{p_{drift}} = \dot{p}_{drift}$ [{p}s^{-1}]. It may change with time t too. This is a description of system properties.

Thus output signals will show effects due to these drift properties. The simplest case is the *constant output signal* y, which drifts in time t. Thus, the signal is not constant any more. The nominal temporal trajectory y_{nom} is superposed by a temporal component $\Delta y_{drift}(t)$, which systematically tends in one direction.

$$y(t) = y_{nom} + \Delta y_{drift}(t) \quad [\{y\}]$$

For a typical *stationary random output signal* y(t) the nominal arithmetic mean value drifts [10]. Thus the random signal is *not stationary* any more.

$$\mu_y(t) = \mu_{y_{nom}} + \Delta y_{drift}(t) \quad [\{y\}]$$

Note that the drift signal $\Delta y_{drift}(t)$ can be a random signal with a drifting mean value too.

So we always distinguish between the nominal trajectory (signal) $y_{nom}(t)$ and the deviating trajectory (signal) drift $\Delta y_{drift}(t)$.

These remarks are true for nondynamic and dynamic systems. Questions about the effects of system parameter errors $e_p(t)$ on the output signals y(t) or on the output signal errors $e_y(t)$ have to be answered by system sensitivity analysis.

Drifting systems are regularly stated as *instable*. This is not correct in the strict sense of Signal and System Theory. The International Vocabulary of Metrology (VIM) is misleading too [5] with its definition 4.19: "stability of a measuring instrument: property of a measuring instrument, whereby its metrological properties remain constant in time" with a further quantification of this term: "In terms of the change of a property over a stated time interval." This would be the appropriate definition of a *time invariant system* and not of a *stable system* on the one hand and the suitable specification of a *drifting output signal* on the other hand according to system parameter drift.

6. Conclusion

- The term *instable* characterises a system and not a signal (data). Signals of instable systems move *boundlessly*. Instable systems are rare; they are not useful considering practical applications. On the other side, they can be stabilised by control tools.
- Signals are constant and stationary on the one side and drifting, shifting, varying, fluctuating, oscillating, limit cycling, transient as well as nonconstant, nonstationary, unsteady, unbounded, and so on the other side.
- Constant parameters make a system *time invariant*; varying parameters make it *time variant*.
- The term drift characterises a special parameter varying system, whereby drifting signals arise.

The terms discussed above are standard terms of Signal and System Theory and Stochastics and Statistics. However, their treatment in everyday use and the statements concerned in International Guides and Standards of other fields are more or less inconsistent.

References

[1] Oppenheim, A. V.; Schafer, R. W.: (2010; 3rd ed.); Discrete-time signal processing; Prentice Hall
[2] Oppenheim, A. V.; Willsky, A. S.; Hamid N. S.: (1997; 2nd edition); Signals and Systems; Prentice Hall
[3] Papoulis, A.; Pillai, S. U.: (2002); Probability, random variables, and stochastic processes; McGraw-Hill
[4] ISO : (2008); Evaluation of measurement – Guide to the expression of uncertainty in measurement (GUM); JCGM 100
[5] ISO : (2008; 3rd edition); International vocabulary of metrology – Basic and general concepts and associated terms (VIM); JCGM 200
[6] ISO : (2010); Statistics - Vocabulary and symbols - Part 2: Applied statistics; ISO 3534-2
[7] Frank, P. M.: (1978); Introduction to System Sensitivity Theory; Academic Press
[8] NIST: Engineering Statistics Handbook; http://www.itl.nist.gov/div898/handbook/ppc/section4/ppc45.htm
[9] Ruhm, K. H.: (2010); Process and System - A Dual Definition, Revisited with Consequences in Metrology; 13th IMEKO TC1-TC7 Joint Symposium, London, Journal of Physics: Conference Series 238 (2010) 012037; http://iopscience.iop.org/1742-6596/238/1/012037
[10] Ruhm, K. H.: (2005); Averaging One Variable; http://www.mmm.ethz.ch/dok01/d0000547.pdf
[11] Ruhm, K. H.: (2011); Drift; http://www.mmm.ethz.ch/dok01/e0000603.pdf

Journal of Physics: Conference Series **459** (2013) 012045 doi:10.1088/1742-6596/459/1/012045

Wireless Sensing Node Network Management for Monitoring Landslide Disaster

S.Takayama, J.Akiyama, T. Fujiki, N.A.B.Mokhtar
Dept.E.E.Engng., Fac.Sci.&Engng., Ritsumeikan Univ.BKC
Noji, Kusatsu, Shiga 525-8577, JAPAN

E-mail: s-tkym@se.ritsumei.ac.jp

Abstract. This paper shows the network management and operation to monitor landslide disaster at slop of mountain and hill. Natural disasters damage a measuring system easily. It is necessary for the measuring system to be flexible and robust. The measuring network proposed in this paper is the telemetry system consisted of host system (HS) and local sensing nodes network system (LSNNS). LSNNS operates autonomously and sometimes is controlled by commands from HS. HS collects data/information of landslide disaster from LSNNS, and controls LSNNS remotely. HS and LSNNS are communicated by using "cloud" system. The dual communication is very effective and convenient to manage a network system operation.

1. Introduction

Natural disasters are terrible and so important concerns for us. They destroy social infrastructures, loss human lives, and damage economic condition of people, companies, organizations and governments. It is very hard for measuring system to monitor landslide disasters. Measuring instruments are broken easily by the disaster. By reason of dangerousness, measuring person cannot be with the system. Measuring system must monitor at a wide area not only a point. As landslide disaster is occurred suddenly, the system waits and senses it at a moment after long time monitoring.

Wireless sensing node network system (WSNNS) is one of effective solution to monitor the landslide disaster. The network formation is mesh. The distributed sensing nodes operate cooperatively. Then the WSNNS becomes possible to measure acceleration, soil moisture and land position at a wide area for long time. As sensing nodes are supporting each other, the WSNNS can continue the measurement even if some nodes have been destroyed by landslide disaster. The sensing node analyses measuring parameters to find a sign of landslide disaster and to estimate the condition of system and sensor devices. The WSNNS operates autonomously as a local system (LSNNS). Then the measurement data and estimated information are transmitted to Host system (HS) at other place where measuring person monitors and collects them safely. There are several kinds of communication method between WSNNS and HS. "Cloud" system presents so useful services to realize a dual way communication method between them. As HS does not demand measuring data transmission from LSNNS in real time, the communication is enough asynchronously. The storage and mail system of "Cloud" realize an effective and convenient communication to transmit the data/information from LSNNS and the command from HS.

In monitoring landslide disasters, the total measurement network system constructed by LSNNS, HS and the communication method using "cloud" system realizes so useful and convenient measurement system. In following, the network management is shown.

Journal of Physics: Conference Series **459** (2013) 012045 doi:10.1088/1742-6596/459/1/012045

2. Measurement Network System for Monitoring Landslide Disaster

Fig.1 shows the system construction of total measurement network to monitor landslide disaster. The system is consisted with Host System (HS), "Cloud" system, Local Sensing Node Network System (LSNNS).

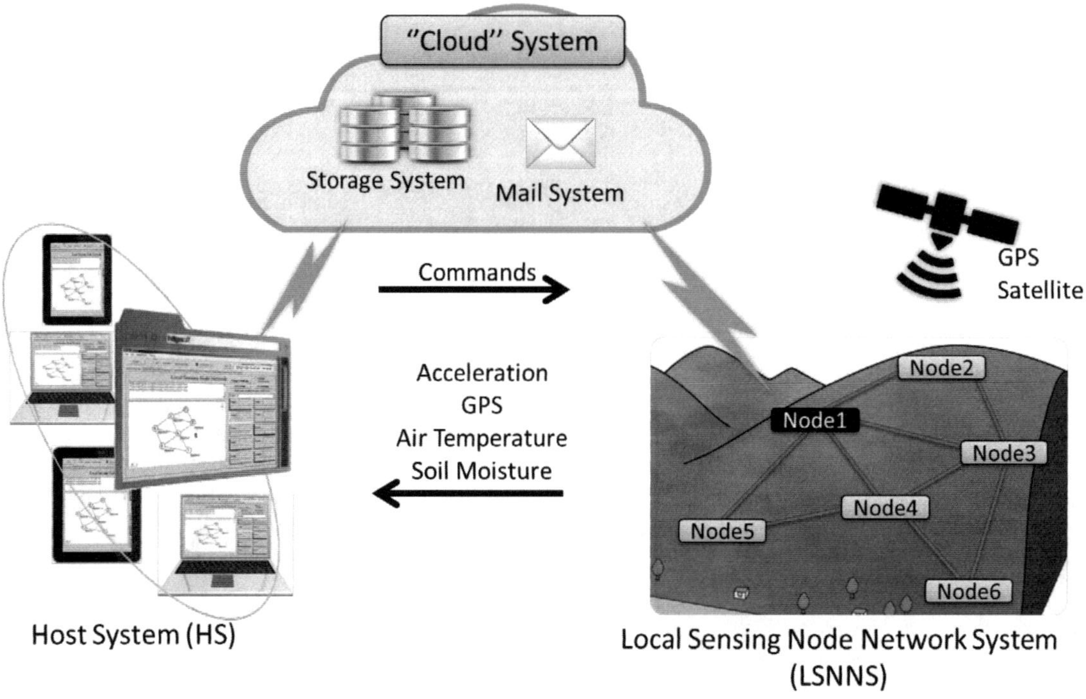

Figure 1. System construction of total measurement network to monitor landslide disaster

2.1. Local Sensing Node Network System

LSNNS is the local measuring network which is consisted of plural distributed sensing nodes connected with near field wireless communication at around slope of mountain and hill. One of them is a top node which is the gateway of LSNNS for data and command. The top node mounts a smart phone to communicate data and command with "Cloud" system. Each sensing node mounts some sensors, control board and wireless communication device (Figure 2, Figure 3). The sensors are 3D acceleration, GPS, temperature and soil moisture. The 3D acceleration sensor is used to monitor ground acceleration by landslide disaster. The occurrence and types (slide down, rolling down) of landslide are recognized by continuous analyse of 3D acceleration angles and power spectrum of acceleration signal for 1 sec.. GPS is used to know the ground position and movement of sensing node. The position precision is within 5 m. By information of ground positions of sensing nodes, HS can monitor the topology of LSNNS which becomes the basic information to construct the routing table of LSNNS. The ground movement of sensing node is referred to find a sign of very slow landslide. The wireless communication device connects sensing nodes in near field. The operation frequency is 1.2GHz. The transmission baud late is 9600 bps. The communication distance in wood is about 50-60 m. The control board takes some kinds of role to manage the communication control, to decode and operate commands from HS, to acquire measuring data from sensors and to estimate information of landslide disaster and system condition by analysing the data.

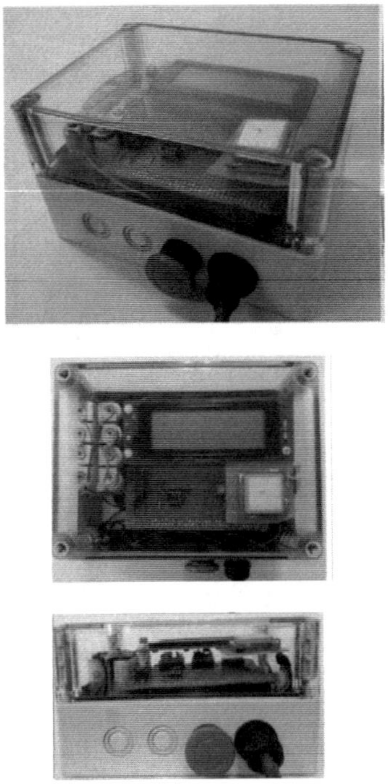

Figure 2. A sensing node

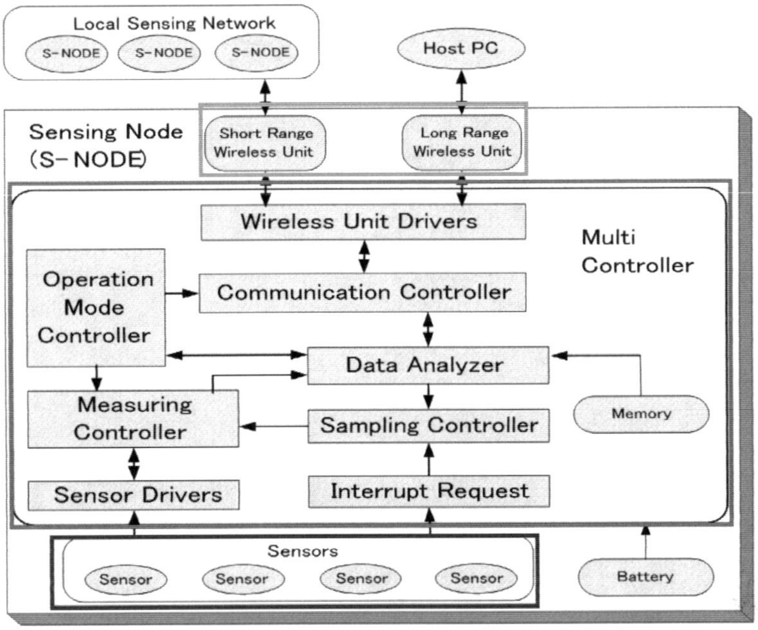

- Wireless communication unit : MU-1-1216
- Control board : SH3DSP
- Sensors : Acceleration, GPS, Temperature

Figure 3. System construction of sensing node

LSNNS has three kinds of operation mode which are initialization, normal and urgency mode. In the initialization mode, sensing nodes recognize the condition of mutual connections each other by beacon test signal transmissions. By these connection data, HS estimates the topology of LSNNS and constructs the routing table. In the normal mode, the sensing node operates on demand by HS. According to the command from HS, the sensing node replaces system parameters like sampling time. And that returns measuring data, estimated information and system conditions to HS. In the urgent mode, the sensing node recognized the occurrence of landslide disaster send urgent information to neighbour nodes. The nodes known the urgent information also send it new neighbour nodes. Repeating this transmission between sensing nodes, the urgent information is influence node to node. Finally, that information reaches to the top node of LSNNS. The top node sends the information to HS via "Cloud" system urgently.

2.2. "Cloud" System

"Cloud" is a collection system of several kinds of service (Storage system, Mail system and so on). The system is so convenient to exchange data and information from everywhere. Storage capacity in commercial "Cloud" system is 25GB freely. This measurement network system uses this "Cloud" system to communicate commands and data/information between HS and the top node of LSNNS by the mail system. They are packaged in a mail message each other. Now, the top node of LSNNS can send 13 messages per minute maximally which is depended on the characteristic of smartphone mounted in top node. As the communication is asynchronously, HS and top node of LSNNS check mail messages regularly. Figure 4 shows a series of measuring data transmitted by mail messages. Figure 5 shows the situation of growth of local sensing node network which is informed by measuring data. Data transmission rate will be increased more by using direct access technology to storage system. But in actually, as the analysis of continuous measuring data like acceleration is processed at sensing nodes in LSNNS (transformation to information), data message via "Cloud" system is limited to temperature and absolute time/position by GPS. Then huge number of mail messages is not occurred in this measurement network system.

Figure 4. A series of measuring data transmitted

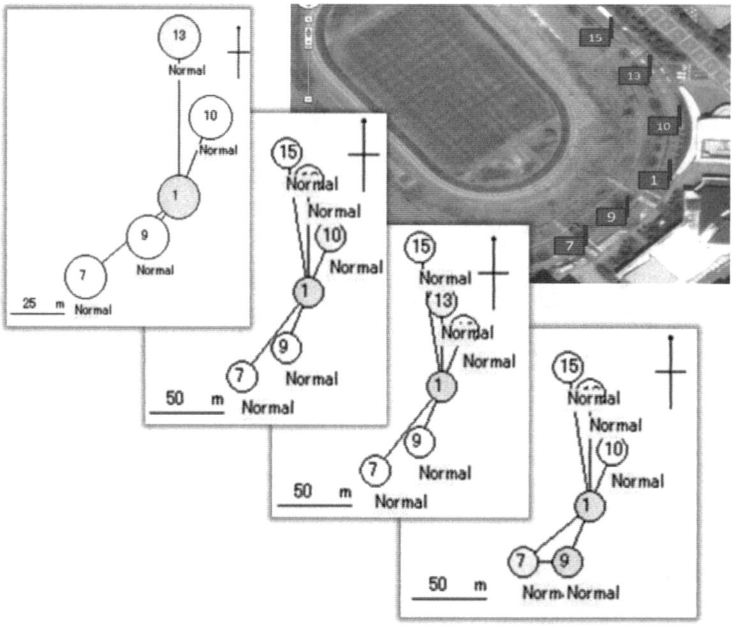

Figure 5. Situation of growth of local sensing node network which is informed by measuring data

2.3. Host System

HS takes some roles for system management with measuring person far from LSNNS. As HS is designed to exchange data/information to "Cloud" system with wireless communication network, the hardware is various like desktop, mobile, PC, tablet and any other which can access to internet. HS checks mail arrivals regularly in "Cloud". At finding, it takes data/information from mail messages and distinguishes them according to their characteristics. In HS, they are displayed sequentially as graph and table to measuring persons. At getting the information of occurrence of landslide disaster, HS alarms it to them autonomously. Using the measuring data of absolute position by GPS, HS forms some kinds of routing table of LSNNS. The routing table is a map of ways which commands and data are transmitted in LSNNS. The plural routing tables are prepared as risk management for local broking of sensing nodes in LSNNS. HS sends several kinds of command to LSNNS, which change the operation modes (initialization and normal), request to any nodes to send measuring data and operation conditions.

Using human interface windows in HS, Measuring person can monitor a list of measuring data (acceleration, GPS, temperature)(Figure 4), status of network topology(Figure 5), status of communication and time transition of measuring data. Figure 6 shows a command operation window. At left area, operator can select (1) target sensing nodes in LSNNS, (2) measuring data of GPS, acceleration sensor, temperature sensor which he wants to collect. HS converts these requests to a command message in an edit box shown at left top. By click "Send" button of right side of the command line box. HS sends the command to LSNNS. The command is displayed in right above box. At receiving the responses from sending nodes, the message lines are shown in right below box.

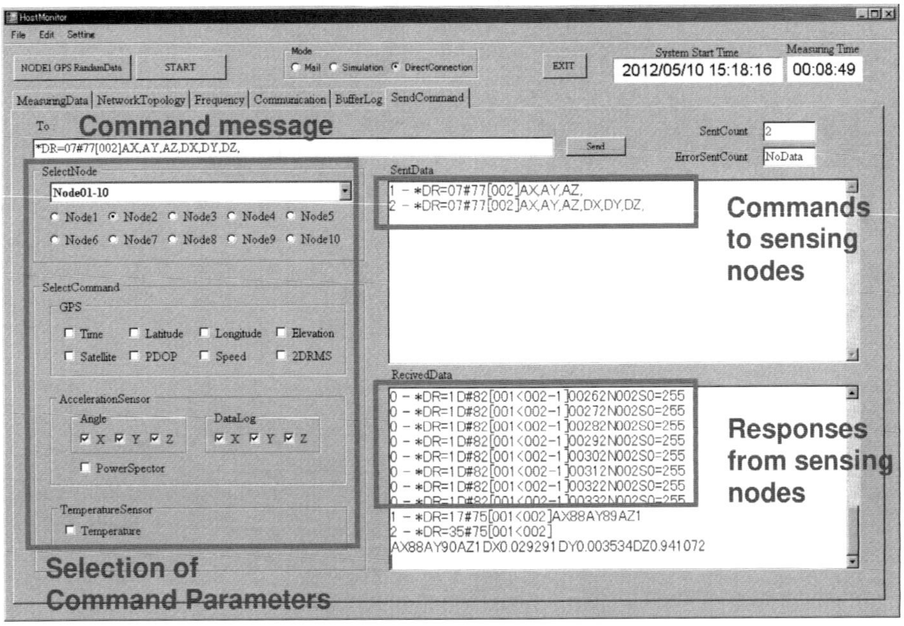

Figure 6. Command operation window in HS

Figure 7 shows the monitoring window of sensing node conditions. The conditions (3D body inclination, absolute position, inside temperature and so on) are recognized autonomously by the sensing node itself. The information is requested and collected by commands from HS. In Figure 7, body inclinations of a sensing node are shown.

Figure 8 shows the information flow of landslide disaster detected by the analysis of acceleration signal measured at a sensing node. To avoid communication conflictions in LSNNS, usually,

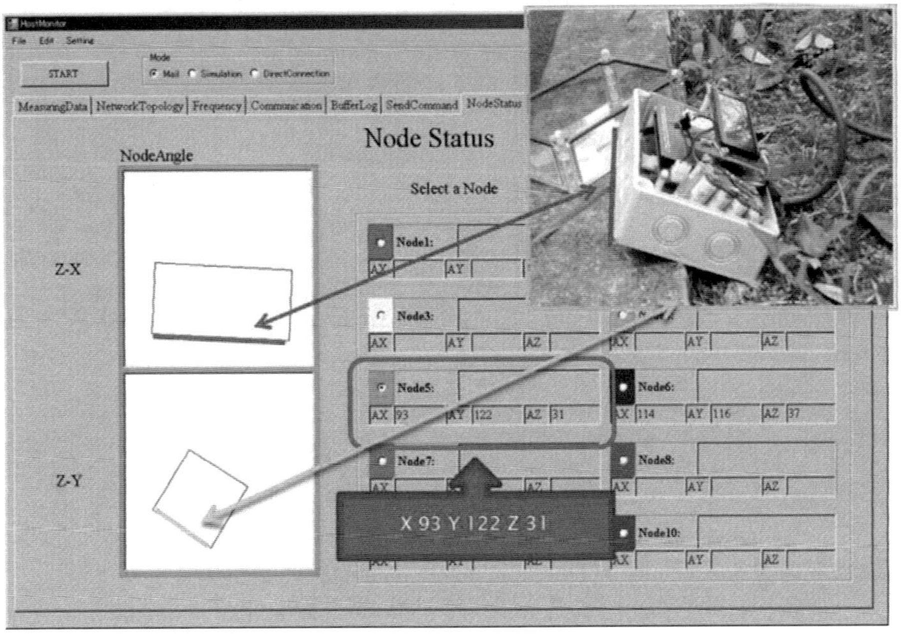

Figure 7. Monitoring window of sensing node conditions in HS

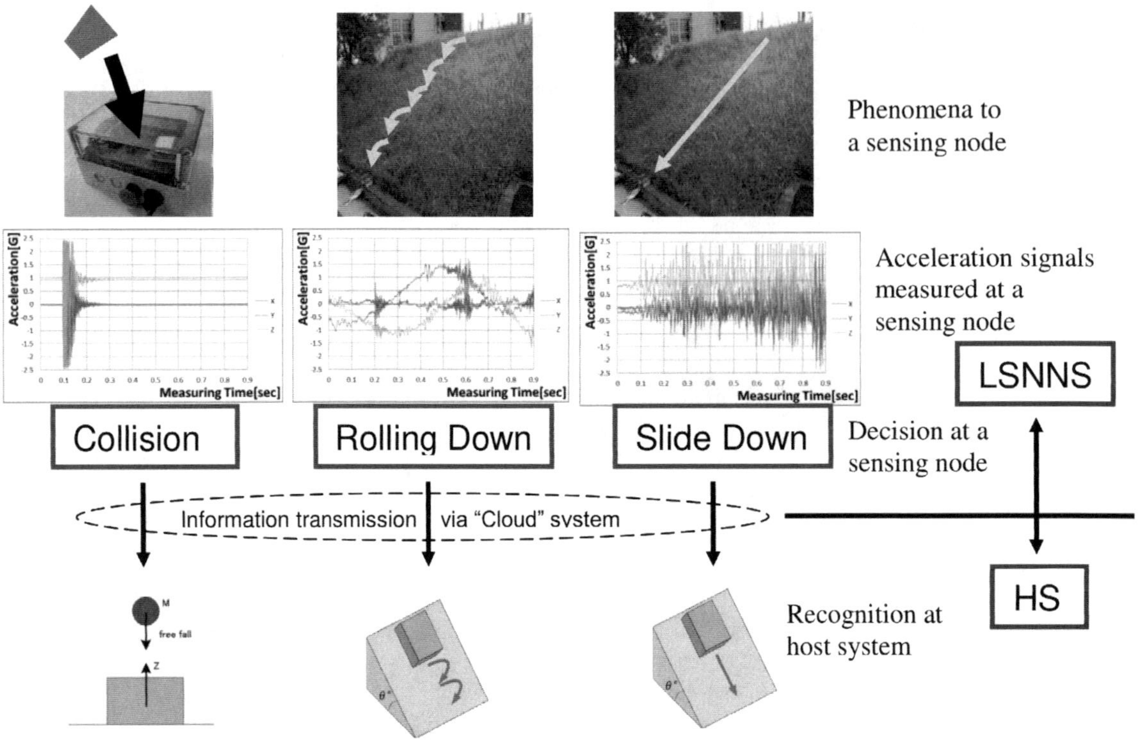

Figure 8. Information flow of landslide disaster detected by the analysis of acceleration signal measured at a sensing node

measurement data is analysed at a sensing node. And the information as results is transmitted from LSNNS to HS via "Cloud" system. By reconstruction of HS software as WEB software, measuring person can operate measurement network system and monitor landslide disasters by any PC and Tablet from everywhere.

3. Conclusion

In this paper, the system construction of total measurement network to monitor landslide disaster, which is consisted with Host System (HS), "Cloud" system, Local Sensing Node Network System (LSNNS). The measurement system to monitor natural disasters is needed robust, flexible, useful and reliable operation management. LSNNS is so robust, flexible system for landslide disaster. That reforms the network routing autonomously. "Cloud" system presents usefulness to access the system from everywhere and the flexibleness to communicate data/information and command between HS and LSNNS. In addition to them, HS as WEB software realizes portability of total measurement network system for measuring persons.

References

[1] S.Taenaka, Y.Watanabe and S.Takayama, "Construction of Wireless Sensing Network System for Landslide Disaster Monitoring", ACTA IMEKO 2012, pp.TC19-P-5, 2012.

[2] Rahman, F.; Kumar, A.; Nagendra, G.; Gupta, G.S., "Network approach for physiological parameters measurement", IEEE Trans. Instr. and Meas., Vol.54, pp. 337 - 346, 2005.

[3] S.Taenaka, J.Akiyama, T.Fujiki and S.Takayama, "Design of Flexible Wireless Sensing Network System for Landslide Disaster Monitoring", Proc. of SICE International Conference 2012, pp.1443-1459, 2012

Journal of Physics: Conference Series **459** (2013) 012046 doi:10.1088/1742-6596/459/1/012046

Problems in choosing methods and tools for conformity assessment of prepackages at the manufacturer

J. Dobiliene[1], A. Meskuotiene, E. Raudiene
Metrology Institute of Kaunas University of Technology, Studentu str. 50-454, Kaunas, Lithuania
E-mail: justina.dobiliene@ktu.lt

Abstract. In the article is a systematic analysis of the prepackaged products conformity assessment that could help in practice by applying the general provisions of the Directives to conformity assessment of prepackaged products. Manufacturers have to choose the method of conformity assessment that best suits the needs of his/her firm by choosing the optimal packing structure, methods and tools, based on the recommendations of the Article. Procedures for the maintenance of measuring instruments (calibration and verification) depend on the composition of packing line, level of automation. They shall be chosen only after the accurate identification of packaging line. When choosing the measuring instruments for packing manufacturers must provide traceability capabilities of measuring instruments used in weighing process, properly using calibration and verification results of the measuring instruments. Measurement results have demonstrated that for evaluation of average mass of prepackages procedures, mentioned below, can be applied. Statistical analysis procedure is more accurate than random sample procedure. It was proved that accuracy (repeatability and systematic error) of measurement results for sticky products is more than 2 times bigger than for powdery products (especially for low mass). Manufacturer shall pay attention to this fact while adjusting packing line for sticky products.

1. Introduction

Conformity assessment and approval is one of the conditions to implement the policy of free movement of goods. Legal basis is Council Directive 76/211/EEC of 20 January 1976 on the approximation of the laws of the Member States relating to the making-up by weight or by volume of certain prepackaged products.

Conformity assessment of prepackages is related with legal metrology, thus it provides consumers with confidence that requirements for the quantity of products are met. Extremely strict requirements are set for placing prepackages on the market and their surveillance. Conformity assessment is related with actual quantity of content of a product in prepackage, with statistical evaluation of a prepackage batch. The packer or importer shall be responsible for ensuring that prepackages meet the requirements of the Directive [1].

In the area of conformity assessment and enforcement it is possible to choose them from two or more alternatives. Such a choice should be based on scientific research. Chosen test methods and

1 Corresponding author: E-mail: justina.dobiliene@ktu.lt (Justina Dobiliene), Phone: +370 37 300776, fax: +370 37 351 252.

tools, possibilities to ensure their traceability must be optimized and fully meet the objective set for them.

Difficulties occur, and particularly manufacturers are facing the problems, in choosing forms and methods for conformity assessment of prepackaged products, measuring instruments, in providing maintenance of measuring instruments.

1. Choice of optimal structure for the product packaging

Actual quantity of product contents in prepackages (package and product) general imply the mass or volume of the product. Packaging process at the manufacturer or packer and its control are realized in four different stages (Figure 1).

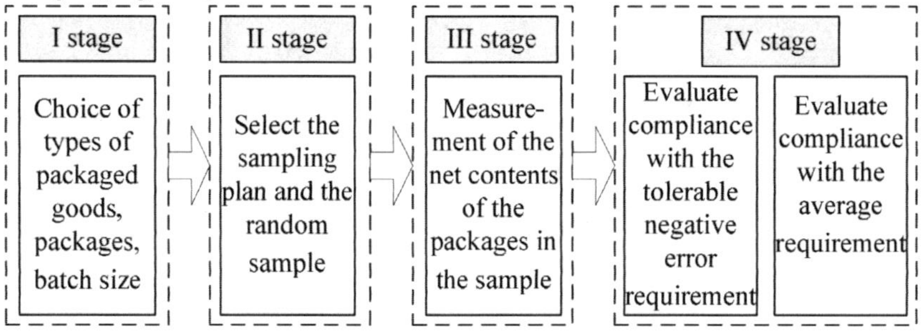

Figure 1. Product packaging stages

In the first stage types of packaged goods, weight packages, and batch size are chosen. In the second stage sampling plan and random sample are selected. It is possible to use sampling plan and random sample presented in Directives. In the third stage method that will be used to assess the net amount of contents should be chosen. The last stage comprises control, i.e. a comparison of the net quantity of individual prepackaged products with the tolerable negative error and compliance with the average requirement (AQC) [2].There are three general rules to follow:

- the average net content of the package in a batch may not be less than the declared quantity;
- the number of non-conforming prepackages must not exceed 2,5%;
- the prepackages having a negative error greater than twice the tolerable negative error must not be in a batch. [2]

These three rules must be implemented by the manufacturer during the product packaging and quantity control. It is therefore important to choose the product packaging structures, methods and tools properly.

Generally packing structure depends on:

- type of prepackaged product (product contains additives or not, glazed frozen product, product having liquid), its physical characteristics (liquid, powdery), and batch size for a production;
- number of measuring instruments in the packing line (automatic checkweighers, automatic gravimetric filling instruments, non-automatic weighing instruments);
- status of measuring instruments (being verified (legal metrology), calibrated).

In consideration of the above mentioned criteria product packaging organizational structures can be identified as those that are shown in Figure 2.

These criteria are related to one another. Type of packaged goods or packages and batch size determine the level of automation of packaging process. Accordingly, this level of automation determines the level of automation of chosen dosing devices and measuring instruments. Planning of packaging process itself influences the surveillance types of instruments in use.

Automatic checkweighers and automatic gravimetric filling instruments are usually used to dose and pack powdery, liquid products. Non-automatic weighing instruments are used very often for weighing hard lumps products, such as fish. Measuring instruments shall be assigned to legal or industrial metrology depending on how the weighing is carried out (individual weighing or automatic) and at what stage. This determines the choice of surveillance type – verification or calibration.

Thus, the organizational structure depends on the composition of the packing line. Certain case is when prepacking and weighing of a product is done at the same time. Another case is when a product is prepacked only by a coating, for example, by a plastic film. In a case of powdery products they are almost always prepacked and weighed at the same time. In a case of hard lumps products (such as fish) prepacking is done by automatic machine using a film coating and package sealing. Mass is measured individually by non-automatic weighing instruments. Then the non-automatic weighing instruments shall be assigned to packing line. In this case several identical non-automated packaging lines can be included in the product packing line, where each product is weighed by non-automatic weighing instruments and packing machine performs only a package sealing function. Similarly, if packing is done manually by placing the product into the package with sealing it and using non-automatic weighing instrument to weigh every prepackaged product; the non-automatic weighing instruments are assigned to the packing line.

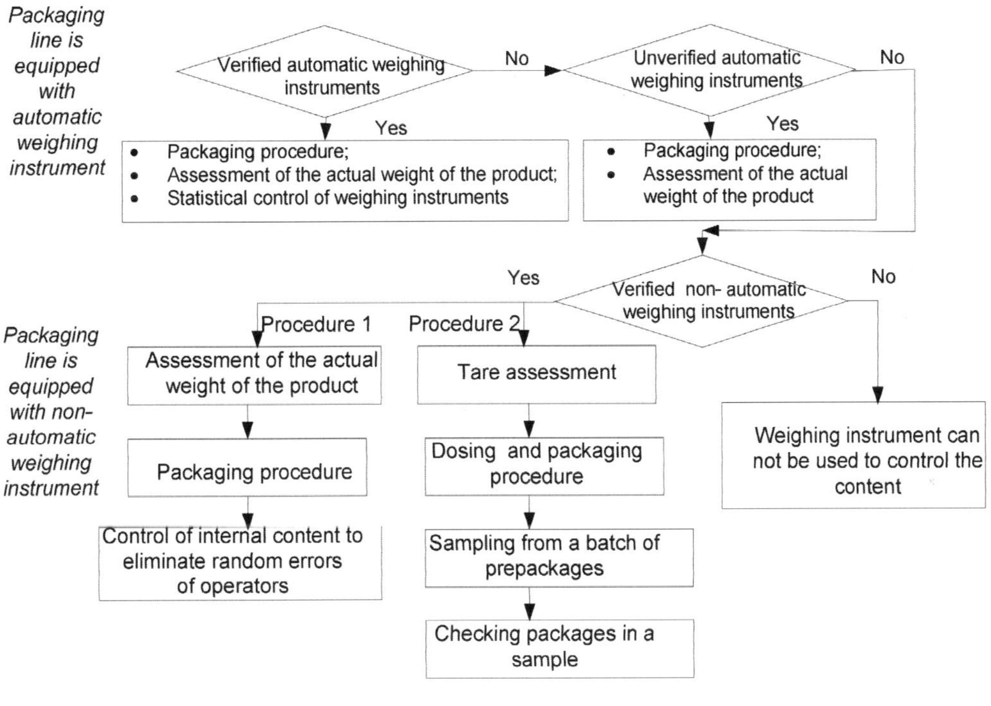

Figure 2. Organizational structure of the product packaging

In order to be able to show that the e-marked prepackages comply with the requirements, the packer must maintain his measuring equipment. The measurements of the actual content of prepackages, the density of liquid product, the weight of packaging materials and other relevant measurements must be carried out by means of a legal and suitable measuring instrument. Measuring instruments that are used directly for measuring process of quantity are assigned to legal metrology. They are subject to verification. Generally control measuring instruments are used: a non-automatic weighing instrument, a checkweigher, an automatic gravimetric filling instrument, a volumetric measure. However, in the case of non checked automatic weighing instruments, that are involved in dispensing and packaging process the weight control can be proven by checked non-automatic weighing instruments. Other measuring instruments that are not directly used for evaluation the quantity of the product in the package, for example, measuring instruments used for the monitoring of environmental conditions are not subject to legal metrology. For those measuring instruments a different maintenance and calibration regime is necessary. The packer shall organize this regime in line with the conditions of use and behavior.

2. The adaptability of the procedures

Check of prepackages in situ can be carried out by two methods: 1) using a random sample check with selection of samples during the packaging at least once per hour, evaluating the average of actual amount, and recording defective items and 2) carrying out selective statistical check by sampling method, i.e. by sampling from the batch of prepackages. Having chosen any of these methods, decision regarding approval of prepackages or rejection of a batch is taken after whole batch is manufactured. After a rejection of batch produced it must be quarantined. To quarantine means: the prepackages will not be put onto the market.

Using the first of the above method for investigation the prepackages batch, when the average is calculated, it is important to separate an hour's production (to identify it) because of when the average of the measuring results that relate to an hour's production is less than the nominal quantity, the hour's production must be quarantined. In this way the only part of the production would be quarantined, not the whole batch. Besides, choosing the batch classification into the hourly production, when the average of a sample falls below a warning limit, packers can undertake corrective action to bring the average back on target by adjusting the filling machine. It is a good practice for the packer to carry out a check on every (production) hour. Checks might be carried out at longer intervals, provided that the quantity control system is set up appropriately to take into account drift in the packing process. This also applies for monitoring prepackages below the TU1 (TU1-the nominal quantity minus the tolerable negative error) and TU2 (TU2 the nominal quantity minus twice the tolerable negative error)-limits.

3. Comparative analysis of methods

Methods using a random sample checking or statistical analysis procedures were investigated and shown in Figure 3.

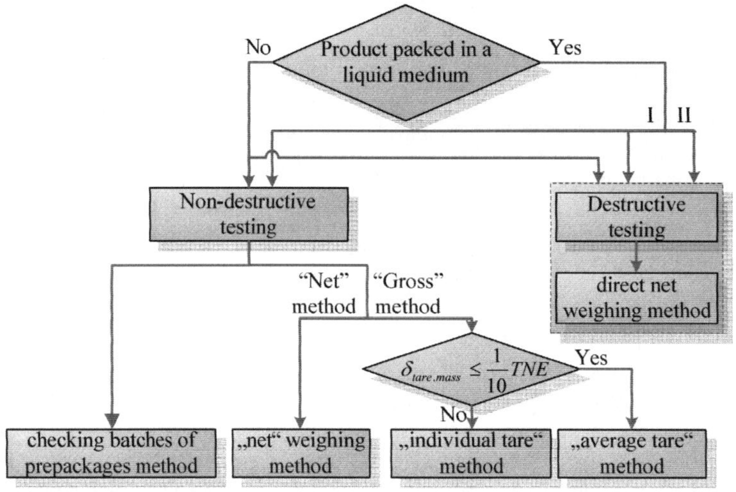

Figure 3. Algorithm for choice of methods of product mass determination in prepackage

The choice of measurement method for product actual quantity depends on the product type, packaging type, and the product batch size. When using non-destructive testing to control the quantity of prepackages, random sampling testing or statistical analysis procedures can be used. When a selective statistical verification is carried out, testing method of the prepackaged products batch is used. It is convenient to use the method as the measurement of actual content of the product is carried out after packing the whole batch. When choosing the test of random sample during the manufacturing process a packer has to take samples of the product at least once per hour. Using „net" weighing method the actual content of prepackage is measured directly: weighing the contents of prepackage without package. If this method is chosen quantity control is carried out by taking samples from the production line before the product packaging operation. One relevant thing is that no product can get "lost" when weighing and packing. Measuring the content itself is the best but often not possible. Otherwise, the mass of package has to be determined beforehand. Consequently, there are two ways for determination of actual mass of product in prepackage indirectly (thereinafter tare) [3]:

- Gross mass of prepackage minus mass of individual tare (thereinafter "individual tare" method);
- Gross mass of prepackage minus average mass of tare (thereinafter "average tare" method).

The essence of all methods is calculation of prepackaged products with a negative tolerance of the product contents higher than the permissible negative tolerance. Experiments were carried out with powdery (nuts of various sorts) and sticky (dried plums, dates) products for the same batch of similar nominal mass (from 40g to 1500g). Two main characteristics were calculated, i.e. the average quantity of prepackages and repeatability of measurement results. Typical examples of the results are presented in Figure 4.

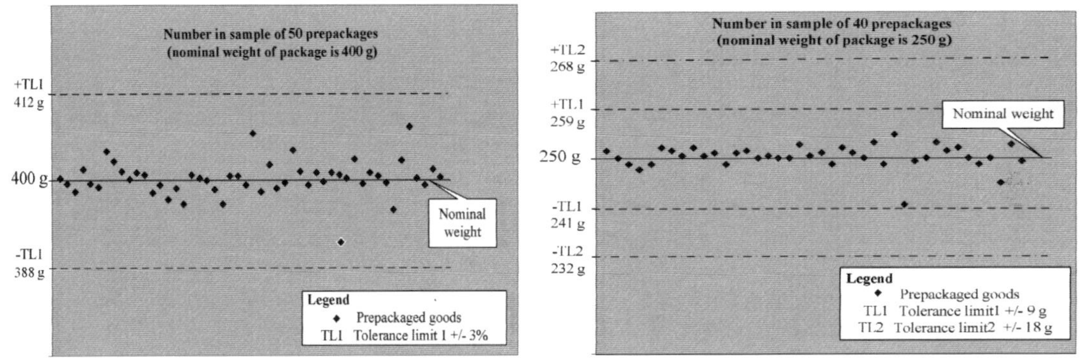

Figure 4. Typical examples of the results of the statistical analysis and of the random sample testing procedures ("average tare" method)

Summarized results of procedures (repeatability of measurement results for different nominal masses) are shown in Figure 5.

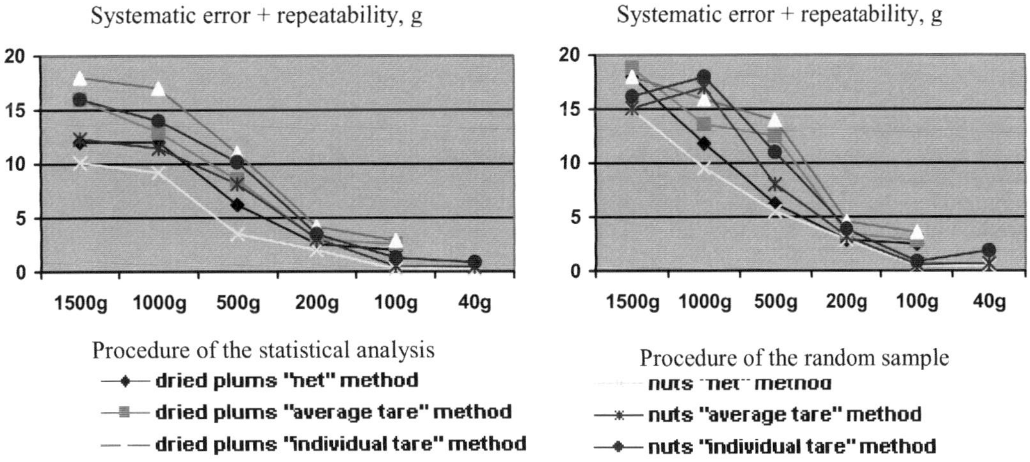

Figure 5. Results (repeatability) of the of the statistical analysis and random sample testing procedure for different methods

Results that are given in Figure 5 show that "average of plums" is usually negative and "average of nuts" is usually positive for different procedures even though the conditions and adjustment of automatic weighing instrument are the same. That means that manufacturer have to reject whole line of products because of the one of rules (that is mentioned in chapter 1) is not satisfied.

4. Conclusions

1. Manufacturers have to choose the conformity assessment method that best satisfies the needs of his/her firm by choosing the optimal packing structure, methods, and tools, based on recommendations of the Article.
2. Procedures for the maintenance of measuring instruments (calibration and verification) depend on the composition of packing line, level of automation. They shall be chosen only after the accurate identification of packaging line.
3. When choosing the measuring instruments for packing manufacturers must provide traceability capabilities of measuring instruments used in weighing process, by properly using calibration and verification results of the measuring instruments.
4. Measurement results have demonstrated that both procedures are suitable for evaluation of average mass of prepackages from the point of view of accuracy. Evaluating trueness and precision of separate methods it can be concluded that "net" method is the most accurate. But its application at manufacturer is limited. Statistical analysis procedure is more accurate than random sample procedure. It was proved that systematic error + repeatability of measurement results for sticky products is more than 2 times bigger than for powdery products (especially for low masses up to 200 g). Manufacturer has to pay attention to this fact while adjusting packing line for sticky products.

References

[1] Council Directive 76/211/EEC of 20 January 1976 on the approximation of the laws of the Member States relating to the making-up by weight or by volume of certain prepackaged products.
[2] Welmec 6.6 2003 *Guide for recognition of procedures* (Vienna: BEV) p 24
[3] Meškuotienė A and Raudienė E 2009 *Investigation of Product Quantity Evaluation Models in Prepackages* vol 2(44) (Kaunas: Technology Matavimai) pp 14-19

Journal of Physics: Conference Series **459** (2013) 012047 doi:10.1088/1742-6596/459/1/012047

A test procedure for energetic and performance analysis of cold appliances for the food industry

F. Armani, A. Boscolo
APL, Engineering and Architecture dept. University of Trieste, via Valerio 10, Trieste, Italy

E-mail: francesco.armani@phd.units.it

Abstract. In this article we present a novel approach for the characterization of cold appliances and in particular of refrigerators based on the standard vapour compression cycle with a reciprocating on/off compressor. The test procedure is based on a virtual instrument that perform both the stimulus and the data acquisition on the device under test. Acquired data is elaborated to fit a semi-empirical model based on the energetic balances between thermal and electrical sub systems and the heat exchanged with the environment. This approach results in a simple method to calculate useful parameters of the refrigerator, such as energetic performance, cooling effect and limit values of thermal loads. The test procedure requires only a few temperatures and the electric power consumption to be monitored, resulting in a low impact on the refrigerator. Preliminary tests showed a good estimation of parameters and prediction of energy consumption and heat extraction capacity of the refrigerator under test.

1. Introduction

Domestic and professional refrigerators are a large energy demanding appliances in both emerging and developing countries. In particular household refrigerators cover the 15% of the domestic power needs[1,2]. Their capillary spread over the entire food chain, combined with the 24 hours a day use, justify this energetic demand and make important even the smallest energy efficiency improvement.

Performance enhancement of refrigerators is being addressed by different points of view[3]. Thermal insulators, high heat capacity materials, novel refrigerant fluids, stirling[4] and linear compressors, thermo-acoustic[3] and magneto-thermal[5] effect are all subjects of great interest as promising means to improve the efficiency of refrigerators. From the other point it looks clear that the energetic efficiency of a refrigerator can also be increased by users with their own actions. Several studies have underlined that a 10%–20% of energy consumption in the domestic sector could be saved by modifying the sole behaviour of the occupants[6,7]. A real–time feedback on refrigerator energetic needs have been identified as an essential information to help individuals improve energy–use behaviour[8,9].

Useful information on the state of the refrigerator such as electrical power consumption, performance or heat extracted from the cabinet are difficult to measure directly without a complex acquisition system. An estimation of these values can be determined with the support of a behavioural model of the refrigerator tuned on its load characteristics. The components to be taken into account to characterise refrigerators and in general cold appliances are heat pump (generally compressor, evaporator and isenthalpic valve), cabinet or room to be cooled and external environment (temperature and humidity). To characterise the entire refrigerator the layout of the components in the cabinet and their interconnection must be taken into account as well because the overall performance is strongly related to the practical assembly of the system. This implies that the real behaviour can't be defined by the sole combination of the characteristics of its components but require an overall performance characterisation. Moreover cold appliances are in general non-linear systems, thus their characterisation requires multiple points of load and environmental conditions to be acquired.

Standard procedures to characterize a refrigeration compressor are mainly based on the measurement of its values of temperature, power consumption, pressures and extracted heat at different points of its load curve. The approach here proposed draws from this method, in which the refrigerator is characterized as a whole over its entire range of loads and ambient temperatures. This method is associated to a mathematical model which is based on the balances between the different energy components that describe the refrigerator. In the stationary state heat extracted by the thermodynamic cycle compensates thermal load of food and insulation losses. The load curves for different ambient temperatures are used to tune the model which is then able to emulate the behaviour of the refrigerator and provide useful information such as cell thermal conductivity, amount of net heat extracted from the cell and electrical power consumption. The simplified thermodynamic equations are derived from the steady state semi empirical approach proposed by Gonçalves et al.[10] and Hermes et al.[11].

This article presents this approach to the characterisation of a professional refrigerator and provide a first experimental validation of the underlying model. The tests were performed in a climatic room and with a controlled heat source positioned in refrigerator cell to achieve the entire range of loads and temperatures defined by the main test standards[12,13] while maintaining the stationary condition.

2. Experimental setup

The semi-empirical model emulates the steady state behaviour of a vapour compression cycle refrigerator, in which the reciprocating compressor is controlled by an on/off type thermostat. The mathematical approach is similar to those proposed by Gonçalves et al. and Hermes et al., in which the refrigerator is divided into sub-systems (Figure 1), namely compressor, evaporator, heat exchanger, capillary valve, condenser and refrigerator cell.

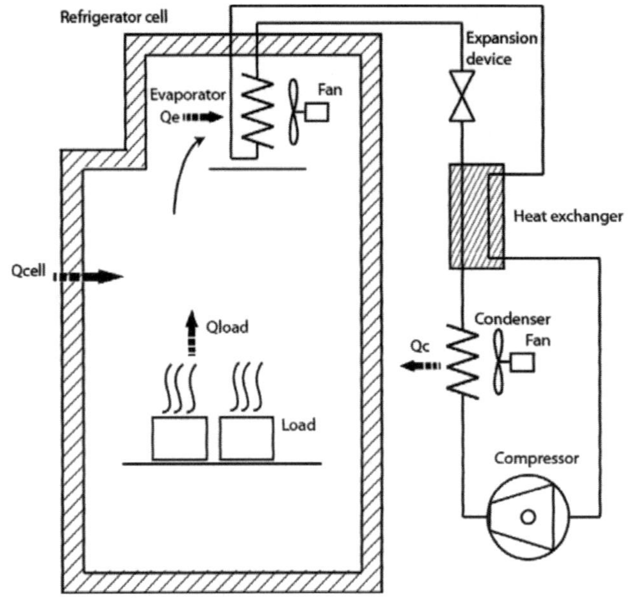

Figure 1. Simplified diagram of a vapour compression cycle refrigerator.

Journal of Physics: Conference Series **459** (2013) 012047 doi:10.1088/1742-6596/459/1/012047

In this work the thermal load absorbed by the evaporator is divided in three contributions:

- Heat introduced in the cell by insulation losses
- Heat produced by electric equipment located in the cold compartment (evaporator fan, defrost heater, door heater and cell lamp)
- Heat released by the food present in the cold compartment, including pans, grids and moist air introduced when the door is opened.

Each sub-model is described by its simplified thermodynamic behaviour; refrigerant enthalpy and mass flow rate are the state functions that connect these subsystems.

The experiments were designed to achieve two different tasks: sub models verification and model fitting. First a large amount of data was collected to verify that each part of the overall model was coherent with the real behaviour of refrigerator's related element. The second task was to provide the model with sufficient data to allow parameter fitting procedure. A second set of data was used to evaluate model prediction capabilities.

All of the experiments have been performed under stationary conditions of ambient temperature, while in the climatic chamber humidity was kept at values below 20%. In order to emulate food contribution to the thermal load, a controllable constant heat source was placed in the refrigerator cell. The methodology has been tested with a standard 500l, single door professional refrigerator with a R134a vapour compression cycle, a GL90TB reciprocating compressor and fin tube evaporator and condenser. Ambient temperature and humidity were imposed by the use of a climatic room with a ±2°C temperature error. Temperatures were acquired using a set of T-type thermocouples connected to a NI thermocouple data acquisition system (±0,1°C error). The pressure transducer mounted on the suction side of the compressor had a 0,1Bar resolution. Power measurements were performed with a WT500 and a Microvip3 electrical power meters (±0,1W and ±0,5W error respectively).

Thermal load supplied to the refrigerator's cell through electric resistors is equivalent to extracting heat from a constant heat source. In this way electrical/thermal dummy loads are intended as a useful way to emulate any kind of food and its load profile (humid air could be included as well, with proper assumptions on the heat exchange performance of the evaporator). The net thermal load was introduced in the refrigerator cell by turning On and Off the electrical power supplied to a rack of 4 power resistors. Resistor values were calculated to dissipate a nominal power of 25W, 50W, 100W and 200W. In such way it was possible to supply 16 values of power between 0W and 375W, with a step resolution of 25W.

Each experiment comprised of a preliminary phase to allow transients due to ambient temperature or power variations to be settled. The acquisition phase was set to 2 hours with a sample rate of $1Sas^{-1}$. Two sets of experiments were performed. The first one comprised 270 experiments of 2 hours each. This data was then used to fit model's parameters. A second set of 20 experiments was then performed and used as validation data. These experiments covered the range of thermal loads between 0W and 250W and ambient temperatures between 14°C and 43°C.

A system of virtual instruments, developed in Labview environment, performed tasks of data acquisition and control of thermal load supplied by the heaters. A first VI generated the stimulus to control the climatic room and the heat source inside the refrigerator's cell. A second VI collected the data from the various measurement units. After each test, acquired data was elaborated with another VI to calculate the enthalpy of the refrigerant in points of interest in the thermodynamic diagram. Values were calculated for the compressor On state only; in the Off phase it was assumed there was no refrigerating effect, as evaporator's fan was stopped, thus reducing significantly the air flow through the evaporator.

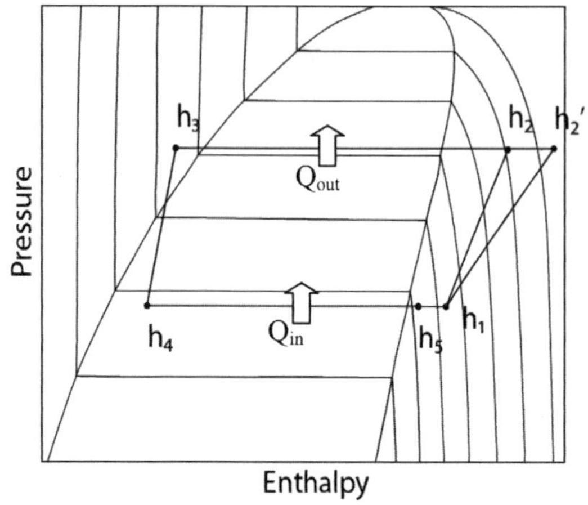

Figure 2. Simplified P-h diagram of the refrigeration cycle.

Look-up tables derived from the NIST database were used to relate R134a refrigerant enthalpy and entropy to its temperature and pressure in different points of the P-h vapour compression cycle diagram (Figure 2). Activation period, duty cycle, on state, average electrical power and average value of temperatures were calculated. Average values of thermodynamic properties such as isentropic compression enthalpy and extracted and rejected heat were provided as well. Concluding, each set of data, related to a 2hrs test, was then reduced by averaging each parameter. Values were split in two groups, associated with the state of the compressor. Thus thermodynamic properties, temperatures and electrical powers were averaged for both the ON and OFF states of the compressor.

3. Results

A refrigerator that works on a stationary condition can be defined by mainly four parameters, these are ambient temperature, electrical power consumption, heat load in the refrigerated cabinet and compressor's duty cycle.

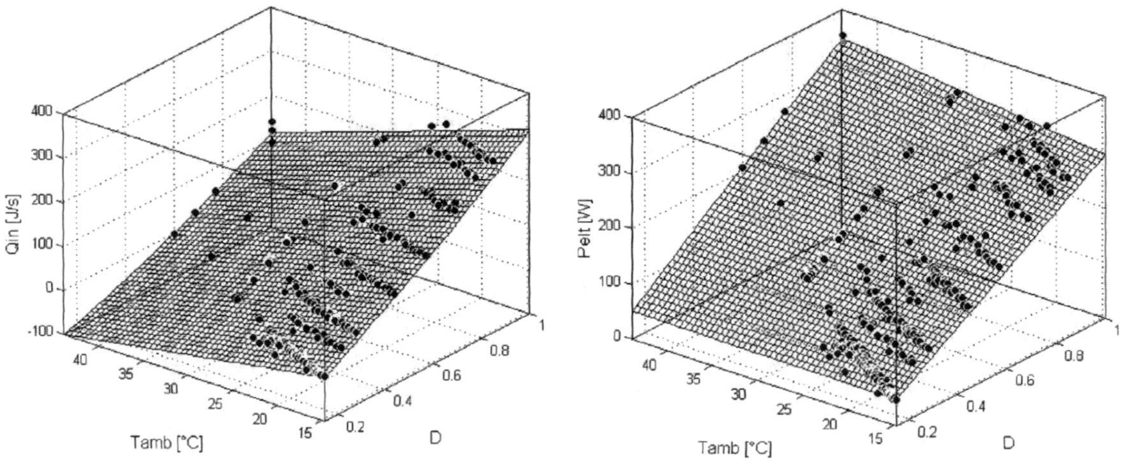

Figure 3. Net heat load (left) and electrical power consumption (right) as a function of ambient temperature and duty cycle.

In Figure 3 are reported experimental data (dots) and model response (surfaces) of heat load and electrical power consumption as functions of ambient temperature and duty cycle.

A preliminary validation of the model is reported in Figure 4. The measured and predicted heat load in the refrigerated cell and the electrical power consumption are compared. These show good agreement between the model and the device under test. An important coefficient derived from the model is the thermal insulation of the refrigerator cell. For the DUT its estimated value was 2.4 W/°K, that is in good agreement with the 2.5 W/°K stated by the constructor of the refrigerator.

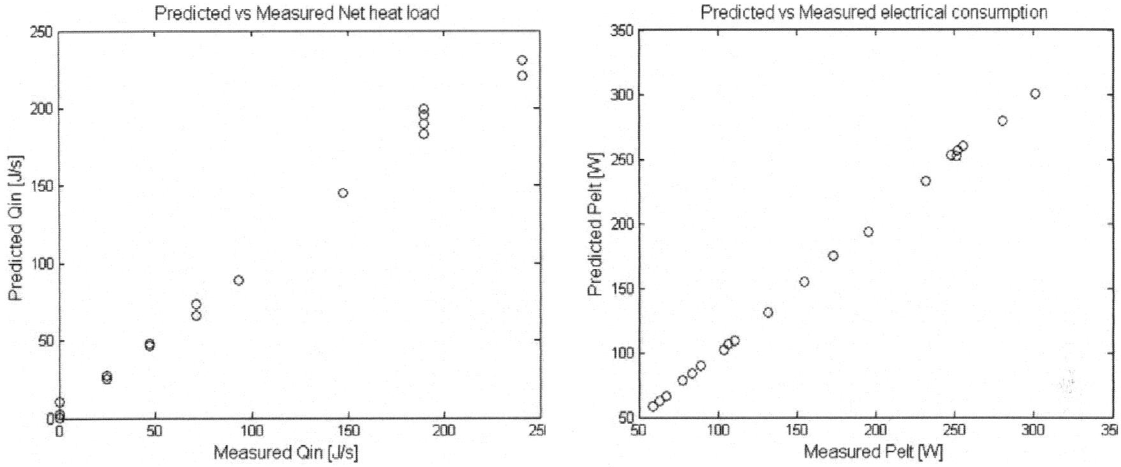

Figure 4. Comparison between predicted and measured heat load (left) and electrical power consumption (right).

4. Conclusions

A novel method for the characterisation of professional cold appliances has been proposed. A first evaluation has underlined its capabilities to assess the performance and maximum load of a refrigerator, with minimal computational requirements and with minimal effort on the device under test. Although a large amount of data has been collected, the method requires only three temperature sensors and a power meter to monitor electrical consumption.

The same method can be used to generate a model with good capabilities in the prediction of the power consumption of the refrigerator and its thermal load response over the entire range of ambient temperatures indicated in the international standards for professional refrigeration. This approach could represent a simple solution for the characterisation of the refrigerator in function of other parameters, such as door opening events, defrost cycle etc. and for the estimation of the thermal load associated to different foods. It is also suitable to be embedded in the refrigerator in order to provide the user with real time estimation of power consumption and cooling capability.

Ageing of components of the refrigerator is a topic that needs to be inspected as well to assess whether a duty cycle-based characterisation of the machine can be used over the entire life-time of the appliance.

Although this method based on steady state energetic balance has been proposed for refrigerators it can be adapted to other kind of cold appliances such as blast chillers and in general HVAC. The advantages could be the low impact on DUTs and the capability to evaluate their entire load curve in a limited set of experiments.

References

[1] IEA 2008 Worldwide trends in energy use and efficiency: key insights from IEA indicator analysis, IEA/OECD, Paris, www.iea.org/papers/2008/indicators_2008.pdf
[2] Bertoldi P, Atanasiu B 2009 Electricity and Consumption Efficiency Trends in European Union

- Status Report. EU Commission Joint Research Centre ISBN 978-92-79-13614-6, EUR 24005 EN

[3] Tassou S A, Lewis J S, Ge Y T, Hadawey A, Chaer I 2010 A review of emerging technologies for food refrigeration applications *Applied Thermal Engineering* **30** 263-276

[4] Thombare D G, Verma S K 2008 Technological development in the Stirling cycle engines *Renewable and Sustainable Energy Reviews* **12** 1-38

[5] Yu B, Liu M, Egolf P W, Kitanovski A 2010 A review of magnetic refrigerator and heat pump prototypes built before the year 2010 *International Journal of Refrigeration* **33** 1029-1060

[6] Owens J, Wilhite H 1988 Household energy behavior in Nordic countries-an unrealized energy saving potential *Energy* **13** 853-859

[7] Nielsen L, 1993 How to get the birds in the bush into your hand. Results from a Danish research project on electricity savings *Energy Policy* **21** 1133-1144

[8] Mansouri I, Newborough M, Probert D 1996 Energy consumption in UK households: Impact of domestic electrical appliances *Applied Energy* **54** 211-285

[9] Wood G, Newborough M 2003 Dynamic energy-consumption indicators for domestic appliances: Environment, behaviour and design *Energy and Buildings* **35** 821-841

[10] Gonçalves J M, Melo C, Hermes C J L 2009 A semi-empirical model for steady-state simulation of household refrigerators *Applied Thermal Engineering* **29** 1622-1630

[11] Hermes C J L, Melo C, Knabben F T, Gonçalves J M 2009 Prediction of the energy consumption of household refrigerators and freezers via steady-state simulation *Applied Energy* **86** 1311-1319

[12] Bansal P K 2003 Developing new test procedures for domestic refrigerators: Harmonisation issues and future R&D needs - A review *International Journal of Refrigeration* **26** 735-748

[13] Mahlia TM I, Saidur R 2010 A review on test procedure, energy efficiency standards and energy labels for room air conditioners and refrigerator-freezers *Renewable and Sustainable Energy Reviews* **14** 1888-1900

[14] Ding G L 2007 Recent developments in simulation techniques for vapour-compression refrigeration systems *International Journal of Refrigeration* **30** 1119-1133

Journal of Physics: Conference Series **459** (2013) 012048 doi:10.1088/1742-6596/459/1/012048

Experimental techniques for ballistic pressure measurements and recent development in means of calibration

L Elkarous[1,2,3], F Coghe[1], M Pirlot[1] and J C Golinval[2]

[1] Royal Military Academy, Renaissance avenue 30, B-1000 Brussels, Belgium
[2] University of Liege, Chemin des chevreuils, 1 B52/3, B-4000 Liège, Belgium
[3] Military Academy, 8012-Nabeul, Tunisia

E-mail: elkarouslamine@yahoo.fr

Abstract. This paper presents a study carried out with the commonly used experimental techniques of ballistic pressure measurement. The comparison criteria were the peak chamber pressure and its standard deviation inside specific weapon/ammunition system configurations. It is impossible to determine exactly how precise either crusher, direct or conformal transducer methods are, as there is no way to know exactly what the actual pressure is; Nevertheless, the combined use of these measuring techniques could improve accuracy. Furthermore, a particular attention has been devoted to the problem of calibration. Calibration of crusher gauges and piezoelectric transducers is paramount and an essential task for a correct determination of the pressure inside a weapon. This topic has not been completely addressed yet and still requires further investigation. In this work, state of the art calibration methods are presented together with their specific aspects. Many solutions have been developed to satisfy this demand; nevertheless current systems do not cover the whole range of needs, calling for further development effort. In this work, research being carried out for the development of suitable practical calibration methods will be presented. The behavior of copper crushers under different high strain rates by the use of the Split Hopkinson Pressure Bars (SHPB) technique is investigated in particular. The Johnson-Cook model was employed as suitable model for the numerical study using FEM code

1. Introduction

The aim of interior ballistics is to obtain the required velocity to hit the target and inflict the desired damage. The maximum pressure generated by the burn of the propellant inside the ammunition shell must not exceed the yield strength of the weapon barrel. Therefore, it's very important to know the peak pressure value to ensure a safe use of the weapon. Earlier, until the mid of the 1960s, the crusher gauge method, invented in 1860s by Nobel, was the only commonly used and standardized method for gas chamber pressure measurement [1]. A copper or lead cylinder is compressed by a piston fitted to a piston hole into the chamber of the barrel. Under direct effect of the gas pressure generated by the burning of the gunpowder on the base of the piston, the crusher is permanently deformed. The deformed length of the crusher is measured and compared to a conversion table provided by the supplier with each lot of crushers to estimate the peak pressure. The value of maximum pressure is expressed in Copper Units of Pressure (CUP) or Lead Units of Pressure (LUP).
Major drawbacks of this technique are that it has a limited accuracy and gives only the chamber peak pressure value. However, it is still frequently used, since it is simple (there is no further

instrumentation needed), cheap and considered as accurate enough to obtain a rapid estimation of the chamber peak pressure, e.g. in ammunition testing.
Since 1960, piezoelectric transducers have superseded crusher gauges. The discovery of the piezoelectric phenomena was done by Pierre and Jacques Curie in 1880 [2]. Some crystals such as quartz have the capability to produce electric charges when they are squeezed by a mechanical force. This charge is characterized by high impedance. That's why, the use of the piezoelectric technique began effectively in 1935 when a successful application was done by Oerlikon Bührle of Switzerland which was described by Dr. Werner Gohlke in his book published in 1959 on Piezoelectric Technology. The use of piezoelectric pressure measurement in the field of interior ballistics knew a great progress since the development of the charge amplifiers by W.P. Kistler in the 1950s [3].
Today, research and development of piezoelectric pressure measurements are shaped by three main worldwide organizations: NATO (North Atlantic Treaty Organization), C.I.P. (Commission Internationale Permanente pour l'épreuve des armes à feu portatives) and SAAMI (Sporting Arms and Ammunition Manufacturers' Institute). These last two organizations develop proof standards mainly intended to the evaluation of the ammunition for civilian uses. The major differences between the measurement methods defined by these organizations are the transducer fixture points and the measuring techniques. NATO recognizes the direct gas measurement method according to the EPVAT (Electronic Pressure Velocity and Action Time) which it means that the transducer has to be mounted at the case mouth. Both SAAMI and C.I.P. use chamber pressure measurement but C.I.P. uses a transducer which require a hole drilled into the cartridge case and fired by a specific prepared barrel. However, SAAMI employs the so called conformal measurement where the transducer has a similar part as the NATO and C.I.P. sensors except an additional piston is cut in the side of the chamber to conform the cartridge case [4,5,6].
In this paper, a study carried out with the commonly used experimental techniques of high pressure gas chamber measurement is presented. The comparison criteria were the peak chamber pressure, its standard deviation inside specific weapon/ammunition system configurations. Furthermore, a particular attention has been devoted to the problem of calibration. Calibration of crusher gauges and piezoelectric transducers is paramount and an essential task for a correct determination of the pressure inside a weapon. This topic has not been completely addressed yet and still requires further investigations.

2. Experimental techniques for ballistic pressure measurement

2.1. Piezoelectric transducers and measuring chain

Kistler type 6215 and PCB type 117B104 are two different types of piezoelectric transducers often used for pressure measurements inside a barrel of caliber .50 inch [3,7]. As shown in the figure 2 below, the transducer is the first element of a measuring chain which contains other components such as charge amplifier for signal conditioning, data acquisition unit (DAQ) and computer with suitable software for data evaluation and handling.

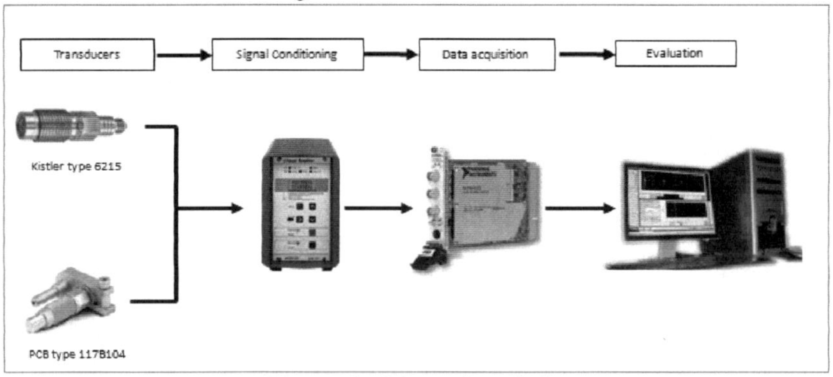

Figure 1. Measuring chain

Calibration of charge amplifier was carried out with a charge calibrator. The data acquisition board consists in a multi-channel device of four high speed digitizers. Each digitizer has two channels in parallel with a resolution of 14 bits and a maximum sampling rate of 100 MHz. Here, the sampling frequency was set to 1 MHz to ensure proper signal measurement. The major conversion occurring in the DAQ board is an analog to digital conversion. Filtering was performed at the frequencies of 10 KHz by a 2nd order Butterworth law pass filter. This filter low-cut frequency allows to obtain the most likely pattern of the temporal variation of the gas pressure without losing too much useful information (good compromise).Signal processing was achieved with the suitable software LabVIEW.

2.2. Firing tests setup

2.2.1. Weapon configuration. Firing tests have been performed with an instrumented .50 inch barrel manufactured according to military standards and mounted on a universal ballistic breech with an interchangeable barrel (BMCI). The major advantage of this caliber is the obtaining of a long ballistic cycle which permits to visualize clearly all the events which occur inside the barrel until the exit of the projectile. The piezoelectric transducers and the copper crusher gauge can be installed in the same mounting position in respect to barrel length but displaced by 90 degrees to each other. The copper crusher gauge had a nominal height of 4.91 mm and a nominal diameter of 3.0 mm.

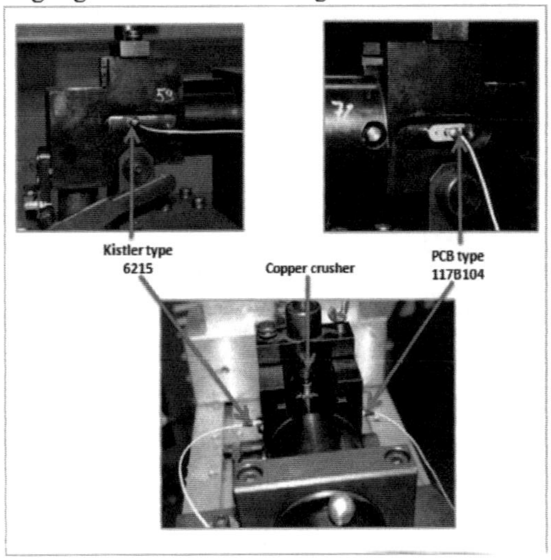

Figure 2. Transducers and crusher installation

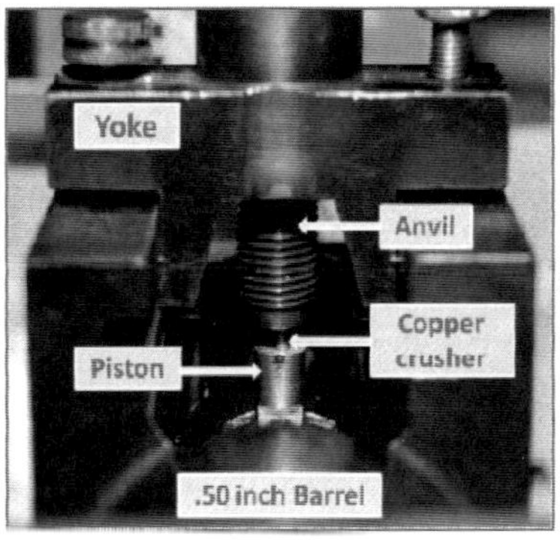

Figure 3. Copper crusher gauge setup.

The triggering was realized using a muzzle flash detector which detects the flash at the exit of the projectile from the muzzle.

*2.2.2. Ammunition configuration. T*he 12.7x99 ammunition filled with the gun powder type .50 inch (WC 860) have been used to achieve the firing tests. A full metal jacket projectile type M33 of 42.5 g is also used for all experiments. The cartridge cases were drilled with a hole of 2.5 mm diameter according to military standards. Obstruction of the hole was done by a heat-resistant adhesive tape.
Four types of ammunition from different manufacturers and lot numbers have been selected to illustrate the influence of cartridge case material: FNB 06, PINDAD 86, IVI 10 and EMZ 87. Indeed, It was expected that the conformal pressure measurement depend strongly onto physical parameters like case thickness and hardness. Ammunition case hardness was evaluated using Vickers Diamond Pyramid Hardness test (table1). The sensitivities of the conformal PCB type 117B104 transducer for each type of ammunition type are also listed in the same table 1.

Table 1. Characteristics of ammunition cases

Ammunition	Hardness Vickers HV 100g	Average Sensitivity (pC/bar[a])
IVI	129.25	1.66
FNB	130.25	1.65
PINDAD	137.75	1.63
EMZ	174.50	1.60

[a] bar is a pressure unit is often used for transducer sensitivity (1bar = 0.1 MPa).

The first series of forty shots of each ammunition type have been carried out to compare only the standard Kistler type 6215 and the conformal PCB type 117B104 transducers. While the second series of ten shots was conducted with different gunpowder masses (12 g, 14 g and 16g) to evaluate the Kistler pressure transducer and the copper crushers. All measurements were taken at 21 ±2°C in a completely enclosed shooting test stand free from weather influences. Results of this series have been used in the second part of this paper to study the dynamic behavior of copper crusher gauges.

2.3. Results and discussion

2.3.1. Kistler type 6215 versus PCB 117B104. Figure 4 below shows typical pressure-time curves for the Kistler and PCB transducers when FNB ammunition was fired. The shape of pressure time curve given by the three sensors are highly similar: rising until a peak value, and then an exponential-like decay to a pressure elevated ambient. As for the Kistler type 6215, the PCB type 117B104 allows to visualize the whole combustion cycle especially the jump of pressure with the departure of the projectile and the two bumps on the top of the curve when the peak pressure is reached which is due to the use of a deterred ball powder. Nevertheless, some differences still exist. This concerns mainly the values of the peak pressure which are slightly different. The conformal sensor exhibits also a slower rate of decay once the peak pressure has been reached. This may be due to the permanent deformation of cartridge case which continues to urge the sensor at the end of ballistic cycle (residual pressure). When the hardness became high, as for the EMZ ammunition case, irregularities appear in the pressure-time curves, especially regarding the descending branch, which can lead to wrong estimation of the projectile velocity. This phenomenon disappears gradually as the hardness decreases.

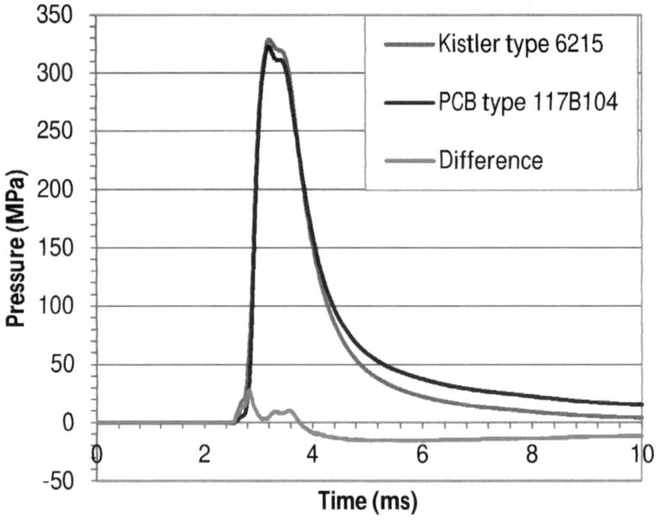

Figure 4. Gas chamber pressure for different transducers

The average peak pressure and its standard deviation for each transducer measurements were determined according to the Guide to the Expression of Uncertainty in Measurement (GUM) [8]. It was assumed that the maximum pressure follows a normal distribution N(μ,σ) which was confirmed by the normality test of Kolmogorov. The estimator of the mean peak pressure value is given by $\bar{x} = \frac{1}{n}\sum_{i=1}^{n} x_i$ and the estimator of the standard deviation is given by $s = \sqrt{\frac{1}{n-1}\sum_{i=1}^{n}(x_i - \bar{x})}$. The Chi-squared distribution $\mathcal{X}^2$ was used to determine the maximum standard deviation σ_{max} which is considered more significant to calculate the measurement uncertainty. We know that:

$$\frac{s^2}{\mathcal{X}_M^2}(n-1) \leq \sigma^2 \leq \frac{s^2}{\mathcal{X}_m^2}(n-1)$$

Then, the maximum standard deviation is given by: $\sigma_{max} = \sqrt{\frac{s^2}{\mathcal{X}_m^2}(n-1)}$

The following table gives the obtained results of the average (Av.) and the maximum standard deviation (Std.) of peak pressures P_{max} (MPa) for each transducer and ammunition type.

Table 2. Mean and maximum standard deviation of peak pressure

	IVI		FNB		PINDAD		EMZ	
	Av. (MPa)	Std. (MPa)	Av. (MPa)	Std. (MPa)	Av. (MPa)	Std. (MPa)	Av. (MPa)	Std. (MPa)
Kistler type 6215	343.45	9.75	319.86	14.37	333.39	13.21	324.08	12.00
PCB type 117B104	333.00	10.78	311.77	14.12	320.69	16.62	310.28	16.73
Average peak pressure difference (%)	3		2.5		3.8		4.25	

The PCB type117B104 always gives lower peak pressure values than the Kistler type 6215. The average peak pressure difference usually exceeds 3 %. To compare statistically the results given by the transducers, hypothesis tests were carried out. The tests were based on the use of the Chi-squared distribution $\mathcal{X}^2$ to compare the standard deviation as shown in table 2. Hence, it was concluded that measurements of the two transducers are not statistically different.

Moreover, the expanded uncertainty U (maximum) is given by $U = k\sigma_{max}$. Considering an interval with a confidence level approximately 95%, the relative uncertainty is computed by:

$$U_{rel} = \frac{2\sigma_{max}}{P_{am}}$$

Where, P_{am} is the average of the peak pressure values. The relative uncertainties are given in the table 3 below.

Table 3. Relative uncertainties

	Relative uncertainty (%)			
	IVI	FNB	PINDAD	EMZ
Kistler type 6215	5.68	8.98	7.92	7.41
PCB type 117B104	6.47	9.06	10.37	10.78

The results of the conformal transducer depend strongly on the characteristics of the ammunition cases. Indeed, for the conformal PCB type 117B104, the relative uncertainty increases with the ammunition case hardness. The influence was sometimes remarkable, especially when the hardness is relatively high. This can explain the values of the growth of the relative uncertainty which exceeds 10 % with PINDAD and EMZ ammunition cases.

2.3.2. Kistler type 6215 versus copper crusher gauge. Table 4 gives the experimental average and the standard deviation of the peak pressure for the piezoelectric Kistler type 6215 and the copper crusher gauges when FNB ammunitions are fired with three different gunpowder masses (12 g, 14 g and 16 g). Copper gauge peak pressures originate from the conversion table provided by the supplier.

Table 4. Results of pressure measurements for different powder masses

	12 g		**14 g**		**16 g**	
	Kistler 6215	Copper Crusher	Kistler 6215	Copper Crusher	Kistler 6215	Copper Crusher
Average peak pressure (MPa)	159.15	127.65	251.47	223.27	319.56	291.92
Std. deviation peak pressure (MPa)	14.15	25.51	19.74	21.03	20.64	25.25
Average peak pressure difference (%)	19.8		11.2		8.65	

The peak pressure given by the crusher is always lower than the measured by the Kistler piezoelectric transducer. The difference between values may reach 20 % as for the 12 g powder mass but it is less than 10 % for the higher 16 g powder mass. These significant differences are mainly related to the used supplier conversion table.

Under the effect of the pressure generated by the combustion gas, it deforms plastically under different high strain rates. However, peak pressure values provided by the conversion table are obtained by a quasi-static calibration (10^{-3}-10^{-1} /s) which do not take into account the dynamic behavior of copper crusher. This statement was also verified experimentally. Therefore, recommendations were made to develop a suitable model dynamic calibration method for the copper crusher gauge. The purpose of this method was to develop a model for the conversion of the deformation in crusher length to maximum pressure. This task was achieved by the investigation of the behavior of the copper crusher at different high strain rates with the Hopkinson Split Pressure Bar (HSPB).

The accuracy of the developed model was verified by a numerical finite element modeling. In this regards, the gas chamber pressure measurements of the transducer Kistler type 6215 was applied to the copper crusher as reference pressure.

3. Dynamic calibration of copper crusher gauges

3.1. Crusher material modeling

Two different strength models were considered for the copper material of the crushers. An experimental study was performed in order to obtain the necessary parameters of each.

A first considered model was the Zerilli-Armstrong model [9], which in its version for face-centered cubic materials (FCC) like copper is given by:

$$\sigma = C_0 + C_2\sqrt{\varepsilon_p}e^{-C_3T + C_4\ln(\dot{\varepsilon})T}$$

With σ the stress, ε_p the plastic strain, $\dot{\varepsilon}$ the strain rate, T the temperature, and finally C_0, C_2, C_3, C_4 the material parameters fitted to the real material behavior.

The second considered strength model was the Johnson-Cook model [10], which originally was developed to model the mechanical behavior of body-centered cubic materials (BCC), as steels for instance. Nevertheless, the Johnson-Cook model is a very versatile model that has often been used outside its original scope, typically with satisfactory results. Its mathematical expression is given by:

$$\sigma = \left[A + B\varepsilon_p^n\right]\left[1 + C\ln\left(\frac{\dot{\varepsilon}}{\dot{\varepsilon}_0}\right)\right]\left[1 - \left(\frac{T - T_r}{T_m - T_r}\right)^m\right]$$

With A, B, C, n and m the material parameters to determine, and $\dot{\varepsilon}_0$ a reference strain rate (chosen as 1 s^{-1}). T_m and T_0 are respectively the melt temperature of the considered material and a reference temperature. Based on the melt temperature of pure copper, T_m was chosen at 1357 K, whereas the reference temperature T_0 was chosen at room temperature (294 K).

The material parameters for the Johnson-Cook model described above were determined using the SHPB. It is one of the few known setups nowadays for studying material behavior at high strain rates [11,12]. A compressive longitudinal wave is produced by a projectile impacting the input bar. This wave is transported through the bar and once it reaches the specimen interface, a part of the waves is reflected, and a part is transmitted. These three waves are recorded with strain gauges glued on the input and outpt bar.

The ABAL laboratory setup consists of 2 m long input and output bars with a diameter of 30 mm and made out of high-strength steel. A double set of strain gages was used on both the input bar and the output bar to eliminate any influence of parasitic bending of the bars. The signals coming from the strain gages were sampled at 10 MHz and recorded using a digital acquisition system. The final stress-strain curves of each test were obtained using the typical relationships between the strains in the input and output bar and the deformation history of the sample, which are integrated in a custom data analysis software implemented in a LabVIEW environment.

The cylindrical samples used for the experiments were made out of the same material as the original copper crushers used in the ballistic experiments, but had slightly different dimensions in order to have a good signal-to-noise ratio for the SHPB testing. The samples had respectively a height and diameter of 10 mm and 7.2 mm.

In total 11 dynamic compression tests were performed in a strain rate range of approx. 650 s^{-1} up to 1400 s^{-1}. The final strains varied correspondingly from approx. 0.1 up to 0.25, comparable to the strains achieved in the copper crushers during a typical pressure measurement. All samples had an initial temperature corresponding to room temperature.

A typical stress-strain curve (true strain rate 1181 s^{-1}) is shown in Figure 5. The pseudo-elastical part of the curve was removed using a 0.2 % offset criterion as used for quasi-static tensile and compression testing.

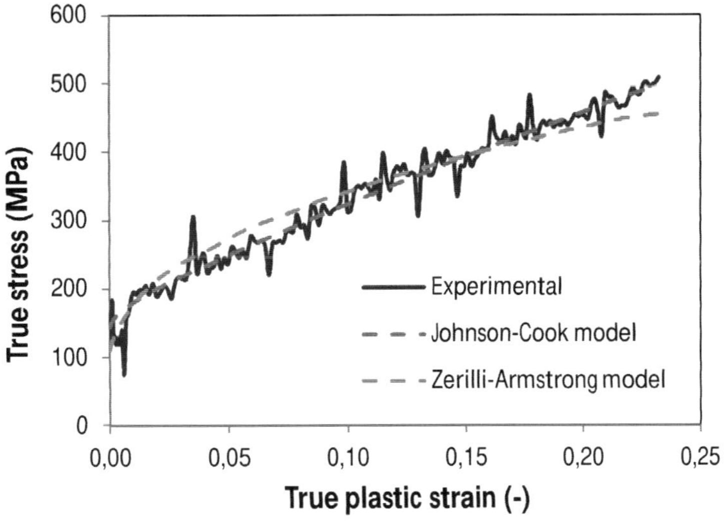

Figure 5. Comparison between an experimental stress-strain curve and the modeled results

The resulting parameters for both models are given in Table 5 and compared with literature values [13] for pure copper (OFHC). It can be observed that the crusher material shows a much more pronounced strain hardening than OFHC copper, which could most likely be explained by the crusher material being a copper alloy (instead of the initially assumed pure copper material). As in the case of the Johnson-Cook model the temperature sensitivity exponent, m, did not converge to any meaningful physical value (which can be attributed to the fact that all testing was done at room temperature), its value was set arbitrarily to 1. A representative comparison between an experimental stress-strain curve and the modeled curves is given in Figure 5 as well. Even if it was developed for BCC materials, the Johnson-Cook model fits the experimental curve better than the Zerilli-Armstrong model, especially for the beginning and the end of the curves. The Johnson-Cook strength model was hence preferred to the Zerilli-Armstrong model for the subsequent finite element modeling.

Table 5. Johnson-Cook and Zerilli-Armstrong model parameters

Johnson-Cook model					
	A (MPa)	**B (MPa)**	**C (-)**	**n (-)**	**m (-)**
Own result	119	935	0.049	0.89	1
Literature	90	292	0.025	0.31	1.09
Zerilli-Armstrong model					
	C_0 (MPa)	**C_2 (MPa)**	**C_3 (-)**	**C_4 (-)**	
Own result	107	853	0.0054	0.000681	
Literature	65	890	0.0028	0.000115	

3.2. Numerical Modeling

The finite element modeling was carried out with the ANSYS Workbench 14 [14]. If the strength model is accurate enough, the permanent deformation of the crusher Δl_{sim} could be compared to the reference permanent deformation Δl_{ref} corresponding to the initially applied peak pressure P_{peak} (using the conversion table), and to the experimentally obtained average residual deformation Δl_{exp}. If the simulated permanent deformation corresponded to the reference permanent deformation, the developed model could be assumed to be correct.

Both piston and crusher were modeled using a Lagrangian mesh and implementing the previously fitted Johnson-Cook model as the strength model. The piston's inertia could be hence evaluated. The measured pressure-time curve was taken as the input "true" pressure. The equation of state was obtained from the material models for OFHC copper, as available in the ANSYS material library. No failure model was considered.

The final deformations of the copper crusher as a function of the powder mass are given in table 6.

Tableau 6. Comparison between reference, simulated and experimental permanent deformation.

Powder mass (g)	P_{peak} (MPa)	$P_{crusher}$ (MPa)	Δl_{ref} (mm)	Δl_{sim} (mm)	Δl_{exp} (mm)
12	159.15	127.65	0.69	0.64	0.49
14	250.15	218.4	1.35	1.29	1.1
16	319.6	292	1.80	1.73	1.6

With increasing powder mass, the residual deformation increases as well, as expected. It is also clear that the simulated residual deformations correspond better to the reference deformations than to the average experimental deformations. However, they are still less than the reference deformations. The reasons of these differences are certainly two: the piston's inertia and the strength model which originate from the developed calibration method. Indeed, the Johnson-Cook model was determined at different strain rates. The question was then raised about the choice of the required strain rate of the deformation of copper crusher. One of the possible solutions consists of the estimation of this parameter from the pressure-time curve. Once this has been done, the appropriate stress-strain curve has to be used to convert the deformation in the length of the crusher to a maximum pressure.

Journal of Physics: Conference Series **459** (2013) 012048 doi:10.1088/1742-6596/459/1/012048

4. Calibration of piezoelectric pressure transducers

4.1. Calibration of ballistic pressure transducers

Calibration is a procedure to find the relationship between a known input e(t) and a measured output s(t) in well-defined conditions [15], which means the determination of the transfer function.

In static calibration, the ratio of output and input is constant and gives the sensitivity of the sensor. In the dynamic calibration, the ratio leads to a complex function H(f) which is defined as the ratio of the Fourier transforms of the output S(f) and the input E(f) [16,17,18]:

$$H(f) = \frac{S(f)}{E(f)} = \frac{\int_0^\infty s(t)e^{-j2\pi ft}dt}{\int_0^\infty e(t)e^{-j2\pi ft}dt}$$

In the most of cases, the measured signals are sampled and non-periodic. Discrete Fourier Transform (DFT) is then used to compute the transfer function, also known as the Frequency Response Function (FRF). The FRF allows the full characterization of pressure transducers both in the static and the dynamic regimes. Dynamic calibration involves the determination of several properties of the pressure sensor such as the FRF (amplitude and phase), resonant frequency, damping ratio, rise time and overshoot [16]. In 1972, the ASME published a guide for the dynamic calibration of pressure transducers [19]. A revised version of this guide has been available since 2002 published by the ISA where a description of the methods employed for dynamically calibrating pressure transducers is given [20]. A diversity of periodic and aperiodic generators is described. These systems allow the characterization respectively of the harmonic and transient response of pressure sensors. Periodic pressure signals have low amplitude and limited bandwidth while aperiodic generators produce pressure signals with high amplitude and large bandwidth (several hundred kHz). Aperiodic generators include shock tube, quick-opening valve devices, explosive devices (close bombe), and the drop weight devices (hammer and ball). They create step-like pressures or single pressure pulses similar to a half-sine wave [16,18,20]. Each pressure generator has its range of operation. The choice of the required device is then determined by the conditions of use of the sensor to be calibrated.

Until today, a pressure primary standard for the dynamic calibration of high pressure sensors does not exist yet. Few national institutes of metrology such as LNE, NPL and NIM have developed a series of shock tubes to achieve this challenge but no one of these systems was able to satisfy entirely the needs [17,18].Further work is still required because the generated pressure step signals have some limitations especially in amplitude (several tens MPa) and the low limit of frequency which remains significantly high. Also a reference pressure sensor still needed to measure the output in most of the cases which is generally calibrated statically.

In the field of interior ballistics, the need of reliable dynamic calibration techniques for piezoelectric high pressure sensors has been increasing in the last years since accurate dynamic measurements of gas chamber pressure became paramount in many applications. The best way still a static calibration followed by a dynamic verification. First of all, only the working standard is calibrated statically, by the use of a pressure balance as primary standard, which allows a step-wise calibration. Then, a quasi-static calibration of the piezoelectric transducer is fulfilled by a high continuous pressure generator. Finally, its dynamic behavior is checked against the "dynamic" working standard. These procedures are explained in the figure 6 below [21].

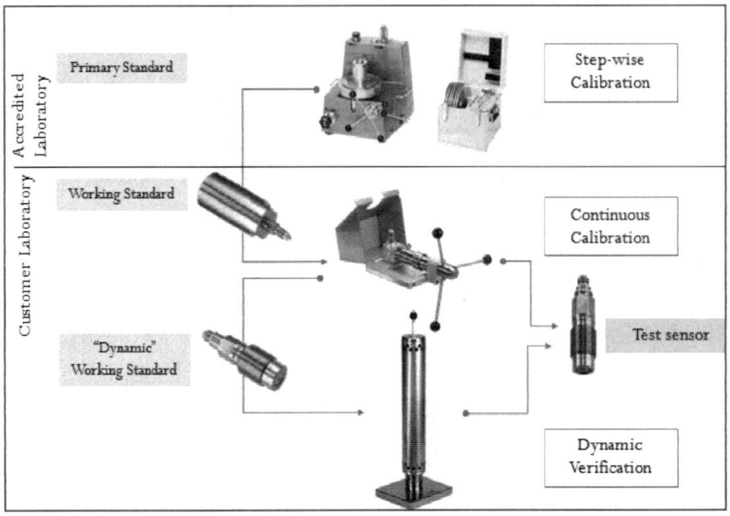

Figure 6. Calibration of piezoelectric pressure transducer

4.2. Development of dynamic calibration generator

The aperiodic pressure pulses generator produces a single half-cycle sine wave with high amplitude (several hundred MPa). However, the frequency bandwidth of this system is not wide enough to characterize the high pressure transducers. This limitation is essentially due to the big width of the generated pressure pulse. In theory, an ideal system will generate an ideal impulse signal. The transfer function of the sensor can be directly derived by computing the Fourier transform of the measuring chain response. A system like this does not exist. However, it is possible to produce a pseudo-Dirac function which allows to reach high frequencies.

A solution based on the use of a piston in contact with oil-filled pressure chamber has been developed and tested. The high pressure was obtained by impacting the piston with projectiles launched by an air gun. The aim of the setup was to create a system which may generate narrow pressure pulses with high accuracy. The change of projectile length leads to create pulses with different width. Figure 7 gives an overview of the obtained results. A maximum pressure of 500 MPa can easily be reached but the narrowest generated pressure pulse has a width of 0.4 ms.

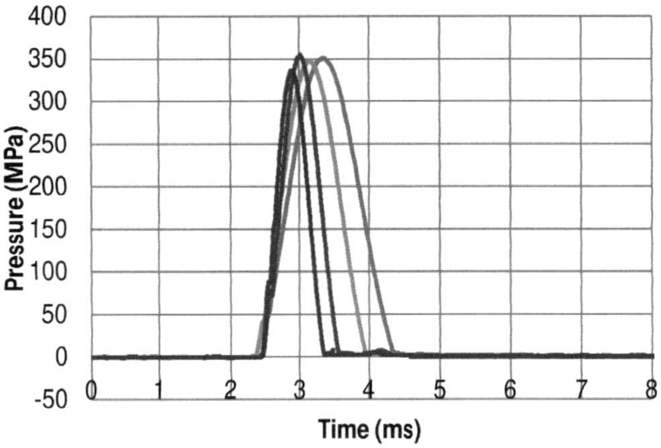

Figure 7 : Different pressure pulses

The amplitude and the phase of the FRF are estimated by the following expression:

$$H(f) = \frac{G_{SS}(f)}{G_{SE}(f)}$$

Journal of Physics: Conference Series **459** (2013) 012048 doi:10.1088/1742-6596/459/1/012048

Where $G_{SS}(f)$ and $G_{SE}(f)$ are respectively the Power spectrum of the measuring chain response of the reference transducer and the Cross power spectrum of the measuring chain response of the tested and the reference transducers.

Results show that the system has a limited frequency bandwidth. The low limit frequency and the high limit frequency are respectively 100 Hz and 4 kHz which are obtained for a maximum deviation of 2.5 % of the amplitude ratio. As the FRF was estimated using a reference transducer calibrated statically to measure the input pressure pulse, this method is still also a comparison calibration method.

The way to improve the developed system was identified by the use of Hokinson pressure bar device as shown in the figure 8 below.

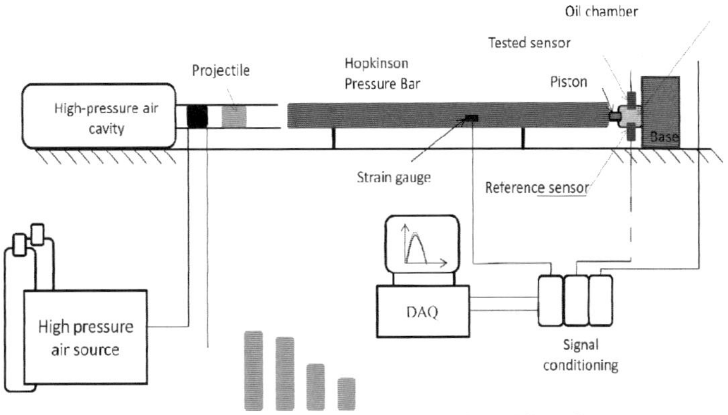

Figure 8. Setup of the dynamic calibration

This dynamic pressure generator produces a pseudo-Dirac pressure wave. It allows to reach a quit high frequency and has wide frequency bandwidth as it was approved in the recent study made by Zhang et al. [22]. Finite element modelling was also carried out to develop a numerical model for the described setup.

5. Conclusion

Gas chamber pressure measurement is one of the most demanded measurements in ballistic testing. Experimental method usually used to measure this physical quantity were reviewed and compared to elucidate on the typical measurement differences one can find between the results of not only crusher and piezoelectric measurements, but also between the different piezoelectric measurement techniques. Furthermore, as the accuracy of pressure measurement depends strongly on the calibration method, a big amount of this work was dedicated to the problems of dynamic calibration to develop in-situ reliable techniques for both copper crusher and piezoelectric transducers. Further works still needed to improve the accuracy of the behavior model for copper crusher gauges and the generated pressure pulses frequency bandwidth for dynamic calibration of piezoelectric high pressure transducers.

References

[1] NATO *Pressure Measurement by Crusher Gauges NATO Approved Tests for Crusher Gauges* AEP-23 Ed. 2, NATO Standardization Agency, 2005.
[2] Pierre Curie *Œuvres de Pierre Curie* Société française de physique, 1984.
[3] Kistler Instrumente AG *Technical documents.*
[4] NATO *Combination Electronic Pressure, Velocity And* Action Time (EPVAT) Test Procedure PFP (NAAG-LG/3-SG/1) D1 Ch. 12, NATO.
[5] C.I.P. *Edition Synthétique des décisions C.I.P. en vigueur* Bureau Permanent C.I.P. Brussels, 2011
[6] SAAMI Technical Committee *SAAMI/ANSI standards* available at http://www.saami.org
[7] J. Lally and R. Rhen *Confromal Sensor Measures Ammunition through shell Case* PCB Piezotronics, 1999
[8] Norme AFNOR NF ENV 13005 *Guide pour l'expression de l'incertitude de mesure* AFNOR, 1999

[9] Zerilli F. and Armstrong R. *Description of tantalum deformation behavior by dislocation mechanics based constitutive relations* J. Appl. Phys., 68, pp. 1580-1591, 1990
[10] Johnson, G. R., and Cook, W. H., 1983 A Constitutive Model and Data for Metals Subjected to Large Strains, High Strain Rates and High Temperatures Proceedings of the 7th International Symposium on Ballistics, The Hague, The Netherlands, pp. 541-547
[11] Kolsky H. *An Investigation of the Mechanical Properties of Materials at Very High Loading Rates* Proc. *Phys. Soc. London*, B62, pp. 676, 1949
[12] Hopkinson B. *A Method of Measuring the Pressure Produced in the Detonation of High Explosives or by the Impact of Bullets* Roy. Soc. Phil. Trans., A213, pp. 437-456, 1914
[13] Banerjee B. *An evaluation of plastic flow stress models for the simulation of high temperature and high-strain-rate deformation of metals* Dept. of Mechanical Engineering, University of Utah, USA, 2005
[14] ANSYS Workbench, http://www.ansys.com
[15] ISO *Vocabulaire international des termes fondamentaux et généraux de métrologie* 2e édition, 1993
[16] J.L. Schweppe, L.C. Eichenberger, D.F. Muster, E.L. Michaels and G.F. Paskusz *Methods for the Dynamic Calibration of Pressure Transducers* NBS Monograph 67, December 12, 1963. .
[17] V.E. Bean *Dynamic pressure metrology* Metrologia, vol. 30, pp. 737-741, 1993.
[18] J.P. Damion, "Means of dynamic calibration for pressure transducers", Metrologia, vol. 30, pp. 743-746, 1993.
[19] ASME *A Guide for the Dynamic Calibration of Pressure Transducers* ANSI B88.1-1972 (R1995), 1995.
[20] ISA *A guide for the dynamic calibration of pressure transducers* ISA-37.1601-2002, 2002
[21] C. Cater *Calibration of Piezoelectric Sensors for Combustion Pressure Analysis* KIAG, 2006
[22] Y. Zhang, J. Zu, H.Y Zang *The Dynamic Calibration Method of High-Pressure Transducer under based on quasi-static function excitation source*, Measurement 45 (2012), 1981-1988.

Protein Aggregation Measurement through Electrical Impedance Spectroscopy

A. Affanni[1], A. Corazza[2], G. Esposito[2], F. Fogolari[2], M. Polano[2]

[1]University of Udine, Dept. of Electrical, Management and Mechanical Engineering, via delle Scienze 206, 33100 Udine (Italy)

[2]University of Udine, Dept. of Medical and Biological Science, piazzale Kolbe 3, 33100 Udine (Italy)

antonio.affanni@uniud.it

Abstract. The paper presents a novel methodology to measure the fibril formation in protein solutions. We designed a bench consisting of a sensor having interdigitated electrodes, a PDMS hermetic reservoir and an impedance meter automatically driven by calculator. The impedance data are interpolated with a lumped elements model and their change over time can provide information on the aggregation process. Encouraging results have been obtained by testing the methodology on K-casein, a protein of milk, with and without the addition of a drug inhibiting the aggregation. The amount of sample needed to perform this measurement is by far lower than the amount needed by fluorescence analysis.

1. Introduction

Protein aggregation leading to amyloid fibrillation is responsible for several serious human diseases. Neurodegenerative diseases such as Alzheimer's Disease (AD), Parkinson's Disease (PD), Huntington's Disease (HD), Amyotrophic Lateral Sclerosis (ALS) are some of the important diseases associated to the presence of protein amyloid deposits. Amyloid aggregates consist of fibers containing misfolded proteins with substantial extent of beta-sheet conformation, arranged in the so-called cross-beta structure. The complexity and dynamics of protein folding present unique challenges for elucidating the molecular mechanisms involved in protein aggregation and designing effective amyloid inhibitors. Continuous development of creative approaches to identify an ultimate solution for controlling protein amyloid aggregation in biopharmaceuticals and clinical pathology is clearly required.

K-casein is arguably the best characterized milk protein and constitutes over 70–80% of total bovine milk protein. Recently it has been demonstrated that the reduced and carboxymethylate K-casein (RCM-k-CN) can be used as first trial for screen compounds with putative inhibitor or aggregation blocker activity [1]. RCM-k-CN is an amyloidogenic protein that readily undergoes nucleation-dependent aggregation and amyloid fibril formation via a similar pathway to disease-specific amyloidogenic peptides like Amiloid β (Aβ) , which is associated with Alzheimer's disease [2].

Currently the measurement of RCM-k-CN fibrillar aggregation is performed by means of optical fluorescence. A thioflavin-T (ThT) fluorescence binding assay is used. When ThT binds to amyloid aggregates, it exhibit fluorescence at 485 nm following excitation at 440 nm. By monitoring the time-course of this ThT fluorescence, the fibril formation kinetics can be assessed [3].

In the present paper we investigate a new method, based on the Electrical Impedance Spectroscopy, for monitoring the aggregation of proteins into fibrils. We developed a bench composed by a sensor with interdigitated gold electrodes, a Poly Dimethyl Siloxane (PDMS) hermetic and removable reservoir, a LCR meter and a control panel to perform automatic measurements. Since the protein and the ThT are dissolved in a buffer whose electrical properties are mainly conductive, when the fibrils of the protein drop on the electrode surface there is a local change of charge distribution, that modifies the impedance spectroscopy response. The preliminary results suggest that electric impedance can be a new valuable method for analyzing the fibrillar aggregation of proteins for screening fibrillar aggregation inhibitors without the presence of ThT molecule as marker.

2. Bench Description and Measurements Setup

The developed system is represented in Figure 1. A fully programmable control panel (described in detail in [4]), has been developed in order to automatically acquire impedance data from the Agilent E4980A LCR meter via USB interface.

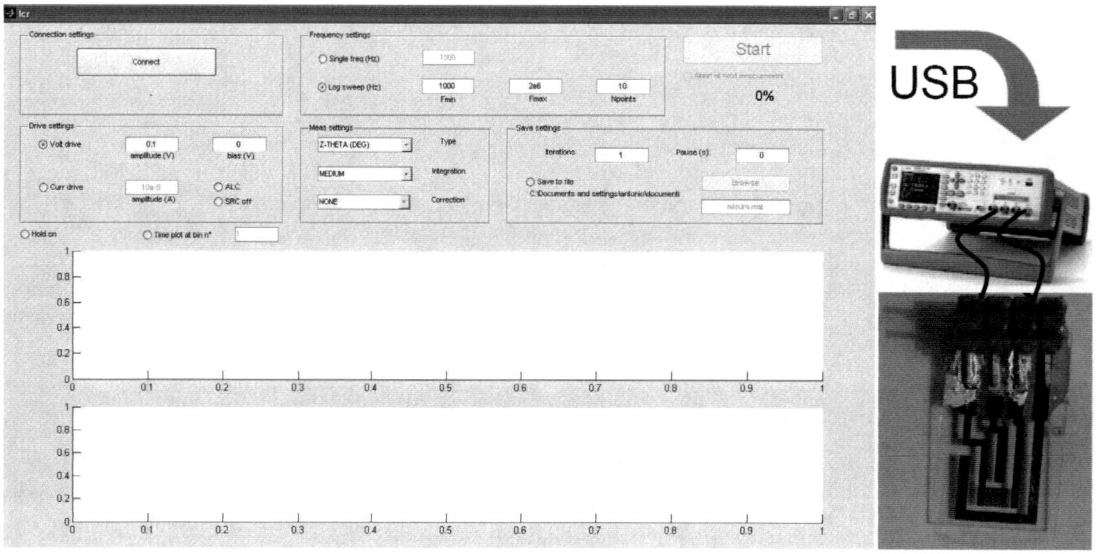

Figure 1: Developed bench is composed by a control panel, a LCR meter and an intgerdigitated electrodes sensor.

To the LCR meter it is connected a sensor having gold interdigitated electrodes sputtered on quartz; on the sensing electrodes we posed a removable PDMS reservoir which is topped by a removable polycarbonate slide. The three layers are held together with a clamp.

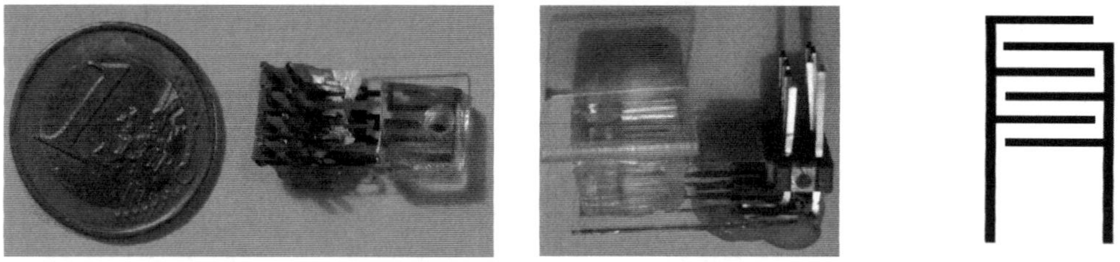

(a) (b) (c)

Figure 2: (a) Top view and (b) side view of the sensor with reservoir; (c) electrodes arrangement.

The sensor is composed by 65 couples of interdigitated electrodes; the dimensions of each finger are 10 μm × 2 mm and the space between fingers is 5 μm. The PDMS cylindric reservoir is 3 mm in diameter and 8 mm in height, yielding to a sample volume of 60 μl, the assembly of the sensor is shown in Figure 2 (a) and (b).

2.1. *Measurements Setup*

The solution of RCM-k-CN freshly dissolved in 50 mM sodium phospate buffer pH 7.2 10 μM ThT at final concentration of 0.5 mg/ml (27 μM) is poured in the reservoir and hermetically closed with the top of the sensor; the sensor is then posed at controlled temperature in a thermostated chamber at 37 °C.

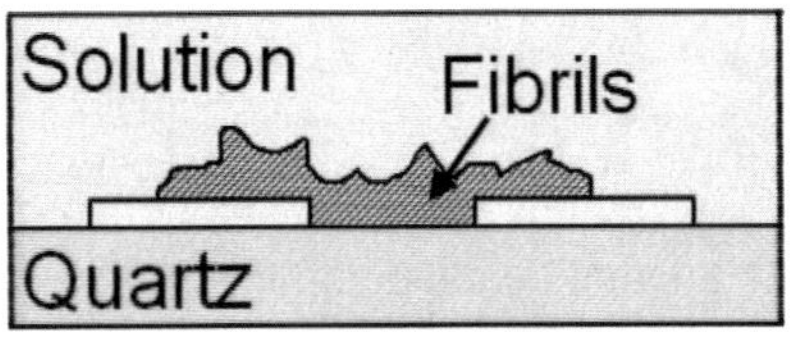

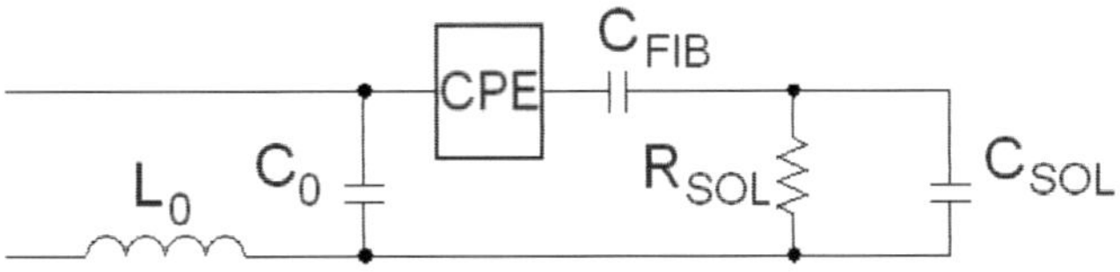

Figure 3: (a) Fibrils precipitation on gold electrodes; (b) equivalent circuit

Figure 3 (a) shows the precipitation of protein fibrils on the gold electrodes and Figure 3 (b) shows the equivalent circuit considered for assessment of fibril deposition. The elements C_0 and L_0 have been compensated performing the "open" and "short" compensations before measuring the sample properties. The bench has been set to perform impedance measurements with a drive voltage of 100 mV (to avoid redox reactions) with 24 points frequencies in the range [20, $2 \cdot 10^6$] Hz logarithmically spaced. Sweeps are performed every 5 minutes for a total duration of 18 hours. The Constant Phase Element CPE in Figure 3 (b) represents the effect of double layer due to interaction between the solution and the electrode [5], its impedance is

$$Z_{CPE} = \frac{1}{(j\omega)^{\alpha} C_{DL}} \quad (1)$$

being C_{DL} the double layer capacitance and $0 \le \alpha \le 1$ its phase shift. The quantities R_{SOL} and C_{SOL} represent the impedance behaviour of the solution while the measurand is represented by the change over time of the capacitance C_{FIB} due to the fibril precipitation on the electrodes.

3. Experimental results

Figures 4 (a) and (b) show, respectively, the absolute value and the phase of measured impedance (after the open-short compensations) during a 18 hours experiment assaying fibril precipitation on the

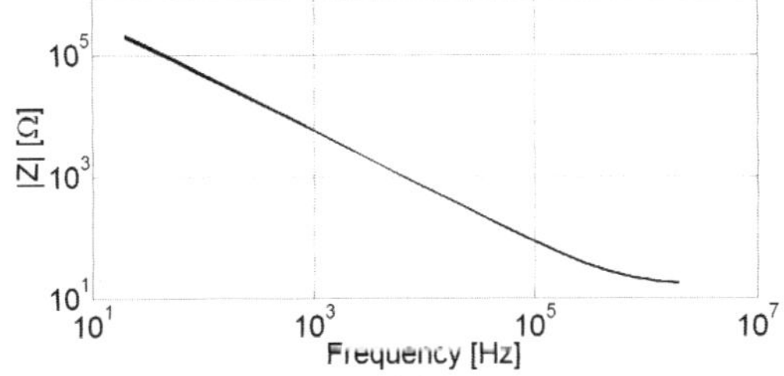

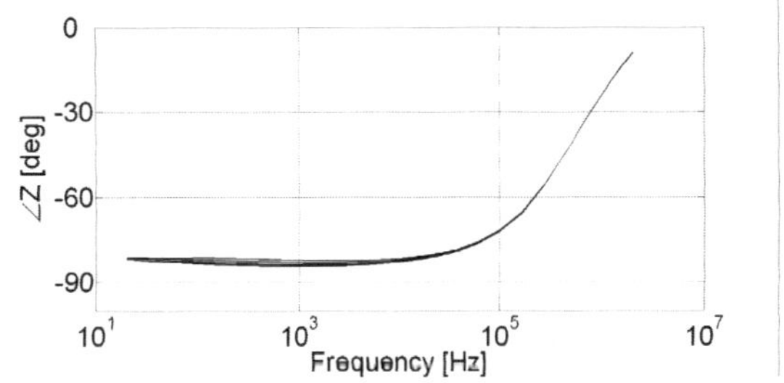

Figure 4: (a) Impedance magnitude and (b) phase behaviour.

electrodes. The accuracy of acquired impedance data after "open" and "short" compensations is 0.1% [6].

As a first comment it is clear that the double layer (CPE) plays a significant role and it strongly masks the impedance changes over time; it is also possible to see that the quantity C_{SOL} is too small and does not intervene in this frequency range.

In order to better distinguish the parameters of the model which are varying over time, the spectroscopy impedance data have been interpolated with an iterative fitting software [7]. The interpolated data provided $R_{SOL} \approx 10\ \Omega$, $C_{DL} \approx 70$ nF, $\alpha \approx 0.9$, the accuracy of interpolated data is 5%. The quantity C_{FIB} cannot be directly estimated by the fitting software since in the model it is in series to a CPE having a high value of α. For this reason we performed the impedance measurement of the buffer solution in the same conditions, with data acquired every 5 minutes over 18 hours.

The C_{FIB} value can be derived considering the double layer changes over time of the protein with respect to the double layer changes of the buffer:

$$C_{FIB,PROTEIN} = \left(\left. \frac{(j\omega)^{1-\alpha}}{C_{DL}} \right|_{PROTEIN} - \left. \frac{(j\omega)^{1-\alpha}}{C_{DL}} \right|_{BUFFER} \right)^{-1} \tag{2}$$

In (2) we considered the double layer impedance for protein as the sum of the double layer contribution plus the C_{FIB} contribution and we subtracted the double layer contribution of the buffer; to verify our data we repeated the experiment three times in the same conditions.

In order to validate our approach we added to the buffer blank and to the protein solution the drug doxycycline at concentration of 81 μM. This doxycycline concentration is known to abolish Aβ fibril formation. As in (2) we evaluated the quantity

$$C_{FIB,PROTEIN+DRUG} = \left(\left. \frac{(j\omega)^{1-\alpha}}{C_{DL}} \right|_{PROTEIN+DRUG} - \left. \frac{(j\omega)^{1-\alpha}}{C_{DL}} \right|_{BUFFER+DRUG} \right)^{-1} \tag{3}$$

The experiments with and without drug were repeated three times in the same conditions and results proved encouraging.

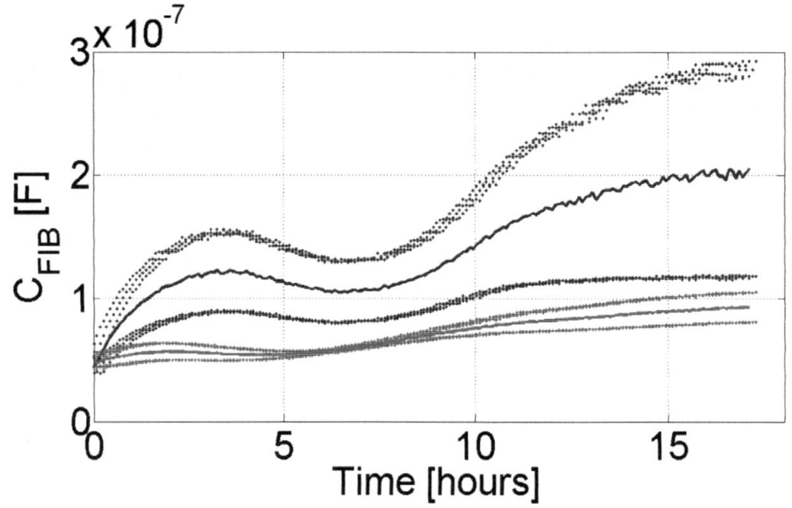

Figure 5: C_{FIB} estimation with (red line) doxycycline inhibitor addition and without (blue line) doxycycline inhibitor addition. The points above and below the curves represent the experimental standard deviation of the mean that was extracted from three experiments.

Figure 5 shows the estimation of C_{FIB} when the protein solution fibrillates (blue line) and when doxycycline is added (red line). The points surrounding the curves represent the accuracy of the measurements estimated as standard deviation of the mean of the three experiments performed.

4. Discussion and Conclusions

Figure 5 shows that the behaviours of the protein with and without the tested drug are quite different, especially in the first eight hours of the experiments. This is in good agreement with the doxycyline principle of operation; this drug in fact mainly inhibits the initial formation of oligomers which can lead to long organized protein fibrils.

Doxycicline does not necessarily prevent the amorphous precipitation of the protein that could be related to the residual C_{FIB} increase observed over long times. It is expected, however, that the impedance characteristics of the solution could be sensitive to the type of precipitating aggreagates, thereby enabling the assessment of the actual process undergoing in solution. This issue, along with the very relevance of the precipitation, compared to the formation of the very anisotropic soluble adducts that occurs in solution upon at the early stages of fibrillogenesis, requires further investigation. In addition the effects on amyloidogenic protein solutions of other perturbations such as agitation or ultrasound application could be interestingly monitored by impedance change measurements.

The results that have been reported in this first preliminary work show that the Electrical Impedance Spectroscopy could be a good method for measuring fibril formation for evaluating new drug efficiency. With this method the experiments can be carried out on the protein solution without the addition of any contrast medium (like ThT) thus providing information on the real aggregation kinetics; the addition of ThT in fact can modify the fibril kinetics since ThT binds to the fibrils and may therefore compete with or even interact the tested drug.. Moreover, with the presented device, measurements can be performed with a very limited amount of solution and therefore the method efficiency is coupled to sensitivity advantage.

References

[1] Carver, J. A., Duggan, P. J., Ecroyd, H., Liu, Y., Meyer, A. G. & Tranberg, C. E. "Carboxymethylated-k-casein: A convenient tool for the identification of polyphenolic inhibitors of amyloid fibril formation", Bioorganic & Medicinal Chemistry, 18 (1), 2010, 222-228.

[2] Ecroyd, H., Koudelka, T., Thorn, D. C., Williams, D. M., Devlin, G., Hoffmann, P. & Carver, J. A. "Dissociation from the oligomeric state is the ratelimiting step in amyloid fibril formation by k-casein.", J. Biol. Chem. 283, 2008, 9012–9022.

[3] H. LeVine III, "Thioflavine T interaction with synthetic Alzheimer's disease β-amyloid peptides: Detection of amyloid aggregation in solution", Protein Science 2 (3), 1993, 404-410

[4] A. Affanni, R. Specogna, F. Trevisan "Measurement Bench for Impedance Tomography during Hemostasis Process in whole Blood", IEEE International Symposium on on Medical Measurements and Applications (MeMeA 2011) Proceedings, Bari (Italy), 30-31 May 2011, pp. 80-83

[5] E. Barsoukov, J. R. Macdonald, "Impedance Spectroscopy Theory, Experiment, and Applications", 2nd ed. J. Wiley and Sons, 2005, 30 – 90

[6] Agilent E4980A Precision LCR Meter Users Guide, 7th Edition, Agilent Technology, Santa Clara, CA, Manufacturing No. E4980-90070, Jun. 2008.

[7] http://www.jrossmacdonald.com/levminfo.html

[8] K.S. Cole, R.H. Cole, "Dispersion and absorption in dielectrics: 1. Alternating current characteristics", J. Chem. Phys. vol. 9 pp. 341-351,1941.

[9] J. R. Macdonald, "Impedance Spectroscopy", Annals of Bionedical Engineering, vol. 20, pp. 289-305, 1992

Study on the AFM Force Spectroscopy method for elastic modulus measurement of living cells

A Demichelis[1], S Pavarelli[1], L Mortati[1], G Sassi[1 2] and M Sassi[1]

[1] Istituto Nazionale di Ricerca Metrologica I.N.Ri.M., Torino, Italy
[2] Dipartimento di Scienza dei Materiali e Ingegneria Chimica - Politecnico di Torino, Italy

E.mail: a.demichelis@inrim.it

Abstract. The cell elasticity gives information about its pathological state and metastatic potential. The aim of this paper is to study the AFM Force Spectroscopy technique with the future goal of realizing a reference method for accurate elastic modulus measurement in the elasticity range of living cells. This biological range has not been yet explored with a metrological approach. Practical hints are given for the realization of a Sylgard elasticity scale. Systematic effects given by the sample curing thickness and nanoindenter geometry have been found with regards of the measured elastic modulus. AFM measurement reproducibility better than 20% is obtained in the entire investigated elastic modulus scale of 10^1 - 10^4 kPa.

1. Introduction

The measurement of mechanical properties of living cells and extracellular components is becoming increasingly important in different field like cancer and developmental biology. The cell elasticity gives information about its pathological state and metastatic potential [1], the extra-cellular matrix elasticity influences cell lineage specification [2]. Accurate cell elasticity measurements can be performed using the AFM Force Spectroscopy method. It allows to quantify the cell Young's modulus E through measures of the cantilever deflection when approached to the cell surface [3].

In a typical force-indentation curve obtained by the AFM measurement the force values on the y-axis are calculated from the measurement of the cantilever vertical deflection V [Volts], the system sensitivity S [nm/V] and the cantilever constant k [N/m]. The indentation values δ [nm] on the x-axes are calculate from the measurement of the cantilever piezo height z [m]. S depends on the positioning of the cantilever over the cantilever holder. The Young's modulus E of the sample is calculated fitting the extend part of the force-indentation AFM curve with the classical Hertzian model for a spheric indenter [3].

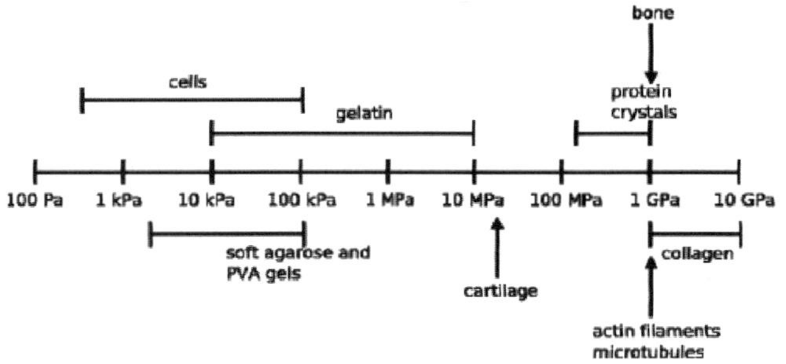

Figure 1. Elastic modulus of biological materials [4].

The aim of the present work is to study the AFM nanomechanical analysis in an elasticity range of 10^1 - 10^4 kPa. The future goal is to make this method a reference method, in the cell elasticity *E* range of 10^{-1} – 10^2 kPa (figure 1) not yet explored with a metrological approach. To make the AFM method a "reference method" an accuracy budget must be evaluated, with the aid of a stable and homogeneous *E* sample scale. To reach this prefixed objective the following steps have been done in this work: realization of a stable and homogeneous *E* scale in the interesting biological range, realization of the suitable nanoindenter, optimization of the AFM measurement parameters, AFM measurement of the *E* scale, analysis of the AFM measurement uncertainty, comparison with other literature methods.

2. Materials and methods

2.1. The preparation of the working E scale

Sylgard 184 has been chosen as materials for the realization of the working *E* scale (10^1 - 10^4 kPa), this because it presents a tunable *E* varying the base/curing ration, it allows to realize very low *E* materials (down to few kPa), it presents at a microscopic level a very homogeneous surface and let to construct mechanically stable samples. Sylgard 184 is a viscoelastic polymer of cross-linked polydimethylsiloxane PDMS chains that can be prepared curing short PDMS chains (the Sylgard 184 base agent) with hydrogenated-PDMS chains (the Sylgard 184 curing agent). The chemical curing reaction (hydrogen addition to the vinyl ends of PDMS chains, catalyzed by Pt and heat) causes the internal re-arranging of the random-distributed PDMS chains that expose to the surface idrofobic –CH_3 groups.
Thick Sylgard 184 (Dow Corning) cylinders, 1.5 cm diameter per 1.5 cm height, have been realized weighing in a plastic can the Sylgard components with nominal base/curing ratio of 2.5, 5, 7.5, 15, 25 and 50 by weight and stirring them for 5 minutes. Samples have been realized let curing the Sylgard blend in a 24-well plate.
Thin cylinders, 2.5 cm diameter per 0.1 cm height with nominal base/curing ratio of 5, 15, 30, 60 have been prepared, curing them in rings lean on a glass slides. 20 days curing time at 20°C has been waited leaving the Sylgard cylinders in the well plate submerged by deionized water and leaving the cylinder in the ring exposed to the air.
For the AFM measurements the thick cylinders are indented on both surfaces: for the upper surface the nanoindenter has been immerged in the corresponding well, for the botton surface (flat and not-water and air contaminated) the cylinders are extracted from the plastic plate, turned upside down and fixed with Vaseline on a petri dish filled with deionized water. The thin cylinders (attached at the botton to the glass slide) instead are analyzed immerging the entire system in the petri dish filled with deionized water.

2.2. The preparation of the nanoindenter

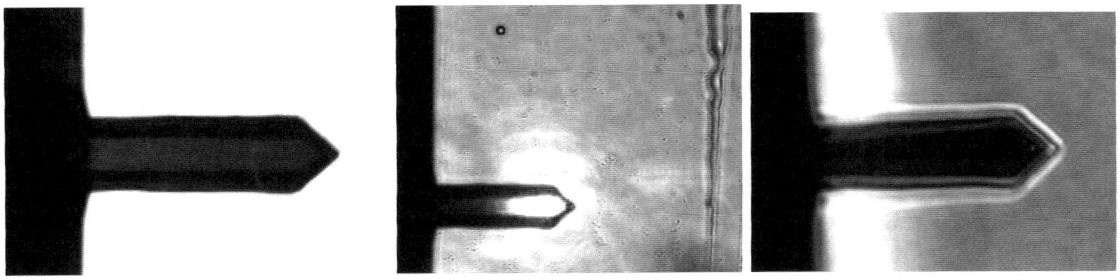

Figure 3. Nanoindenter construction procedure. From left to right: the clean and new tipless cantilever; cantilever with the laser spot on it, the glue at the right and the SiO_2 sphere in the top; the realized indenter with the attached sphere.

As nanoindenter geometry a SiO_2 sphere with a known diameter has been attached to the top of a tipless rigid AFM cantilever. A very small quantity of glue (Dymax OP-29), a drop with a diameter similar to the SiO_2 sphere diameter (GmbH microparticles o.d. 7.75 ± 0.29 µm), has been load on the top of a tipless cantilever (Nanosensors TL-NCH) acting on the setpoint parameters of the approach and force curve measurement of the AFM instrument (JPK Nanowizard II). In a second moment the sphere has been attached performing a force curve on its top. In figure 3 is reported the optical images step of the construction procedure of the nanoindenter.
A 40 N/m value has been used in this work for the elasticity constant of the nanoindenter (nominal manufacturer value). A 25 nm/V value has been used in this work for the sensitivity of the piezo-positioning of the AFM instrument.
Two geometries of nanoindenter are realized, following named indenter 1 and 2.

3. Results

3.1. Optimization of the AFM indentation speed

To optimize the AFM measurement the E value has been analyzed in function of the indentation speed. A petri dish with a thin Sylgard sample is placed on the AFM stage and force curve measurements are acquired in liquid, in order to avoid the "jump-to-contact" effect, in an indention speed range from 0.1 to 10 µm/s. 1 V has been set as preload setpoint of cantilever during the approach and force spectroscopy measurement, 5 µm is set as total measurement z length.
As a result (figure 2), a speed such that the obtained E lies in the plateau between low indentations (where the E is underestimated because of the closeness to the contact point) and high indentations (where the E is overestimated because of the sensing of the substrate stiffness or sample viscous effects) has been chosen. Moreover in the E plateau the speed that maximize the work of adhesion of the sample to the indenter (in correspondence of the maximum of the adhesion forces F_{AD}) is chosen, obtaining the optimum value of 1 µm/s (measurement z length of 5 µm in 5 seconds extend time).

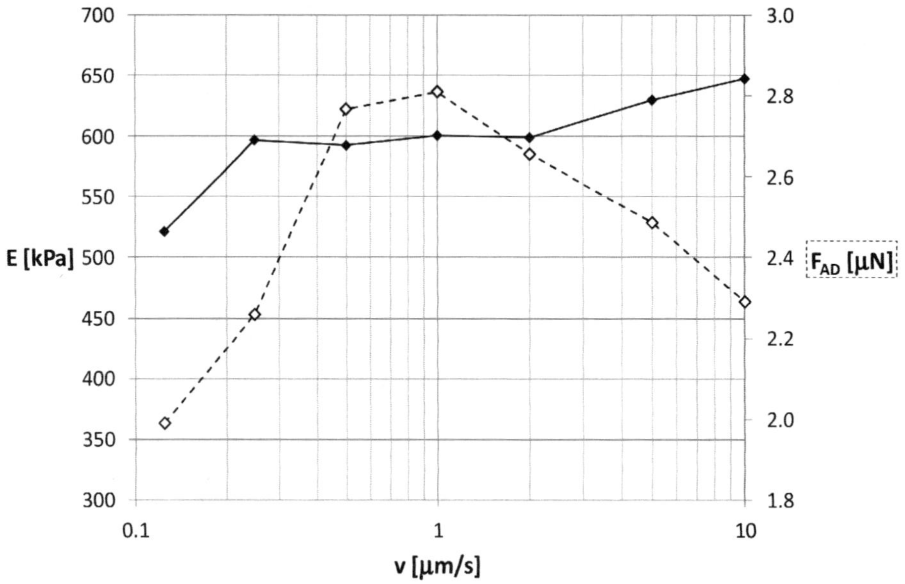

Figure 4. Indentification of the optimum indentation speed for Sylgard samples.

3.2. Study of the AFM measurement uncertainty

In figure 5 are reported the *E* measurements done on the Sylgard reference thick cylinders (contaminated and uncontaminated surface), on the thin cylinders and on the thick cylinders tested with nanoindenter 2. The error bars represents the obtained 20% reproducibility of the thick cylinders measurements with indenter 1.

In figure 6 are reported the AFM measurement of this work in comparison with the unconfined compression measurement done over the same thick Sylgard cylinders. In addition AFM measurement done by Carillo et al. [5] with a sample curing thickness similar to our thick samples, and microfluidic measurements done by Gutierrez et al. [6] are reported.

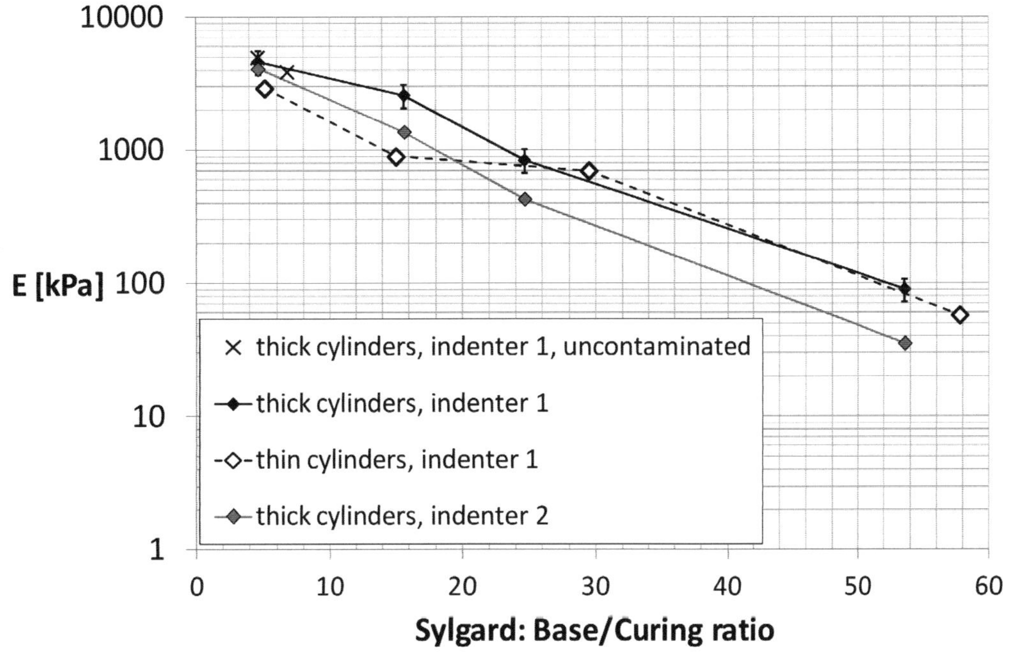

Figure 5. Analysis of the AFM results in function of the nanoindenter geometry and sample thickness.

4. Discussion and Conclusion

In this work has been metrologically analyzed the AFM method for Young' modulus measurements on a wide *E* range comprising the cellular elasticity range. A Sylgard working *E* scale in this range has been prepared in different curing and contamination conditions. AFM reproducibility parameters have been optimized allowing to obtain a total measurement reproducibility of 20% along all the tested *E* measure range. This reproducibility value considers a 10% homogeneity of the Sylgard surface elasticity and a 10% measurement repeatability of the approach and nanoindentation process.

A systematic *E* effect due to the sample curing thickness has been observed for the less elastic cylinders (*E* decreases for thin 5 and 15 cylinders). It follows that the shape of the curing container plays an important role on the surface elasticity, at constant curing time and temperature and base/curing ratio. This can be explain considering that the reducing of the confinement space of the polymer (i.e. the well-plate in respect to the open flat ring) increase the curing reaction efficiency leading to stiffer sample. In fact in more confined space the conformational entropy of the polymeric chains decreases, the chains free-energy increases, the thermal stress energy increase and the chains are more attracted between each other.

Another systematic *E* effect due to the geometry of the nanoindenter has been observed for the more elastic cylinders (*E* decreases for 15, 25, 50 cylinders tested with nanoindenter 2). This can be due to the significant variation of the cantilever elasticity constant once realizing a new nanoindenter.
No effects of the Sylgard surface contamination (by water and air) are found on the surface elasticity, therefore seems not to be necessary to preserve the Sylgard cylinders surface during storage.

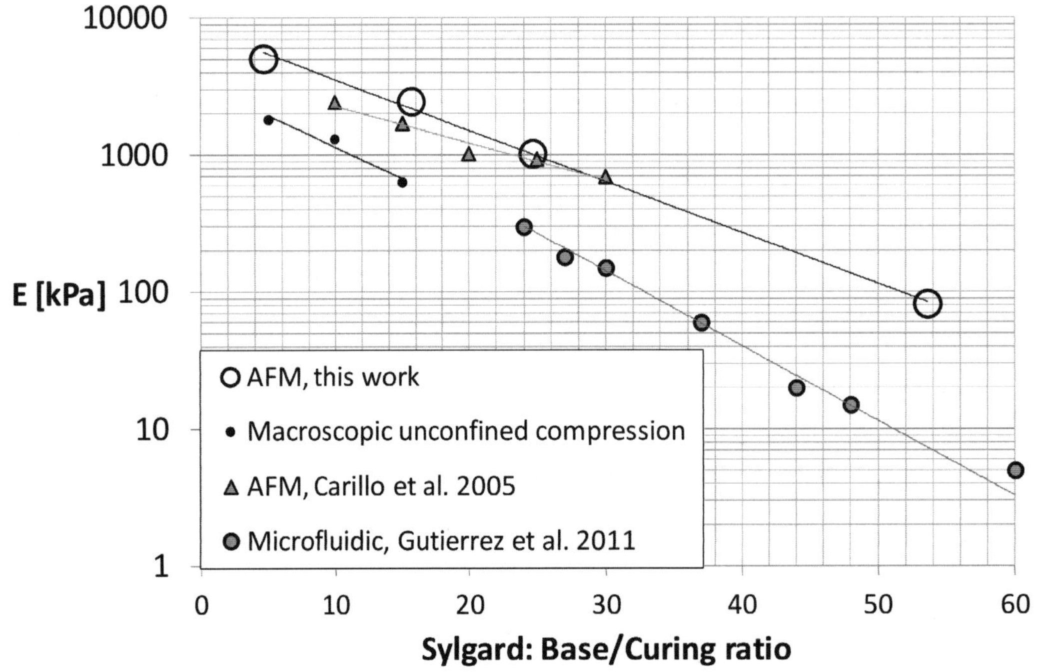

Figure 6. Comparison of the Sylgard *E* measures of the present work with literature values.

The *E* measures of this work are resulted in accord with the literature AFM values done by Carillo et al. (figure 6), but are resulted different from the same measurements done with other techniques (macroscopic compression and microfluidic). This stress the fact that an accuracy budget of the reported methods are needed in order to let a comparison.
In our opinion accurate AFM measures can be realized quantifying the uncertainty of the following influence quantities: the sensitivity of the AFM piezo-positioning, the elasticity constant of the nanoindenter, the diameter of the indentation sphere and considering the adhesion forces of the Sylgard surface in the contact mechanics model (the Hertzian model underestimated the effective contact area). Moreover if Sylgard cylinders are intended to be used for the metrological characterization of the AFM instrument the sample curing thickness must to be kept constant in order to realize stable measuring samples.
Exploring these metrological issues will give a right way of performing accurate elastic modulus measurement on living cells, where further expedients should be then taken with regards to the inhomogeneous and instable softer surface.

References

[1] Cross S E, Jin Y S, Rao J and Gimzewski J K 2007 Nanomechanical analysis of cells from cancer patients *Nature Nanotechnology* **2**, 780 - 783

[2] Wenger M P E, Bozec L, Horton M A and Mesquida P 2007 Mechanical Properties of Collagen Fibrils *Biophysical Journal* **93** 4 1255-1263

[3] Lekka M and Wiltowska-Zuber J 2009 Biomedical applications of AFM *Journal of Physics: Conference Series* **146** 012023

[4] Alonso J L, Goldmann W H 2003 Feeling the forces: Atomic force microscopy in cell biology *Life Sciences* **72** 2553-2560

[5] Carrillo F, Gupta S, Balooch M, Marshall S J, Marshall G W, Pruitt L and Puttlitza C M 2005 Nanoindentation of polydimethylsiloxane elastomers: Effect of crosslinking, work of adhesion, and fluid environment on elastic modulus *J. Mater. Res.* **20** 10

[6] Gutierrez E and Groisman A 2011 Measurements of Elastic Moduli of Silicone Gel Substrates with a Microfluidic Device *Plos One* **6** 9 e25534

Journal of Physics: Conference Series **459** (2013) 012051 doi:10.1088/1742-6596/459/1/012051

Uncertainty analysis of cell counting by metabolic assays

C.Divieto[1], L.Revel[1], G.Sassi[1,2] and M.P.Sassi[1]

[1] Istituto Nazionale di Ricerca Metrologica INRIM., Torino (Italy)
[2] Department of Applied Science and Technology, Politecnico di Torino, Torino (Italy)

c.divieto@inrim.it

Abstract. Cell counting is a fundamental procedure in living cell culture-based experiments and protocols in which the cell number quantification is required. The number of cells is one of the parameters necessary to investigate several cell culture features requiring to be monitored as function of time, such as cell viability, proliferation, growth, fitness and metabolism. Aim of this paper is contributing to declare a comprehensive uncertainty budget for cell counting through metabolic assays according to the EURACHEM/CITAC Guide *Quantifying Uncertainty in Analytical Measurement.*

1. Introduction

The aim of this paper is to contribute to declare a comprehensive uncertainty budget for cell counting by metabolic assays according to the EURACHEM/CITAC Guide *Quantifying Uncertainty in Analytical Measurement* [1].

The lack of traceability of diagnostic and clinical measurement justifies the consideration on the quantification of cells number as a measurement service requested by National Metrology Institutes (NMIs). NMIs raise the issue of quantification of cell population in both bi-dimensional (2D) cell cultures and three-dimensional (3D) cells systems mimicking real tissues [2].

Cell counting is a fundamental procedure in living cell culture-based experiments and protocols in which the cell number quantification is required. The cell number is one of the parameters necessary to investigate several cell culture features such as cell viability, proliferation, growth, fitness and metabolism which require a monitoring as function of time [3]. Cell counting is the quantification of the quantity of cells/organisms in a bioreactor which is any limited volume where a cell culture evolves. Cell number can be determined by removing cells from the bioreactor and counting them with either a hemocytometer or an electronic cell counter, or by cell image analysis, or by assaying a metabolic activity of the cells within the bioreactor (e.g. intracellular enzymes activity) [4], when the activity is proportional to the number of cells explicating that activity.

Metabolic assays measure a metabolic activity accounting for vital cells only and quantify the number of cells under the hypothesis that all the active cells show a metabolic activity and that the mean of these cell activities is stable. There are at present some metabolic reaction-based methods for monitoring animal cell number (giving information on cell proliferation) over time [5, 6]. However, most of them requires the use of substances that are cytotoxic [5], or results in cell lysis to measure the metabolic product [5, 6, 7]. The method chosen to be analyzed in this work is a metabolic method based on the reduction of resazurin into the end product resorufin. It is not cytotoxic and do not require any cell damaging steps [8]. The cell counting methods based on metabolic activities are

Journal of Physics: Conference Series **459** (2013) 012051 doi:10.1088/1742-6596/459/1/012051

indirect methods and require a high reproducibility in experiments to calibrate the method and in experiments with samples. The identification of standard measurement units for fluorescence intensities and absorbance are certainly a significant prerogative for methods unification in order to increase their traceability, regardless of the adopted measurement instrument.

This paper presents a brief description of the measurement method, states a realistic model equation and evaluates the uncertainty of a CellTiter-blue® proliferation assay for use to evaluate the cell number and estimate the cell proliferation in 2D cell culture.

2. Materials and Measurement method

2.1. *Cell seeding and cell culture medium*

The human mesenchymal stem cells (hMSCs) were purchased from Lonza Group Ltd (Basel, Switzerland). They are bone marrow derived-hMSCs from donor. hMSCs were expanded and maintained in non-differentiating cell growth αMEM (alpha modified Minimum Essential Medium - Lonza, Wokingham, UK) with FBS 10% v/v (Fetal Bovine Serum, Lonza, Wokingham, UK), L-glutammine 2 mM, penicillin 100 U/ml and streptomycin 100 g/ml (Lonza Wokingham, UK). Cells were grown in monolayers in Petri dish (plastic vessels) for cell culture and fresh medium was replaced every 3-4 days until cells reached about the 80% of confluence(area occupied by cells/total area). Cells were then washed once with 1X Phosphate Buffer Saline (PBS), detached with 0.25% Trypsin – 0.53 mM Ethylenediaminetetraacetic acid (EDTA) solution, counted by means of a hemocytometer and suspended at several concentrations (cells/ml) in 100 μl non-differentiating αMEM growth medium to be seeded and cultured statically in 96 well microplates. Then, cells are placed in an incubator at 37°C with 5% CO_2 over night to let them to adhere to the bottom of the wells. Cells were managed in laminar flow hood (class II) under sterile conditions.

2.2. *CellTiter-Blue® proliferation assay*

The CellTiter-Blue (CTB) proliferation assay was purchased from Promega Corporation. It is a fluorimetric and metabolic assay used for determining the number of living cells in a biological *in vitro* system. It is based on a redox reaction: the reduction of resazurin (dark blue in color, redox dye with a slight intrinsic fluorescence) into the end product resorufin (pink, highly fluorescent molecule, $\lambda_{ex/em}$ 579/584 nm), made by redox enzymes. Only living cells can perform this reaction because nonviable cells rapidly lose metabolic activity and do not generate any fluorescent signal [8]. O'Brien et al. in 2000 proposed the resazurin dye, the original name of the Alamar Blue dye, to assess the mammalian 2D cell cultures cytotoxicity [9]. CellTiter-Blue reagent is a buffered solution containing highly purified resazurin (substrate of the reaction). The resaruzin molecule can penetrate cells by passing the cell membrane and into the cytoplasm is reduced by cytosolic, microsomal and mitochondrial redox enzymes producing the fluorescent resorufin (reaction product). Resorufin diffuses out of cells back to the culture medium which alone does not reduce resazurin [10]. After cell adhesion, resazurin is added as 10% v/v to the cell culture growth medium in the well containing the cells (adherent on the bottom of the well). Fluorescence intensity of the resorufin was measured after a contact time (0.5 to 6 hours) of resazurin with cells.

Important to note that in the specific case the entire liquid phase (cell culture growth medium containing resazurin) volume was used to measure fluorescence intensity, i.e. 100 μl. However, cell culture can be establish in a different vessel (such as a Petri dish) where the liquid phase volume is much higher. In that case, the volume of resorufin used for fluorescence intensity measurement (typically 100 μl) is taken from the liquid phase (previously mixed to be homogeneous) and placed in a well of a multiplate (typically a 96 well plate) for the measurement. Two volumes are therefore distinguished: the total liquid phase volume (V_{LP}) in which the reaction occurs and the volume of the aliquot of liquid phase placed in the well (V_w) in which the measurement is made. In the following measurand definition, both volumes V_{LP} and V_w will be taken into account, even if in the specific measurements they are the same entity.

2.3. *Fluorescence measurement system*

The GloMax®-Multi Microplate Multimode Reader (Promega Corp.) was used to measure fluorescence intensity. Excitation was performed at 525 nm wavelength while emission was measured in the range 580-640 nm by selecting a pre-set program for CTB. Microplates with 96 wells were used.

3. Measurand definition

The cell culture, of which the number of cells (*N*) is measured, is in a certain volume of liquid phase (made by growth medium in which the resazurin is diluted) in cell culture vessel, such as a well of a microplate.

The number of cells *N* in the cell culture liquid phase volume is proportional to the molar concentration of the reaction product (resorufin) in the liquid phase $C_{Prod,t}$ available in the well at contact time τ and to the volume of the liquid phase in the cell culture vessel V_{LF}.

In the following equation, K_1 defines the specific metabolic activity of the cell and depends by the contact time (τ) between resazurin and cells

$$N = K_1(\tau) \cdot C_{Prod,t} \cdot V_{LF} \tag{1}$$

The product concentration at the contact time τ in the liquid phase is proportional to the fluorescence intensity emitted by the molecules of product $I_{f,Prod,t}$. K_2 is influenced by the liquid phase volume in the measurement well (V_W).

$$C_{Prod,t} = K_2(V_W) \cdot I_{f,Prod,t} \tag{2}$$

The cell culture medium and the reaction product resorufin could emit in the same spectrum region and lead to modifications of the fluorescence intensity. It is, hence, necessary to measure the blank (b) in absence of cells but with times and treatments analogous to those of the samples (s) with cells.

$$I_{f,Prod,t} = (I_{f,s,t} - I_{f,b,t}) \tag{3}$$

Hence, the measurand equation can be written as:

$$N = K(I_{f,s,t} - I_{f,b,t})V_{LF} \tag{4}$$

Where $K = f\,(K1,K2)$ and is therefore influenced by τ, V_W and from the metabolic activity of the cells.

4. Analysis of quantities influencing the measurement

Each input quantity x_i appearing in equation (4) (*V*, *K*, *I* and the dilution in the measurement procedure) has been considered as quantitiy influencing the measurement. The estimation of the associated standard uncertainties $u(x_i)$ and of the systematic effects have been given, as shown in table 5.1. A detailed description of each input quantity follows.

4.1. *Fluorescence*

The fluorescence intensity is determined by the instrument used for the measurement, hence the fluorescence measurement needs to be characterized with respect to the adopted measurement system. Here the GloMax®-Multi Microplate Multimode Reader was adopted.

The fluorescence measurement can be influenced by: repeatability and reproducibility of the measurement, the position of the well within the plate, the plate re-positioning into the measuring system and the sample volume loaded in the well.

For the sample volume loaded in the well, an analysis of the fluorescence intensity sensitivity needs to be carried out to evaluate the sensitivity coefficients.

The repeatability of the fluorescence measure in a single well has been evaluated lower than 0.5% over the whole fluorescence intensity range of interest. The reproducibility evaluated among the 96 wells on the same plate has been found lower than 1%, even though results from 5-10% of wells were outliers. Thus, it is necessary to work in triplicate in order to eventually identify outliers.

The plate placement into the plate reader could contribute to the reproducibility. Hence, the standard deviation of repeated measurements has been calculated.

The system measures the fluorescence intensity due to excitation produced by an incident ray entering the well. The effective optical path length of the incident beam is influenced by the liquid level in the well, thus the fluorescence intensity has been measured as function of the volume filling the well V_W. The fluorescence intensity is highest in the range around 100 µl. This volume minimizes the contribution to the uncertainty of the filling-the-well liquid volume fluorescence intensity.

To summarize, the fluorescence intensity uncertainty has a component of repeatability of instrument measure (0.5%) and a component of reproducibility due to the well positions in the plate (1%). A further contribution is due to the volume of liquid filling the well and can be evaluated separately as function of the available sample volume. Sensitivity coefficients of sample and blank fluorescence intensity ($I_{f,s,2.5h}$ and $I_{f,b,2.5h}$) are calculated by derivatives of eq. (4) with respect to $I_{f,s,2.5h}$ and $I_{f,b,2.5h}$.

4.2. *Liquid volumes*

The type of liquid phase, namely the cell culture medium, is one of the most influencing factors that can interfere on the reaction rate by increasing or decreasing the cell growth. Hence, two different cell culture media added each one with two different FBS quantities have been analyzed. The difference of fluorescence intensity of the different media as function of the contact time τ, i.e., contact between resazurin and cells, for several different cell concentrations (cell number/ml) seeded on wells, has been evaluated. The volume of the liquid phase in the cell culture vessel V_{LF} and the volume filling the measurement well V_w have been measured by a calibrated micropipette. V_{LF} is affected by the residual volumes during liquid replacements, evaporation during the contact time and micropipette uncertainty. In the specific tests the cell culture vessel was the well itself, thus V_W and V_{LF} are the same volume. A 3% total uncertainty has been calculated. The sensitivity coefficient of V_{LF} has been calculated by the derivative of eq (4) with respect to V_{LF}.

4.3. *Slope*

The slope of calibration curve (*K*) have been calculated by the regression analysis of fluorescence intensity experimental data and the nominal value of cells in culture. The nominal value of cells in culture is estimated from the cell seeding concentration measured by detaching and counting the cell by means of an hemocytometer, e.g., Neubauer chamber.

The linear regression minimizes the objective function at the estimated *K* value [11]:

$$K = \frac{\sum_{i=1,n} \frac{N_i\left(I_{f,s,t,i}-I_{f,b,t,i}\right)V_{LF,i}}{\sigma_{y,i}^2}}{\sum_{i=1,n} \frac{\left[\left(I_{f,s,t,i}-I_{f,b,t,i}\right)V_{LF,i}\right]^2}{\sigma_{y,i}^2}} \tag{5}$$

The uncertainty of K [11] is:

$$u(K) = \sqrt{\frac{1}{\sum_{i=1,n} \frac{\left[\left(I_{f,s,t,i}-I_{f,b,t,i}\right)V_{LF,i}\right]^2}{\sigma_{y,i}^2}}} \tag{6}$$

where $\sigma_{y,i}$ is the uncertainty of each experimental point and combines the uncertainty of all the measured quantities, i.e., number of cells, volume and fluorescence intensities. Cell counting relative uncertainty, evaluated as the reproducibility of the Neubauer chamber's cell counting, has been calculated to be 5%.

The K uncertainty decreases when the ratio between cell number N and available liquid volume V_{LF} increases and depends on the number of experimental points. The experimental conditions in which K is calculated should be as similar as possible to the experimental conditions in which the unknown samples are measured. The K uncertainty has been calculated to be 2.2% for time of contact larger than 1 hour and 3% at 0.5 hours of contact. The sensitivity coefficient of K has been calculated by derivative of eq (4) with respect to K.

4.4. Dilution

The fluorescence intensity detection limit of the GloMax®-Multi Microplate Multimode Reader is $5\cdot10^5$ fluorescence units (or counts). Consequently, the maximum ratio between the number of cells and the volume of available liquid has a limit and for values above the detection limit a dilution of the samples is necessary. For different τ, the limit value has been analyzed. The optimal dilution is with the blank solution. However, the available volumes of the blank solution are rarely sufficient for each dilution. Hence, it has been necessary to dilute also the blank solution to obtain a correct linearity of the calibration curve. The dilution of sample and blank solutions with a solvent similar to the blank, i.e., fresh medium, has given a good response.

4.5. Correlation between input quantities

In order to keep the evaluation simple we did not consider correlation between none of the input quantities.

5. Uncertainty budget

The uncertainty budget of the cell count by CTB assay is reported in table 5.1. It has been calculated for a contact time of 2.5h and low cell number at the experimental conditions of the tests. The Significance Index (SI) is the ratio between the contribution of the analyzed influence quantity and the maximum contribution to the uncertainty provided by one of the influence quantities. SI lower than 1% indicates that the contribution to the uncertainty is negligible, SI upper that 10% indicates that the contribution is relevant, SI=100% indicates the most relevant contribution.

The most relevant contribution was given by K. The K uncertainty can be reduced by increasing the number of measurement points of calibration curve; in addition, the fluorescence intensity may be enhanced calibrating the system by fluorophores standard solutions.

Table 5.1. Uncertainty budget for cell number quantification by CTB assay

Component	*Unit*	*Value*	*Source of uncertainty*		*Standard uncertainty*	*Relative standard uncertainty*	*Sensitivity Coefficient*	*Uncertainty contribution*	*SI**
x	$[x]$	x_i			$u(x_i)$	$u(x_i)/x_i$	$\lvert c_i\rvert$	$[u(x_i)\cdot c_i]^2$	*see above
$I_{f,s,2.5h}$	counts	$3.8\cdot10^5$	reproducibility	$3.8\cdot10^3$	$4.2\cdot10^3$	1.1%	$2.7\cdot10^{-2}$	$1.3\cdot10^4$	35%
			repeatability	$1.9\cdot10^3$					
$I_{f,b,2.5h}$	counts	$5.0\cdot10^4$	reproducibility	$5.0\cdot10^3$	$5.6\cdot10^3$	11.2%	$2.7\cdot10^{-2}$	$2.3\cdot10^4$	60%
			repeatability	$2.5\cdot10^3$					
V_{LF}	dm^3	$1.0\cdot10^{-4}$	pipette calibration	$1.0\cdot10^{-6}$	$1.0\cdot10^{-6}$	1.0%	$8.9\cdot10^7$	$7.9\cdot10^3$	21%
K	dm^{-3}	270	regression	5.9	5.9	2.2%	$3.3\cdot10^1$	$3.8\cdot10^4$	100%
Cell number (N)		**8910**	***Combined standard uncertainty***		**286**	**3.2%**			

Journal of Physics: Conference Series **459** (2013) 012051 doi:10.1088/1742-6596/459/1/012051

6. Discussion and conclusions

Cell counting methods based on the monitoring of metabolic activities are indirect methods and require high reproducibility in experiments to calibrate instruments and to analyze unknown samples. The identification of standard measurement units for fluorescence intensities is certainly a significant prerogative for comparing the results obtained by using a certain investigation methodology (i.e. CTB assay) in order to increase its traceability, independently from the adopted measurement instrument.

Results of this work on the evaluation of cell number measurement uncertainty by CTB assay in 2D cell cultures, revealed that the main contributions to the uncertainty are: repeatability and reproducibility of the measurement system and a contribution due to the contact time between the substrate of the metabolic reaction and the cells, the measured sample volume and the cellular metabolic activity. Standard uncertainty was found to be around 3.5% in the tested experimental conditions. All the influence quantities gave a relevant contribution to the total uncertainty, it means that to reduce uncertainty by one order of magnitude, the uncertainty of all the quantities must be reduced.

This work is a first approach to the uncertainty evaluation of a metabolic assay in conventional 2D cell cultures and can be the basis for the innovative and recently applied 3D cell culture systems where several other components can influence the measurement of the cell number in order to increase the traceability of those measurements, independently from the adopted measurement instrument.

References

[1] Ellison L R, Rosslein M and Williams A, 2000 EURACHEM/CITAC Guide, Quantifying Uncertainty in Analytical Measurement, Second Edition

[2] Marriott J, O'Connor G and Parkes H, March 2011 Study of measurement service and comparison needs for an international measurement infrastructure for the biosciences and biotechnology: input for the BIPM work programme *Rapport BIPM 2011/02, LGC/R/2011/123*

[3] Gillies R J, Didier N and Denton M 1986 Determination of cell number in monolayer cultures *Analytical Biochemistry* **159** 109-13

[4] Simonson M S and Dunn M J 1987 Direct measurements of cell number using computer-aided video microscopy" *Analytical Biochemistry* **167** 106-12

[5] Berridge M V, Tan A S, McCoy K D and Wang R 1996 The Biochemical and Cellular Basis of cell Proliferation Assays that use Tetrazolium Salts *Biochemical* **4** 14-19

[6] Marshall N J, Goodwin C J, Holt S J 1995 A critical assessment of the use of microculture tetrazolium assays to measure cell growth and function *Growth Regulation* **5** 69-84

[7] Crouch S P M, Kozlowski R, Slater K J and Fletcher J 1993 The use of ATP bioluminescence as a measure of cell proliferation and cytotoxicity *Journal of Immunological Methods* **160** 81-88

[8] CellTiter-Blue™ Cell Viability Assay Technical Bulletin # TB317, Promega Corporation, 2009

[9] O'Brien J, Wilson I, Orton T and Pognan F 2000 Investigation of the Alamar Blue (resazurin) fluorescent dye for the assessment of mammalian cell cytotoxicity *European Journal of Biochemistry* **267** 5421-26

[10] Gonzalez R J, Tarloff J B 2001 Evaluation of hepatic subcellular fractions for Alamar blue and MTT reductase activity *Toxicology In Vitro* **15** 257-59

[11] Taylor J R 1982 An introduction to error analysis *University Science Books*

The importance of risk-aversion as a measurable psychological parameter governing risk-taking behaviour

P. J. Thomas

Risk Management, Reliability and Maintenance Group, School of Engineering and Mathematical Sciences, City University London, Northampton Square, London EC1V 0HB, United Kingdom

p.j.thomas@city.ac.uk; pjt3.michaelmas@gmail.com

Abstract. A utility function with risk-aversion as its sole parameter is developed and used to examine the well-known psychological phenomenon, whereby risk averse people adopt behavioural strategies that are extreme and apparently highly risky. The pioneering work of the psychologist, John W. Atkinson, is revisited, and utility theory is used to extend his mathematical model. His explanation of the psychology involved is improved by regarding risk-aversion not as a discrete variable with three possible states: risk averse, risk neutral and risk confident, but as continuous and covering a large range. A probability distribution is derived, the "motivational density", to describe the process of selecting tasks of different degrees of difficulty. An assessment is then made of practicable methods for measuring risk-aversion.

1. Introduction

"Fear of failure explains all those rogue trader scandals, most of which started with a cover up", according to former banker turned best-selling author, Robert Kelsey [1], [2]. Kelsey says "It wasn't that I was unable to take the risks that are part and parcel of an investment banker's role ... I could take ludicrous risks in some of the most volatile trading environments on the planet ... It was my judgement of risk that was undermined by my fears [as was] my ability to sort good fear from bad fear [which] can lead to either paralysis ... or nonsensical leaps."

Such extreme behaviour by fearful people has been known to psychologists for many decade. As John W. Atkinson noted: "The tendency for anxious persons to set either extremely high or very low aspirations has been noted over and over again in the literature on level of aspiration" [3]. Atkinson was setting out to explain the behaviour observed with five-year-olds playing the game of hoop-the-peg, where greater rewards were on offer the farther back the child stood. Here children who had been found by independent tests to be highly achievement-motivated took up a throwing position that was a challenging but realistic distance from the peg. But the children found previously to have low achievement motivation were found to act differently. Either they stood almost on top of the peg or else, perplexingly, so far back that failure was almost certain.

Atkinson devised a simple, but powerful mathematical model in order to explain this apparently contradictory behaviour. But it is possible to improve on his explanation by developing a utility function with risk-aversion as its governing parameter. Risk-aversion is a well established parameter from the field of economics and decision science, and the fact that it can be defined precisely in mathematical terms means that Atkinson's results may be generalised, thus offering the potential for the accurate measurements of risk-aversion.

2. John W. Atkinson's mathematical insights

Atkinson needed to devise a mathematical model in order to explain the apparently contradictory behaviour described above. For any task, he saw the "motivation to succeed", D_s as the product of three factors: the incentive to succeed, I_s, the "expectancy" or probability of success, p_s, and the "motive to succeed", M_s. This would be counterbalanced by the "motivation to avoid failure", D_f, found by multiplying the incentive to avoid failure, I_f, by both the "expectancy" (= probability) of failure, p_f, and the "motive to avoid failure", M_f. The resultant motivation to achieve, D_r, would then be the difference between the motivations to succeed and to avoid failure:

$$D_r = D_s - D_f = M_s p_s I_s - M_f p_f I_f \qquad (1)$$

Moreover, he modelled the incentive to succeed as the complement of the probability of success:

$$I_s = 1 - p_s = p_f \qquad (2)$$

reasoning that the more difficult a task was, the greater would be the gain in both personal satisfaction and kudos when it was achieved. Atkinson foresaw this model carrying over to more general tasks faced in life, commenting that "the ranking of occupations according to their prestige ... clearly suggests that occupations accorded greater prestige are also more difficult to obtain".

Atkinson accorded failure a psychological weighting that was different from that accorded to lack of success, which might attract only a null weighting. He took losing as being painful in its own right, with the dissatisfaction and embarrassment at failing with an easy task being greater than those associated with failure in a difficult task. He considered the incentive to avoid failure proportional to its probability of success, with a unity constant of proportionality:

$$I_f = p_s \qquad (3)$$

The greater the task's chance of success, the greater would be the dissatisfaction and shame if the subject failed at that task. (The link between degree of shame and the extent to which individuals fear failure is explored further in [4]).

Nevertheless, the model Atkinson produced had some deficiencies. Literally interpreted, it pointed to the anxious individual preferring a task where success was either completely certain or completely impossible, but still having no net motivation to undertake either. To cover this, Atkinson invoked a further influence, "e.g. social pressures", as necessary to persuade the individual to take on either the easiest available or the hardest available task, but did not include this in the model. Moreover, Atkinson's model was unable to distinguish between degrees of anxiousness beyond just three states, which might be characterised as risk averse, risk neutral and risk confident. Yet common experience suggests that people vary in their levels of anxiety in a more gradual way than this ternary system would suggest, underlining the need to have a fuller model. This paper combines Atkinson's insights with utility theory to construct the more general mathematical model outlined in the next section.

3. Modelling the hoop-the-peg game using utility theory

The reward associated with success in a task may be described by a utility function of which the argument is the incentive for success. The utility of success is then modelled using a Power utility function [5], with risk-aversion, ε, as parameter:

$$u_s = a_s (1 - p_s)^{1-\varepsilon} \qquad (4)$$

where a_s is a constant that reflects the fact that the utility function is correct to a positive multiplier. Its explicit inclusion will facilitate the later comparison with the utility of failure which will be considered next. The utility function of equation (4) becomes identical to Atkinson's incentive for success, $u_s = I_s$, when $\varepsilon = 0$ and $a_s = 1$, in much the same way that a utility function applied to money becomes either the money or a multiple of it when $\varepsilon = 0$.

The penalty associated with failure in a task may be described by a disutility function, the argument of which will be the incentive to avoid failure. In choosing the form of the disutility function, we note that, while utility functions are usually assumed concave, reflecting satiation with good things, disutility functions may be taken generally to be convex, reflecting an accelerating growth in dissatisfaction as bad things accumulate [6]. While successive increments of utility can be expected to fall in size under the addition of equal increments of wealth, the increments of disutility for successive increments of debt are likely to increase in absolute value, implying that the disutility of being £20,000 in debt is more than 10 times worse than being indebted to the tune of £2,000. Accordingly, a suitable form for disutility is the Power disutility function, which has the same form as the Power utility function, except that the risk-aversion parameter, ε, is preceded by a plus rather than a minus sign. It is assumed that the individual's utility and disutility functions will use the same value of risk-aversion. Thus the disutility of failure, v_f, is given by:

$$v_f = a_f p_s^{1+\varepsilon} \tag{5}$$

where the constant, a_f, reflects the fact that the utility function is correct to a positive multiplier.

The disutility function of equation (6) becomes identical to Atkinson's incentive to avoid failure, $v_f = I_f$, when $\varepsilon = 0$ and $a_f = 1$, in a similar way to that in which a disutility function applied to debt becomes simply the debt when $\varepsilon = 0$. The utility of failure is then the negative of the disutility of failure: $u_f = -v_f$.

Let us apply the utility model developed above to a task with success probability, p_s. The expected utility, $y(\varepsilon, p_s)$, for a person with risk-aversion, ε, is then

$$y(\varepsilon, p_s) = E[U(\varepsilon)] = p_s u_s + (1-p_s)u_f = a_s p_s (1-p_s)^{1-\varepsilon} - a_f (1-p_s) p_s^{1+\varepsilon} \tag{6}$$

While Atkinson's model of the hoop-the-peg game assigns a discrete uniform probability distribution to tasks with success probabilities ranging from 0.1 and 0.9, the model may be generalised by assuming there to be a continuum of tasks with success probability, p_s, conforming to a uniform distribution between 0 to 1. The probability density for success probability, $f(p_s)$, will then obey simply: $f(p_s) = 1$ for $0 \le p_s \le 1$. Hence the expected utility, $y_T(\varepsilon)$, across all tasks for a person with risk-aversion, ε, will be:

$$y_T(\varepsilon) = \int_{p_s=0}^{1} y(\varepsilon, p_s) f(p_s) dp_s = a_s \int_{p_s=0}^{1} p_s (1-p_s)^{1-\varepsilon} dp_s - a_f \int_{p_s=0}^{1} (1-p_s) p_s^{1+\varepsilon} dp_s = \frac{a_s}{(2-\varepsilon)(3-\varepsilon)} - \frac{a_f}{(2+\varepsilon)(3+\varepsilon)} \tag{7}$$

In betting, a fair game is one where the expected monetary gains are offset exactly by the expected monetary losses. By analogy, we may define a fair game when the rewards may be other than monetary as one where the expected net reward is zero for a risk-neutral person. Since $\varepsilon = 0$ for a risk-neutral person, a fair game implies, from equation (7):

$$y_T(0) = \frac{1}{6}(a_s - a_f) = 0 \tag{8}$$

This will be satisfied when $a_s = a_f$. The risk-neutral person will regard taking part in such a game with equanimity, since, if he undertook each of the tasks within the game, he would not expect overall to lose, even if he would not expect to win overall either. The idea of a fair game may be expanded to encompass the notion of a "fair game *according to his lights*" if we replace the condition $y_T(0) = 0$ by $y_T(\varepsilon) = 0$, which will imply that a person with risk aversion, ε, would regard the game as fair in that if he undertook each of the tasks in it, he could expect to break-even in the sense that his psychological pain would be matched by his psychological gain. This condition may be satisfied by setting the weighting of utility of success (arbitrarily) at $a_s = 1$ and varying a_f. Solving equation (7) under these conditions gives

Journal of Physics: Conference Series **459** (2013) 012052 doi:10.1088/1742-6596/459/1/012052

$$a_f = \frac{2+\varepsilon}{2-\varepsilon}\frac{3+\varepsilon}{3-\varepsilon} \tag{9}$$

Substituting back into equation (3) gives the expected utility for each task for someone with risk-aversion, ε, under the condition that he will regard the game as fair according to his lights, so that he will be prepared to play it:

$$y(\varepsilon, p_s) = p_s(1-p_s)^{1-\varepsilon} - \left(\frac{2+\varepsilon}{2-\varepsilon}\frac{3+\varepsilon}{3-\varepsilon}\right)(1-p_s)p_s^{1+\varepsilon} \tag{10}$$

It may be noted that this equation depends solely on the task's success probability and the individual's risk-aversion.

4. The choice of task for an individual of given risk-aversion

Plotting equation (10) shows that the choice of task is unimodal for a risk-confident individual, who will have a negative risk-aversion (Fig.1), while it is bimodal for a risk-averse individual (Fig. 2).

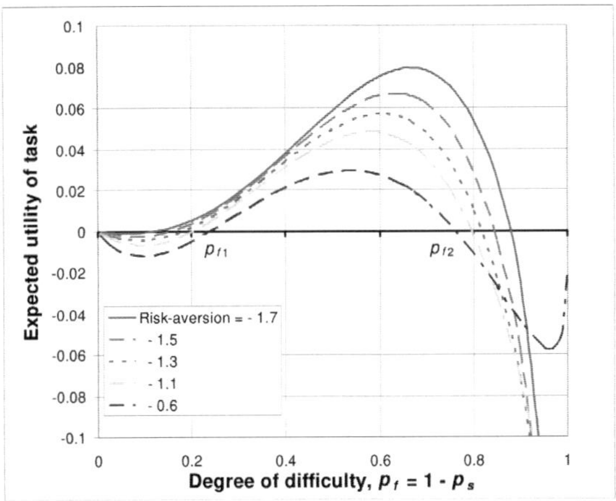

Figure 1. Risk confident individual, $\varepsilon < 0$

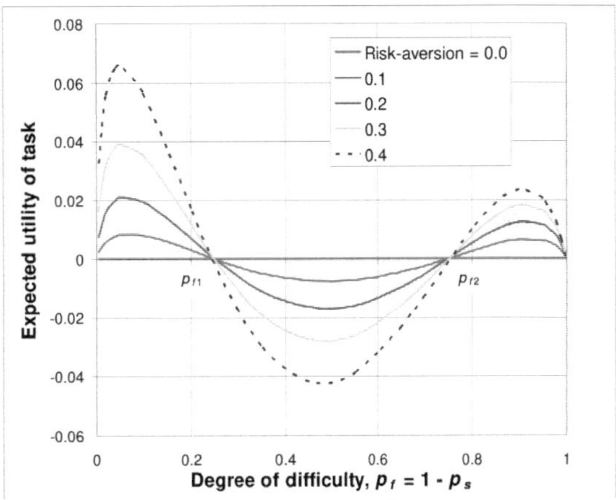

Figure 2. Risk averse individual, $\varepsilon > 0$

The curves are plotted against the degree of difficulty of the task, the failure probability, $p_f = 1 - p_s$. The curve at the risk-neutral value, $\varepsilon = 0$, coincides with the p_f-axis. For all other values of risk-aversion in the range defined by the model, $-2 \le \varepsilon \le 1$, the curve starts and finishes on the p_f-axis and also cuts it twice, at the values, p_{f1} and p_{f2}, given by:

$$p_{f1} = \frac{1}{2} - \frac{1}{2}\sqrt{1 - 4\left(\frac{2+\varepsilon}{2-\varepsilon}\frac{3+\varepsilon}{3-\varepsilon}\right)^{-\frac{1}{\varepsilon}}} \; ; \; p_{f2} = 1 - p_{f1} \tag{11}$$

It is striking that, while risk-confident individuals are predicted to choose tasks that are challenging, but not impossible, the model suggests that people who are risk-averse may choose either a very easy task, or else one with a high degree of difficulty. This is exactly the behaviour reported by Atkinson.

The situation may be generalised by assigning a probability density for the chosen degree of difficulty that is proportional to the expected utility associated with that degree of difficulty provided the expected utility is positive, and zero otherwise. This yields a "motivational density", $h(\varepsilon, p_f)$, applicable to all tasks where the reward is the esteem and kudos associated with the completion of the task rather than a specific, direct benefit; such a benefit may well accrue later, but cannot be included in the calculation of utility because it is not known. Such a situation applies widely to tasks in life, including many associated with people's professional life. As graphed in Figures 3 and 4 against degree of difficulty, p_f, the motivational density is given by:

$$h(\varepsilon, p_f) = \begin{cases} \max\left\{0, \dfrac{(2+\varepsilon)(3+\varepsilon)p_f(1-p_f)^{1+\varepsilon} - (2-\varepsilon)(3-\varepsilon)(1-p_f)p_f^{1-\varepsilon}}{D}\right\} & \text{for } \varepsilon < 0 \\ \max\left\{0, \dfrac{(2-\varepsilon)(3-\varepsilon)(1-p_f)p_f^{1-\varepsilon} - (2+\varepsilon)(3+\varepsilon)p_f(1-p_f)^{1+\varepsilon}}{D}\right\} & \text{for } \varepsilon > 0 \end{cases} \quad (12)$$

$$D = (1-p_{f2})^{2-\varepsilon}((2-\varepsilon)p_{f2}+1) - (1-p_{f1})^{2-\varepsilon}((2-\varepsilon)p_{f1}+1) + (3+\varepsilon)p_{f2}^{2+\varepsilon} - (3+\varepsilon)p_{f1}^{2+\varepsilon} + (2+\varepsilon)p_{f1}^{3+\varepsilon} - (2+\varepsilon)p_{f2}^{3+\varepsilon} \quad (13)$$

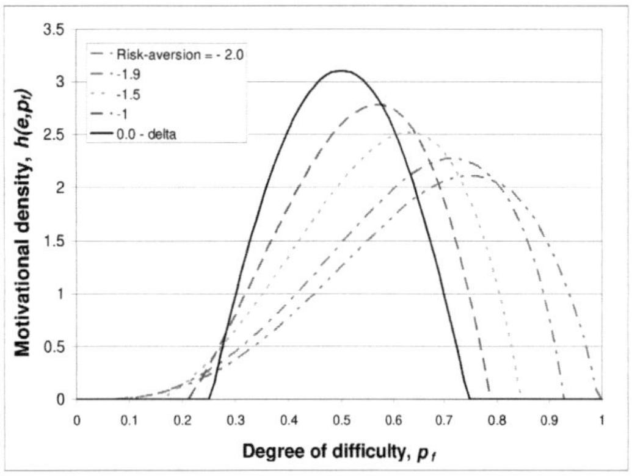

Figure 3. Risk confident individual, $\varepsilon < 0$

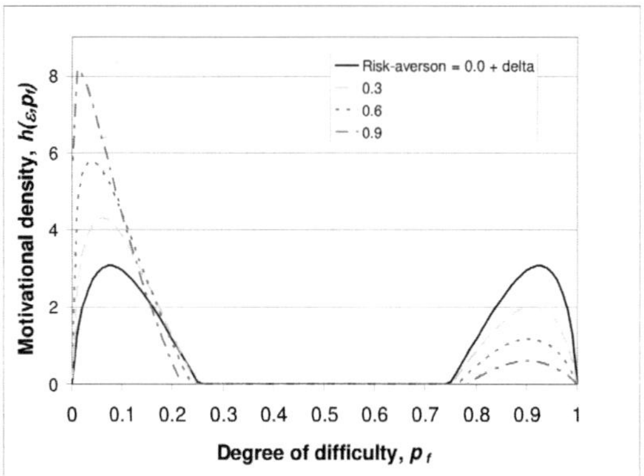

Figure 4. Risk averse individual, $\varepsilon > 0$

It is clear from Figure 3 that for risk confident people, their chosen degree of difficulty increases as risk-confidence grows (risk-aversion becomes more negative). However, the situation is more complex for risk averse individuals, where the distribution is bimodal. From Figure 4, people with a positive but very low risk-aversion will choose either a very easy task, with a probability of failure of 25% or less, or else a very hard task, with a probability of failure of 75% or more. As they become more risk averse, they will tend to prefer easier tasks, but their tendency to choose very difficult tasks persists, as shown by the lobe in the motivational density curve centred around the high degree of difficulty, $p_f \approx 0.9$.

Accurate measurement of risk-aversion becomes possible as soon as the motivational density has been defined mathematically. Setting a series of generally similar tasks where the degree of difficulty needs to be chosen will enable an average chosen degree of difficulty to be calculated, from which the individual's risk-aversion may be deduced. Figures 5 and 6 show the relationship between risk-aversion and the mean for both risk confident and risk averse subjects, with Figure 5 showing also the mode for risk confident persons. These figures show that the average degree of difficulty tends to 0.5 as risk-aversion, ε, approaches zero both from above and below. Even though Figures 3 and 4 make clear that the mode divides into two as ε increases past 0.0, Figures 5 and 6 show that, on average, more risk averse individuals prefer easier tasks, confirming the intuitive interpretation of that parameter. A rational individual will experience different levels of risk-aversion depending on the decision facing him [7], [8], but nevertheless psychological tests interpreted via equations (12) and (13) may provide a ranking of the risk-aversions of different individuals facing similar decisions.

5. Conclusions on the importance of the risk-aversion parameter

It has been observed previously that risk-aversion, acting as the governing parameter of a utility function, can provide a mathematical explanation for panic: at a very high level of risk-aversion, the decision maker may act rashly because he is unable to distinguish between acting to enhance safety or else behaving in such a way as to diminish it [7], [8]. This study extends the field in which a utility

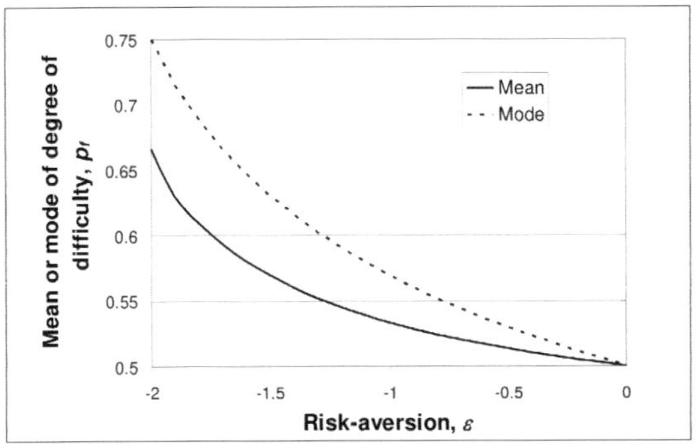

Figure 5. Risk-confident individual, $\varepsilon < 0$

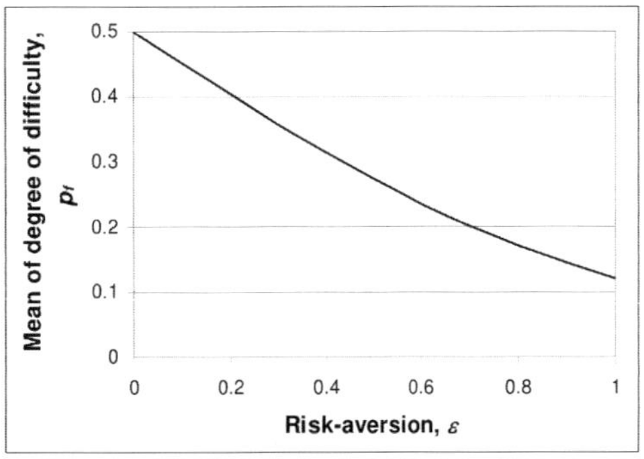

Figure 6. Risk-averse individual, $\varepsilon > 0$

function with risk-aversion as parameter can provide an insight into human behaviour, in this case into actions where the potential rewards are not monetary (at least in the first instance), but satisfaction, esteem and kudos. It has been shown once again how individuals who are highly risk averse may act in the rash way observed in practice. Methods of applying the newly derived motivational density as the basis for measuring risk-aversion have been discussed, while further methods will be discussed in a companion paper [9].

6. Acknowledgements

This work was carried out as part of the NREFS project, Management of Nuclear Risk Issues: Environmental, Financial and Safety, led by City University London and carried out in collaboration with Manchester, Warwick and Open Universities as part of the UK-India Civil Nuclear Power Collaboration. The authors acknowledge gratefully the support of the Engineering and Physical Sciences Research Council (EPSRC) under grant reference number EP/K007580/1. The views expressed in the paper are those of the author and not necessarily those of the NREFS project.

References

[1] Kelsey, R., 2012a, "Tears for fears", *Square Mile*, ISSN 1752-9956, squaremile.com, 74 – 76.

[2] Kelsey, R., 2012b, *What's stopping you? Why smart people don't always reach their potential and how you can*, 2nd Ed., Capstone, London.

[3] Atkinson, J. W., 1957, "Motivational determinants of risk-taking behaviour", *Psychological Review*, Vol. 64, No. 6, 359 – 372.

[4] McGregor, H. A. and Elliot, A. J., 2005, "The shame of failure: examining the link between fear of failure and shame", *Personality and Social Psychology Bulletin*, Vol. 31, No. 2, 218 – 231.

[5] Thomas, P. J., 2010, "An absolute scale for measuring the utility of money", Proc. 13th IMEKO TC1-TC7 Joint Symposium *"Without Measurement no Science, without Science no Measurement"*, September 1-3, City University, London, UK, IOP Publishing, Journal of Physics: Conference Series 238 (2010) 012039 doi:10.1088/1742-6596/238/1/012039.

[6] Ramsey, F.P., 1928, 'A Mathematical Theory of Saving', *Economic Journal*, Vol. 38, Issue 152, pp. 543 – 559, Dec.

[7] Thomas, P. J., Jones, R. D. and Boyle, W. J. O., 2010, "The limits to risk aversion: Part 1. The point of indiscriminate decision", *Process Safety and Environmental Protection*, Vol. 88, No. 6, November, pages 381 – 395.

[8] Thomas, P. J., Jones, R. D. and Boyle, W. J. O., 2010, " The limits to risk aversion. Part 2: The permission point and examples", *Process Safety and Environmental Protection*, Vol. 88, No. 6, November, pages 396 – 406.

[9] Thomas, P. J., 2013, "Methods for measuring risk aversion: problems and solutions", Proc. 2013 Joint IMEKO TC1-TC7-TC13 Symposium, *Measurement across physical and behavioural sciences*, 4-6 September 2013, Genova, Palazzo Ducale – ITALY

Frequency analysis of DC tolerant current transformers

P Mlejnek[1], **P Kaspar**[2]

[1]Czech Technical University in Prague, University Centre for Energy Efficient Buildings, Nam. Sitna 3105, 272 01 Kladno, CZ
[2]Czech Technical University in Prague, Faculty of Electrical Engineering, Technicka 2, 166 27 Prague 6, CZ

mlejnp1@fel.cvut.cz

Abstract. This article deals with wide frequency range behaviour of DC tolerant current transformers that are usually used in modern static energy meters. In this application current transformers must comply with European and International Standards in their accuracy and DC tolerance. Therefore, the linear DC tolerant current transformers and double core current transformers are used in this field. More details about the problems of these particular types of transformers can be found in our previous works. Although these transformers are designed mainly for power distribution network frequency (50/60 Hz), it can be interesting to understand their behaviour in wider frequency range. Based on this knowledge the new generations of energy meters with measuring quality of electric energy will be produced. This solution brings better measurement of consumption of nonlinear loads or measurement of non-sinusoidal voltage and current sources such as solar cells or fuel cells. The determination of actual power consumption in such energy meters is done using particular harmonics component of current and voltage. We measured the phase and ratio errors that are the most important parameters of current transformers, to characterize several samples of current transformers of both types.

1. Introduction

This paper deals with wide frequency range behaviour of current transformers (CT) that are used in energy meters. In present day, the static energy meters replace the traditionally used Ferraris energy meters. The static energy meters are based on digitalization of voltage and current waveforms. Then the energy is computed in microcontroller.

This principle allows, due to high performance of today's microcontrollers, getting more information of voltage and current distortion. Moreover it allows measuring power of higher harmonics components in distorted (non-harmonic) power networks, which is the typical case today. In this application the phase shift between voltage and current is crucial. The knowledge of phase shift between corresponding harmonics allows designing new generation of energy meters where actual power consumption will be computed from particular harmonics and it allows measurement of deformation power [1].

For analysis of higher harmonic components of measured current it is necessary to know about behavior of CT in wider frequency range.

CTs used in static energy meters must comply with European and International Standards e.g. EN 62053-21, EN 50470-3 etc. These standards define several requirements and among them the accuracy and DC immunity that are affected mainly by CT in energy meter. Moreover, another requirement that

Journal of Physics: Conference Series **459** (2013) 012053 doi:10.1088/1742-6596/459/1/012053

is not defined in Standards is the price of current sensor and whole device that is important for manufacturers and their customers.

According to these requirements two main types of CT are used: single core linear CT and double core CT. Linear CT consists of the core made from amorphous or nanocrystalline material with very linear and narrow B-H loop and high saturation magnetic field strength H_s. This arrangement brings non-zero but almost constant phase error that can be easily compensated.

Double core CT consists of two parallel or concentric ring cores that are together wound by the secondary coil. High permeability core ensures very low ratio and phase error and low permeability core with high saturating magnetic field strength ensures high DC immunity. This design is cheaper than linear CT; therefore manufacturers are interested in it. On the other side double core CT introduces some problems under special condition e.g. measurement of consumption of non resistive (reactive) load together with a load that is powered through a diode. Generally, when the DC magnetic flux appears in the core the phase error rapidly increases because of the high permeability core saturation. For purely active load (with cos φ close to 1) this is not such a big problem thanks to the flat maximum of cosine function. For reactive load, even small phase error causes large error of power measurement. More about these problems can be found in our previous works [2], [3] and in [4].

As stated above some additional functions can be implemented in the energy meters to increase their functionality – e.g. monitoring of energy quality, spectrum analysis etc. For this is necessary to understand the measurement errors of CT in wider frequency range than the nominal 50/60 Hz [5].

2. Theoretical Analysis of Errors

Commonly used equations are derived from simplified model of current transformer (Figure 1) and they define amplitude and phase error.

$$\varepsilon_I = \frac{R}{\omega L} \cdot \sin\delta \tag{1}$$

$$\varphi_I = \arctan\left(\frac{R}{\omega L} \cdot \cos\delta\right) \tag{2}$$

where δ represents the core losses angle, $R = R_{Cu} + R_b$ is a sum of winding resistance and burden resistance, L is main inductance and $\omega = 2\pi f$, where f is frequency of current.

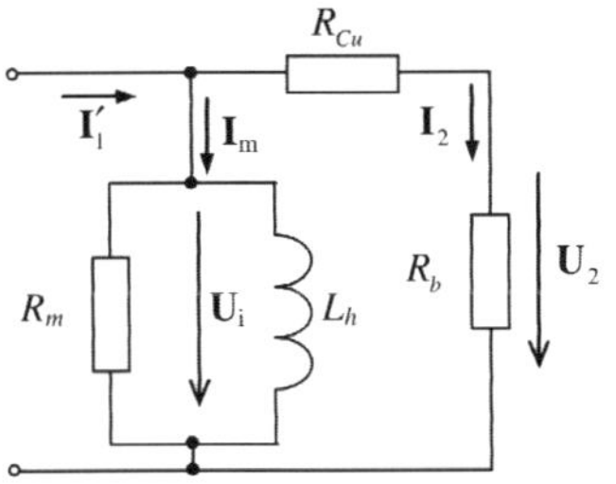

Figure 1. Simplified model of current transformer (1 turn primary winding, no secondary stray induction).

From the equations above can be seen that both errors decrease with increasing frequency (ω).

The amplitude error (ratio error) is very low even for low frequencies (tenths of percentage) and therefore it is not so important to measure its dependency on frequency.

Moreover, the change of the phase error can cause large measurement error especially at measurement with low power factor (PF), see [2], [3], [4].

These errors apply only to an ideal current transformer. There are other properties of CT that modify this dependence and it is very difficult to calculate the error while taking all these properties into account.

3. Real Measurements of the Errors

We arranged the experimental measurement setup that consists of generator Agilent 33120, amplifier Krohn-Hite KH75 and precision power meter ZES Zimmer LMG500 (see block diagram on Fig. 2).

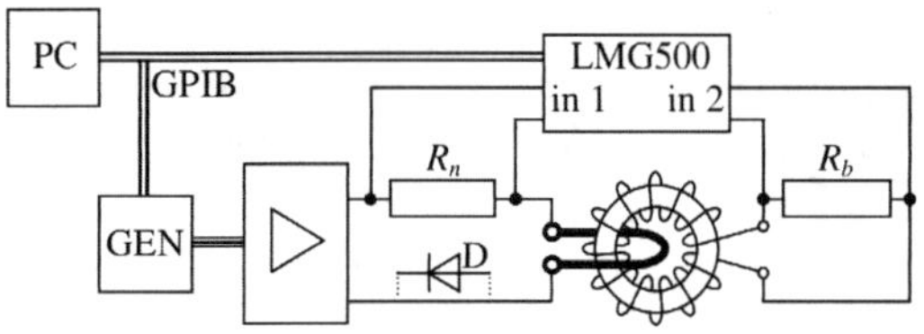

Figure 2. Block diagram of measurement setup (R_n – internal sensing resistance of LMG500 power meter, D – Schottky diode for half-wave rectified current measurement)

The generator and power meter were controlled by PC via GPIB. The frequency varied from 50 Hz up to 5 kHz and the waveforms of primary and secondary currents were sampled. Primary current was measured on built-in shunt resistor in LMG500. The secondary current was measured on burden resistor specified for the CT by manufacturer and the current sensor input of LMG500 was used for sampling.

Table 1. Specification of used current transformers.

Manufacturer	Vacuumschmelze	Magnetec	Taehwatrans	Oswell
Type	4626X131	MB-361-01	TD120L	DCT103
Design	linear	linear	linear	dual core
Current range	100 A	100 A	510 A	100 A
DC tolerance	100 A	100 A	120 A	100 A
Turns ratio	1:2500	1:2500	1:2500	1:2500
Phase error	3.3 deg	5.1 deg	3.3 deg	< 0.33 deg
Burden	7.5 Ω	7.5 Ω	1 Ω (7.5 Ω)	12,5 Ω

Since some measurements have been done with non sinusoidal current, the phase error in this case was defined as a phase shift of 1st harmonic – see equation (3). The LMG500 power analyzer uses simultaneous sampling of all channels so the FFT analysis of both waveforms provides the magnitudes and consistent relative phases of particular harmonics. The phase error was calculated from these results.

$$\varphi_I = \varphi_{I\ sec\ 1^{st}\ harm} - \varphi_{I\ prim\ 1^{st}\ harm} \tag{3}$$

Estimation of uncertainty of phase error calculation is not easy and it exceeds the scope of this paper. In our first measurement we compare the phase error of CT at 50 Hz with specification from manufacturers and it fits well.

We measured two different types of CT in our tests. Dual core CT and precise linear CT from different manufacturers (see Table 1 for the parameters).

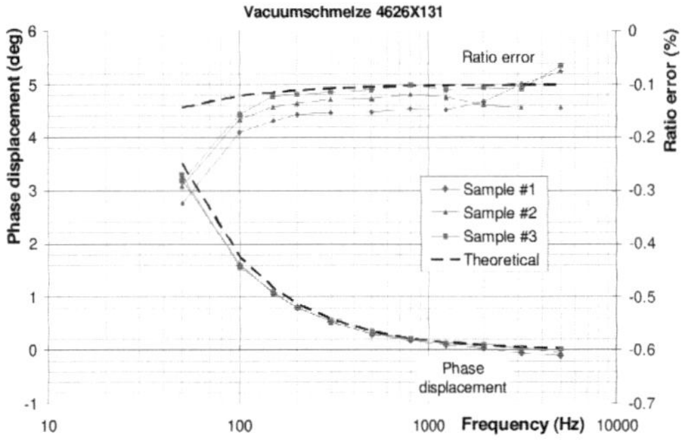

Figure 3. Errors of 4626X131 (I_{prim} = 10 A, L = 3.2 H, R_{Cu} = 54.1 Ω, R_b = 7.5 Ω, δ = 0.4°)

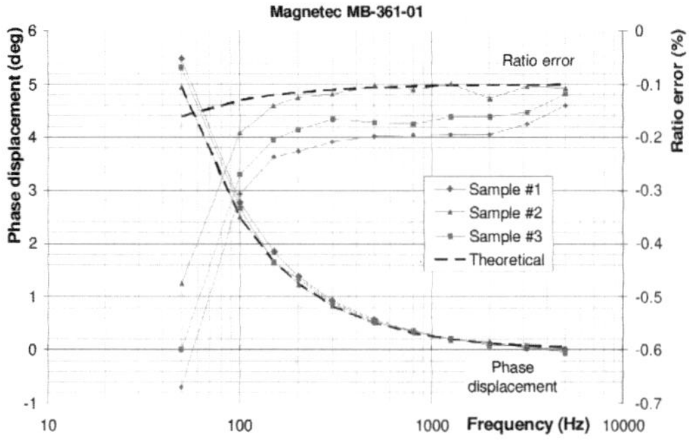

Figure 4. Errors of MB-361-01 (I_{prim} = 10 A, L = 2.0 H, R_{Cu} = 47.8 Ω, R_b = 7.5 Ω, δ = 0.4°)

Figure 3 shows the frequency dependency of phase displacement and ratio error of three samples of linear CT VAC 4626X131. Results well correspond with theoretically computed values that are depicted by black dashed line. Similar measurement was done for next three samples of CT MB-361-01 (Figure 4) and for TD120L (Figure 5). Despite of recommended burden we used the 7.5 Ω even for the TD120L due to the same sensitivity and higher output voltage.

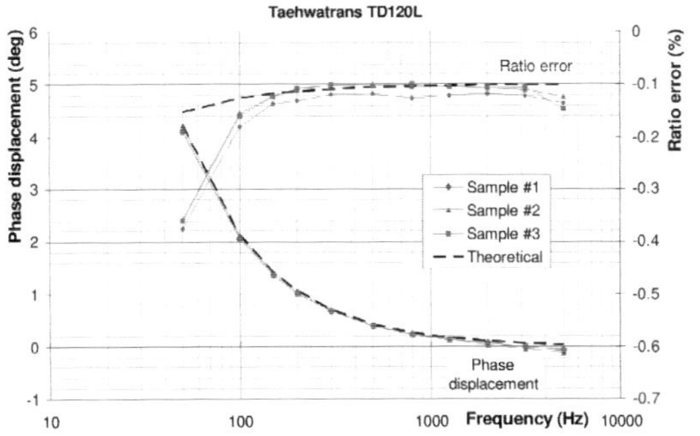

Figure 5. Errors of TD120L (I_{prim} = 10 A, L = 2.7 H, R_{Cu} = 55.9 Ω, R_b = 7.5 Ω, δ = 0.4°)

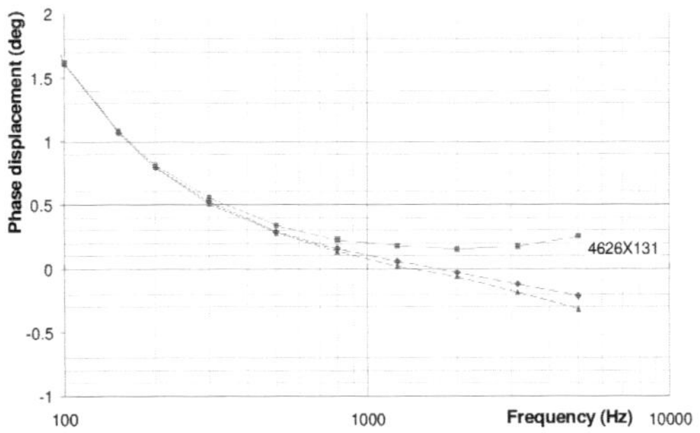

Figure 6. Detail of phase displacement curves for several position of primary turns

Perceptible variation at higher frequency (>2 kHz) is caused by parasitic properties of CT (stray field and core losses, capacitances etc.), which are strongly dependent on position of primary conductor with respect to the CT. This frequency is very close to resonant frequency (around 1.5 kHz). Some types of phase displacement of same sample above resonant frequency can be seen in detail in Figure 6.

4. Measurement with Non-sinusoidal Current

It is important to measure the phase displacement even for non-sinusoidal current. Half-wave rectified current is an important type of primary current because it is a frequent case of DC load seen today in applications of energy meters. A Schottky diode (with very low forward voltage drop and thus very low power loss) has been inserted in primary circuit for this measurement.

As expected the phase displacement remain unchanged for all linear CT due to the constant permeability in wide range of magnetic field strength. The DC component of measured current moves the operating point along the B-H loop. The only differences were at higher frequencies due to resonant effects (see above).

On the other hand there was significant change for double core CT as can be seen in Figure 7 (cp. red and green lines). This change is due to the saturation of high permeability core by DC component of primary current. Thus magnetic flux flows through low permeability core with significant and non-constant phase displacement (about 3 deg).

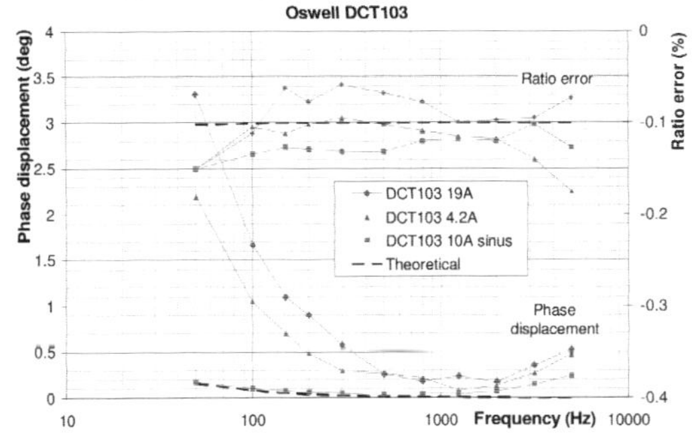

Figure 7. Errors of DCT103 (L = 135.5 H, R_{Cu} = 117 Ω, R_b = 12.5 Ω, δ = 0.4°)

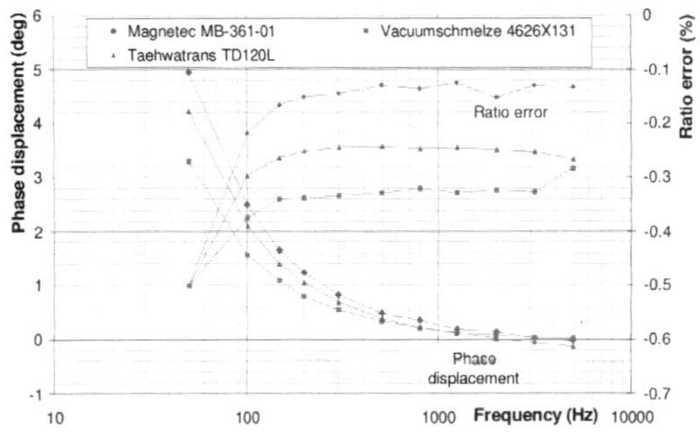

Figure 8. Comparison of phase displacement and ratio error of different current transformers

The measurement with half-wave rectified primary current has been performed for two amplitudes of primary current. In contrast to linear CT (constant permeability of core in wide range of DC magnetizing) the phase displacement of dual core CT varies with amplitude changing. It is due to the low quality of low permeability core (in order to be cheaper than the linear CTs) and causes variation of phase displacement with changing DC magnetization of CT. The results of measurement for varying amplitude of primary current can be also seen in Figure 7 (see blue and green lines).

5. Conclusion

In present day the manufacturers of static energy meters would like to implement new features to their products in order to get advantage against their competitors. Some of these features can be measuring of energy quality or analysis of power of higher harmonics. These features allow using the energy meters in Smart Grids which is required by Directive of the European Parliament and Council Directive 2009/72/EC.

It is very important to select proper current sensor for the current measurement in energy meter or other device that can measure also non-sinusoidal current. Cheap double core CT can be used in these devices, but the problems with non-constant phase displacement have to be considered. Especially in application with non-resistive loads such CT should be replaced with high-quality linear CT.

Current transformers for energy meters are specially designed for the single frequency but they can be used even in wider frequency range after phase displacement correction as was shown in the text above. The frequency dependency of the errors of a current transformer can be very well predicted according to the basic parameters – inductance L and resistance of winding R_{Cu} as was shown in figures 3, 4 and 5. Fig. 8 compares the results for all tested linear types CT.

References

[1] Sedlacek M and Stoudek Z 2011 Active Power Measurements - An Overview and a Comparison of DSP Algorithms by Noncoherent Sampling *Metrology and Measurement Systems* **18**(2) 173–184

[2] Mlejnek P and Kaspar P 2009 Drawback of Using Double Core Current Transformers in Static Watt-Hour Meters *Sensor Letters* **7**(3) 394–396

[3] Mlejnek P and Kaspar P 2008 The Phase Displacement of DC Tolerant Current Transformers under DC Biasing *Journal of Electrical Engineering* **59**(7) 94–96

[4] Hermann F 2007 Untersuchung von Zählern mit Mischkern-Stromtransformatoren *ew—das magazin für die energie wirtschaft* **16** 33–36

[5] Draxler K and Styblikova R 1999 Use of Instrument Current Transformers in Wider Frequency Range *IMEKO - XV World Congress* **4** 113–117

Journal of Physics: Conference Series **459** (2013) 012054 doi:10.1088/1742-6596/459/1/012054

Modelling adequacy of calibration curves and dynamic characteristics of measuring instruments

V A Granovskii,[1] **T N Siraia**
General Research Fellow, Leading Researcher,
Concern CSRI Elektropribor, JSC,
30, Malaya Posadskaya St., 197046 St. Petersburg, Russia

E-mail: vgranovsky@eprib.ru / valgr39@mail.ru

Abstract. Calibration curves and dynamic characteristics are widely used as metrological models of measuring instruments properties; but the initial concepts and formal representations have not been clearly formulated so far. This paper focuses on the construction of a formal framework, which should enable introduction and estimation of adequacy as a characteristic of the model quality. The formal system is constructed, that is similar to the representative theory formalization. So, an initial model for calibration curve is a complex $\{U, C_0, \Phi\}$, containing a set of dependences U, a space of continuous monotonous functions C_0, and monotonic homomorphism $\Phi: U \rightarrow C_0$. Construction of calibration curve is started from an initial table of experimental data, and transfer to the analytical form of calibration curve is analysed. It is demonstrated that adequacy parameters should be expressed by functionals of both measurement accuracy and frequency range. The analogy of measurement and modelling of metrological characteristics should be extended up to the construction of system ensuring the traceability of calibration curves and dynamic characteristics. This system can be created using the experience in constructing a system ensuring the measurement traceability.

1. Introduction

Metrological characteristics of measuring instruments – such as calibration curves and dynamic characteristics – are widely used as examples in metrological publications devoted to modelling problems. Moreover, these characteristics usually influence deeply on the quality of measurements. However, the modelling problem in this area has not been clearly formulated so far, thus not allowing correct definition or estimation of the result adequacy.

This paper focuses on the construction of a system of initial representations, which should enable correct introduction and estimation of adequacy as a characteristic of quality of the modelling procedure and its results.

It seems appropriate to construct a formal system which is similar to the system of measurement formalization provided by the representative measurement theory. It may be based on the formal scheme of representative theory, and on the main stages of measurement, and also on the adequacy analysis as applied to measurement modelling.

---

[1] To whom any correspondence should be addressed.

2 General Formalization

2.1 Formalization of Measurement Procedure

As is known, in representative theory the measurement is interpreted as a complex (ordered triple) [1]:

$$\{X, R, \Phi\}, \tag{1}$$

containing: X – set of object physical characteristics (empirical system with relations),
R – complete numerical set with relations,
Φ: $X \rightarrow R$ – homomorphic mapping of the empirical system into numerical set.

As applied to a certain measurement, this general complex may be defined concretely, and it is rational to realize in compliance with the main stages of measurement [2]. Therefore, at the stage of a measurement problem statement this complex is realized as the triple

$$\{X_0, R, \Phi_0\}, \tag{2}$$

where X_0 - the considered characteristic of the object under study;
Φ_0 - operation of comparison with the corresponding unit.

After the preparation measurement stage the complex takes the form:

$$\{X_1, R, \Phi_1\}, \tag{3}$$

where X_1 is the measurand, Φ_1 is the measurement method.

The complex, corresponding to the stage of measurement experiment, is of the form:

$$\{x_1, R, \varphi_1\}, \tag{4}$$

where x_1 is the input signal, φ_1 is the measurement procedure.

Finally, the measurement result may be presented in the form

$$r = \varphi_1 (x_1). \tag{5}$$

The error (or uncertainty) of this result can be expressed as follows:

$$\Delta (r) = \varepsilon (\Phi_1, \ \varphi_1) (X_1)\,, \tag{6}$$

where the operation $\varepsilon (\Phi_1, \ \varphi_1)$ denotes the discrepancy between operations φ_1 and Φ_1, which is applied to measurand X_1.

As the error as expressed by (6) is estimated by experimental data, the expression

$$\Delta_1(r) = \varepsilon (\Phi_1, \ \varphi_1)\, x_1 \tag{7}$$

can serve as an estimate of error (uncertainty) $\Delta (r)$.

2.2 Formalization of Calibration Curve

While proceeding from the measurement to modelling of metrological characteristics, particularly, the calibration curve, we may construct a formal scheme, which is similar to (1) – (4). First of all, we have an initial general complex (triple) of the form:

$$\{U, C_0, \Phi\} \tag{8}$$

where U - a set of all monotonic dependences (between two physical quantities),
C_0 - a space of continuous monotonous functions,
Φ: $U \rightarrow C_0$ - a order preserving function (homomorphism).

It is also rational to proceed in compliance with the main stages of modelling procedure [3]. So, at the stage of the modelling problem setting the complex becomes as follows

$$\{U_0, C_0, \Phi_0\}, \tag{9}$$

where U_0 - the required dependence,
Φ_0 - two-dimensional comparison with the units of two quantities (mapping onto coordinate plane).

Further, a modelling preparation step stage yields the complex

$$\{ U_1, C_0, \Phi_1 \}, \tag{10}$$

where U_1 is the modelled dependence, Φ_1 is the modelling method.

After obtaining the initial experimental data, i.e., the basic data table

$$u_1 = \{(x_i, y_i), i=1...n\}, \tag{11}$$

we get the complex

$$\{u_1, C_0, \varphi_1\}, \tag{12}$$

where φ_1 is the table generation procedure.

Based on the introduced representations, the modelling adequacy and its estimate can be defined by analogy with (6) – (7), and these may be presented as follows

$$\alpha = \delta(\Phi_1 , \varphi_1) (U_1), \qquad \alpha_1 = \delta(\Phi_1 , \varphi_1) (u_1), \tag{13}$$

where $\delta(\Phi_1 , \varphi_1)$ denotes the discrepancy between operations Φ_1 and φ_1.

3 Analysis of Calibration Curve Adequacy

3.1 Factors of Calibration Curve Adequacy

Definition of adequacy concept makes it possible to analyze it as a qualitative characteristic of modelling results of calibration curves and dynamic characteristics.

We start with the calibration curve by saying that it can be represented in tabular, graphic, or analytical form. So the table (11) is an initial form, which is quite suitable for practical application. However, the issue of proceeding to other representation form, i.e., to continue modelling, is relevant. Therefore, adequacy analysis is inseparably connected with the validation of the curve form.

If Table (11) is used in itself, the objective meaning contained in experimental data is saved in complete and non-changed form. To construct a table, the primary data are processed for each point, and estimates of errors (uncertainties) are also obtained. Thus, we have an extended table:

$$u_e = \{(x_i, \Delta x_i; y_i, \Delta y_i), i = 1...m\}. \tag{14}$$

The expected way of adequacy analysis is to realize the modelling adequacy estimates (13) as applied to Table (11), which is a model of the required input-output dependence. At the same time, we come to recognize that adequacy of the table as a model of dependence is not merely conditioned by the measurement accuracy Δ, but it is also determined by the level of the table detailing. Here the accuracy Δ is expressed by data error set $\{ (\Delta x_i; \Delta y_i), i = 1...m \}$, and the level of detailing H is expressed by the number of point m or, in other words, the frequency range of the model.

Therefore the measurement accuracy Δ and frequency range H are the two factors (two coordinates) of adequacy, whose increase, if assumed independent, should result in higher model adequacy. The factors can be changed independently; however, the effect on the model adequacy depends on the level of the other factor. For instance, with fixed sampling interval h (sample size m) the increasing accuracy enhances the adequacy in limited range only depending on h. On the other side, with the fixed accuracy Δ, if h grows from the low initial level, it will lead to adequacy increase, and above the certain level, to its decrease. Therefore adequacy parameters should be expressed by functionals of both Δ and h.

3.2 Analysis of Calibration Curve Representation

If one need to convert a calibration curve from the tabular (11) to the analytical form, such as

$$Y = f(X) = f(X, a_1, ..., a_k), \tag{15}$$

the additional a priori data are required. These may be obtained from analysis of physical principles of measuring instruments, or some previous experiments. On the other side, during the conversion some primary information is inevitably lost, and further it would be irretrievable. Additionally, revision of Table (11) may be needed irrespective of the form of calibration curve.

If particular importance in constructing the calibration curves is given to adequacy, then selection of the curve form becomes less significant, especially with modern IT capabilities.

However, obvious drawbacks of the tabular form should be mentioned as follows:

- complexity of obtaining and storage of calibration curves;
- possible uncertainty of curve usage.

The last issue has two main aspects, which are the formal one (poor conditioning of data processing task) and the technical one, that is required calibration accuracy depends on application conditions of measuring instruments.

For instance, if the measurements in calibration are accurate, but the measurements while using of calibration curve are inaccurate, then the considerable uncertainty in determination of input can occur. Vice versa, inaccurate calibration and accurate using of calibration curve can result in appearance of gaps, where the input cannot be determined.

Traditionally, the analytical form (15) is preferred, i.e., representation of calibration curve as a continuous function $\boldsymbol{f}$, which depends on a small number of estimated parameters $\boldsymbol{a}_1, \ldots, \boldsymbol{a}_k$. However, this representation is not always possible or rational.

Advantages of analytical form include the following: easy use and tests of curve, compactness of representation, and possibility of estimating the results accuracy. This form is achieved first of all by applying physical principles, which should enhance the adequacy of calibration curves. There are, however, some drawbacks, since the objectivity of data contained in initial Table (11) is decreased.

Thus, the following requirements are placed upon the analytical model of calibration curve:

- optimization of resources during the curve construction and use;
- providing uncertainty during the use;
- possibility of estimating and validating the curve accuracy.

It follows from above that the decision on analytical modelling should be taken before the data are obtained, at the stage of experiment planning. Actually the data obtaining (initial table) should be planned and organized depending on the modelling problem. Generally speaking, the dimension of analytical model should be correlated at least with sampling interval h of initial Table (11): for example by transition from h to the upper frequency of model spectrum by sampling theorem.

3.3 Analysis of Analytical Form for Calibration Curve

The basic steps of analytical modelling are in general similar to the measurement stages. At the initial step, the model general structure is formed based on the physical principles of measuring instrument. Then, criteria to select the model order are formulated based on conditions of inversibility, i.e., effective use of calibration curves in practice.

Accuracy is interpreted in a specific way as applied to the analytical model of calibration curve. Unlike the general measurement problem, here the object under consideration is a function; therefore in the context of calibration curves it would be natural to use the notion of adequacy characteristic instead of error. The proposed notion can be interpreted in two ways: as uncertainty of the curve analytical form, or as inaccuracy of the curve parameters. The first type of uncertainty is surely more significant for the analysis of model adequacy.

Selection of the analytical form is illustrated as a stepwise search of embedded sequence using the stopping criteria [3]. The criteria should be based on physical principles of conversion implemented in the sensor or device, and on admissible deviations from them in real devices. Consequently the model adequacy should be coordinated with the practically required measurement accuracy.

On the whole, modelling of calibration curve is a cyclic process, which is conditioned by the above described principal uncertainty of the dependence inherent in the measuring instrument (device). Therefore, design of the calibration experiment and selection of criteria play the key role.

Mathematical design methods also gain some specific features in this context. Frequently optimal plans obtained using formal methods prove unpractical or physically unrealizable.

A simple example is a construction of linear calibration curve

$$\boldsymbol{Y = a + bx} \tag{16}$$

on the specified interval (x_{min}, x_{max}). If the input valuts $\boldsymbol{x_1, \ldots, x_m}$ can be selected, and the curve is constructed using the least squares method (LSM), the formal optimal plan prescribes [4] that the measurements are made at the extreme points, x_{min} and x_{max}, with the equal number of observations. With the calibration curves, however, this plan is not rational, because in practice it is also important to determine nonlinearity of the curve, and also to investigate the middle part of interval in detail; besides, physical realization often requires time or space separation of the points. Frequently the uniform distribution of points x_i over the interval (x_{min}, x_{max}) proves practical; it is also convenient from the computational viewpoint [2].

Since the construction of calibration curve is a cyclic process, validation of transition from one model to the other, and selection of transition direction are of great importance. Initial requirements to the curve features are commonly minimal: continuity and monotoneness, and sometimes smoothness (for physical considerations). In most cases it is natural to start from a linear calibration curve (16).

There are many options to extend the linear curve form, such as:

a) polynomial curves of low degree;

b) nonlinear curves linearized by transformation of variables;

c) linear-fractional curves;

d) nonlinear curves as presented by linear combinations of the known functions;

e) piecewise-linear curves obtained by several linear segments.

Each of extension options has its advantages and disadvantages. Polynomial extension, for instance, allows the use of LSM, and LSM-based analysis. On the other hand, a fortunate transformation of variables (option b) can significantly enhance the calibration adequacy, and provide more physically illustrative curve.

Modelling of calibration curves involves two closely connected aspects: physical principles and computation aids. Two formal languages corresponding to these aspects are transformation or group-oriented, and functional-computational language.

The most important classes of calibration curves are the most narrow, linear class, and the widest, containing all the continuous monotonous functions; both are the transformation groups. It means that these classes of functions $\boldsymbol{\Omega}$ have the following properties:

a) ***closed composition***: if all functions $f_i(x)$, $i = 1\ldots n$, belong to class $\boldsymbol{\Omega}$, then their composition $f(x) = f_n(f_{n-1}(\ldots f_2(f_1(x))\ldots))$ also belongs to Ω;

b) ***inversibility***: if function $f(x) \in \boldsymbol{\Omega}$, then the inverse function $f^{-1}(x)$ belongs to $\boldsymbol{\Omega}$ as well.

The group of linear-fractional transformations meets these conditions, too; it specifies a class of calibration curves suitable for the measurements, which are physically realizable. The linear-fractional curves group is much broader than the linear group, and it is the only extension of the linear group, which does not coincide with the widest group of all monotonous curves [5].

The group structure of calibration curve classes is convenient for application problems, uncluding description of transformation sequences or their inversions. Therefore, linear-fractional calibration curves are very promising for metrology.

4 Analysis of Dynamic Characteristics

As to the dynamic characteristics, everything described above about the calibration curves applies to them as well. For dynamic measurements (including determination of dynamic characteristics) the complex, by analogy with (8), becomes as follows

$$\{\boldsymbol{D, B, \Phi}\}, \tag{17}$$

where $\boldsymbol{D}$ - set of dependences (between input and output quantities), $\boldsymbol{B}$ - space of continuous operators, representing dynamic properties, $\boldsymbol{\Phi}: \boldsymbol{D} \rightarrow \boldsymbol{B}$ - corresponding homomorphism.

Nevertheless, this complex does essentially differ from the analogous sets for static case, such as (1) and (8). Namely, in the case of calibration curves, a structure of the set of dependences $\boldsymbol{U}$ is solely determined by the continous scale of property under investigation. So the elements of the set are not interrelated in formal or physical ways, and the sequence of values $\boldsymbol{x_1, \ldots, x_m}$ is only determined by practical considerations. As opposed to this, in the case of dynamic measurements, the sequence of input values for dependences $\boldsymbol{D}$ is strictly ordered by fundamental time parameter. Thus, the dynamic measurements of the form

$$\{ \boldsymbol{D}_I, \boldsymbol{B}_I, \boldsymbol{\Phi}_I \}, \tag{18}$$

may be only carried out according the rate of the physical process. In its turn, the set of dynamic characteristics $\boldsymbol{B}_I = \{\beta_i\}$ determine the accuracy of the process representation or measurement. So it is rational to determine the characteristics while the more rapid process, than the measurement process. Thus there are clear distinctions between calibration and dynamic measurements, which can be easily presented as relations in temporal and spectral domain.

Also, there exists a more strict and unambiguous classification of dynamic models based on physical properties of material systems. For modelling purposes, the dynamic characteristics are classified into linear and nonlinear, minimum and nonminimum phase, lumped and distributed parameter. Futher, in linear class the dynamic characteristics are represented by typical functional models: rational transfer functions, transient and impulse response in the form of exponential series of various orders, etc.

Limitations of the spectrum of desired dynamic models have clear physical foundations, as contrary to their formal nature with respect to the calibration curves. These specific features do not make the modeling of dynamic characteristics more complicated, but allow a clearer formulation of the modeling problem.

5 Conclusions

Two significant conclusions can be drawn.

First, the adequacy is not reduced to the error (uncertainty), and it is a more complicated (two-dimensional, at least) attribute of the model.

Second, analogy of the modelling of metrological characteristics and the measurement should be extended up to the construction of system ensuring the traceability of calibration curves or dynamic characteristics. This system can be created using the experience in constructing a system ensuring the measurement traceability.

This work is supported by Russian Foundation for Fundamental Researches, grants No. 13-08-00688, 13-08-00778.

References

[1] Finkelstein L Measurement and instrumentation science – Measurement, 1994, **14**, 1.
[2] Granovskii V and Siraya T Methods for Data Processing in Measurement. - Energoatomizdat, Leningrad, 1990 (in Russian).
[3] Granovskii V and Siraya T Model adequacy in measurements: estimation and improvement - AMCTM VIII – Ser. Adv. Math. for Appl. Sci., World Scientific, 2009.
[4] Ermakov S and Zigljavsky A Mathematical theory of optimal experiment. – Moscow, Nauka, 1987 (in Russian).
[5] Krotkov I , Kreinovich V and Mazin V General form of measurement transformations, allowing computation metrological analysis of measuring systems – Izmeritelnaya technica, 1987, **10.**

Journal of Physics: Conference Series **459** (2013) 012055 doi:10.1088/1742-6596/459/1/012055

Conception and development of an optical methodology applied to long-distance measurement of suspension bridges dynamic displacement

L Lages Martins[1], J M Rebordão[2] and A Silva Ribeiro[1]

[1] National Laboratory for Civil Engineering (LNEC), Lisbon, Portugal
[2] Laboratory of Optics, Lasers and Systems (LOLS), Lisbon, Portugal

E-mail: lfmartins@lnec.pt

Abstract. This paper describes the conception and development of an optical system applied to suspension bridge structural monitoring, aiming real-time and long-distance measurement of dynamical three-dimensional displacement, namely, in the central section of the main span. The main innovative issues related to this optical approach are described and a comparison with other optical and non-optical measurement systems is performed. Moreover, a computational simulator tool developed for the optical system design and validation of the implemented image processing and calculation algorithms is also presented.

1. Introduction

In today's society, the mobility of persons and goods is assured by large transport networks where bridges are key elements with relevant impact on the social and economical development of a region or country. Therefore, it implies the need to assure the operational safety of bridges, usually by performing structural observation in order to characterize and analyse their performance in accordance with design safety requirements and behavior historical records, thus supporting maintenance and management decisions.

As in other relevant large structures, bridge observation is based on different field activities such as visual inspection and structural monitoring using measuring instruments placed on critical bridge structural elements. In general, different types of quantities can be measured, namely, temperature, wind velocity and direction, acceleration, rotation, extension and displacement.

In the studied case – long-span suspension bridge observation – the major concern relies in the measurement of three-dimensional displacement of the stiffness girder in the central section of the main span, being a challenge because conventional off-the-shelf instrumentation usually does not provide suitable solutions. A typical constrain, for instance, is found when using displacement transducers since it is not possible to define absolute reference points in the central section region. The use of other measurement methods such as geodesic observation and hydrostatic leveling are also considered inappropriate due to the dynamic structural behavior of the suspension bridge, usually characterized by having low frequency and high amplitude displacements, specially, in the vertical direction of its stiffness girder.

In recent years, research efforts have been seeking the development of non-contact and long-distance measurement systems able to provide adequate solutions for this problem. The three main approaches proposed in this context are: (i) global navigation satellite systems (GNSS) [1]; (ii) microwave interferometry [2]; (iii) optical systems [3-6].

The first approach mentioned – GNSS – is supported on the use of antennas installed near the measurement region receiving signals from available satellites, thus providing absolute three-dimensional displacement measurements without any limitation regarding to the geographical location of the bridge. Although the absence of long-term instrumental drift makes this approach attractive, its accuracy is strongly dependent on several external factors, namely, the number and distribution of the on sight satellites, the atmospheric conditions and the multi-path effect resulting from multiple signal reflections on the bridge's metallic components (see Figure 1). This last effect also compromises the use of interferometric systems, since the emitted microwave signal can be multi-reflected on the metallic stiffness girder of a suspension bridge, thus affecting its measurement accuracy.

Figure 1 - The 25th of April suspension bridge in Lisbon (Portugal).

Optical systems composed of digital cameras and active targets, combined with the use of computer vision techniques, are not affected by the multi-path effect being considered among the most promising and adequate to give appropriate solutions for long-distance and non-invasive three-dimensional displacement measurement. In fact, they allow real-time and long-term monitoring, being able to measure low frequency and high amplitude dynamic displacements in suspension bridges, considering the absence of heavy rain or fog and (still to be assessed) low impact of refraction and turbulence effects.

This paper focus on the conception and development of an optical methodology applied to structural monitoring for three-dimensional dynamical displacement measurement in the central section of the suspension bridge main span. In comparison with other studied optical systems [3-6], the proposed methodology is considered innovative since it is supported on the use of a long focal length digital camera placed in the central section of the main span, oriented towards the tower foundation where a set of four active targets composed of near-infrared LEDs is located.

This optical configuration requires the previous intrinsic parameterization of the applied digital camera, being the experimental procedure a known difficulty when applied to long focal length lenses [7]. In addition, the accuracy of the measurements depends on several optical and mechanical influence quantities, namely, the stability of camera's intrinsic parameters and the mechanical connection to the bridge stiffness girder. This paper provides some solutions in order to support the aim that the variation observed for the camera position can be considered representative of the bridge's three-dimensional displacement in the installation section.

Since the optical measurement system is computationally complex, integrating digital image processing (which identifies and measures the target's position in sensor coordinates) combined with a triangulation operation by an iterative least squares procedure for the bridge's three-dimensional displacement determination, a validation procedure for algorithm implementation is required. For this purpose, a computer simulator was developed – to be described in this paper – allowing the user to generate digital images of active targets with known three-dimensional dynamic displacement of the applied digital camera. Consequently, the simulated target images can be used as input data for the

Journal of Physics: Conference Series **459** (2013) 012055 doi:10.1088/1742-6596/459/1/012055

computational algorithms, allowing a comparison between the output displacement values and the simulated camera displacement.

2. Measurement optical methodology

The conception and development of the optical system was intended to minimize the observation distance between cameras and targets and, being the main displacement measurement of interest located on the central section of the bridge's main span, this implies positioning cameras and targets on that particular region or near the tower foundations, thus reducing the impact of atmospheric phenomena and increasing the system's measurement sensitivity.

Following a photogrammetric approach, cameras placed on a tower foundation should observe not only the monitoring targets placed in the central section of the main span but also reference targets whose coordinates in a world referential are accurately known. However, the significant dynamic behavior of the suspension bridge structural elements implies placing reference targets on static regions, namely, in the tower foundation or anchorages opposite to the cameras. In large structures such as suspension bridges, this situation leads to observation distances ranging from 500 meters up to 1500 meters, implying the use of reduced focal length lenses that compromise the system's sensitivity and the use of unreasonable large targets, when employing simple and affordable off-the-shelf digital cameras.

The proposed optical methodology makes use of a digital camera with a long focal length lens (identical to the one presented in Figure 2) rigidly installed in the central section of the bridge's main span and oriented towards one of the tower's foundation where a set of four active reference targets is placed, being each target composed of several near-infrared LEDs organized in a circular geometrical pattern. This approach requires the accurate knowledge of the relative coordinates between the four targets (world referential) and it is supported on the previous intrinsic parameterization of the applied camera from which its focal length, principal point coordinates and lens distortion corrections [8] are accurately estimated.

Figure 2 – Off-the-shelf digital camera with long focal length lens.

This optical approach results in six unknown variables related to the camera's extrinsic parameters, namely, its position (three coordinates for the corresponding three-dimensional directions) and orientation (three rotation angles or the corresponding rotation quaternion), requiring the observation of, at least, four targets in order to obtain a stable solution in the triangulation process. The observed changes in the position of the camera are expected to be representative of dynamic displacement of the central section of the bridge's main span, if adequate solutions are implemented in response to the two main measurement concerns.

One specific concern is due to the transmission of high vibration frequencies which can compromise the displacement measurement accuracy, therefore, particular care must be given to the mechanical connection established between the camera and the stiffness girder, namely, through the use of vibration isolator materials. Another specific concern is related to the stability of the camera's intrinsic parameters, which must be regularly evaluated in order to evaluate internal damages caused by its continuous exposure to bridge vibration by traffic and wind actions.

The intrinsic parameterization is a key issue of the proposed approach and it reveals to be a challenging task in the case of long focal length cameras while following the perspective or pinhole camera model. Three main parameterization approaches [7] are currently known: (i) the test-field method; (ii) the goniometric method; (iii) and the diffractive optical element (DOE) method. Since all of the mentioned parameterization methods are characterized by an identical accuracy level [7], the DOE method [10] is more attractive to implement due to a less complex experimental setup (stable laser source with accurate wavelength, beam collimator and DOE with known spatial frequency) when compared with the test-field method which requires a high number of targets distributed over a large structure and several observation stations placed far away from the targets (observation distances can be higher than 50 meters), and the goniometric method where an accurate goniometer must be used in order to determine the camera's orientation relative to a collimated light source. In addition, the DOE method is capable of providing stable and accurate estimates based on a single image of the produced diffraction spots on the camera's sensor while the other two methods require processing a high number of digital images.

In comparison with other studied optical measurement approaches, the developed system presents four main advantages: (i) higher sensitivity to the bridge's longitudinal displacement by using long focal length lens; (ii) the world referential is located in a static region where the relative spatial position and orientation of each target is previously characterized in a metrological laboratorial facility; (iii) the number of unknown variables is reduced to six in the triangulation process since the camera's position variation is considered representative of the bridge's three-dimensional displacement; (iv) only one camera is required to perform three-dimensional displacement measurement.

3. Computational simulator tool

The proposed optical methodology has a significant computational component which is a key element for the identification and the calculation of image coordinates of the observed reference targets. In addition to the previous knowledge of the camera's intrinsic parameters, the targets image coordinates are used in triangulation operation to determine its extrinsic parameters (position and orientation), from which the bridge displacement can be obtained.

Due to the complexity of the image processing and calculation operations, the detailed validation of the implemented algorithms is advisable. With this purpose, a computational simulator is used for generating digital image sequences of active targets obtained from a camera with known dynamical displacement. These images can be used as input data for the proposed digital image processing and triangulation algorithms. The obtained output displacement can be compared with the simulated displacement of the camera, as displayed on Figure 3.

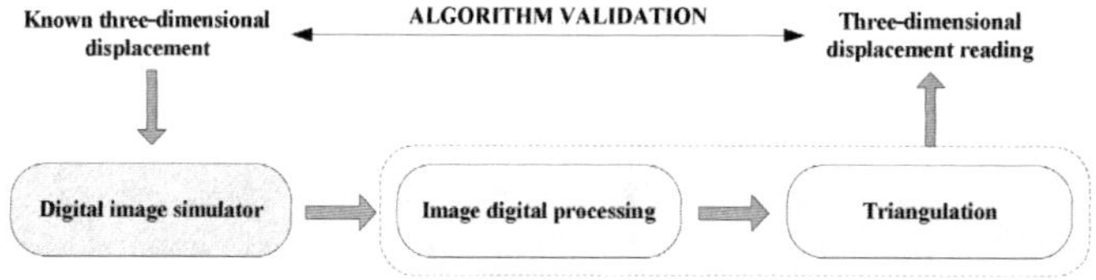

Figure 3 – Schematic representation of the algorithm validation process.

The simulator tool, build in a Matlab® environment, is capable of creating virtual observation scenarios from input variables which are grouped in four categories – camera, target, monitoring conditions and structural dynamics – as displayed in Table 1.

Table 1 – Categories of variables in the simulator tool.

Category	Variables
Camera	Camera type, number of cameras, spatial position and orientation
Target	LED type, number of targets, spatial position and orientation, type (static or dynamic target), shape (cross, rectangular or circular), number of LED lines, number of LEDs per line and LED spacing
Monitoring conditions	Initial and final monitoring instants, image acquisition frequency, turbulence, exposure time
Structural dynamics	Vibration mode, amplitude, frequency and phase

Pre-defined digital cameras and LEDs with specific dimensional, geometrical, radiometric or sensorial characteristics (mentioned in Table 2) can be selected in the simulator through the variables "Camera type" and "LED type".

Table 2 – Subcategories of variables related to camera and LED types.

Subcategory	Variables
Camera type	Focal length, F-number, pixel dimensions, number of pixels, principal point coordinates, focal plane and sensor rotation angle, lens distortion parameters, optical transmittance, quantum efficiency, CCD sensor detection and saturation limits, reading and amplification noise, dark noise, dark signal non-uniformity (DSNU), photo response non-uniformity (PRNU) and quantization noise
LED type	Light intensity, wavelength and viewing angle

Based on the above input data, the simulator tool is able to provide realistic digital images sequences of the four reference targets (see Figure 4), as they were observed from a camera with known dynamic displacement.

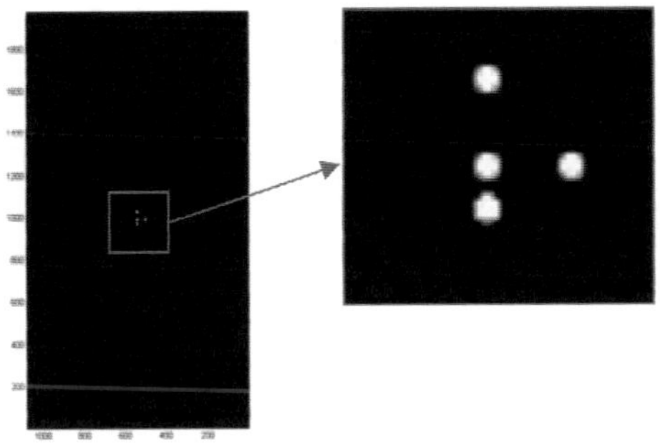

Figure 4 – Reference targets image (normal and amplified) obtained from the simulator tool.

Due to the high variability of observation scenarios (camera and targets position, orientation and mutual distance) and geometrical, radiometric and sensorial characteristics of the main optical components (camera and targets), this simulator tool is also useful in terms of measurement system design, namely, in the dynamical analysis of the camera's field-of-view which, for the proposed optical approach, should always provide a complete image of the four reference targets.

4. Conclusions

This paper describes the main issues related to the conception and development of an optical methodology for long-distance measurement of suspension bridge dynamic three-dimensional displacement in the central section of its main span. By following an optical approach, this measurement method is not affected by the multi-path effect related to other emergent and complementary solutions such as GNSS and interferometric systems but it gathers their main advantages related continuous, real-time, non-contact and long-distance measurement [3-6].

When compared with previous developed optical systems for bridge observation, the described optical approach can be considered innovative in terms of the system's architecture and composition by placing only one digital camera in the central section of the bridge's main span, thus allowing the use of long focal length lens and consequently improving the displacement sensitivity and simplifying the triangulation procedures. In order to achieve an acceptable accuracy level, special attention must be given to the mechanical connection between the camera and the bridge's stiffness girder and to the long-term stability of the camera's intrinsic parameters.

The developed computational simulator tool has a relevant contribution for the validation of the algorithm implementation applied to digital image processing and the triangulation operations which gives the measurement estimates for the three-dimensional displacement. It is also a useful tool for the measurement system design taking into account the high variability of observation scenarios and optical system components.

Being a key issue for the success of the studied optical approach, work related to the DOE intrinsic parameterization of cameras composed of long focal length lens is already underway and results are expected to be obtained soon.

The proposed optical methodology is intended to have an important role in the suspension bridge safety monitoring context, by assuring the accurate measurement of dynamic displacement in its central section. In addition with other complementary measuring instruments and visual inspection activity, it will contribute for early structural failure detection, thus improving the bridge's safety.

The paper also illustrates the interdisciplinary character of the described study and its knowledge interaction with several scientific disciplines, namely, Civil Engineering, Optical Metrology, Photogrammetry and Computer Vision, and their individual contribution for solving this class of measurement problems.

References

[1] Nakamura S, 2000 *Journal of Structural Engineering* **126**, 1413-1419
[2] Pasquale G, Bernardini G, Ricci P, Gentile C, 2010 *IEEE Aerospace and Electronics Systems Magazine* **25**, 19-26
[3] Macdonald J, Dagless E, Thomas B, Taylor C, 1997 *Proceedings Instrumentation Civil Engineers Transport* **123**, 241-248
[4] Olaszek, P, 1999 *Measurement* **25**, 227-236
[5] Wahbeh A, Caffrey P, Masri S, 2003 *Smart Materials and Structures*, **12**, 785-794
[6] Santos C, Costa C, Batista J, 2012 *Strain* **48,** 108-123
[7] Hieronymus J, 2012 *International Archives of the Photogrammetry, Remote Sensing and Spatial Information Sciences* **39**, 595-599
[8] Brown D, 1966 *Photogrammetric Engineering* **32**, 444-462
[9] Stamatopoulos C, Fraser C, Cronk S, 2011 *International Archives of Photogrammetry and Remote Sensing*
[10] Bauer M, Grießbach D, Hermerschmidt A, Krüger S, Scheele M, Schischmanow A, 2008 *Optics Express* 16, 20241-20248

Journal of Physics: Conference Series **459** (2013) 012056 doi:10.1088/1742-6596/459/1/012056

Uncertainty analysis of a stereo system performing ego-motion measurements in a simulated planetary environment

M. Pertile, S. Debei, E. Lorenzini

CISAS G. Colombo, University of Padova, Italy

E-mail: marco.pertile@unipd.it

Abstract. In planetary exploration space missions, motion measurement of a vehicle on the surface of a planet is a very important task. In this work, a non linear vision-based algorithm for ego-motion measurement is described and calibrated using telephoto lenses. Several motion types, including displacement, rotation and their combination, are considered and the evaluated uncertainties are compared, pointing out strengths and weaknesses of employing telephoto lenses for motion measurement.

1. Introduction

In planetary exploration, motion measurement of a vehicle on the surface of a planet should be very accurate in order to track the vehicle also for long paths. The odometric evaluation of vehicle position and attitude performed measuring the rotation of wheels has wide uncertainty due to slippage of wheels on a natural, often sandy or slippery, surface. Moreover, on extraterrestrial planets GPS-like positioning systems are not yet available and inertial navigation sensors exhibit unacceptable drifts. Thus, the need of a reliable and accurate motion instrument is particularly relevant. In this work a vision-based instrument for displacement and rotation is described and calibrated using a simulated rocky scene. Displacement and rotation is measured through the images taken by a vision system. The use of stereo systems for visual-odometry is well known, e.g. see [1] - [9]. Stereo-processing allows estimation of the three dimensional (3D) location of landmarks observed by a stereo-camera. If the same landmarks are acquired and detected by a stereo-system that moves from an initial position to a final one, the two 3D point clouds allow to evaluate the position and orientation of the stereo-system in the final position with reference to the initial one. [5] describes a method for visual odometry based on a stereo camera, providing experimental results gathered during development and flight phases of the NASA's twin Mars Exploration Rovers Spirit and Opportunity landed on the surface of Mars in January 2004. [8] emphasizes the importance of a detailed and correct uncertainty evaluation, and describes a method for visual-odometry that allows to reduce the final measurement uncertainty. However, [8] does not take explicitly into account the uncertainty contribution of the feature detector that allows to find the projection of landmarks in the image plane. In the papers [1]- [8], a detailed calibration of the vision system as a measurement instrument is not present, while in our analysis it is performed according to [10]. Moreover, in the cited literature, the focus is the development of a frame to frame visual odometry, but how the measurement uncertainty changes if one or more frames are skipped is not analyzed. In this work, we calculate the motion of a stereo system and its uncertainty always taking as reference the images acquired in the initial position; thus, we directly analyze what happens when the motion step is increased and the intermediate frames are not used by the algorithm. The methods described in [1] - [9] require that the same 3D landmarks are observed in the initial and final position; this requirement limits the maximum translation and, even more, the maximum rotation that can be measured in one step. For wider ranges, several incremental measurements can be performed and summed together, with the warning that uncertainty may significantly increase with the number of steps, i.e. small errors in each measurement step can eventually cause large errors in the estimated trajectory. This drawback can be solved using more global approaches (e.g. Simultaneous Localization And Mapping, loop-closing, semi-global optimization). However, in the present work the attention is focused on a single step measurement (the same 3D landmarks must be observed in the first and in the final position) and more global approaches, which are beyond the purpose of this work, are not taken into account. Once the intrinsic and extrinsic parameters of a stereo system are carefully determined, one of the main uncertainty sources is the position on the image plane of the features detected and matched in corresponding images. Several feature detectors and descriptors were considered in the preliminary phases of this work, see [11] – [18]. Particularly, [15] compares several different detectors invariant to scale and affine transformations and finds out that the best results are obtained by the Hessian-Affine detector and the Maximally Stable Extremal Regions (MSER) approach. The last one performs well on images containing homogeneous regions with distinctive boundaries. [18] makes a comparison among different descriptors and concludes that the Scale Invariant Feature Transform SIFT and the Gradient Location and Orientation Histogram GLOH descriptors are the best. For the reasons described above, the selected detector is the Hessian-Affine (MSER was discarded since the simulated rocky scenes do not contain homogeneous regions with distinctive boundaries), while SIFT descriptor is chosen.

The present paper gives four main contributions, in terms of novelty and originality: first, it performs a detailed calibration and uncertainty evaluation according to [10] and [19] using an experimental set-up which comprises a simulated rocky scene and reference instruments for both translation and rotation measurements; second, a particularly tricky case is taken into account: the two cameras of the stereo system are equipped with a telephoto lens with a small angle of view; the performed analysis allows to evaluate how the whole measurement algorithm behaves in this difficult case; third, the present work directly analyses what happens when the motion step is increased between two acquired couples of images, i.e. when the whole number of analyzed images is reduced; fourth, the paper describes a slightly new measurement method obtained combining together subroutines and procedures recently published and described in different literature sources. In the following section 2 the measurement algorithm is presented, section 3 describes the calibration procedure and the uncertainty analysis, and in section 5 the experimental results are discussed.

2. Measurement algorithm

In this section, the procedure employed to perform the displacement and rotation measurement of the stereo system is described. The goal is to calculate the displacement and rotation of a calibrated stereo system using the images acquired in an initial position and in a second one. Thus, the input quantities to be measured are a displacement and/or a rotation of the vision system and the output of the indirect measurement is a numerical evaluation of the displacement and rotation vectors. Rotation is described by a sequence of Euler angles around axes X, Y, Z. The procedure begins with the detection of image features (keypoints) that is performed using the Hessian-Affine detector, see [9], [15], [18] for details. After the corresponding 2D features are detected, a triangulation phase allows to compute the 3D coordinates of the physical landmarks acquired by the cameras. The middle point algorithm is used for triangulation, as in [5], for more details see [9].

The uncertainties of 3D points are needed for the evaluation of stereo system displacement and rotation as described in subsection 2.1.2. Uncertainty evaluation for triangulated 3D points becomes an uncertainty propagation task, which is performed by the Kline-McClintock formula, see GUM [10]. This method is selected, instead of the Monte Carlo propagation approach, since the calculation is embedded in the algorithm that calculate the displacement and rotation of the stereo camera, and is performed for all detected features (see subsection 2.1.2). Thus, the use of a Monte Carlo simulation could lead to unacceptable time delays in the position and attitude evaluation of the stereo camera.

2.1. Stereo system displacement and rotation

When the stereo system is calibrated (intrinsic parameters of both cameras and the position and orientation of camera 2 with reference to camera 1 are known), the 3D points ${}^{P1}\mathbf{X}$ can be calculated for all features detected by both cameras when the vision system is in an initial position P1. The notation ${}^{P1}\mathbf{X}$ means that these points are expressed in the reference frame attached to the first camera, when the vision system is in the initial position P1. When the vision system is moved (cameras are rigidly connected) from the initial position P1 to a second position P2, the same procedure can be used to compute the 3D vectors ${}^{P2}\mathbf{X}$ of the same features detected by both cameras in the second position P2 and expressed in the new frame 1 attached to the first camera. For each feature that is detected by both cameras in both positions P1, P2, the following equation can be written:

$$ {}^{P1}\mathbf{X}_i = {}^{P1}_{P2}\mathbf{R} \cdot {}^{P2}\mathbf{X}_i + {}^{P1}\mathbf{P}_{P2,P1} \tag{1} $$

Where ${}^{P1}_{P2}\mathbf{R}$ is the rotation matrix from frame 1 in the second position P2 to frame 1 in the initial position P1; ${}^{P1}\mathbf{P}_{P2,P1}$ is the origin of frame 1 in P2 with reference to the origin of frame 1 in P1 and expressed in P1; i= 1, …, n, with n being the number of common detected and matched features. Vector ${}^{P1}\mathbf{P}_{P2,P1}$ and the Euler angles that define matrix ${}^{P1}_{P2}\mathbf{R}$ are the numerical output values of the whole measurement procedure and, in this work are evaluated in two steps in a similar way as in [5]: first, a less accurate motion is estimated by least squares estimation embedded within a random sample consensus (RANSAC) process to remove outliers; then, a maximum likelihood motion estimation is performed minimizing a non linear problem. The main differences with [5] are: the feature detection and matching algorithms employed in this work are more advanced; the linear least squares approach is derived from [6]; the non linear minimization procedure comprises the Levenberg-Marquardt algorithm; the analysis is performed for motion steps of different sizes and not in a frame by frame way, i.e. we progressively increased the distance (displacement and rotation) between acquired images.

2.1.1. Linear least squares step. In the linear motion evaluation an error vector $\mathbf{e}_i$ is defined for each couple of 3D points ${}^{P1}\mathbf{X}$, ${}^{P2}\mathbf{X}$ detected in both the first P1 and the second P2 position and then the cost function E to be minimized is calculated:

$$ \mathbf{e}_i = {}^{P1}\mathbf{X}_i - {}^{P1}_{P2}\mathbf{R} \cdot {}^{P2}\mathbf{X}_i - {}^{P1}\mathbf{P}_{P2,P1} \qquad E = \sum_{i=1}^{n} \left\| \mathbf{e}_i \right\|^2 \tag{2} $$

In order to separate the evaluation of rotation and translation, the centers of the two point clouds are subtracted from the 3D points. Using the formulas explained in [6], first the rotation matrix is evaluated and then the translation ${}^{P1}\boldsymbol{P}_{P2,P1}$of the stereo system is calculated rearranging equation (2).

2.1.2. Non linear step. In the non linear step also the covariance matrices of the 3D landmarks, calculated as explained above, are taken into account. For the non linear analysis, the errors defined for each 3D feature in equation (2) is weighted using the inverse of a combined covariance matrix. For each 3D point evaluated with the stereo system in the first position, the corresponding covariance matrix is evaluated; then, this evaluation is repeated for the same feature in the second position to obtain the covariance matrix ${}^{P2}\mathbf{C}_i$. ${}^{P1}\mathbf{C}_i$ and ${}^{P2}\mathbf{C}_i$ of the same 3D landmark are combined together and the new cost function E_{nl} (minimized by the Levenberg-Marquardt algorithm) is:

$$ \mathbf{C}_i = {}^{P1}\mathbf{C}_i + {}^{P1}_{P2}\mathbf{R} \cdot {}^{P2}\mathbf{C}_i \cdot {}^{P1}_{P2}\mathbf{R}^T \qquad E_{nl} = \sum_{i=1}^{n} \left(\mathbf{e}_i^T \cdot \mathbf{C}_i^{-1} \cdot \mathbf{e}_i \right) \tag{3} $$

In this way each component of the error vector $\mathbf{e}_i$ of the feature i can have a different weight that takes into account the uncertainty of feature i along the considered direction. This heteroscedastic modeling of uncertainty, i.e. inhomogeneous (it may be different from point to point) and anisotropic (it may be different in each direction) is particularly useful in stereo systems whose baseline (i.e. distance between cameras) is small and the uncertainty along a direction roughly parallel to the optical axes may be much greater than in the other directions. A properly tuned RANSAC algorithm used in the linear phase of motion estimation allows to identify and exclude possible outlier landmarks. Thus, the non linear phase is performed taking into account only 3D points that passed RANSAC algorithm.

3. Calibration procedure and Uncertainty analysis

The first step to calibrate the whole measurement system is the determination of intrinsic and extrinsic parameters of the stereo system. The selected method is described in [19]. After the intrinsic and extrinsic parameters of the stereo system are evaluated, the calibration of the measurement system can be performed. To calibrate the whole measurement system, the stereo cameras are mounted on a rotary motorized stage, and/or on a linear motorized stage. The stereo system position and orientation are changed by known quantities and measured by the encoders of the linear and rotary stages which are used as reference instruments; in each position both cameras acquire an image; the imposed movements are evaluated analyzing images in different positions and orientations. Both cameras are aimed to a simulated planetary scene obtained with crumpled brown paper. In experimental tests the types of movements that are taken into account are: longitudinal displacement along the optical axes of the cameras; transverse displacement orthogonal to the optical axes of cameras; pure rotation; rotation combined with transverse displacement.

Besides the experimental calibration tests, a detailed uncertainty analysis is carried out according to the metrological procedures described in [10] and [19]. A Monte Carlo propagation is used to evaluate the uncertainty of linear displacement and rotation, which are treated as the output quantities of an indirect measurement. In this analysis phase, the aim is to evaluate uncertainties of position and attitude of the stereo system as precisely as possible, and not to calculate displacements and rotations in a fast way. Moreover, the whole measurement algorithm is highly non linear. Thus, a Monte Carlo simulation is employed instead of the propagation formula embedded in the calculation algorithm as described in section 2 and that is an approximated method.

Several uncertainty sources are analyzed and evaluated using experimental tests also with the simulated planetary scene. Particularly, the uncertainty associated with the following quantities are taken into account: intrinsic and extrinsic parameters of the stereo system, whose uncertainties are evaluated using the camera calibration procedure described in [20], and the positions of image features, whose uncertainty is affected by several contributions. In this paper two main contributions are evaluated for image features: the reading uncertainty (image noise) of the cameras, and the lighting variations of the simulated scene. According to [10] and [19], all uncertainty sources are expressed by a probability density function (PDF) and are then propagated to the output displacement of the stereo system using a Monte Carlo simulation. In this way, both the calibration curve and its uncertainty are evaluated for the output measured displacement and/or rotation.

4. Results

The results are the obtained calibration curves of the instrument (figures 1-5). Figure 1 shows the curve and its evaluated uncertainty for longitudinal displacement (substantially parallel to the optical axes), figure 2 for transverse displacement (substantially orthogonal to the optical axes), Figure 3 for rotation, figures 4 and 5 for combined rotation and transverse displacement. For all figures 1-5, uncertainty is depicted with level of confidence (l.o.c.) 95.5% and in the upper graph the vision system measurement is depicted vs. the motion measured by encoders; in the bottom graph the difference between the values measured by the stereo system and the imposed ones (measured by encoders) is depicted for each step amplitude that is analyzed. Comparing figures 1 and 2, the uncertainty evaluated for transverse displacement is larger than that obtained for longitudinal movements, except for the first two motion steps of figure 2. This is an interesting result of the analyzed approach, since the uncertainty of 3D points acquired by the stereo system is wider along the direction parallel to the optical axis than along a transverse direction, due to the small distance between the two cameras. However, this disadvantage along longitudinal direction is compensated by the greater number of

Journal of Physics: Conference Series **459** (2013) 012056 doi:10.1088/1742-6596/459/1/012056

image features correctly matched in the case of axial displacement. Figure 6 depicts the number of matched features vs. motion step amplitude for all the four considered types of motion.

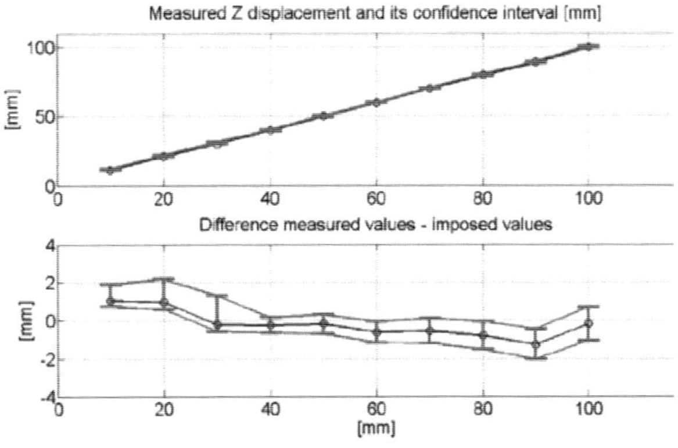

Figure 1. Longitudinal displacement (l.o.c. 95.5%).

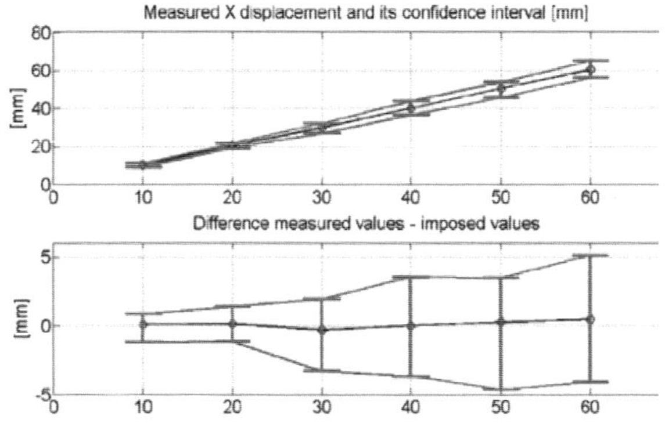

Figure 2. Transverse displacement (l.o.c. 95.5%).

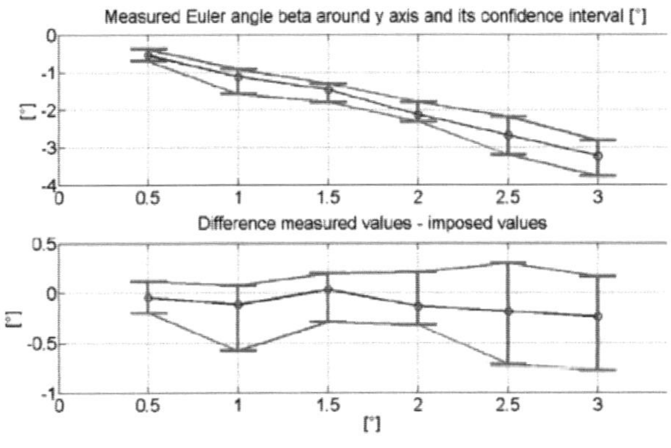

Figure 3. Pure rotation (l.o.c. 95.5%).

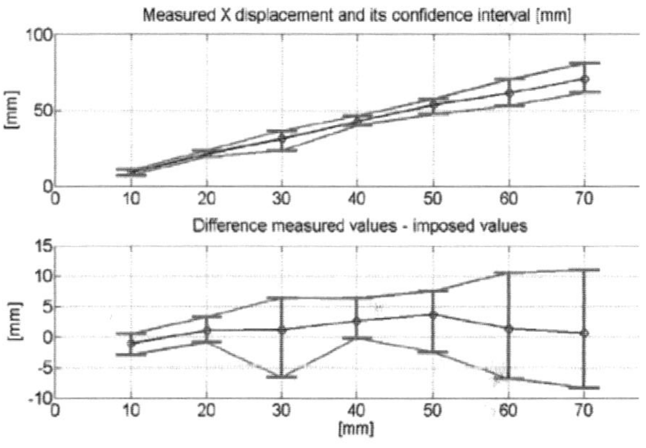

Figure 4. Transverse displacement, with rotation.

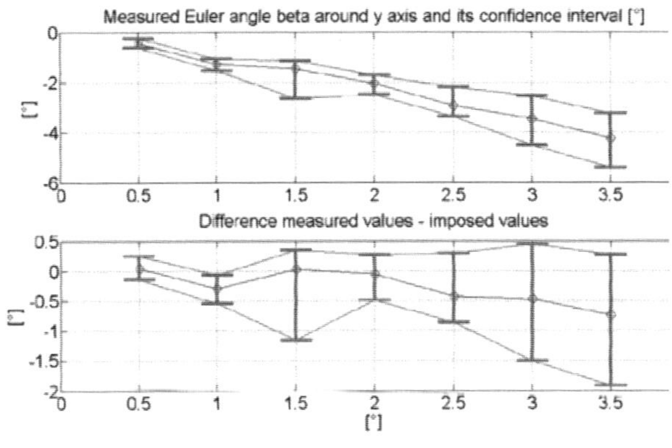

Figure 5. Rotation combined with transverse displacement (l.o.c. 95.5%).

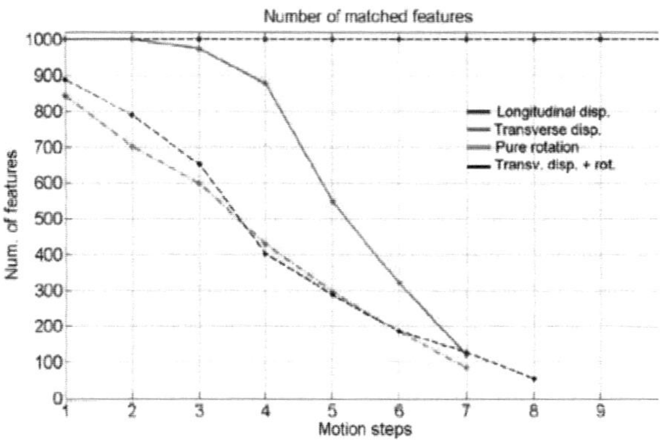

Figure 6. Number of matched features vs. motion step amplitude.

From figure 6 it is clear that in the first two motion steps of the transverse displacement case (figure 2) the algorithm employs the same number of 3D points than in the longitudinal displacement case. This fact may explain the comparable uncertainty obtained. In the other motion steps of the transverse displacement case, the number of matched features rapidly deceases and uncertainty gets wider. Thus, with a photo lens, due to its small angle of view, the variation of the number of matched features seems to prevail over the differences in the uncertainty of 3D points.
Figure 3 shows that the evaluation of small rotation angles with a telephoto lens is not an easy task. There are some physical limiting factors inherent in the stereo system with two telephoto lenses that was employed: the angle of view of the cameras is small (9.80° x 7.85° in our case); the two cameras have substantially parallel optical axes; the depth of field of the cameras is narrow, thus, the acquired 3D points exhibit a small distance along the optical axes of the cameras. The physical factors listed above make the maximum measurable angular step very limited, as confirmed by figure 3. The fourth type of motion that was taken into account is particularly difficult to tackle using a telephoto lens.

In all cases, except for the first one of longitudinal displacement (figure 1), the evaluated uncertainty increases with motion step. A possible reason of this behavior is that the larger the motion step and the fewer the correctly matched features among images; thus, the averaging effect associated with a large number of matched features decreases with step amplitude. This reduction of matched features is not present for movements substantially parallel to the optical axis, see figure 6. Thus, in the first case the uncertainty does not increase with the motion step and longer distances can be measured in this direction than in the transverse one, as it can be seen comparing figures 1 and 2.

5. Conclusion

The calibration of a stereo system as an instrument for displacement and rotation measurement was described. The calibration was carried out using a simulated rocky scene and different types of motion. This procedure allowed to highlight advantages and drawbacks of the considered system which is provided with telephoto lenses. The variation of uncertainty due to a progressive increase of the motion step between acquired images was analyzed.

References

[1] Arun K. S., Huang T. S., Blostein S. D., *IEEE Trans on PAMI*, Vol.9, No.5, 1987, pp. 698–700.
[2] Eggert D., Lorusso A., Fisher R., *Machine Vision and App.*, Vol.9, No.5-6, 1997, pp.272–290.
[3] Umeyama S., *IEEE Trans on PAMI*, Vol. 13, No.4, 1991, pp. 376 – 380.
[4] Olson C. F., et Al., *Robotics and Autonomous Systems*, Vol. 43, 2003, pp. 215–229.
[5] Cheng Y., Maimone M. W., Matthies L., *IEEE Robotics & Automation Magazine*, 2006.
[6] Horn B.K.P., et Al., *Jou. of the Optical Soc. of Am.*, July 1988, Vol.5, pp.1127.
[7] Ma Y., Soatto S., Kosecka J., Sastry S. S., *An invitation to 3-D Vision*, Springer, 2004.
[8] Dubbelman G., Groen F. C. A., *IEEE CVPR*, 2009.
[9] M Pertile, M Magnabosco, S Debei, *Journal of Physics: Conference Series* 238 (2010) 012031.
[10] BIPM, IEC, IFCC, ILAC, ISO, IUPAC, IUPAP, OIML, *Evaluation of measurement data—Guide to the Expression of Uncertainty in Measurement*, Geneva, Switzerland: ISO, 2008.
[11] Harris C., Stephens M., *Proc. Alvey Vision Conf.*, 1988, pp. 147-151.
[12] Mikolajczyk K., Schmid C., *Proc. 8th Int. Conf. Computer Vision*, 2001, pp. 525-531.
[13] Lowe D., *Proc. 7th Int. Conf. Computer Vision*, 1999, pp. 1150-1157.
[14] Lowe D., *Int. J. Computer Vision*, Vol. 2, No. 60, 2004, pp. 91-110.
[15] Mikolajczyk K., Schmid C., *Int. J. Computer Vision*, Vol. 1, No. 60, 2004, pp. 63-86.
[16] Mikolajczyk K., et Al., *Int. J. Computer Vision*, No. 65, 2005, pp. 43–72.
[17] Matas J., et Al., *Image and Vision Computing*, Vol. 22, 2004, pp.761–767.
[18] Mikolajczyk K., Schmid C., *IEEE Trans On PAMI*, Vol. 27, No. 10, 2005.
[19] BIPM, IEC, IFCC, ILAC, ISO, IUPAC, IUPAP, OIML, *Evaluation of measurement data—Supplement 1 to the Guide to the Expression of Uncertainty in Measurement*.
[20] Zhang Z., *IEEE Trans on PAMI*, vol. 22, n.11, 2000.

Journal of Physics: Conference Series **459** (2013) 012057 doi:10.1088/1742-6596/459/1/012057

Quantifying Human Response: Linking metrological and psychometric characterisations of Man as a Measurement Instrument

L R Pendrill[1]
SP Technical Research Institute of Sweden, Measurement Technology, Box 857, SE-50115 Borås (SE), phone: +46 767 88 54 44,
mailto:leslie.pendrill@sp.se

William P. Fisher, Jr.,
BEAR Center, Graduate School of Education, University of California, Berkeley, CA (USA) & Principal, LivingCapitalMetrics Consulting
mailto:william@livingcapitalmetrics.com

Abstract. A better understanding of how to characterise human response is essential to improved person-centred care and other situations where human factors are crucial. Challenges to introducing classical metrological concepts such as measurement uncertainty and traceability when characterising Man as a Measurement Instrument include the failure of many statistical tools when applied to ordinal measurement scales and a lack of metrological references in, for instance, healthcare. The present work attempts to link metrological and psychometric (Rasch) characterisation of Man as a Measurement Instrument in a study of elementary tasks, such as counting dots, where one knows independently the expected value because the measurement object (collection of dots) is prepared in advance. The analysis is compared and contrasted with recent approaches to this problem by others, for instance using signal error fidelity.

1. Introduction

Advances in technology and information processing are allowing the enhancement of various human functions, be it machine learning [1] to assist in mining the ever increasing amounts of information available in society or aiding a disabled, ill or elderly person to cope better with care as well as everyday tasks [2]. To ensure the quality and promote innovation of these diverse technological services to Mankind, better understanding is needed of how a human being perceives quantities such as information content and service quality, not merely in technical terms but in even more 'human' quantities such as comfort, pleasure, beauty [3] and so on.

When characterising the human measuring instrument [4], be it with the five senses or even taking account of the full physiological, mental, cognitive and behavioural richness of human perception, formulation of the commensurate metrological concepts as established in traditional engineering is as yet in its infancy for qualitative [5,6] and subjective measurement [7,8].

It is still common, for instance, to find incorrect analysis of the scores obtained with questionnaires and similar instruments often used to measure human response, where the challenge [9] is that several of the most common tools of statistics – such as calculation of a mean or standard deviation - cannot be used to characterise the location and dispersion of qualitative measurements on ordinal scales [10] typical in such measurements.

Additionally, some method of metrological traceability for measurements based on ordinal observations is needed when the ability of a person to perform a task of classifying an entity of given reference level is to be determined. In some areas of rating person ability in this way – for instance, in the medical sector – it may be possible to promote the same kind of quality-assurance as is routine in

[1] To whom any correspondence should be addressed

more quantitative measurement, that is, in terms of metrological traceability and with declared levels of measurement quality as uncertainty. Patient health, for example, is increasingly rated by health clinics on ordinal scales linearized via a log-odds transformation, and appropriate treatment is decided on by the doctor by comparing the actual ratings with corresponding pattern of 'typical' health ratings for similar patients from earlier studies. Obviously, the comparability of such ratings has to be reliable to a sufficient degree of accuracy if the patient is to be treated appropriately. Given that such accuracy can be be regularly obtained, the observational framework could be redesigned to omit the observed ordinal scores and to incorporate a metrologically traceable unit reported alongside an uncertainty term [7,11,12].

2. Rationale for a potential expansion of the metrological framework

As an example of earlier work, Fisher [7] assesses the feasibility of developing and deploying a universal metric through studies of the quantitative stability of a physical disability construct across a range of measurement instruments and samples. One seeks estimates of the person and instrument measures that are "not affected by the abilities or attitudes of the particular persons measured, or by the difficulties of the particular survey or test items used to measure". Fisher draws parallels with traditional metrological concepts in engineering and physics with corresponding application of these concepts in psychometrics. He argues that this reflects an on-going transition in health care from "local economies of disease-crisis management to regional, national, and international economies of population-based, preventive health management". He points out that the development and commensurate demands for scale-free measurement providing accountability and comparability in healthcare are being enabled by readily accessible computational and communicational resources through internetworked personal computers.

The measurement system concept from the perspective of psychometrics where a human being is, in one way or another, a critical element of the measurement system has recently been considered [13]. The operation of any measurement system includes three steps, recalled by Mari [14]: Calibration; Data acquisition; Data presentation. Examples of recent studies of psychophysical scaling [15] where perceptual intensity is related to stimulus intensity include an attempted explanation of the Weber-Fechner law in terms of signal error fidelity [16].

The present work offers an approach which attempts to link a traditional engineering metrological characterisation of measurement systems to commonly used techniques in behavioural and psychometric studies, where the ability of a human being to perform tasks is characterised over a range of levels of challenge employing the well-established approaches in psychological measurement developed by Rasch and his students [17-19]. Studies of situations are made where a human acts as an instrument for elementary tasks – counting dots [20-21] – where a key observation will be that human performance, described in terms of person ability and level of challenge for different tasks, can in fact be related to invariant measures of location and dispersion on an interval scale when Man acts 'metrologically' as a Measurement Instrument [22, 23]. In such elementary cases, one knows independently the expected value since the measurement object (collection of dots) is prepared in advance, thus allowing calibration and assessment of measurement uncertainty which can then be compared with the ability of the human to perform the task.

3. Grounding measuring in counting

The Mundurucu are an Amazonian indigenous people with little access to Western-style educational resources [21]. Though they lack such technologies as rulers, graphs, weight scales, and numbers greater than five, the Mundurucu have sophisticated, though approximative and nonverbal, conceptions of space and number. Research investigating the conceptual link between number and spatially distributed dots suggests that the Mundurucu intuitively employ a logarithmic transformation of impressions of varying amounts, meaning that "larger numbers require a proportional larger difference in order to remain equally discriminable" (Weber's law) [21].

This difference between intuitive sense impressions and measurement results was first noted by Fechner for the human senses, and has been documented in recent neurological research [24]. This "Gaussian tuning curve" functions like an internal biological slide rule, compactly collapsing several orders of magnitude into a portable system of relatively constant imprecision [21]. In showing the logarithmic proportions between sensations and stimuli, Fechner provided a basis for Thurstone to shift the focus of human measurement from psychophysical to psychological, economic, and social phenomena [25]. Rasch later independently developed ideas along the same lines [26].

Further, real things, from the sides of triangles to rocks to performances, are not identical, and so cannot conform completely to the parameters in a scientific law or measurement model. The measurement of counting ability, like the measurement of weight or length, requires the definition of an invariant unit amount that will not correspond directly with empirically observed counting numbers or things counted [22]. Again, the natural logarithm plays an important role in providing experimentally verifiable criteria for demonstrating linear proportional relationships among parameters in a model.

4. Human performance related to invariant measures on interval scale

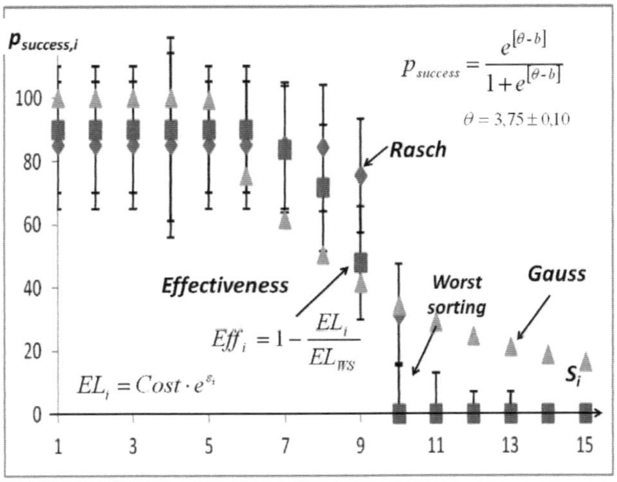

Fig. 1. Comparison of effectiveness (4), Rasch (1) and Gaussian (3) modelling of counting range 1 – 10 for the Munduruсu (w = 0,17)

Two alternative metrological approaches to deriving how the probability of successful counting depends on measures of location and dispersion of perceptive judgments over a range of stimulus values will be compared here in terms (resolution and bias) as one would use when characterising a measurement instrument (in this case, the human counter) (Fig. 1).

4.1 Probabilistic approaches to subjective differences

The Rasch approach [17-19] models individual probabilities of success in counting as the difference between person ability θ and level of challenge b::

$$p_{success} = \frac{e^{[\theta - b]}}{1+e^{[\theta - b]}} . \qquad (1)$$

No distributional assumptions are required, counts of observed success in the assigned task are sufficient statistics [28-29], and individual ability measures are expressed as the average difference between the person's location and each task's location on the log-odds continuum. For dichotomous observations, the model can be written to include an error term as follows:

$$X_{n,i} = P_{n,i} \pm \sqrt{P_{n,i} \cdot (1 - P_{n,i})} \quad (2)$$

Here $X_{n,i}$ is the scored response of person n to item i, and $P_{n,i}$ is the probability of success given in (1). The binomial error distributions for dichotomous scores approximate Gaussian distributions when accumulated across all the observations, as they are in the estimation process [30].

The second approach expresses the ability to distinguish ('resolve') two counts with stimulus values, $S_i = n_1$ and n_2, in terms of the distance, $D(P,Q)$, between two distributions, $P = N(\mu_1, \sigma_1^2)$ and $Q = N(\mu_2, \sigma_2^2)$, each centred at the respective mean count and the standard uncertainty in the difference is given[2] by $u = w\sqrt{\frac{(n_1^2 + n_2^2)}{12}}$. This approach has been used in studies of the acuity of "number sense" where values of the Weber constant, w, range from 1 to 0.1 for the acuity of "number sense" for people of different counting abilities (from infants to educated adults) [27]. The counting efficiency, $p_{success}$, is then expressed in terms of an error rate defined as the area under the overlap in the two count distributions. Assuming Gaussian distributions:

$$p_{success} = 1 - erfc\left[\frac{|n_1 - n_2|}{\sqrt{2} \cdot u}\right] \quad (3)$$

4.2 Classification effectiveness in terms of counting bias on ordinal scale

Perceived values, P_i, such as counts, normally can only be referred to an ordinal scale, since exact perceptive distances between different points are not known. An approach to overcoming the traditional limitations of statistical measures of location and dispersion on such scales is by the inclusion of a cost function as a distance metric. Following this approach, a measure of the effectiveness, *Eff*, of sorting on an ordinal scale has been proposed [31-32] as: $Eff_i = 1 - \frac{EL_i}{EL_{WS}}$ (4)

where EL_i is the expected loss associated with incorrect classification at level i and EL_{WS} is the worst-case loss. In this second approach, the expected loss, and thereby the classification efficiency, associated with incorrect counts are expressed as functions of the 'bias' $\varepsilon_i = S_i - P_i$ for each count (rather than using the difference in adjacent counts used in the first approach above).

5. Results

Despite the differences in these approaches, reasonably good agreement (Fig. 1) is obtained between estimates of the probability of success (at counting in the present example) if both the expected loss in expression (4) and the level of challenge, b_i, for each count in expression (1) are modelled with exponential functions of, respectively, the count bias, ε_i, and stimulus value, S_i. In both approaches, the worst-case loss (most challenging task) is taken ('anchored') to be the count i = 10. The Gaussian psychophysical modelling, for its part, does not rely on these assumptions, so the discrepancies between the three curves of Figure 1 could indicate an increase slightly milder than exponential in the level of challenge across the range of counts from 1 to 10.

[2] Assuming a uniform (rectangular) distribution of the difference in means

We will also compare our results with other recent approaches [16], where the logarithmic relationship in the Weber–Fechner law [33-34] is couched in terms of perception characterised in information theory by 'quantization' followed by efficient coding where fundamental limits of compressibility are expressed in terms of entropy [35]. Fechner's law can be derived by cumulating a measure of dissimilarity of a stimulus from its "immediate" neighbour value where, according to [36], the distance, *D(P,Q)*, can be expressed more generally in terms of the symmetric Kullback-Leibler divergence information measure.

6. Discussion

Historically, psychometrics extended the Weber-Fechner law from the constructs of psychophysics to the measurement of psychological and social constructs observable only in ordinal terms. Linear comparisons effected via application of the log-odds transformation to observed scores unexpectedly offered several additional highly practical advantages. Foremost among these were new capacities for qualitative meaningfulness and for systematically accounting for missing data.

The first of these followed from the reproducible and invariant hierarchies in the questions asked [37]. Numbers ceased being mere digits but were instead strongly associated with substantive variation in the measured construct, such that the increasing difficulty of the questions asked could be understood in terms of developmental sequences. Comparisons of individual measures were now not expressed only in numeric terms, but in terms of performances relative to invariantly ordered learning progressions [38]. Additional new efficiencies are being realized within psychometrics as empirical estimation gives way to strongly predictive theories of the measured constructs [39-40].

The second new practical advantage is the equating of measures made from different sets of items from tests, surveys, and assessments measuring the same thing [39]. Just as new capacities for qualitative meaningfulness were realized from the repeated emergence of the same item hierarchies across samples, so, too, did identifiable commonalities in the item hierarchies emerge across item sets. Data-based equating methods are giving rise to theory-based methods as understanding of the cognitive and performative bases of these hierarchies improves [40]. Enhanced interpretability and the equating of different instruments bode well for future metrological standards and traceability.

7. Conclusion

In well-designed experimental contexts, counts of observed behaviours or decisions may provide evidence that invariant unit amounts exist, and this evidence may be useful in calibrating instrumentation and in devising a theory of the construct. The value of theory in this context is realized to the extent that the construct can effectively be synthesized in the laboratory. Given a theory capable of predicting 90% or more of the observed variance in the counts providing empirical evidence for experimental evaluation, a provisionally satisfactory understanding of the construct can be claimed. Following Feynman's point that, "What I cannot create, I do not understand" [41], judicious use of a combination of empirical, instrumental, and theoretical techniques may ultimately lead to new horizons in psychology and the social sciences [42]. Metrological traceability to reference standard units will be of central importance in this process.

5. References

[1] Zhang W, Yang Y, Wang Q, Shu F 2011 *Proceedings of the 23rd International Conference on Software Engineering & Knowledge Engineering* (**SEKE'2011**), Miami Beach, USA, July 7-9 2011 , http://www.ksi.edu/seke/Proceedings/seke11/51_Wen_Zhang.pdf

[2] Farbrot A, Abbas S, Nihlstrand A, Dagman J, Emardson R, Kanerva S, Pendrill L R 2013 *The Simon Foundation for Continence's Innovating for Continence Conference Series*, Chicago (US), April 2013

[3] Schmidhuber J 2009 *J SICE*, **48**(1):21-32, PDF
[4] Berglund B, Rossi G B, Townsend J, Pendrill L R 2011 *Theory and methods of measurements with persons* (New York: Taylor & Francis)
[5] Mari L 1999 *Measurement* **25** 183 – 92
[6] Pendrill L R 2011 *AMCTM 2011 International Conference on Advanced Mathematical and Computational Tools in Metrology and Testing*, Göteborg June 20-22 2011, http://www.sp.se/AMCTM2011
[7] Fisher W P Jr 1997 *J Outcome Meas* **1**(2) 87-113
[8] Pendrill L R, Berglund B *et al.* 2010 *NCSLi Measure* **5** 42-54
[9] Svensson E 2001 *J Rehab Med* **33** 47-48
[10] Stevens S S 1946 *Science* **103** 2684 677-680
[11] Wright B D 1997 *Physical Med & Rehabil State of the Art Reviews* **11** 261-288
[12] Davis A M, Perruccio A V, Canizares M et al. 2008 *Osteoarthritis Cartilage* **16** 551-559
[13] Wilson M 2011 *Joint International IMEKO TC1 + TC7 + TC13 Symposium*, August 31st – September 2nd, Jena, Germany, urn:nbn:de:gbv:ilm1-2011imeko-005:8
[14] Mari L 2000 *Measurement* **27** 71-84
[15] Berglund B 2011 Chapter 2 in ref [4]
[16] Sun J Z, Wang G I, Goyal V K, Varshney L R 2012 *J Math Psychol* http://dx.doi.org/10.1016/j.jmp.2012.08.002
[17] Wright B D 1999 pp 65-104 in *The new rules of measurement: What every educator and psychologist should know*, ed S E Embretson and S L Hershberger (Hillsdale, New Jersey: Lawrence Erlbaum Associates) pp. 65-104
[18] Rasch G 1961 pp. 321–334 in *Proceedings of the Fourth Berkeley Symposium on Mathematical Statistics and Probability*, IV. (Berkeley, California: University of California Press)
[19] Andrich D 2004 *Medical Care* **42** 1-16
[20] Burro R, Sartori R, Vidotto G 2011 *Qual Quant* **45** 43-58, http://dx.doi.org//10.1007/s11135-009-9282-3
[21] Dehaene S, Izard V, Spelke E, Pica P 2008 *Science*, **320** 1217 – 1220
[22] Cooper G, Fisher W P 2011 Session on Fundamentals of measurement science. International Measurement Confederation (IMEKO). Jena, Germany, August 31 to September 2. http://www.db-thueringen.de/servlets/DerivateServlet/Derivate-24494/ilm1-2011imeko-019.pdf
[23] Wright B D 1989 *Rasch Measurement Transactions* **3** 62 http://www.rasch.org/rmt/rmt32e.htm
[24] Nieder A, Miller E K 2004 *Proc. Natl. Acad. Sci. U.S.A.* **101** 7457
[25] Masin S, Zudini C V, Antonelli M 2009 *J History Behavioral Sciences* **45** 56-65
[26] Andrich D 1978 *Ap Psychol Meas* **2** 449-460
[27] Halberda J, Feigenson L 2008 *Dev Psychol* **44** 1457-1465
[28] Andersen E B 1977 *Psychometrika* **42** 69-81
[29] Fischer G H 1981 *Psychometrika* **46** 59-77
[30] Linacre JM 2010 *Rasch Measurement Transactions* **23** 1238
[31] Bashkansky E, Dror S, Ravid R, Grabov P 2007 *Quality Engineering* **19** 235-244
[32] Bashkansky E, Gadrich T 2010 *Accred Qual Assur* **15** 331-336 DOI 10.1007/s00769-009-0620-x
[33] Dehaene S. 2003 *Trends in Cognitive Sciences* **7** 145–147
[34] Nieder A, Miller E K 2003 *Neuron* **37** 149–157
[35] Cover T M, Thomas J A 1991 *Elements of information theory* New York, John Wiley & Sons
[36] Dzhafarov E N 2011 Chapter 9 in ref [4]
[37] Wright B D, Masters G N 1982 *Rating scale analysis* Chicago, MESA Press
[38] Wilson M R 2009 *J Res Science Teaching* **46** 716-730
[39] Wright B D, Stone M H 1979 *Best test design* Chicago, MESA Press
[40] Stenner A J, Stone M H 2003 *Rasch Measurement Transactions* **17** 929-930
[41] Hawking S W 2001 *The universe in a nutshell* New York, Bantam Books
[42] Stenner A J, Fisher W P, Stone M, Burdick D 2013 Causal Rasch models *Frontiers in Psychology* (in review)

IMEKO 2013 TC1 + TC7 + TC13 IOP Publishing
Journal of Physics: Conference Series **459** (2013) 012058 doi:10.1088/1742-6596/459/1/012058

The experimental verification of metrological properties of direct current Watt-hour meter

P Otomański[1], **P Zazula**[2]
[1] Poznan University of Technology, Institute of Electrical Engineering and Electronics, Piotrowo 3A, 60 – 965 Poznań, POLAND
[2] Regional Verification Office, Electrical Quantities Measurement Laboratory, Krakowska 19, 61-893 Poznań, POLAND

E-mail: otoman@et.put.poznan.pl

Abstract. The paper presents a method for evaluating the accuracy of indications of the direct current Watt-hour meters in a designed and constructed measuring system. Such measuring system is composed of two multi-function calibrators, and a specialised high-voltage attachment dedicated for this system, which makes it possible to generate direct voltages in a required range up to 4 kV with a suitably high precision. The authors described in detail particular elements of the measuring system together with the results of its calibration. They showed the results of experiments carried out on a large and representative number of LE3000plus Watt-hour meter items.

1. Introduction

The evaluation of metrological parameters of devices dedicated for settlements among enterprises and market contributors is a complex measurement problem. In practice, calibration is frequently assumed as the proper form of such evaluation. The values obtained as result of calibration, associated with the uncertainty of determining them, should be defined in specified measuring points, according to respective standards and other supplementary documents. The paper presents a method for evaluating the parameters of DC energy meter LE3000plus. Meters of that type find application in measuring the power and energy of direct current in the railway traction network. The currently used meters are modern measurement devices, usually microprocessors, equipped with an additional measurement interface that allows the remote transmission of registered measurement results.

The evaluation of meter will be in this paper focused on the determination of the accuracy of measuring the voltage, current, and import/export energy of the supply network.

The evaluated Watt-hour meter of type LE3000plus consists of: high-voltage, measuring part LE3000plus_HVM, low-voltage, communication part - LE3000plus_KOM, measuring shunt, fiber optic cable connecting the high-voltage and low-voltage parts, GSM/GPS antenna, together with antenna wires. A detailed description of all the functions and potential of this meter is included in [1].

2. Measuring system for the evaluation of the metrological properties of Watt-hour meters

A measuring system was prepared for calibrating the evaluated meter whose construction was described in the previous chapter. It has a direct current supply in a range of 0 V - 4 kV, and a signal

---

[1] To whom any correspondence should be addressed.

of current shunt corresponding to the flow of a current with values between a few hundred and over a thousand amperes.
The measuring system consists of two multi-function calibrators, and a specialised high-voltage attachment dedicated for this system, which allows the generation of direct voltages in a demanded range of up to 4 kV with suitably high precision.
Fig. 1 presents a diagram of the prepared measuring system for the testing of direct current Watt-hour meters.

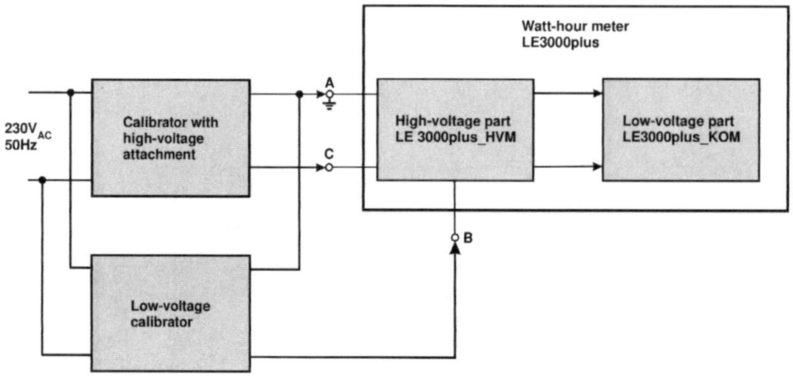

Figure 1. Diagram of measuring system for testing the direct current Watt-hour meters.

The signal between terminals A and B corresponds, in working conditions, to the voltage drop on the external measurement shunt, and the signal between terminals A and C is the voltage corresponding actually to a high voltage in electrical traction. In the developed measuring system voltage U_{AB} attained small values, and therefore the costs of insulation between internal blocks of the measuring system were minimized. Terminal C of the developed measuring system is located on a high minus potential, of the order of 4 kV. Proper configuration and insulation of this terminal guaranteed safe operation of the whole measuring system. The developed measuring system provides appropriate metrological parameters, minimum energy consumption from the supply network, which is relevant in view of the research costs, as well as adequate separation of components.
The determined relative value of expanded uncertainty of this part of the measuring system equals 0.06 % and it is stated in the current calibration certificate. In the original version of the system, during the determining of electrical energy values, time was measured with mechanical stopwatch. At the measurement of time of the order of 30 min. with such stopwatch, the value of expanded uncertainty at a level of 0.12 s was achieved. A detailed description of these problems can be found in work [2].
Essential modifications were introduced in the present version of measuring system, which enabled the energy measurement inaccuracies to be significantly limited. The modifications allowed the error related to the reaction time of the person conducting the measurements to be eliminated. High voltage obtained from the voltage calibrator with attachment is connected to the investigated meter for the whole duration of measurements. During normal work of a meter in railway traction network the load current is measured by the meter indirectly, through a voltage drop on the external shunt. In the measuring system the load current measured by the meter is simulated by the low-voltage calibrator, the output signal of which simulates the voltage drop from the shunt. The calibrator is equipped with a precise time distributor with a range of settings 1 ÷ 85 min. It generates a signal during a strictly specified time interval; in standard version it is a 15-minute interval. The construction of this calibrator makes it possible to generate an impulse with the longest possible rise time, which is identical to the shortest possible time of switching the impulse on and off. This calibrator, together with the time distributor that is its integral part, was calibrated in order to evaluate the inaccuracies of time and voltage measurement.

The results of calibrating the time distributor were compared to the state standard of units of time and frequency measurement. For this purpose, a digital frequency-time meter of type 53132A was used as the control device, synchronized with the model frequency from the state standard of units of time and frequency measurement. The low-voltage calibrator was tested in lab with an ambient temperature of (21.9 ÷ 23.0) °C and relative humidity (38.3 ÷ 41.3) %. The calibration results were compared to the state standard of units of direct electrical voltage measurement by using a digital multimeter. The measurement uncertainties were determined according to the recommendations in document [3].
The calibrator with high-voltage attachment as well as the low-voltage calibrator work with an inner feedback loop. Therefore, the output signals of both devices are stable in time. The value of expanded uncertainty of time measurement in this measuring system, compared to the measuring method with a mechanical stopwatch, is over fifty times smaller.
A detailed description of these problems can be found in work [4].

3. Experimental research

In the measuring system presented in Fig. 1, the Watt-hour meters of direct current LE 3000plus were calibrated. During this process, the indications of the same meter were compared with the settings on control devices. The readings were made from the display of the very meter, with resolution equal to 0.1 kWh. Because, in case of meters installed in electrical locomotives, the energy imported from and exported to the network is in standard version registered in 15-minute records, it is purpose-full and essential that also the measurement time of the meter's parameters equals 15 minutes. Therefore, it is possible to compare the results displayed on the meter with the values set on the measuring devices.

3.1. Voltage measurements

The meter under evaluation was installed in the measuring system – its diagram is shown in Fig. 1. The tests were carried out for nominal voltage U_n = 3300 V, and also for voltages differing from the nominal voltage by ± 10 %. For each of these voltages the measurements were repeated many times. The measurements were carried out in lab with an ambient temperature of (22.5 ÷ 22.9) °C and relative humidity (41.1 ÷ 45.3) %. Examples of the results achieved are shown in Table 1.

Table 1. Results of calibrating LE3000plus meter at voltage measurements

U_n = 3300 V								
Test point U = 2970 V			Test point U = 3300 V			Test point U = 3630 V		
W_{wsk}	W_{zad}	Δ	W_{wsk}	W_{zad}	Δ	W_{wsk}	W_{zad}	Δ
V	V	V	V	V	V	V	V	V
2970	-2971.00	-1.00	3300	-3300.80	-0.80	3630	-3630.80	-0.80
2970	-2971.20	-1.20	3300	-3301.00	-1.00	3630	-3631.00	-1.00
2970	-2971.00	-1.00	3300	-3300.60	-0.60	3630	-3630.60	-0.60
2970	-2971.20	-1.20	3300	-3301.00	-1.00	3630	-3631.00	-1.00
2970	-2971.00	-1.00	3300	-3300.60	-0.60	3630	-3630.80	-0.80
2970	-2971.20	-1.20	3300	-3301.00	-1.00	3630	-3630.80	-0.80
2970	-2971.00	-1.00	3300	-3300.60	-0.60	3630	-3630.60	-0.60
2970	-2971.20	-1.20	3300	-3301.00	-1.00	3630	-3630.80	-0.80
2970	-2971.00	-1.00	3300	-3300.60	-0.60	3630	-3630.60	-0.60
2970	-2971.20	-1.20	3300	-3300.80	-0.80	3630	-3630.80	-0.80

Correct values W_{zad} of voltage measurement were given according to the setting adopted on the control equipment. Because according to the diagram shown in Fig. 1, the system works with the so-called "inverted mass", in this case the indication error Δ takes a form according to equation (1):

$$\Delta = W_{wsk} - (W_{zad}) \tag{1}$$

where W_{wsk} is the value indicated by the evaluated meter.

3.2. Current measurements

Similar meter evaluation was carried out with regard to current changes. The measurements were done in seven test points for the following current values: 50 A, 100 A, 250 A, 500 A, 800 A, 1000 A, and 1500 A, and the value of nominal current for this meter was I_n = 1000 A. For each of these currents, the measurements were repeated many times. The measurements were carried out in lab at the same values of temperature and humidity variation as in the case of voltage measurements. Examples of test results achieved during the measurements for 3 test points are presented in Table 2. Similarly as for the voltage measurements, the value of current indication error Δ was determined according to equation (1).

Table 2. Results of calibrating LE3000plus meter at current measurements

I_n = 1000 A								
Test point I = 50 A			Test point I = 1000 A			Test point I = 1500 A		
W_{wsk}	W_{zad}	Δ	W_{wsk}	W_{zad}	Δ	W_{wsk}	W_{zad}	Δ
A	A	A	A	A	A	A	A	A
49.9	-50.0	-0.1	999.4	-1000.0	-0.6	1498.9	-1500.0	-1.1
49.9	-50.0	-0.1	999.5	-1000.0	-0.5	1498.8	-1500.0	-1.2
49.9	-50.0	-0.1	999.4	-1000.0	-0.6	1498.9	-1500.0	-1.1
49.9	-50.0	-0.1	999.4	-1000.0	-0.6	1498.9	-1500.0	-1.1
49.9	-50.0	-0.1	999.5	-1000.0	-0.5	1498.9	-1500.0	-1.1
49.8	-50.0	-0.2	999.4	-1000.0	-0.6	1498.8	-1500.0	-1.2
49.9	-50.0	-0.1	999.4	-1000.0	-0.6	1498.9	-1500.0	-1.1
49.8	-50.0	-0.2	999.5	-1000.0	-0.5	1498.9	-1500.0	-1.1
49.9	-50.0	-0.1	999.4	-1000.0	-0.6	1498.9	-1500.0	-1.1
50.0	-50.0	0.0	999.5	-1000.0	-0.5	1498.8	-1500.0	-1.2

4. Uncertainty budget

Final measurement result is complete only when it contains both the measurand value and the measurement uncertainty attributed to that value. According to document [5], measurement uncertainty is a non-negative parameter characterizing the dispersion of the quantity values, being attributed to the measurand, calculated on the basis of information used. Such parameter may be e.g. standard deviation, called standard measurement uncertainty (or its specified multiple), or a half of the width of an interval, having a specified extension probability. Usually measurement uncertainty comprises a lot of components. The values of some of them can be determined with Type A evaluation of measurement uncertainty, which consists in determining the value of a component of measurement uncertainty by means of the statistical analysis of the measurand values, obtained in defined measurement conditions.

According to document [6] – with the assumption that the best estimation of the measurand value, defined on the basis of a series of n observations, is arithmetic mean $\bar{x}$ – the value of the standard deviation estimator for the mean is equal to the standard uncertainty determined with Type A evaluation and given with equation (2):

$$s(\bar{x}) = u_A = \sqrt{\frac{1}{n(n-1)}\sum_{i=1}^{n}(x_i - \bar{x})^2}. \tag{2}$$

Other components, with values determined with Type B evaluation of measurement uncertainty, can also be characterized by means of standard deviation, determined with the probability density function based on experience or other information. It consists in determining the value of a component of measurement uncertainty by means of instruments different from those applied in Type A evaluation [5]. There is not one stated algorithm for calculating standard uncertainty with Type B evaluation. A typical situation encountered in metrological practice is when we know only the variation limits of the values around a given result, e.g. the limiting error of measuring equipment Δ_g. It is possible to assume, then, that a value unbiased with the rounding error in each point of the rounding interval is equally probable, i.e. uniform distribution can be assigned to this interval. The above assumption on uniform distribution is rather safe because for such distribution, we obtain the biggest uncertainty value from among all symmetrical, limited, unimodal distributions, with the probability density function decreasing at the ends of the domain.

Statement of all identified component standard uncertainties estimated with A and B Type methods is called uncertainty budget. The aim of creating the uncertainty budget is to prove that the combined uncertainty of measurement result u_c, was estimated in a matter-of-fact, penetrating and verifiable way. With a correctly constructed budget, we can prove that all elements of uncertainty have been analysed, also those that were not the greatest parts of the combined uncertainty. It means that no element was rejected arbitrarily, based on believes or traditional manner of conduct. Moreover, no assumption was made that some element was negligibly small. Thanks to the analysis of uncertainty budget, we can answer the question: which component uncertainties are the greatest parts of the combined uncertainty? We can also indicate directions for improving the measurement, which could lead to significant decrease in the combined uncertainty. On that account, all the component uncertainties taken into consideration, also those with the value estimated as zero, should be listed in the budget. It is essential to assign a suitable type of probability distribution to particular component uncertainties in the uncertainty budget.

For the measurement results presented in Tables 1 and 2, the uncertainty budgets were prepared according to the rule described above.

Below, in Table 3, the uncertainty budget at the voltage measurement for test point U = 3300 V is presented.

Table 3. Uncertainty budget of LE3000plus meter for evaluation of selected voltage measurement

Quantity symbol	Quantity estimate	Standard uncertainty $u(x_i)$	Probability distribution	Sensitivity coefficient c_i	Part of combined uncertainty
U_{wsk}-U_{zad}	-1.1	3.33E-02	Normal	1	3.33E-02
$\Delta\delta_U$	2.13 E-01	2.00E-02	Normal	- 1	2.00E-02
$\delta\Delta_{rU}$	0	2.89E-01	Rectangular	1	2.89E-01
Δ_U	-1.313	-	-	-	2.91E-01

The quantities influencing the value of combined uncertainty, put in the above table, denote respectively:

- U_{wsk}-U_{zad} – difference of indicated value and given value. Because for this factor normal distribution was assumed, uncertainty was determined with A Type method,
- $\Delta\delta_U$ – control device error, read out from a current calibration certificate of calibrator with high-voltage attachment. To this factor also normal probability distribution was assigned,
- $\delta\Delta_{rU}$ – correction resulting from the calibrated meter's resolution. To this factor rectangular distribution was assigned, therefore uncertainty was determined with B Type method,
- Δ_U – factor determined according to equation $(U_{wsk} - U_{zad}) - \Delta\delta_U + \delta\Delta_{rU}$.

Table 4 shows the determined values of expanded uncertainty U, for the results of voltage measurements in Table 1.

Table 4. Results of calibrating voltage measurements for nominal voltage U_n = 3300 V

Voltage measurement (U_n = 3300 V)			
Indicated value U_{wsk}	Correct value	Indication error Δ_U	Measurement uncertainty U
V	V	V	V
2970	-2971.31	-1.3	0.6
3300	-3301.34	-1.3	0.6
3630	-3631.72	-1.7	0.6

For the current measurements contained in Table 2, the uncertainty budget was prepared in a way similar as for voltage measurements. The results for all test points are shown in Table 5.

Table 5. Results of calibrating current measurements for nominal current I_n = 1000 A

Current measurement (I_n = 1000 A)			
Indicated value I_{wsk}	Correct value	Indication error Δ_I	Measurement uncertainty U
A	A	A	A
49.9	-50.00	-0.11	0.09
99.8	-100.00	-0.16	0.09
249.7	-250.00	-0.34	0.11
499.8	-500.00	-0.18	0.11
799.5	-800.00	-0.47	0.12
999.4	-1000.00	-0.56	0.12
1498.9	-1500.00	-1.13	0.14

5. Conclusion

The authors of the paper attempted to discuss complex problems of experimental research on Watt-hour meters for voltage of the order of 4 kV. A designed and constructed measuring system was presented, consisting of two multi-function calibrators, a specialised high-voltage attachment dedicated for this system, in order to generate direct voltages in a range of up to 4 kV. The publication contains the results of experiments carried out on a numerous and representative set of direct current Watt-hour meters LE3000plus items, in the conditions of accredited laboratory. The testing of so large population of devices permitted the authors to draw reliable and objective conclusions. When selecting the measurement points of a meter tested in the measuring system, the authors took into account the conclusions drawn from the testing of meters in working conditions, during their normal work.

6. References

[1] Operational and Maintenance Manual 2011 Direct current Watt-hour meter 3 kV LE3000plus
[2] Otomański P Zazula P 2008 Metrological characteristics of measuring position for direct current Watt-hour meter research in range to 4 kV Electrical Review 2008 315-317
[3] EA-4/02 1999 Expression of the Uncertainty of Measurement in Calibration European co-operation for Accreditation
[4] Otomański P Zazula P The measuring system for the evaluation of the metrological properties of direct current Watt-hour meters Measurement Automation and Monitoring 2012 764-766
[5] JCGM 2008 *International vocabulary of metrology – Basic and general concepts and associated terms (VIM)* JCGM 200:2008
[6] JCGM 2008 *Evaluation of measurement data – Guide to the expression of uncertainty in measurement* JCGM 100:2008

New Concept of Laboratory Exercise on Temperature Measurements Using Thermocouple

Oldřich Slavata[1,2], **Pavel Souček**[1], **Jan Holub**[1,2]

[1] Czech Technical University, Faculty of Electrical Engineering, Dept. of Measurement K13138, Technicka 2, 166 27 PRAGUE 6, Czech Republic

[2] Czech Technical University in Prague, University Center for Energy Efficient Buildings, Nam. Sitna 3105, Kladno, Czech Republic

slavaold@fel.cvut.cz

Abstract: The proposed exercise is focused on the measurement of temperature using thermocouple, students acquire theoretical knowledge of the Seebeck effect, the design and application of thermocouples including the differences between their various types. The students measure the voltage at the thermocouple by various methods: directly with a compensation box, using operational amplifiers and USB module. The exercise explains also general principles of low voltage measurements, error compensation and uncertainty evaluation.

1. Introduction

1.1. Exercise positioning

Laboratory exercise thermocouple, measuring its voltage, junction temperature compensation is part of Labs on Sensors and Measurement. The course is compulsory for students of disciplines: Robotics, Sensors and Instrumentation and Control Systems and a Bachelor's degree Cybernetics and Robotics. Students of other disciplines may enroll as an optional subject.

1.2. Exercise history

Scope of Sensors and Measurements follows the past subject Sensors and Transducers. One of its laboratory exercises used to be the temperature measurement using thermocouples [4]. In this task, the students were asked to measure the temperature in a furnace in the range of 20-100 ° C using a K-type thermocouple. One thermocouple was connected to a transducer 101 XTR manufacturer Burr-Brown (USA), the converter converted the thermocouple voltage to current in the range of 4 - 18 mA. The task of the students was to suggest the resistor values so that the output current range corresponded to the range of measured temperatures. The second thermocouple was connected to the measuring module Janascard AD 232, module containing 18bit integrating AD converter and isothermal terminal block, the data were transmitted to a computer via RS232 bus.

For the needs of a new course the laboratory exercise was substantially upgraded with emphasis on measurement of small voltages (tens of μV) and on the uncertainty calculation of such measurements. Also, multiple thermocouple types have been added.

2. Exercise description

The task of the students in this exercise is to measure the voltage at the thermocouple by different methods and to evaluate the accuracy and uncertainty of these measurements. List of thermocouples is given in Table 1.

Table. 1 List of thermocouples

Type	Sensitivity [μV/°C]	Temperature range	Material	Attachment
E	68	-100 - +1000	Chromel/Constantan	screw
J	55	0 - +760	Iron/Constatntan	sticker
K	40,8	-200 - +1370	Chromel/Alumel	screw
T	43	-160 - +400	Copper/Constantan	magnetic

The advantage over the previously taught exercise [4] is the use of multiple types of thermocouples with different mountings to the object to be measured. In similar tasks taught at the University at Buffalo [5], or the Polytechnic Institute of New York University [7] only one type of thermocouple is used. A similar exercise with multiple types of thermocouples is taught in Portland State University [6].

The measured object is a table lamp shade in temperature range between 0-70 ° C.

2.1. Measurements using a digital multimeter

In this part of the task, students will gain awareness of the size of the output voltage of the thermocouple and the size and speed of change of output voltage with temperature changes. This is an advantage compared to tasks that use only measurements using the AD converter with subsequent data processing machine [5, 6]. Procedures for measuring low voltage that students learn are quite universal and applicable in other structures than the thermocouple measurements. A similar task using basic measuring instruments is taught at the Faculty of Engineering and Technology, University of Jodan [8].

2.1.1. Direct measurement. Task is to measure the voltage at the thermocouple directly using a multimeter manual Mastech MY-64 and laboratory Agilent U3401A multimeter. Wiring diagrams is shown in Figure 1. The junction temperature (the temperature in the laboratory) measured by means of bimetal thermometer must be taken into account when calculating the temperature.

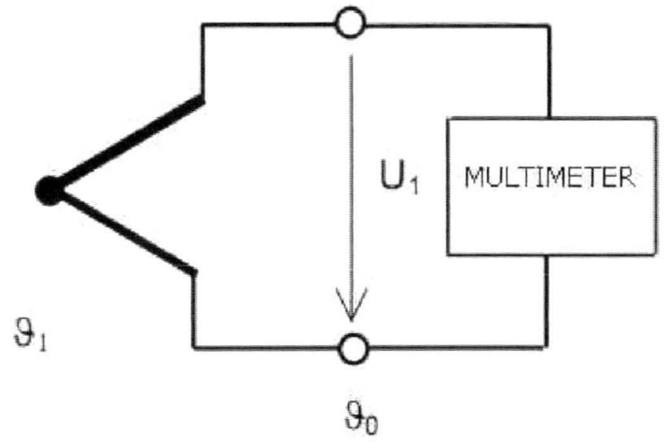

Figure 1. Wiring diagram of direct measurement

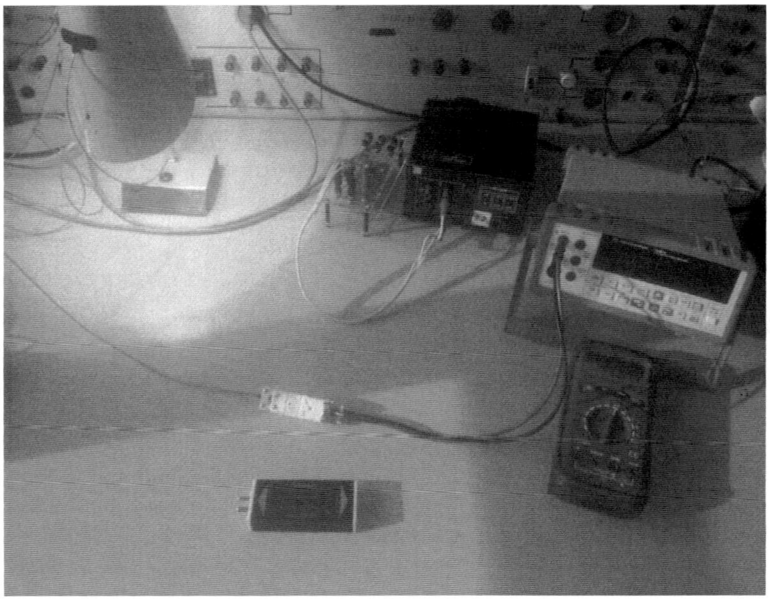

Figure 2. Direct measurement of the output voltage of the thermocouple with optional compensation box OMEGA CJ

2.1.2. Measurement with compensation box OMEGA CJ. Previous measurements are repeated but involving compensation box OMEGA CJ where the thermocouple output voltage corresponds to the comparator at the end of the temperature of 0 ° C. The diagram is shown in Figure 3.

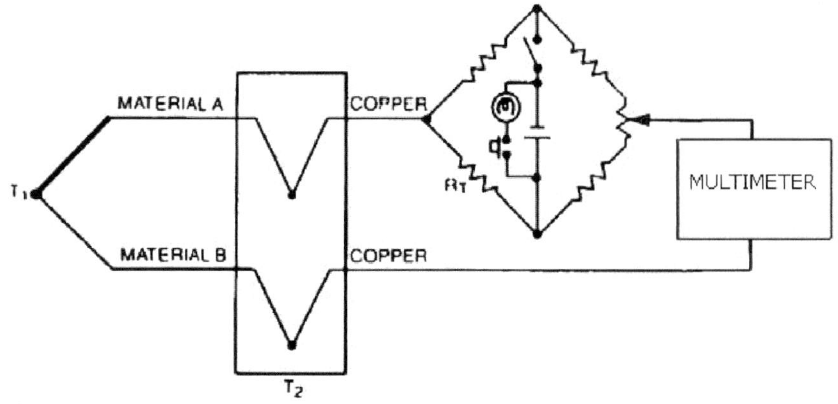

Figure 3. Wiring diagram of measurement with compensation box OMEGA CJ

2.2. *Measurements using an operational amplifier*

The student task in this part of the exercise is to design a circuit with operational amplifier (OA) that amplifies the thermocouple output voltage by the factor of 100. The circuit input resistance must take into account when designing the circuit type and feedback resistors value. The value of the feedback resistor also influences a thermal noise. Students will then verify and deploy their knowledge of other subjects. In comparison to converter U / I 101 XTR used in previously exercised role [4], the circuit with operational amplifier is easier for students to understand. The diagram is shown in Figure 3. The OA offset can be determined by commutating the thermocouple amplifier and compensated consequently.

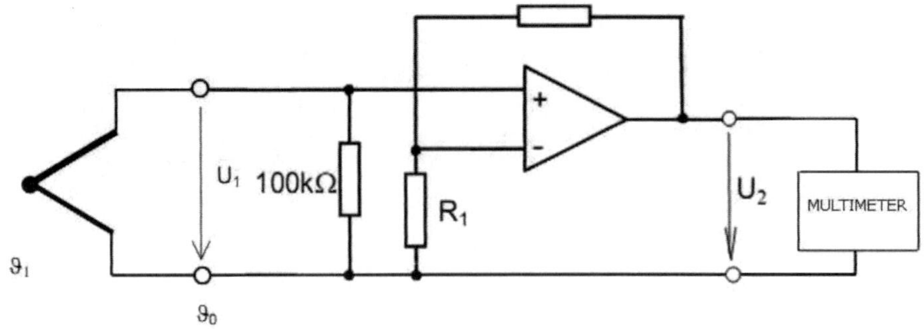

Figure 4. Wiring diagram of measurement with operational amplifier

2.3. *Measurement with USB module with isothermal terminal block*

Automatic measurement and data acquisition with post-processing on the computer plays an irreplaceable role in contemporary measurements. Usually, the part of the temperature measurement task is a measuring module connected to the computer via the bus. The measured data is then processed in Matlab [7], or other specialized software [5, 6].

New laboratory exercise replaced the outdated module Janascard AD 232 by Janascard AD24USB module (Figure 5.) that contains 24bit AD converter with 15nVpp noise which connects isothermal terminal block of 8 differential inputs with a range of 0 - 10V and semiconductor temperature sensor. The module is connected to the computer via the USB bus. Temperature compensation and linearization is performed in software program AD24control (figure 6.). The program allows you to set parameters of the connected sensor and different measurement and display of data.

The task of the students in this section is to record and display the temperature graphically on the screen when the lamp bulb is on or off. Measurements are repeated for different thermocouple types, and different locations on the lamp. Unlike [5, 8], students are not required to determine time constants of the thermocouples as it is part of the instruction in other subjects.

Figure 5. USB module with isothermal terminal block

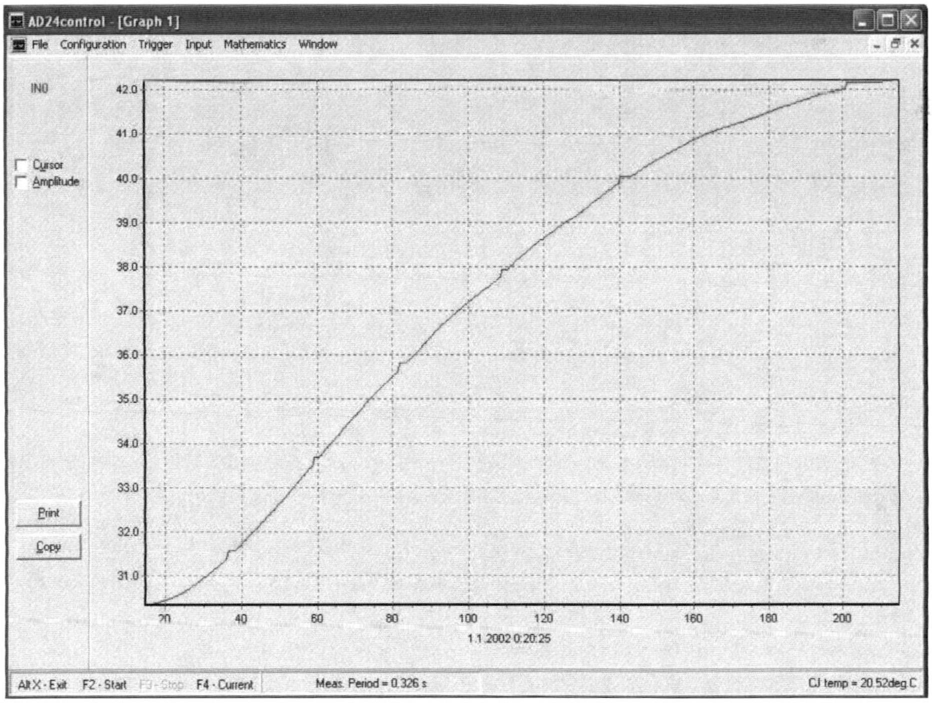

Figure 6. Software program AD24control

3. Summary

The task of the students in the exercise is to measure the thermocouple voltage using different methods and based on that, to calculate the object temperature. In the measurement protocol students are asked to compare different methods in terms of accuracy and to calculate the measurement uncertainty. The students make themselves acquaint with the theory of the Seebeck effect, the temperature dependence of the Seebeck coefficient and the basic compensation procedures. In comparison with the previous exercise version, methods of small voltage measurements and problems associated with this measurement are emphasized. The recap of operational amplifier and related feedback theory is another benefit of this exercise.

4. References

[1] Ripka, P., Ďaďo, S., Kreidl, M., Novák, J.: Senzory (opr. dotisk). Skripta ČVUT FEL, Praha 2007

[2] Haasz, V., Sedláček, M.: Elektrická měření. Přístroje a metody. ČVUT, Praha 2003

[3] http://measure.feld.cvut.cz/en/education/courses/AE3B38SME

[4] http://measure.feld.cvut.cz/cs/system/files/files/cs/vyuka/predmety/x38szp/uloha5.pdf

[5] http://www.eng.buffalo.edu/Classes/mae334/labs/Lab1.pdf

[6] http://web.cecs.pdx.edu/~wernc/ME411W13/files/Experiment%201%20Thermocouples%20W13.pdf

[7] http://mechanical.poly.edu/faculty/slee/ME3511/Exp4.pdf

[8] http://www2.ju.edu.jo/faculties/FacultyofEngineering/Mechatronics/Documents/Measurements%20and%20Control%20Lab/Exp5%20Thermocouple.pdf

IMEKO 2013 TC1 + TC7 + TC13 IOP Publishing
Journal of Physics: Conference Series **459** (2013) 012060 doi:10.1088/1742-6596/459/1/012060

Selection of excitation signals for high-impedance spectroscopy

Michal Kowalewski

Gdansk University of Technology, Faculty of Electronics, Telecommunications and Informatics, Department of Metrology and Optoelectronics, Gdansk, Poland

E-mail: michal.kowalewski@eti.pg.gda.pl

Abstract. The method of fast impedance spectroscopy of technical objects with high impedance ($|Zx| > 1$ GΩ) is evaluated in this paper. An object is excited with a signal generated by digital-to-analog converter (DAC) located on the U2531A DAQ module. Response signals proportional to current flowing through and voltage across the measured object are sampled by analog-to-digital converters (ADC) in the DAQ module. The object impedance spectrum is obtained with use of continuous Fourier transform on the basis of acquired signals. Different excitation signals (square, triangle, sawtooth and sinc) were compared in order to estimate accuracy of impedance spectrum evaluation. The method is evaluated by means of simulation and practical experiment in a measuring system on an exemplary object in the form of multi-element two-terminal RC network modelling of an anticorrosion coating impedance.

1. Introduction

Diagnosis of technical [1] and biological [2] objects and materials [3] with use of impedance spectroscopy is still an open research field. This is due to a fact, that there is no universal method capable to diagnose different objects in comparable, short time and also with high accuracy. Usually higher accuracy of impedance spectroscopy requires more time.

Conventional measurement technique used for impedance spectroscopy is based on single sine technique (SST). An object is excited with a harmonic signal and two signals are measured: voltage across and current through the object. Measurements are repeated for different frequencies in order to obtain impedance spectrum in several decades. For high impedance objects ($|Zx| > 1$ GΩ), e.g. high-thickness anticorrosion coatings or dielectric materials, this leads to very long measurement times (of an order of hours), because measurements must be performed for very low frequencies (1 μHz, 1 mHz). This kind of diagnosis is acceptable only in laboratory conditions and can not be used directly in the field, e.g. on bridges, pipelines or flood banks.

In order to shorten time of measurements different methods were proposed. These methods use multi-sine technique [4, 5] or other measurement signals [6-10]. A promising technique depends on usage of a square pulse for excitation of an object and measuring signals proportional to voltage across and current through the object with ADCs [11]. The object impedance spectrum is then obtained with use of continuous Fourier transform on the basis of linear approximations between samples of discrete response signals. Simulations and experimental research of this method had been performed in [11, 12], but in this paper the method is evaluated by means of different excitation signals (square, triangle, sawtooth and sinc). Accuracy of impedance spectrum evaluation for exemplary model of high impedance anticorrosion coating is compared for each excitation.

2. Description of the method

In the presented method a measured object Z_x is excited with an arbitrary waveform $u_o(t)$ and signals proportional to voltage $u_x(t)$ across and current $i_x(t)$ through the object are measured. The excitation signal is generated by DAQ module U2531A and applied to the object from DAC through A_{out}. The programmed resistor R_L is used at the output of the amplifier A_1 in order to limit the maximum value of current flowing through the object. Response signals $i_x(t)$ and $u_x(t)$ are converted to signals $u_1(t)$ and $u_2(t)$ and applied to ADCs in the DAQ module through A_{in1} and A_{in2}. Current $i_x(t)$ is converted to voltage u_1(t) by current-to-voltage converter (CVC) realized with amplifier A3. Range of the CVC can be changed with resistor R_R in order to fit the measured signal u_1(t) to the range of the ADC.

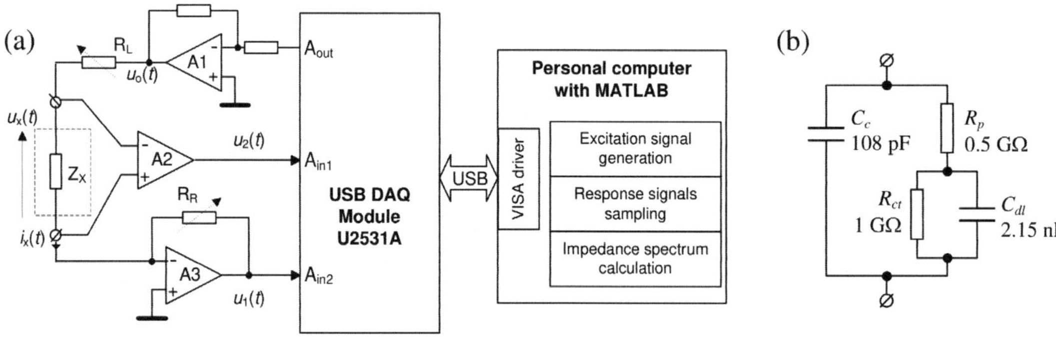

Figure 1. Block diagram of the measuring system (a), the equivalent circuit of anticorrosion coating impedance (b).

Impedance spectrum of the object is calculated from the formula

$$Z(\omega)=\frac{\mathrm{Re}U_2(\omega)+j\,\mathrm{Im}U_2(\omega)}{\mathrm{Re}U_1(\omega)+j\,\mathrm{Im}U_1(\omega)}R_R\,, \tag{1}$$

where $U_1(\omega)$ and $U_2(\omega)$ are continuous Fourier transforms of signals $u_1(t)$ and $u_2(t)$ obtained as follows

$$U_i(j\omega)\approx\sum_{n=1}^{N-1}\int_{t_n}^{t_{n+1}}\tilde{u}_i(t)\exp(-j\omega t)\,dt\,,\quad i=1,2. \tag{2}$$

Response signals $u_1(t)$ and $u_2(t)$ are discrete, while continuous Fourier transform requires continuous signals, hence linear approximations $\tilde{u}_i(t)$ between samples are used in (2).

The method is useful for diagnosis of high-impedance objects, eg. anticorrosion coatings. An example of the equivalent circuit of anticorrosion coating is shown in figure 1b.

In the previous research of the method a unipolar square pulse signal was used as an excitation. However it is interesting to investigate efficiency of the method, when measured object is excited with different arbitrary waveform. The following signals were chosen for comparison: bipolar square, triangle, sawtooth and sinc (figure 2a).

Important problem during selection of an excitation signal is analysis of its spectrum (figure 2b). A shape of the spectrum determines allowable range of measurement frequencies and accuracy of object's impedance calculations for each frequency. First experiment [11] assumed usage of a unipolar square pulse. Spectrum modulus $|U_0|$ for this signal is given by a function of type $\sin(x)/x$, where $x=\pi\cdot k\cdot\tau/T$ (k - spectral line number, τ - square pulse duration, T - acquisition time). If the argument x reaches the value $m\cdot\pi$, then modulus $|U_0|$ equals zero. For this excitation moduli $|U_1|$ and $|U_2|$ of signals $u_1(t)$ and $u_2(t)$ are also given by a function of type $\sin(x)/x$ and it is unable to obtain impedance of an object for frequencies k/τ. Also near these frequencies very high impedance calculation errors are obtained. The limitation of the method was minimized in [12] by division of measurement frequencies in the required range from f_{min} to f_{max} ($f_{min}<1/\tau<f_{max}$) into 2 groups:

- logarithmically spaced frequencies within the range ($f_{\min}$, $0.75/\tau$),
- linearly spaced frequencies ($1.5/\tau$, $2.5/\tau$, …, $f_{\max}$).

For the object shown in figure 1b optimal value of unipolar square pulse duration equals $\tau = 5$ s [11], hence the following set of 15 measurement frequencies placed within 3 decades from $f_{\min} = 1$ mHz to $f_{\max} = 1$ Hz were selected: 1 mHz, 1.7 mHz, 2.7 mHz, 4.5 mHz, 7.4 mHz, 12.2 mHz, 20.2 mHz, 33.4 mHz, 55.1 mHz, 90.9 mHz, 0.15 Hz, 0.3 Hz, 0.5 Hz, 0.7 Hz, 0.9 Hz.

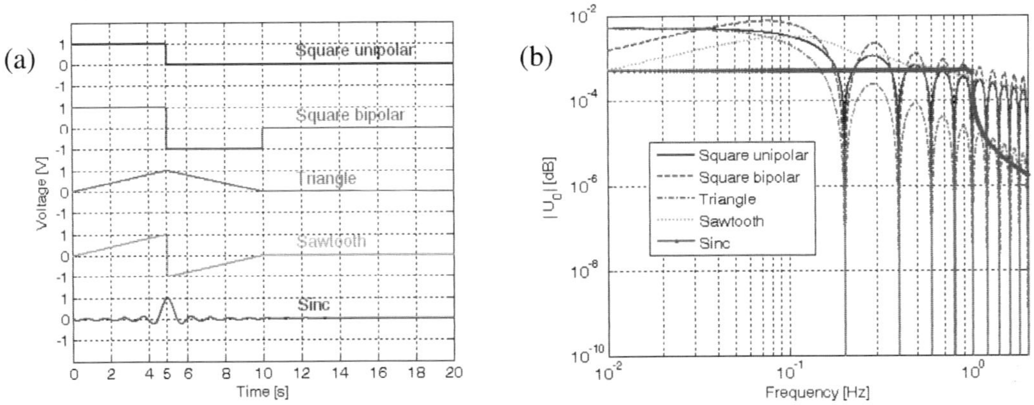

Figure 2. Excitation signals (a) and their spectra (b).

As is shown in figure 2b, the same limitation of the method occurs for triangle excitation, while for sawtooth and sinc signals modulus $|U_0|$ is non-zero for all frequencies in the assumed range and there is no limitation in measurement frequencies selection. However for comparison purposes the same set of measurement frequencies previously obtained for unipolar square pulse was used in simulations.

3. Simulation results

Efficiency of the method of high impedance spectrum calculation for different excitation signals was verified on the model of anticorrosion coating shown in figure 1b. The following parameters of the method were selected: constant sampling frequency ($f_s = 10$ kHz), $T = 100$ s, resolution of ADC (14-bits). Excitation signals with parameters shown graphically in figure 2a were used. Obtained impedance modulus and argument calculation errors are presented in figure 3.

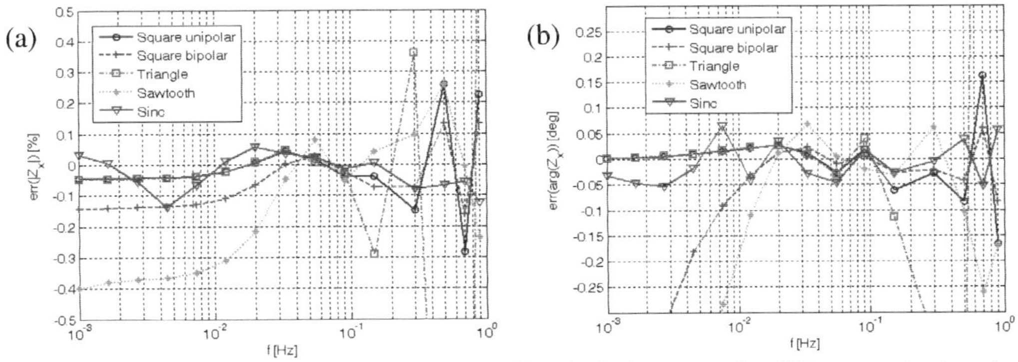

Figure 3. Impedance modulus (a) and argument (b) calculation errors for different excitation signals.

The replacement of an unipolar square excitation with its bipolar form insignificantly improve accuracy of impedance calculations for higher frequencies due to greater values of modulus of bipolar square signal, but worsen accuracy for lower frequencies, especially for impedance argument. Errors increased up to ± 1 °.

Results obtained for triangle and unipolar square excitations are the same for lower frequencies. This is comprehensible, because in this case moduli of both signals have the same values. However for higher frequencies modulus of triangle excitation has significantly lower values, which causes increase of impedance spectrum calculation errors up to ± 3 % for modulus and ± 3 ° for argument.

Efficiency of usage of sawtooth excitation is in the middle between square pulse and triangle. Maximum calculation errors are contained in ranges ± 0.4 % for modulus and ± 2.6 ° for argument.

The best results were obtained for sinc excitation. Due to constant value of modulus of that signal in the whole range of measurement frequencies, dispersion of impedance calculation errors is uniform and changes in the range ± 0.14 % for modulus and ± 0.06 ° for argument. Since usage of a unipolar square pulse gives better results only for lower frequencies, maximum errors for frequencies closer to 1 Hz grow for that signal up to ± 0.3 % for modulus and ± 0.17 ° for argument. Important benefit of sinc excitation also results from the fact that it doesn't have steep edges and there is no need to use irregular sampling method (higher sampling frequencies just after steep edge of an excitation), as in case of square pulse signal. The problem of irregular sampling was discussed in [11, 12].

4. Optimization of the method for sinc excitation

A characteristic feature of the sinc function is that its continuous Fourier transform is a rectangular function. In order to guarantee constant modulus value of that function for measurement frequencies up to $f_{\max}$ a following formula can be used

$$u_0(t) = U_0 \operatorname{sinc}\left(2\pi f_{\max}\left(t-\tau\right)\right), \tag{3}$$

where: U_0 - maximum amplitude, τ - time shift, $f_{\max}$ – maximum frequency in impedance spectrum.

It was assumed that U_0 = 1 V and $\tau = T/2$, hence maximum value of $u_0(t)$ occurs in the middle of acquisition time $u_0(\tau)$ = 1 V.

Important point is determining proper value of T, because starting generation of sinc signal for $T \neq kf_{\max}^{-1}$, (k = 1, 3, 5, ...) causes occurrence of violent changes in values of signals $u_0(t)$ and $u_1(t)$ at the beginning of the sampling process (t = 0). To omit necessity of usage of high sampling rates it is justified to select optimal value of T from the range $T = kf_{\max}^{-1}$.

A simulation focused on selection of optimal value of T for the model of anticorrosion coating was performed, which indicated that maximum calculation errors decrease below ± 1 % for modulus and ± 1 ° for argument when T > 17 s. Obtained calculation errors are presented in figure 4.

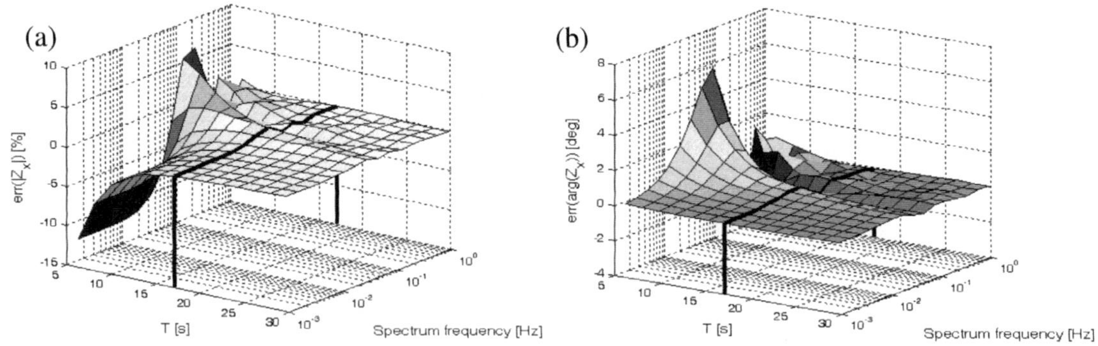

Figure 4. An influence of acquisition time T on impedance modulus (a) and argument (b) calculation errors.

Simulations with use of sinc excitation indicated also that constant sampling frequency can be used and influence of its value on the accuracy of the method is negligible (with assumption that at least 1000 samples are acquired).

Journal of Physics: Conference Series **459** (2013) 012060 doi:10.1088/1742-6596/459/1/012060

5. Experimental results

Measuring system shown in figure 1a was used to supply sinc excitation and obtain impedance spectrum of the model of anticorrosion coating in the form of 4-element RC network shown in figure 1b. In the first step a following values of parameters of the method were assumed: $U_0 = 1$ V, $\tau = 8.5$ s, $T = 17$ s, $f_s = 10$ kHz, $f_{min} = 1$ mHz, $f_{max} = 1$ Hz, which gave significant errors (several times greater than in simulations, especially for low frequencies). It was caused by omitting condition $T \geq f_{min}^{-1}$ required to fulfill in order to evaluate impedance spectrum from frequency f_{min}. Hence, in the second step it was assumed that $T = 1000$ s. In that case impedance spectrum calculation errors decreased insignificantly, because of very high dynamic range of signal $u_0(t)$ - response signals $u_1(t)$ and $u_2(t)$ were measurable only in time range $|t - \tau| < 20$ s.

Performed experiments led to the conclusion that in order to significantly increase accuracy of impedance spectrum calculations in the method with sinc excitation, it is required to decrease range between f_{min} and f_{max}, which will cause that the magnitude $|U_0|$ of that signal rises (figure 5b). This will lead to greater robustness against measurement disturbances and decrease dynamic range of excitation and measured signals. Different shapes of sinc excitation destined to evaluate impedance spectrum in one, two or three decades from $f_{min} = 1$ mHz are shown in figure 5a.

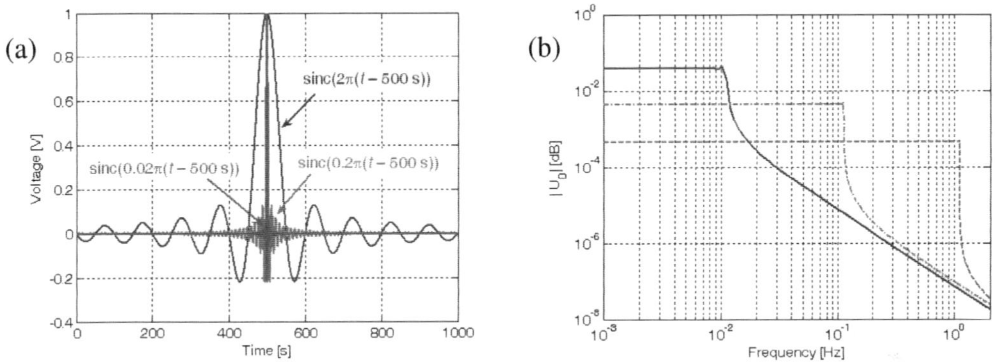

Figure 5. Exemplary sinc excitation signals destined to evaluate impedance spectrum in one (*blue*), two (*red*) or three (*green*) decades (a) and their spectra (b).

In the last step frequency range was limited to only one decade from $f_{min} = 10$ mHz to $f_{max} = 100$ mHz and the following parameters of the method were assumed: $U_0 = 1$ V, $\tau = 50$ s, $T = 100$ s, $f_s = 1$ kHz. Statistical analysis for a series of 10 measurements was performed. Obtained results, shown in figure 6, are satisfactory. Mean values of impedance calculation errors are in the range ± 2 % for modulus and ± 1 ° for argument, while standard deviations are comparable in the entire range of measurement frequencies ($\sigma_{|Zx|} < 1$ %, $\sigma_{arg(Zx)} < 0.6$ °).

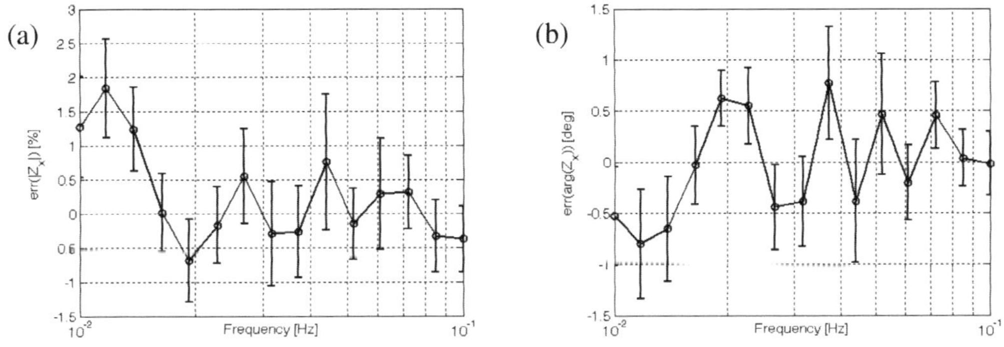

Figure 6. Mean and standard deviation of impedance modulus (a) and argument (b) for a series of 10 measurements of the model of anticorrosion coating.

6. Conclusions

The method of fast impedance spectroscopy of high impedance objects with use of continuous Fourier transform requires selection of an optimal excitation signal. Previous research of the method was performed with use of unipolar square pulse. Due to steep edges of the excitation and response signals there was a need to use irregular sampling methods (division acquisition time into segments with different sampling frequencies or usage of logarithmically changed sampling frequency).

In this paper different excitation signals were compared by taking into account their spectra and impedance spectrum calculation errors. Performed research shows usefulness of sinc excitation, because this signal is free of steep edges, can be developed to have constant spectrum in assumed range of measurement frequencies and also guaranties to obtain lowest object's spectrum calculation errors for all analyzed excitations. There is no need to use high sampling frequencies as in case of a square pulse. Practical experiments in measuring system led to the conclusion that it is justified to decrease range between $f_{\min}$ and $f_{\max}$ in order to increase accuracy of impedance spectrum calculations.

Replacement of the square pulse with sinc excitation offers better conditions to implement the method with continuous Fourier transform in portable instruments designed for diagnosis of technical objects located directly in the field.

References

[1] Skale S, Doležek V and Slemnik M 2008 Electrochemical impedance studies of corrosion protected surfaces covered by epoxy polyamide coating systems *Prog. Organic Coat.* **62** pp 2456–60

[2] Xu Z, Neoh K G and Kishen A 2008 Monitoring acid-demineralization of human dentine by electrochemical impedance spectroscopy *Journal of Dentistry* **36** 1005–12

[3] Srinivas K, Sarah P and Suryanarayana S V 2003 Impedance spectroscopy study of polycrystalline BI6FE2TI3O18 *Bulletin of Material Science* **26** 247–53

[4] Uchiyama T, Ishigame S, Niitsuma J, Aikawa Y and Ohta Y 2008 Multi-frequency bioelectrical impedance analysis of skin rubor with two-electrode technique *Journal of Tissue Viability* **17** 110–14

[5] Sanchez B, Bragos R and Vandersteen G 2011 Influence of the multisine excitation amplitude design for biomedical applications using impedance spectroscopy *33rd Annual International Conference of the IEEE EMBS (Boston, USA, 30 August – 3 September 2011)* pp 3975–78

[6] Min M, Ojarand J, Märtens O, Paavle T, Land R, Annus P, Rist M, Reidla M and Parve T 2012 Binary signals in impedance spectroscopy *34th Annual International Conference of the IEEE EMBS (San Diego, USA, 28 August - 1 September 2012)* pp 134–37

[7] Lindahl P A, Cornachione M A and Shaw S R 2012 A time-domain least squares approach to electrochemical impedance spectroscopy *IEEE Transactions on Instrumentation and Measurement* **61** 3303-11

[8] Mejia-Aguilar A and Pallas-Areny R 2008 Electrical impedance measurement using pulse excitation 2008 *Proc. of 16th IMEKO TC4 Symp. (Florence, Italy, 22-24 September 2008)* pp 567–72

[9] Park S and Yoo J 2002 Apparatus and method for measuring electrochemical impedance at high speed *United States Patent No. 6339334, 15 January 2002*

[10] Yoon C, Barsoukov Y and Kim J H 2004 Laplace transform spectrometer and its measurement method *US Patent 6687631, 3 February 2004*

[11] Hoja J and Lentka G 2011 Method using square-pulse excitation for high-impedance spectroscopy of anticorrosion coatings *IEEE Transactions on Instrumentation and Measurement* **60** 957–64

[12] Kowalewski M, Lentka G and Hoja J 2012 Evaluation of the fast impedance spectroscopy method in the laboratory measurement system *IEEE International Instr. and Meas. Technology Conf. (Graz, Austria, 2012)*

Journal of Physics: Conference Series **459** (2013) 012061 doi:10.1088/1742-6596/459/1/012061

An investigation on probabilistic relations in measurement

F. Crenna[1], **G. B. Rossi, V. Belotti**
Laboratorio di Misure
Università degli Studi di Genova - DIME , Via all'Opera Pia 15A,
16145 GENOVA, Italy

E-mail: crenna@dimec.unige.it

Abstract. In the representational approach, measurement is defined on the basis of primitive (in the context) notion of empirical relation. To account for measurement uncertainty, it is quite natural to consider probabilistic (or, more generally, non deterministic) empirical relations. Here the notion of probabilistic relation is investigated. After a brief historical outline, we point out that probabilistic relations appear in everyday measurement situation. To demonstrate this, a simple illustrative experiment has been developed, whose results are presented and discussed.

1. Probabilistic empirical relation

In a formal approach, for defining measurement, it is convenient to refer it to a weaker notion. In the representational approach, measurement is defined on the basis of the concept of empirical relation, which is a primitive one in the context [5][2] notion of empirical relation. In most of the literature concerning such a theory, empirical relations are regarded as "deterministic", in that they are associated to object according to a "yes-no" ("true-false"), classical logic [12]. To account for measurement uncertainty, it is instead quite natural to consider probabilistic (or, more generally, non deterministic) relations [7, 9, 11, 12]. This approach is more radical to the one provided by the early theory of errors [1], and was probably firstly conceived by Fechner himself [3]. The idea was fruitfully developed in psychophysics. Thurstone provided a (genial) representation of (inner) sensations in term of probabilistic (or random) variables[3]. Link notes that Thurstone's idea was already somewhat present, *in nuce*, in Fechner's thought. Experimental work in psychophysics provided wide empirical support to the need, at least in that field, of conceiving empirical relations as non-deterministic [4-8]. We suggest that the idea of probabilistic relation can be applied to measurement in general, including physics and engineering. We have discussed elsewhere the theoretical aspects of this position. Here we present and discuss a (relatively) simple experiment, concerning voltage measurement, whose result can anyway be extended to situations where the resolution of the observation device interact with the fluctuations of the phenomenon under observation.

---

[1] To whom any correspondence should be addressed.

[2] References are here listed in a chronological order, to provide a feeling of historical development of the subject.

[3] We definitely prefer the term probabilistic to (the more common) random variable, since probability does not necessary imply randomness [10].

2. Experimental set up

In order to provide some experimental data in support to the theoretical considerations, we have designed and performed an experiment in which two objects are compared to order them in respect of some property of interest. To obtain the input - output relationship, we need to have some control on one of the two. If one of them constitutes a fixed reference item, s, the entity of the other , a, can be varied either continuously or in steps smaller than the resolution of the observation device. Then we can evaluate the probability of observing the relations of equal, greater than, or less than, for small variations of the second object. We also measure the reference and the observed objects with an independent, high accuracy instrument in order to obtain additional information that will be used for displaying the results. The whole set-up is shown in Figure 1.

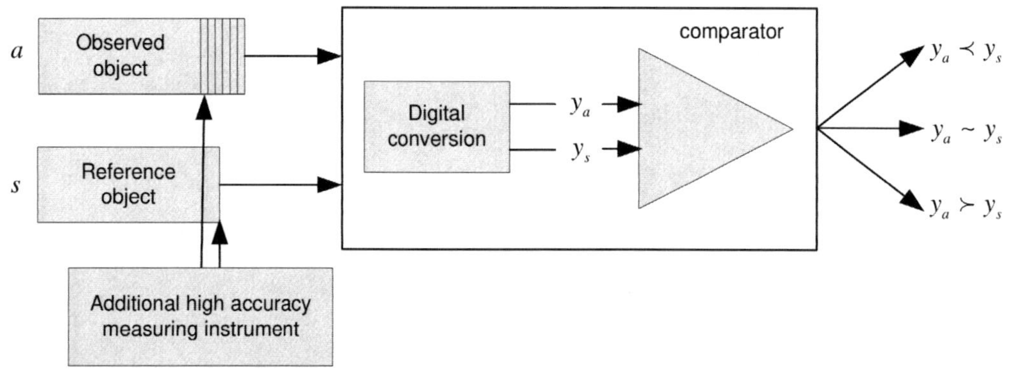

Figure 1. Generic discrete comparison experiment.

In principle it is possible to design such an experiment with any property, such as mass or voltage, given a proper comparator having a limited resolution. On the other hand there are practical implementation issues to be faced such as reference quantities stability and fine control on the variable item. In this regard voltage has been considered a convenient choice, in term of flexibility and ease of implementation. We have designed the voltage comparison experiment shown in Figure 2.

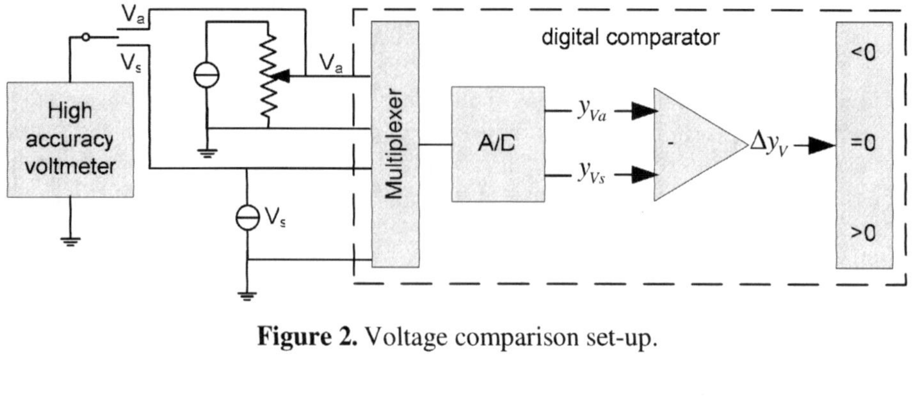

Figure 2. Voltage comparison set-up.

A stable fixed voltage source, V_s , acts as reference object s .

A second stable voltage source can be adjusted by acting on a voltage divider, to obtain the desired series of voltages states, V_a .

Journal of Physics: Conference Series **459** (2013) 012061 doi:10.1088/1742-6596/459/1/012061

The voltage comparator consists in a standard multiplexed data acquisition board, with a proper control software. Through the multiplexer the board connects an input channel at a time to the data conversion chain, so the measurement chain for the two voltages is the same and it performs as a comparator that is moved from one sample to the other.
The resolution of the comparator is the resolution of the data acquisition chain that includes a programmable gain amplifier to manage some flexibility in the input range. During the experiment the gain was set to obtain a ±10 V range and, since the analogue to digital converter has a 12 bits range, the overall resolution is set to 4,883 mV.
A higher accuracy bench voltmeter (Agilent 34401) is connected in parallel to the comparator, so that it is possible to measure the two voltage levels independently. During the experiment the reference voltage V_s was stable in a ±0,1 mV range, while it was possible to control the variable voltage source V_a , down to 0,1 mV. The sampling frequency was set to 500 Hz, and 500 samples were acquired for each channel for a one second time history.
The software evaluates the difference $\Delta y_V = y_{Va} - y_{Vs}$ for each couple of samples ad a logical operator evaluates if Δy_V is equal, larger than or less than zero. The output of each sample is recorded obtaining a sequence of 500 logical output values was recorded.

Table 1. Comparator logical output values

Objects	Readings	Comparator	Output
$a \prec s$	$y_{Va} < y_{Vs}$	$\Delta y_V < 0$	-1
$a \sim s$	$y_{Va} = y_{Vs}$	$\Delta y_V = 0$	0
$a \succ s$	$y_{Va} > y_{Vs}$	$\Delta y_V > 0$	1

So the overall set up, including the software data processing, realises a voltage comparator with known resolution and a logical (yes-no) output, for each of the investigated relations. For practical reasons such results have been conventionally codified with the values -1, 0 and +1.
Since the comparator is subjected to some noise, as the two voltage sources, even in presence at the input of two strictly equal voltages, the comparator output will present a distribution of logical values centred in zero but with considerable probability of lower/higher values.

3. Experimental procedure

After a one hour warm up of the voltage sources to reach a stable condition, the reference voltage V_s is measured and the variable voltage source V_a is set to a value lower than the reference with a nominal difference equivalent to more than 3 bits or 17 mV. A sequence of 500 samples is acquired and the comparison performed for each sample couple. In this situation, even in presence of some noise, the probability to obtain -1 at the output is almost one, due the large difference between the inputs.
Then the variable voltage source is increased in small steps with an increment in voltage that is less than the resolution of the comparator, and the test is repeated for all the values. The voltage sequence as measured at the input of the comparator by the bench voltmeter for all the 16 tests performed is presented in Figure 2.

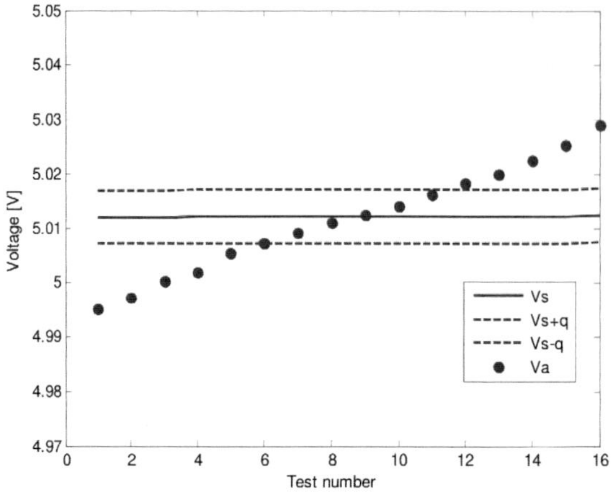

Figure 2. Voltage steps at the input of the comparator.

It is worth noting that the amplitude of each step of V_a is about half the resolution of the comparator, as shown by the dashed lines indicating a voltage variation corresponding to ± 1 bit across the reference level V_s.

The probability for each logical value at the output of the comparator is evaluated at each voltage step and it is possible to obtain a probability curve as a function of the ΔV at the comparator input, as presented in Figure 3.

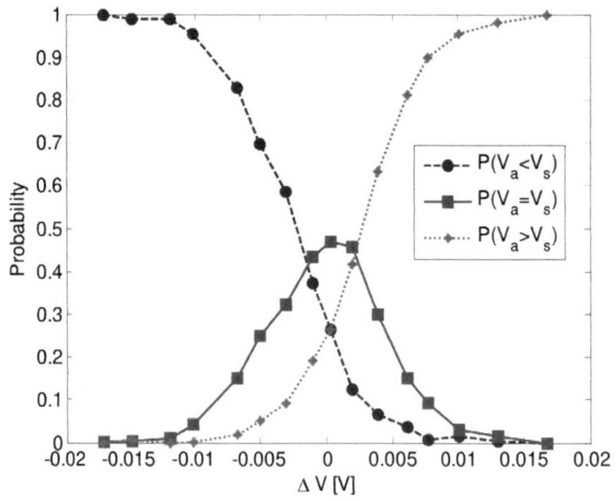

Figure 3. Probability distribution for comparison results, as a function of the voltage difference at the input.

In fact, the basic idea under this experiment, is quite general and it can be applied to a generic quantity, given a discrete comparator with a given resolution. In order to have a clearer view of this, it is possible to redraw the graph presented in Figure 3, considering a generic quantization process, instead of using specific voltage quantities. Scaling the voltage steps to the specific voltage resolution of the comparator that is used in this specific experiment we can find the graph presented in Figure 4

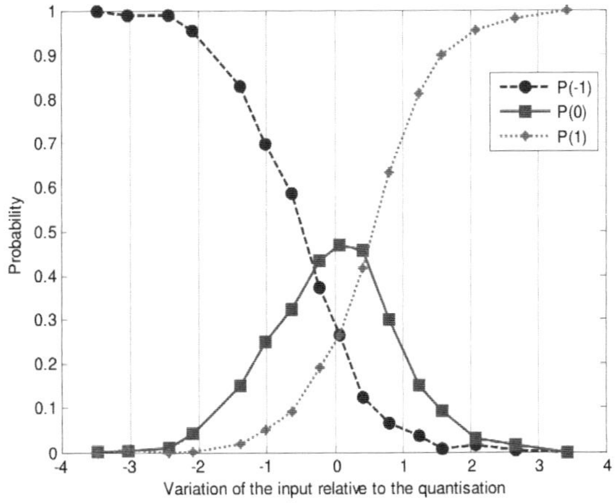

Figure 4. Probability distribution for comparison results, as a function of the input difference relative to quantisation

4. Discussion

Probability distributions presented in Figure 4 show that even when at the input of the discrete comparator two quantities are almost equal, the probability to obtain an equal result by the comparator is less than one (green squares), and there are not negligible probabilities to obtain less than (blue circles) or greater than (red stars) results. This situation is related to the characteristics of the overall comparison chain, in particular the noise level and the resolution. The former is due to the contribution of both the quantities involved and the comparator, while the latter is typical of the comparator device. A higher signal to noise ratio and a smaller resolution would result in more steep changes in the distribution curves, and of course in a more accurate comparison, but even in this more favourable case the result will not be deterministic, since investigating the behaviour between two, even small, quantisation intervals all the three possible results could appear at the output, so the most appropriate description would be in any case a probability distribution.

5. References

[1] Gauss C F 1809 Theoria motus corporum coelestium in sectionibus conicis solem ambientium, Hambourg
[2] Thurstone L L 1927 Psychological Review,
[3] Nowell Jones F 1974 History of psychophysics and judgement, in E C Carterette, M P Friedman (Eds.) Handbook of perception, vol. 2, Academic Press, New York
[4] Baird J C, Noma E 1978 Fundamentals of scaling and psychophysics, Wiley, New York, 1978
[5] Finkelstein L 1982 Theory and philosophy of measurement, in: P.H. Sydenham (Ed.) Handbook of measurement science, vol. 1, Wiley, Chichester, pp. 1-30
[6] Link S W 1994 Psychological science 5 335- 340
[7] Regenwetter M 1996 Journal of mathematical psychology, 40, 219-234
[8] Zwicker E, Fastl 1999 Psychoacoustics, Springer Verlag
[9] Rossi G B 2006 Measurement 39 34-50
[10] Costantini D 2011 Verso una visione probabilistica del mondo, GEM, Padova
[11] Rossi G B 2012 IEEE Transactions on Instrumentation and Measurement, 61 2097-2106
[12] Rossi G B, Crenna F 2013 Probability as a logic for measurement representation, Proceeding of the 2013 Joint IMEKO TC1 TC7 TC13 Symposium, at press.

Journal of Physics: Conference Series **459** (2013) 012062 doi:10.1088/1742-6596/459/1/012062

Conductivity measures coupled with treatment with ion-exchange resin for the assessment of sodium concentration in physiological fluids: analyses on artificial solutions

A Tura[1], S Sbrignadello[1], E Mambelli[2], P Ravazzani[1], A Santoro[2], G Pacini[1]

[1]Institute of Biomedical Engineering, National Research Council, Padova, Italy

[2]Nephrology Dialysis and Hypertension Unit, Policlinico S. Orsola-Malpighi, Bologna, Italy

E-mail: tura@isib.cnr.it

Abstract. In humans, sodium is essential for the regulation of blood volume and pressure. During hemodialysis, sodium measurement is important to preserve the patient from hypo- or hyper-natremia Usually, sodium measurement is performed through laboratory equipment which is typically expensive, and requires manual intervention. We propose a new method, based on conductivity measurement after treatment of dialysate solution through ion-exchange resin. To test this method, we performed in vitro experiments. We prepared 40 ml sodium chloride (NaCl) samples at 280, 140, 70, 35, 17.5, 8.75, 4.375 mEq/l, and some "mixed samples", i.e., with added potassium chloride (KCl) at different concentrations (4.375-17.5 mEq/l), to simulate the confounding factors in a conductivity-based sodium measurement. We measured the conductivity of all samples. Afterwards, each sample was treated for 1 min with 1 g of Dowex G-26 resin, and conductivity measured again. On average, the difference ε in the conductivity between mixed samples and corresponding pure NaCl samples (at the same NaCl concentration) was 20.9%. With treatment with the resin, it was 9.9%, only. We conclude that ion-exchange resin treatment coupled with conductivity measures may be a possible simple approach for continuous and automatic sodium measurement during hemodialysis.

1. Introduction

In humans, sodium is essential for the regulation of blood volume and blood pressure, as well as osmotic equilibrium and pH. In advanced renal failure, sodium balance becomes positive and the extracellular volume expands, and this leads to adverse cardiovascular consequences in patients under hemodialysis (HD) treatment. Thus, the maintenance of the sodium balance is an essential role of chronic HD treatment. As a consequence of a negative or positive sodium gradient between dialysate and plasma, patients can receive an even excessive removal of plasma sodium (leading to hypotensive events), or on the contrary may otherwise experience sodium gain [1]. Thus, reliable measurement of sodium concentration during HD is clinically relevant.

Currently, there are two main approaches for sodium measurement during HD: flame photometry or ion-selective electrode method [2]. These methods are based on medical instruments, such as the hemogas-analyzers, which are not integrated with the HD machine. Thus, the measurement of sodium (and, possibly, other variables) always requires manual intervention by an operator. In addition, the measurement performed by these instruments is typically very expensive.

The aim of this study is to propose an approach for possible simple, not expensive, and automatic measurement of sodium during HD. Our approach is based on conductivity measurement coupled with the use of a ion-exchange resin. Conductivity measurement of plasma, dialysate or ultrafiltrate sample alone cannot provide a highly accurate estimation of sodium concentration, though it should be recognized that the main contribution to conductivity of such samples is in fact due to sodium. However, coupling the simple conductivity measurement with treatment of the samples through appropriate ion-exchange resin can lead to improved accuracy, since the effect of the disturbing factors (especially, other cations) is reduced. In this study, we focus on the presentation of our approach applied to artificial samples.

2. Methods

2.1. Samples of sodium chloride and potassium chloride solutions

We prepared some artificial samples, at different sodium and potassium concentration, based on the following considerations. The sodium concentration in plasma is around 140 mEq/l; the concentration of the other cations, which represent the main confounders for conductivity-based sodium measurements, are around 4-5 mEq/l for potassium and calcium, and 1.5 mEq/l for magnesium [3]. In the analysis of the artificial samples, we assume for simplicity that the main confounder is potassium, and neglect the effect of the other cations. Deionized water (18.5 MΩ × cm resistivity, Millipore MilliQ Element system, USA) and sodium chloride (NaCl, >99.5%, Sigma-Aldrich, Italy) was used to prepare a 500 ml sample of NaCl solution at 280 mEq/l, that is, a concentration value certainly including the highest physiological plasma values (being twice the typical value). Afterwards, through diluition process, we obtained other six samples at lower concentrations, reaching the typical concentration values of plasma potassium and calcium (140, 70, 35, 17.5, 8.75, 4.375 mEq/l).

With the same process, with water and potassium chloride (KCl, >99.0%, Sigma-Aldrich) we prepared samples at the same concentration of the sodium samples (i.e., from 280 to 4.375 mEq/l). We also prepared some "mixed" samples (i.e., containing both sodium and potassium), in this case by direct weighting of the appropriate amount of NaCl and KCl for each sample. Specifically, we prepared a total of nine mixed samples: NaCl at 280 mEq/l and KCl at 17.5, 8.75, 4.375 mEq/l; NaCl at 140 mEq/l and KCl at 8.75, 4.375 mEq/l; NaCl at 70 mEq/l and KCl at 8.75, 4.375 mEq/l; NaCl at 35 mEq/l and KCl at 4.375 mEq/l; NaCl at 17.5 mEq/l and KCl at 4.375 mEq/l (see also Table 1). Such mixed samples were prepared to simulate the presence of a confounding factor (i.e., potassium in this case) for the measurement of sodium concentration in the sample. Of course, depending on the sodium concentration value of the samples, we selected some significant but appropriately lower concentrations of the confounding factor, thus somehow mimicking what would happen in biological samples (see typical physiological plasma ion concentrations reported above).

2.2. Conductivity measures and resin treatment

We performed impedance measurement of the samples through a Solartron 1260 impedance analyzer (Solartron Analytical, UK) [4]. Briefly, through the Solartron 1260 we applied a 100 mV r.m.s. voltage to the outer couple of electrodes of the measurement probe (SP06T model, Delta OHM, Italy), which was immersed into the glass tube containing the analyzed sample. The electric current was measured through the inner electrodes of the probe. We analyzed the impedance of the samples in the 10^3–10^7 Hz range, but in this study we simply considered the impedance value at 1 kHz, as done in several commercial conductivity meters. From the impedance, we then derived conductivity, considering that the geometrical factor of the probe (k-factor) was equal to 0.7 cm^{-1}. For each measure, we used 40 ml of the solution sample of interest. For each sample studied, we performed two independent measures, and took the average conductivity values between the two measures. All the impedance measures were performed with the samples at ambient temperature (20 °C with maximum variations of ±0.3 °C). In a first set of experiments, we measured the conductivity, as described above, in all the prepared samples. In a second set of experiments, we introduced the use of a ion-exchange resin, i.e., the Dowex G-26 exchange resin (Dow Chemical Company, USA). This is a uniform particle size, strong acid cation-exchange resin. Such resin acts replacing the cation species in a solution with hydrogen ions, H^+. However, specially strong affinity is expected for the alkali metals, and with sodium in particular. Thus, the use of the resin should determine variations in the characteristics of analyzed solution samples that are related to the sodium concentration. In this second set of experiments, we immersed different quantities of resin in a 40 ml sample of NaCl at 140 mEq/l. Specifically, we tested the effect of 1, 2, 3, 4 g of resin. For each quantity of resin, we stirred the sample solution through an electromagnetic micro-stirrer (Velp Scientifica, Italy) for 1 min. Afterwards, the resin was separated by the sample solution, and conductivity measure of the sample was performed. Subsequently, with 1 g of resin, we tested over the 140 mEq/l NaCl sample the effect of changing the contact time: in fact, we increased the time to 5, 10 and 20 min. Based on the results obtained in the second set of experiments, we proceeded with a third set of experiments. In this phase, each of the prepared samples, which underwent conductivity measure in the first set of experiments, was treated with 1 g of resin for 1 min, and hence conductivity measure was again performed.

3. Results

The conductivity of the NaCl samples (with concentration from 280 to 4.375 mEq/l) was essentially linear (R^2=0.99, P<0.0001), as shown in Figure 1, top . Similar trend was shown by the KCl samples (R^2=0.99, P<0.0001), Figure 1, bottom.

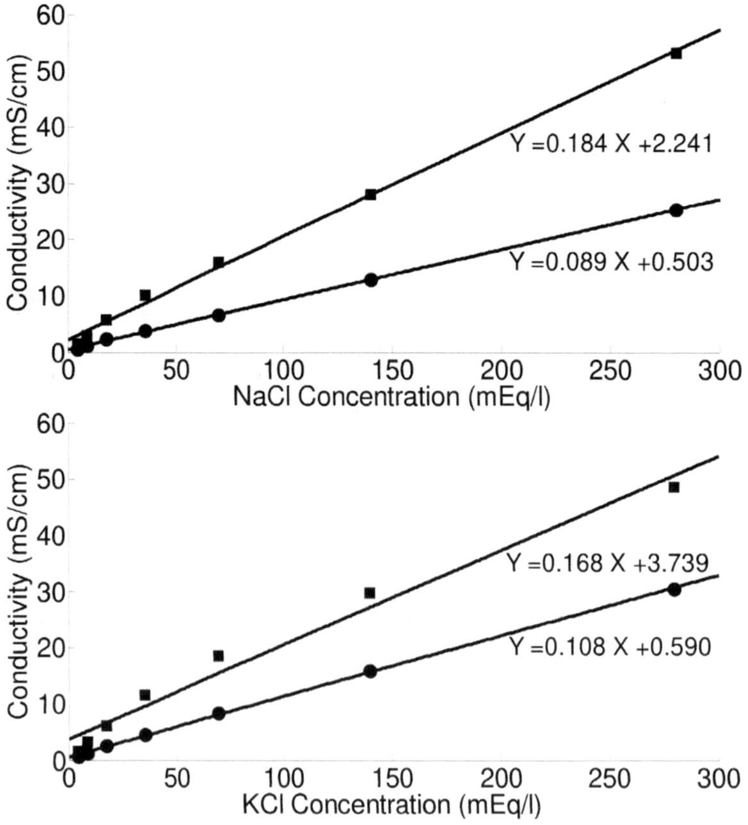

Figure 1. Relationship between concentration and conductivity before (circle) and after (square) resin treatment for NaCl (top) and KCl (bottom). Regression equations are also reported.

In the analysis of the effect of the resin on the sample conductivity, we tested the conductivity variation of the NaCl sample at 140 mEq/l with 1, 2, 3 or 4 g of resin, with a contact time of 1 min. The conductivity difference (before and after resin treatment) was Δ=15.1, Δ=20.9, Δ=24.8, Δ=26.3 mS/cm, for 1, 2, 3, 4 g, respectively. It can be concluded that, for the volume of our sample (40 ml), increasing the quantity of resin over 1 g determines an increase of the resin effect, but the effect is already clearly evident at 1 g. We also tested the changes in resin effect due to variations in the contact time between the resin and the sample. With the NaCl sample at 140 mEq/l, and with the use of 1 g of resin, we found Δ=16.8, Δ=19.4, Δ=20.1 mS/cm, for 5, 10, 20 min, respectively. We concluded that the use of 1 g of resin, with a contact time of 1 min, provided a clear resin effect, thus it was sufficient for the treatment of our 40 ml samples.

The conductivity of the NaCl samples, and of the KCl samples, treated with 1 g of resin for 1 min, was still essentially linear (R^2=0.99 and R^2=0.98, P<0.0001 for both), as reported again in Figure 1. At any concentration, the resin effect was clearly evident for both NaCl and KCl samples. For each concentration value it was possible to quantify the resin effect as the pre-post resin conductivity difference, normalized to the conductivity of the sample before resin treatment. For the NaCl samples, the resin effect was 1.43±0.08 (mean±standard error), whereas for the KCl samples it was 1.21±0.13: this was a demonstration of the stronger effect of the resin on sodium rather than on potassium.

We then analyzed the mixed samples, containing both NaCl and KCl at different concentration values, to simulate the presence of a confounding factor (i.e., potassium) for the measurement of sodium concentration of the sample. For each mixed sample, we computed the percentage difference in conductivity between the mixed sample and the corresponding pure NaCl sample, at the same NaCl concentration (for instance, the conductivity difference between the 280 mEq/l NaCl + 17.5 mEq/l KCl sample and the pure 280 mEq/l NaCl sample, normalized to the latter, etc.). We named such difference ε, since it is an estimation of the error in the conductivity-based estimation of sodium concentration, due to the presence of the confounding factor (potassium). First, we computed ε before the resin treatment (Table 1). Afterwards, we treated the mixed samples with the resin (1g for 1 min), and calculated $\Delta_{pre\text{-}post}$: i.e., the conductivity difference before and after the resin treatment. Again, we computed ε values as explained above, this time based on the pre-post resin conductivity differences (Table 1). It can be noticed that, for each mixed sample, the ε values were always lower when the pre-post resin conductivity differences were considered: in fact, on average, ε without resin treatment was 20.9%, whereas it was 9.9% with resin treatment. This means that, in the analyzed artificial samples, the treatment with resin reduced the confounding effect of other ions (potassium, in this case) on the assessment of sodium concentration, based on conductivity measurement approach.

Table 1. Concentration (mEq/l) of mixed samples (NaCl+KCl), and corresponding difference ε in the conductivity (%) compared to the samples with NaCl alone at the same concentration, without resin treatment, and with resin treatment (pre-post resin conductivity difference).

NaCl → KCl ↓	280	140	70	35	17.5
Pre-treatment					
17.5	ε_{pre} = 17.28	-	-	-	-
8.75	ε_{pre} = 15.27	ε_{pre} = 19.93	ε_{pre} = 31.53	-	-
4.375	ε_{pre} = 12.78	ε_{pre} = 13.09	ε_{pre} = 20.54	ε_{pre} = 24.09	ε_{pre} = 33.13
Post-treatment					
17.5	$\varepsilon_{\Delta pre\text{-}post}$ = 6.35	-	-	-	-
8.75	$\varepsilon_{\Delta pre\text{-}post}$ = 7.45	$\varepsilon_{\Delta pre\text{-}post}$ = 7.07	$\varepsilon_{\Delta pre\text{-}post}$ = 12.83	-	-
4.375	$\varepsilon_{\Delta pre\text{-}post}$ = 2.94	$\varepsilon_{\Delta pre\text{-}post}$ = 0.41	$\varepsilon_{\Delta pre\text{-}post}$ = 17.13	$\varepsilon_{\Delta pre\text{-}post}$ = 23.07	$\varepsilon_{\Delta pre\text{-}post}$ = 11.71

4. Discussion and conclusion

We proposed an approach, based on conductivity measures and resin use, which may contribute to possibly overcome the current limitations in the measure of sodium during HD. Conductivity measure of ultrafiltrate for sodium determination has been recently proposed by the company Bellco, Italy (http://www.bellco.net/public/files/Right_Therapies_News.pdf), based on the hypothesis that the main contribution to conductivity is in fact due to sodium. However, simple conductivity measurement of the ultrafiltrate cannot provide highly accurate estimation of sodium concentration, due to the presence of several disturbing factors: especially, the other cations (potassium, calcium, magnesium), but also glucose [5]. The main novelty of this study is in fact the idea of coupling the simple conductivity measures with treatment of the solution sample with a not expensive (~0.04 € per gram) acid cation-exchange resin, which can reduce the effect of the disturbing factors in sodium determination. In fact, the contact of the solution with the resin determines an "amplification" of the sample conductivity, which depends mainly on the sodium concentration, due to the resin special affinity with sodium. As a consequence, in the conductivity-based determination of the sodium concentration of the sample, resin treatment allows reducing the effect of the disturbing factors. It should be noted that, in our experiments, for simplicity we assumed potassium as representative of any possible disturbing factor. In any case, the other disturbing factors should have lower effect on resin behavior, as previously discussed. In conclusion, we have proposed an approach for possible simple, automatic and not expensive measurement of sodium during a HD session. Further studies, on patients in different clinical conditions, are necessary to prove the actual possibility of integrating the proposed method on a HD machine

References

[1] Hecking M, Kainz A, Hörl WH, Herkner H and Sunder-Plassmann G 2011 Sodium setpoint and sodium gradient: influence on plasma sodium change and weight gain *Am. J. Nephrol.* **33** 39

[2] Worth HG 1985 A comparison of the measurement of sodium and potassium by flame photometry and ion-selective electrode *Ann. Clin. Biochem.* **22** 343

[3] Caduff A, Lutz HU, Heinemann L, Di Benedetto G, Talary MS and Theander S 2011 Dynamics of blood electrolytes in repeated hyper- and/or hypoglycaemic events in patients with type 1 diabetes *Diabetologia* **54** 2678

[4] Tura A, Sbrignadello S, Barison S, Conti S and Pacini G 2007 Impedance spectroscopy of solutions at physiological glucose concentrations *Biophys. Chem.* **129** 235

[5] Tura A, Sbrignadello S, Cianciavicchia D, Pacini G and Ravazzani P 2010 A low frequency electromagnetic sensor for indirect measurement of glucose concentration: in vitro experiments in different conductive solutions *Sensors (Basel)* **10** 5346

Journal of Physics: Conference Series **459** (2013) 012063 doi:10.1088/1742-6596/459/1/012063

A wearable multipoint ultrasonic travel aids for visually impaired

Ilaria Ercoli, Paolo Marchionni and Lorenzo Scalise

Dipartimento di Ingegneria Industriale e Scienze Matematiche (DIISM),
Università Politecnica delle Marche, Via Brecce Bianche, 60131, Ancona, ITALY .

E-mail: l.scalise@univpm.it

Abstract. In 2010, the World Health Organization estimates that there were about 285 million people in the world with disabling eyesight loss (246 millions are visually impaired (VI) and 39 millions are totally blind). For such users, hits during mobility tasks are the reason of major concerns and can reduce the quality of their life. The white cane is the primary device used by the majority of blind or VI users to explore and possibly avoid obstacles; it can monitor only the ground (< 1m) and it does not provide protection for the legs, the trunk and the head. In this paper, authors propose a novel stand-alone Electronic Travel Aid (ETA) device for obstacle detection based on multi- sensing (by 4 ultrasonic transducers) and a microcontroller. Portability, simplicity, reduced dimensions and cost are among the major pros of the reported system, which can detect and localize (angular position and distance from the user) obstacles eventually present in the volume in front of him and on the ground in front of him.

1. Introduction.

In 2010, the World Health Organization estimates that there were about 285 million people in the world with disabling eyesight loss (246 millions are visually impaired (VI) and 39 millions are totally blind). In Italy, the number of VI subjects is estimated in 530.000 and about 50.000 of them are blind. During the last 30 years, with the progress of technology, there has been a growth on the research focusing on novel devices designed for the assistance of visually impaired (VI) people and blind. The first and older device used by VI subjects is a very well known (passive) device: the cane (commonly known as the white cane). The cane is light, portable and cheap and these characteristics certainly have contributed to its success and its dissemination. The most relevant drawback of this device is definitely the inspection range which is limited to a restricted part of the ground (< 1 m) in front of the user's feet; the cane cannot provide information regarding obstacles which are not located on the ground and in the area explored by the cane tip. There are a lot of obstacles located at the trunk and head level which are consequently not individuated and potentially dangerous, such as tree branches, exposed shelves or a propped-open window. Obstacle hits or even just the fear of it can consequently seriously limit blind and VI users quality of life [1-3].
Since early eighties, active devices for obstacle detection have been proposed. Such system are commonly known as Electronic Travel Aids (ETAs) and have been mainly based on use of laser and ultrasonic (US) transducers. Few of them can also be found on the market [2-4]; in the majority of the cases they detect the presence of an obstacle, but they need to be driven by the user and cannot localise the position of the obstacle.

The aim of this paper is to present a multi-sensing, obstacle detection system composed by four ultrasonic sensors; it presents the characteristics to be: portable, standalone, multi-sensing sensors. the proposed system can continuously explore the volume in front of the VI user and provide an aural feedback if one or more obstacles are found; this feature allows the VI subject to anticipate the presence of the obstacles in the explored volume.

Journal of Physics: Conference Series **459** (2013) 012063 doi:10.1088/1742-6596/459/1/012063

Special attention has been dedicated to portability, use of commercially available components and costs (< 250€).

2. Materials and methods

The control of the device is based on the use of a microcontroller (PIC 18F877A) which is used to process input signals from the sensors, to control the temperature an relative humidity conditions and to provide the output information to the user by visual and audio feedbacks (fig.1). Four ultrasonic sensors realize the volume scanning and, eventually, detect the presence of a possible obstacle; their output signals are used by the microcontroller as inputs. The four ultrasonic sensors used are based on a single piezo crystal (XL-MaxSonar-EZ MB 1320) working at 42 kHz; the crystal is firstly excited in order to emit a short wave pack and then it is used as a receiver to collect eventual echoes caused by reflections from objects eventually present in the explored cone (field of measurement: 3 cm to 3 m). Two sensor are used to explore the left and right area (sensor A and B in fig. 1), while one sensor is used to explore the central area (sensor B) and one sensor is used to explore the ground in front of the subject (sensor F). The block diagram of the proposed system is shown in fig.1.

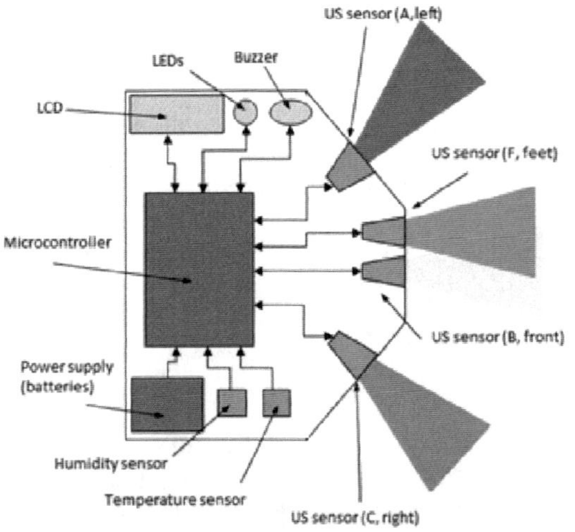

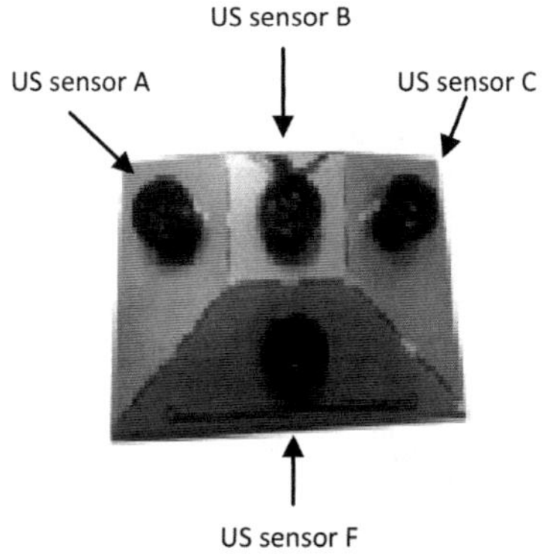

Figure 1. Block diagram of the proposed system (left); frontal view of the system with the four sensor reported (right).

The measurement of the distance between the system and the obstacle is operated by the measurement of the time interval ΔT between the launch of the wave pack and the detection of the obstacle-reflected eco (fig.2). The distance D between the obstacle and the sensor is calculated as:

$$D = \frac{\Delta T v_{air}}{2} \text{ (eq.1)}$$

where v_{air} is the velocity of sound in air.

The velocity of sound in air (v_{air}) is influenced by temperature (T) and the relative humidity (RH) of the air and these quantities can cause uncertainty on the determination of the real sensor to obstacle distance. For this reason a thermo-hygrometer sensor (1 sample/min) has been included in the proposed system in order to reduce uncertainty in the determination of the sound velocity (v_s). T and RH measured values are used to refer to tabled values of the v_{air}.

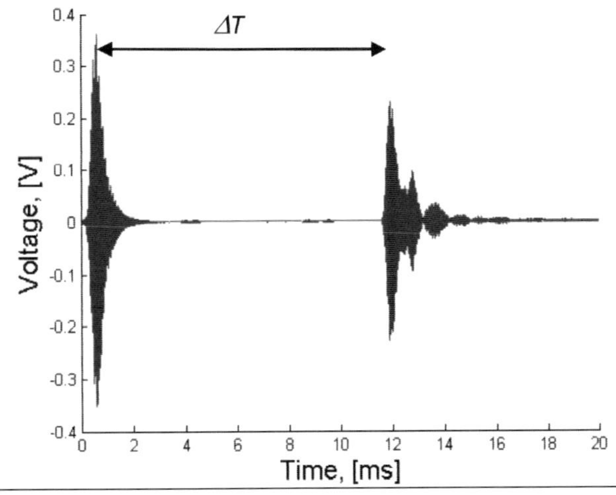

Figure 2. Example of a the signal generated by the ultrasonic sensor (in the case of an obstacle echo).

The presence of an obstacle is communicated to the user by the use of a buzzer; its intensity and frequency is related to the distance and location of the obstacle. The system is worn at the belt buckle (fig. 3) and explores 4 sectors of conical shape (A,B,C and F). Cones A and B are left and right (respect to trunk front direction), oriented respectively at -30° and +30° while cone F is inclined of 25°, toward the terrain. Each cone has an aperture of about 60°. Obstacles eventually present in region A, B or C are signaled as well as obstacles sensed by two sensors (A-B or B-C). Sensor F detects the presence of obstacles on the walking path in the range 1.75 - 2.15 m. Round Robin scheduling algorithm was chosen to extract information by the ultrasonic sensors; time slots of 300 ms are used in order to have a good data refresh. The power supply is provided by a battery (9V) which provide 5V stabilized voltage to the electronics.

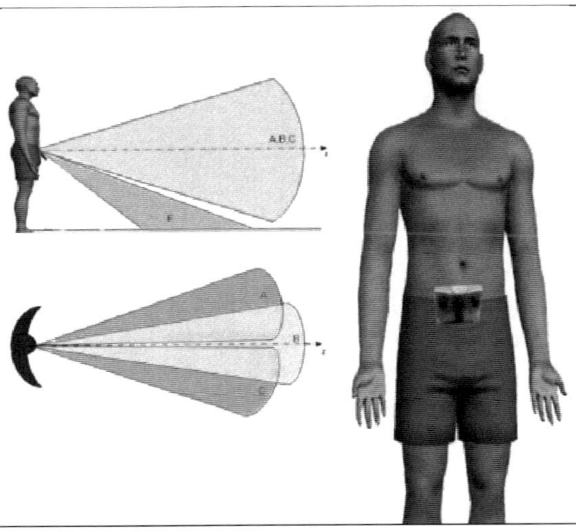

Figure 3. The four conical sectors explored by the US system.

3. Results

The four sensors were firstly calibrated using a scale as reference (resolution 1 mm). The calibration test was carried out on a object of 6X6 cm of area and on the range: 10 – 300 cm. The four sensors have similar behavior and one calibration line is reported in fig. 4 (slope: 1,004 and offset of -0.028 m). The estimated uncertainty, according GUM [5], is ± 0,004 m (k=1).

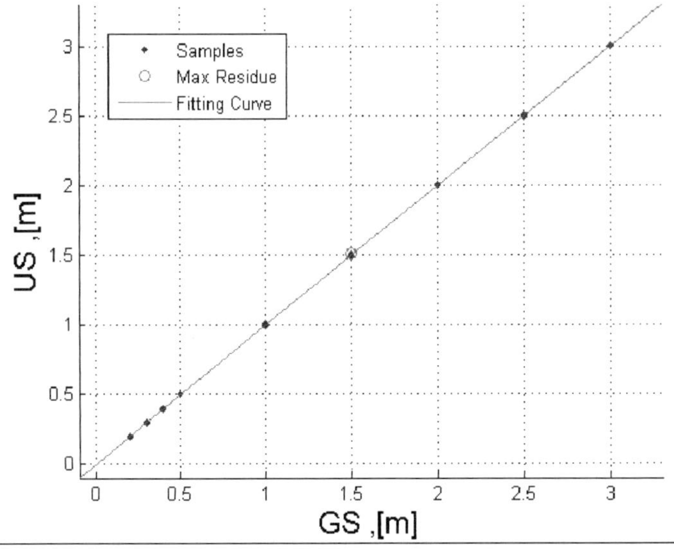

Figure 4. Calibration line for one of the US sensors.

The capability of the sensor in detecting the presence of object of different dimensions, placed at different distances, has been evaluated trough a sensitivity test. Results are reported in fig. 5, where it is possible to observe how the SNR - ratio between the amplitude of the reflected signal (mean of the 4 sensors) and the noise level measured without the obstacle - decreases with distance and with the obstacle area. In particular, it is possible to observe that the smaller obstacle tested (6X6cm) is still detected at the distance of 3m.

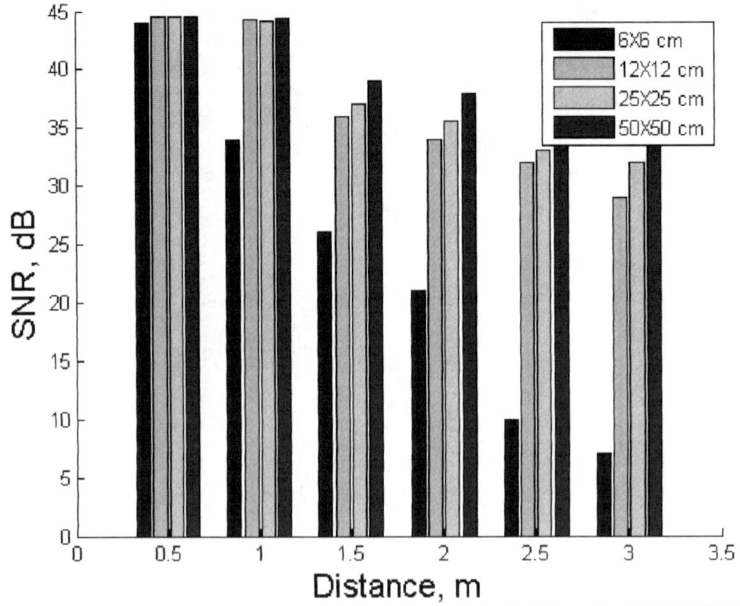

Figure 5. SNR measured for 4 objects of different dimensions.

Journal of Physics: Conference Series **459** (2013) 012063 doi:10.1088/1742-6596/459/1/012063

A Bland-Altman [6] test is operated in order to operate a comparison between the proposed method and the reference. Fig. 7 shows the results of the Bland Altman test and it is possible see that there is not a dependence of the deviation between the two systems by the distance.

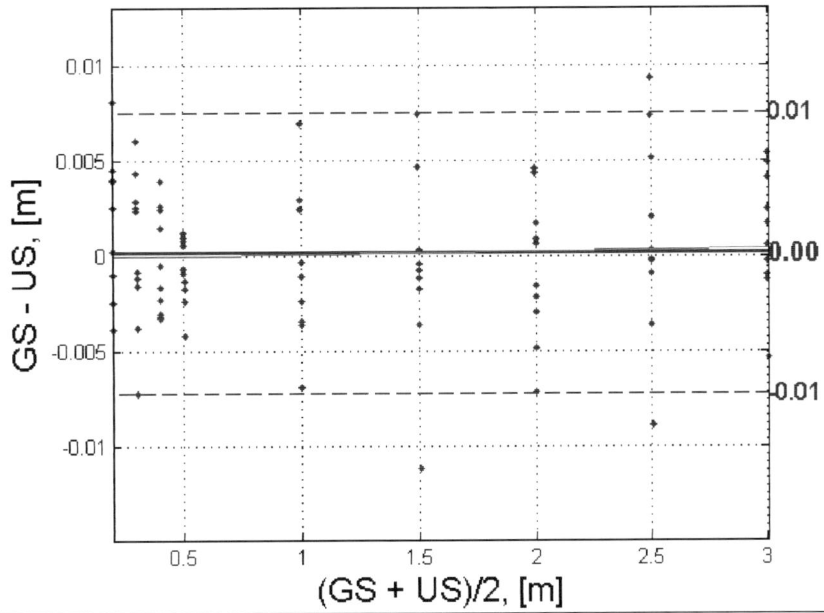

Figure 7. Bland Altman test on the proposed system.

4. Discussion

In this work a novel obstacle detection system, specially devised for blind or visually impaired, is presented. The system is aimed to explore the volume in front of the user. The main advantage respect to the state of the art for ultrasonic system for obstacle detection is the possibility to not only measure the distance from the obstacle, but to be also able to localize the position of the obstacle (left, center up, center down, right). A specific functionality has also been implemented for the exploration of the obstacle at the ground level: the sensor is able to explore the ground surface at about 2 m sending an alert signal to the subject if a ground obstacle (such as a step or a hole) is detected in the area in front of the user feet.

Specific attention has also been put on the portability of the device and on the eventual stigma effect on the VI user. For this reason, the system has been designed in order to be light and easily placed on the belt buckle, living the subjects hands free. Portability is also ensured by the reduced dimensions and the battery supply; while basic system-to-user communication protocol reduce to the minimum the training time.

Finally the use of commercially available components has allowed to reduce the costs needed for the realization of the prototype to < 250 €.

Next steps of the research will aim to:

- improve the transmission of the information to the VI user by vibro-tactile interfaces;
- carry out tests with VI users in order to collecting feedbacks for optimization of the system;
- allow the use of the proposed system in conjunction with the white can (avoiding the interference).

References

[1] WHO (2010) New estimations of visual impairment and blindness.
[2] D. Dakopoulos and N. G. Bourbakis (2010) Wearable Obstacle Avoidance Electronic Travel Aids for Blind: A Survey. IEEE Trans Sys man & Cyb, 40,1.
[3] Communications on vision (1986) Electronic Travel Aids: New Directions for Research, Nat Acad Press.
[4] B. Andò, S. Graziani (2009) Multisensor Strategies to Assist Blind People: A Clear-Path Indicator. IEEE Trans Instr & Meas, 58, 8:2488-94.
[5] JCGM 100:2008. Evaluation of measurement data - Guide to the expression of uncertainty in measurement, Joint Committee for Guides in Metrology.
[6] Altman DG, Bland JM Bland JM, Altman DG (1999). Measuring agreement in method comparison studies. Stat Methods Med Res **8** (2): 135–60.

Journal of Physics: Conference Series **459** (2013) 012064 doi:10.1088/1742-6596/459/1/012064

Metrological reliability of optical coherence tomography in biomedical applications

C M Goloni, G P Temporão and E C Monteiro

Pontifícia Universidade Católica do Rio de Janeiro, Rua Marquês de São Vicente 225, Rio de Janeiro, Brazil

E-mail: `carolg@ele.puc-rio.br`

Abstract. Optical coherence tomography (OCT) has been proving to be an efficient diagnostics technique for imaging in vivo tissues, an optical biopsy with important perspectives as a diagnostic tool for quantitative characterization of tissue structures. Despite its established clinical use, there is no international standard to address the specific requirements for basic safety and essential performance of OCT devices for biomedical imaging. The present work studies the parameters necessary for conformity assessment of optoelectronics equipment used in biomedical applications like Laser, Intense Pulsed Light (IPL), and OCT, targeting to identify the potential requirements to be considered in the case of a future development of a particular standard for OCT equipment. In addition to some of the particular requirements standards for laser and IPL, also applicable for metrological reliability analysis of OCT equipment, specific parameters for OCT's evaluation have been identified, considering its biomedical application. For each parameter identified, its information on the accompanying documents and/or its measurement has been recommended. Among the parameters for which the measurement requirement was recommended, including the uncertainty evaluation, the following are highlighted: optical radiation output, axial and transverse resolution, pulse duration and interval, and beam divergence.

1. Introduction

Optical technology has an extended application in medicine and biology, dating back to the 18th century with the use of microscope. Relatively new, the application in the medical field of optical instruments that take advantage of the coherent properties of light is rising. One technique that has emerged is the optical coherence tomography (OCT), an important and successful application of interferometry in the biomedical sciences. Measuring the reflected light, this technique is capable to provide non-invasive high-resolution cross-sectional images of the internal structures of living tissue.

Originally, due to its low scattering characteristics, OCT was developed for transparent tissue imaging [1]. Later on, the equipment has evolved, allowing the generation of images of non-transparent tissue [2]. Its first clinical application occurred in ophthalmology and the first commercial OCT was released in 1996. Since then, it advanced as a diagnostic and therapeutic technique for different areas. However, risks of adverse events from the use of these technologies are always present. These risks can be avoided with periodical evaluation concerning safety and appropriate performance of the equipment for the referred application, protecting the patient and the medical staff [3].

Every medical electrical equipment must comply with the International Standard published by the International Eletrotechnical Commission (IEC), IEC 60601-1 (Medical electrical equipment - Part 1: General requirements for basic safety and essential performance), and possible existing particular standards. In spite of its established clinical usage, to the present day there is not a specific technical standard published to evaluate optical coherence tomography equipment's conformity to essential safety and performance requirements for biomedical imaging.

Therefore, the present work analyses the IEC particular international standards published for safety and performance evaluation of laser and IPL (Intense Pulsed Light) to identify the parameters also applicable for metrological reliability analysis of OCT equipment. Additionally, specific parameters for OCT evaluation that were not mentioned in these standards were identified, aiming at contributing for the complete list of parameters to be recommended for a possible future development of a specific standard for the non-invasive high-resolution imaging technique by optical coherence tomography.

2. Optical Coherence Tomography

Optical coherence tomography (OCT) is a signal acquisition technique based on the interference between an low-coherence optical field that is divided and later recombined after reflecting back from multiples layers within the sample. The heart of the system is Michelson interferometer illuminated by a low-coherent light source.

One of its advantages is the possibility of non-invasive generation of high resolution imaging of tissue's internal structures, which is of great relevance to the biomedical field. Different from confocal microscopy, which depends on numerical aperture of the optical system, OCT's axial resolution relies only on the coherence of light from the source. This characteristic allows *in vivo* deep tissue measurements non-invasively[4, 5].

Information about the structure of the sample is obtained from the interferogram, the light intensity measured by the photodectetor. Refractive index variations between layers within the sample represent intensity peaks on the interference pattern [6].

3. Parameters for OCT metrological reliability evaluation

This section briefly describes the parameters identified as important to assure metrological reliability of biomedical equipment based on OCT. The suggestions indicated for each evaluation parameter included their measurement and/or the insertion of information of their values in the device's accompanying documents (Table 1). Some of the recommended parameters are already mentioned in the published standards for laser and IPL, as indicated.

3.1. Beam diameter and divergence

IEC 60825-1 standard (Safety of laser products - Part 1: Equipment classification, requirements and user's guide) defines beam diameter as the diameter of the smallest circle which contains 63% of total laser output power. As the laser beam propagates, the beam diameter increases [7]. The IEC 60825-1 standard defines beam divergence as the far field plane angle formed by the beam diameter.

The guarantee that the beam is focused only on the intended area for diagnostic/treatment of the biological tissue is vital to a safe use of the equipment. Therefore, it is suggested to measure the beam divergence of OCT devices. It is recommended to inform the value of the beam diameter on the device's accompanying documents.

3.2. Pulse duration

According to the IEC 60825-1 standard, the pulse duration is the time increment measured between the half peak points of a pulse. The pulse duration must respect the maximum continuous tissue exposure limit, which varies according to each application.

Therefore, in case of a pulsed source, it is recommended not only the pulse duration evaluation but also the inclusion of its value in the accompanying documents.

3.3. Optical radiation output

By the definition on laser's particular standard (IEC 60601-2-22), the output laser can be either the laser energy or the laser power. The laser power is the radiant flux of the work beam and the laser energy is the radiant energy, both incident on the working area [8].

Following the same recommendations of the particular standards for laser and IPL, this parameter should be evaluated for OCT equipment. The accuracy of the radiation output is fundamental to assure successful device-tissue interaction. It is also important to inform the maximum radiation emitted by the light source on the accompanying documents.

3.4. Pulse interval

The interval between pulses is the time from the end of one pulse and the beginning of the next, measured at 50% trailing and leading edges respectively, as the definition of IEC 60601-2-57 states [9].

For devices using pulsed sources, it is suggested to measure and inform the pulse interval value on the accompanying documents.

3.5. Number of pulses in pulse train

A pulse train is a series of pulses where the total pulse time in a solely exposure does not exceeds 0.25 s in the wavelength region between 400 nm and 700 nm. For other wavelengths, this total pulse time exposure changes to 10 s.

As for the other parameters related to pulses, it is suggested to evaluate and inform the value on the device's manual.

3.6. Spectral irradiance

The irradiance (E) is defined by the IEC 60825-1 standard as:

$$E = \frac{d\Phi}{dA} \qquad [W/m^2], \tag{1}$$

where dΦ is the radiant flux incident on an element and A the element's area [7]. For all possible configurations of the device, the spectral irradiance varies according to different wavelengths.

It is recommended to inform the spectral irradiance on the OCT documents, as it is already required by the particular standard for IPL (IEC 60601-2-57).

3.7. Concentricity between aim beam and work beam

The aim beam indicates the area of work and the work beam is the radiation beam emitted by the output laser. The concentricity between them must have a sufficiently small tolerance as to avoid undesirable effects. The maximum lateral displacement allowed between both centers must not be over 50% of the biggest diameter [8].

The concentricity between aim beam and work beam has the same importance as the diameter and divergence of the work beam. To assure the aim marks the real spot where the electromagnetic radiation is applied means guaranteeing the treatment/diagnostic occurs on the correct region. Thus, the measurement of this parameter is recommended to attest the equipment's reliability.

3.8. Exposure time and limit

The exposure time is the duration of continuous laser radiation (or pulse, in case of pulsed sources) incident upon the human body. This value varies according to the application and the type of exposed tissue [9]. As for the exposure limit, the IEC 60601-2-57 particular standard for IPL defines it as the maximum level of eye or skin exposure where adverse biological effects are not expected.

Both these parameters should be known by manufacturers and users to ensure safe application of the equipment in their respective areas. Hence, it is suggested the inclusion of this information in the documentation.

3.9. Ocular and/or skin hazard distance

The particular standard for IPL (IEC 60601-2-57) defines as the distance from the emission aperture where the irradiance in a certain exposure time is equal to the limit exposure value of the eye and skin, respectively, for the parameters ocular and skin hazard distance.

Considering that OCT has ophthalmological and dermatological applications, it is recommended to inform these values on the accompanying documents, according to the equipment's application.

3.10. Central wavelength

The emission wavelength (nominal wavelength) and the source's power determine OCT's depth penetration [4, 5]. The depth penetration of the light is limited by the optical properties of biological tissues, depending on absorption and scattering. In the visible region of the spectrum (400 nm to 700 nm), light absorption by biological tissue is higher compared to the near infrared region (approximately from 800 nm to 2500 nm). The reason for this behavior in the near infrared region is the fact that the wavelength is too long to result in a high number of electrons transitions, but too short to induce vibrational transitions in the water. All biological risks are evaluated around these wavelength values.

In that, it is recommended to inform this parameter on the OCT system's documentation.

3.11. Axial resolution

The axial resolution is an important specification for OCT systems. In most biomedical applications, high axial resolution is necessary to distinguish boundaries and differentiate types of tissue. Also known as longitudinal resolution, it is related to the coherence length of the light source, but also depends on absorption and scattering properties of tissue. This relation with the coherence length is due to the broadband nature of the light, where the interference between optical fields is only observed when the lengths of the optical path on both reference and sample arms correspond to the coherence length of the light [5].

High axial resolution can be achieved regardless of focus conditions of the beam. Considering a source with Gaussian spectrum distribution, the axial resolution Δl is:

$$\Delta l = \left(\frac{2\ln 2}{\pi}\right)\left(\frac{\lambda^2}{\Delta\lambda}\right), \tag{2}$$

where $\Delta\lambda$ is the source band width and λ is the central wavelength.

It is suggested the evaluation of the axial resolution, a crucial parameter to determine imaging potential of OCT systems. The axial resolution values for air and water should be informed on the accompanying documents.

Journal of Physics: Conference Series **459** (2013) 012064 doi:10.1088/1742-6596/459/1/012064

3.12. Transverse resolution

One advantage of the OCT is the total disassociation between the transverse and axial resolution. The transverse resolution Δx, also known as lateral resolution, in OCT systems is determined by the size of the focal point, as in conventional microscopy [10].

$$\Delta x = \left(\frac{4\lambda}{\pi}\right)\left(\frac{f}{d}\right) \quad (3)$$

where d is the spot size at the lens and f is the focal length. High transverse resolution is achieved using large numerical aperture and focusing the beam in a small spot size.

As for the axial resolution, it is suggested both the measurement and inclusion of the transverse resolution value in the documentation.

Table 1. Parameters recommended for reliability evaluation of biomedical equipment based on OCT. The recommendation for the measurement of the parameter and/or inclusion of its information in the accompanying documents is indicated. * parameter mentioned in the standard IEC 60601-2-22 for laser equipment; † parameter mentioned in the standard IEC 60601-2-57 for non-laser light source equipment.

Parameter	Documentation	Measurement
Central wavelength	✓	
Concentricity between aim beam and work beam		✓
Beam diameter	✓	
Ocular hazard distance and/or skin hazard distance†	✓	
Beam divergence*		✓
Pulse or pulse train duration*†	✓	✓
Pulse interval†	✓	✓
Spectral irradiance†	✓	
Exposure time	✓	
Exposure limit	✓	
Number of pulses in pulse train†	✓	✓
Axial resolution	✓	✓
Transverse resolution	✓	✓
Optical radiation output*†	✓	✓

For all the parameters in which measurements are indicated to assess metrological reliability of the optoelectronic equipment, it is recommended to include the estimation of measurement uncertainty.

Considering the necessity of assuring reliable diagnostics and safety during the whole period of biomedical equipment clinical use [3], some of the recommended parameters mentioned on Table 1 should be performed throughout the equipment lifetime, as the concentricity between aim beam and work beam; beam divergence; pulse or pulse train duration; pulse interval; number of pulses in pulse train; axial resolution; transverse resolution; and optical radiation output.

In the literature, it is emphasized another important aspect concerning metrological reliability for this optical technique on continuing expansion in medical diagnostics applications is the method validation, that is currently limited by the lack of comparable techniques and incomplete database regarding optical properties of tissue (refractive index, scattering and absorption of light by the tissue) [11].

4. Conclusion

Although the already well-established clinical use and important perspectives as a diagnostics tool for imaging in vivo tissues, there is no international standard addressed to optical coherence tomography (OCT) for biomedical application.

In this study, the OCT system characteristics are analyzed aiming at identifying parameters that need to be evaluated to ensure metrological reliability for biomedical use of optical coherence tomography devices.

Based on the published particular international standards for laser and non-laser light source equipment intended for biomedical applications [7–9], some common evaluation parameters for OCT and these optical techniques were identified as: ocular and/or skin hazard distance; beam divergence; pulse or pulse train duration; pulse interval; spectral irradiance; number of pulses in pulse train; optical radiation output. Considering the lack of a specific standard for OCT, the evaluation of these parameters in common with other optoelectronic equipment could already be considered for conformity assessment.

Nevertheless, the evaluation of these parameters mentioned in the published standards is not enough to ensure the metrological reliability of OCT. The additional specific parameters, not existent in the laser and IPL standards, but identified in this work as necessary to be evaluated includes: central wavelength; concentricity between aim beam and work beam; beam diameter; exposure time; exposure limit; axial resolution and transverse resolution. The measurement uncertainty estimation is recommended for each measuring parameter suggested.

Considering the great potential of OCT as an imaging technique useful in multiple biomedical applications, it is essential to ensure its metrological reliability. The evaluation parameters suggested in the present work can be a contribution in the case of a future development of a particular standard specifically addressed to the non-invasive imaging technique by OCT intended for clinical applications.

Acknowledgments

This work was financially supported by CNPq and PUC-Rio.

References

[1] Huang D *et al* 1991 *Science* **254** 1178–1181
[2] Brezinski M E and Fujimoto J G 1999 *Selected Topics in Quantum Electronics, IEEE Journal of* **5** 1185–1192
[3] Costa Monteiro E 2007 *Metrologia & Instrumentação* **6** 6–12
[4] Fercher A F, Drexler W, Hitzenberger C K and Lasser T 2003 *Reports on Progress in Physics* **66** 239
[5] Tomlins P H and Wang R K 2005 *Journal of Physics D: Applied Physics* **38** 2519
[6] Saleh B E A and Teich M C 2001 *Fundamentals of Photonics* (Wiley New York)
[7] IEC 2001 *Safety of laser products - Part 1: Equipment classification and requirements, IEC 60825-1*
[8] IEC 2007 *Medical electrical equipment - Part 2-22: Particular requirements for the safety of diagnostic and therapeutic laser equipment, IEC 60601-2-22*
[9] IEC 2011 *Medical electrical equipment - Part 2-57: Particular requirements for the basic safety and essential performance of non-laser light source equipment intended for therapeutic, diagnostic, monitoring and cosmetic/aesthetic use, IEC 60601-2-57*
[10] Schmitt J M 1999 *Selected Topics in Quantum Electronics, IEEE Journal of* **5** 1205–1215
[11] Brownstein M, Hoffman R A, Levenson R, Milner T E, Dowell M, Williams P, White G, Gaigalas A and Hwang J 2007 *Journal of Research of the National Institute of Standards and Technology* **112** 139–152

Journal of Physics: Conference Series **459** (2013) 012065 doi:10.1088/1742-6596/459/1/012065

Non-contact wearable single forearm cardiac biopotential acquisition device

Sérgio Gonçalves[1] and Raul Carneiro Martins[2]
Department of Bioengineering, Instituto Superior Técnico, Lisbon Technical University, Portugal
Instituto de Telecomunicações, Lisboa, Portugal

E-mail: rcmartins@ist.utl.pt

Abstract. In this work the authors propose a novel approach to obtain the electrocardiogram in the forearm using non-contact sensing. This new solution should be at same time portable, ergonomic and robust, enabling its use in different set of applications. A system of four electrodes was used in an adjustable sleeve to be wrapped in the forearm. No additional electrode references were used in other body parts. In order to increase the sensitivity of the system, an harmonium like approach was used in the design of the electrodes. The prototype was then compared with a similar system with a flat conformation. The developed prototype enabled the acquisition of an ECG signal in the forearm and the inclusion of the harmonium like electrode conformation resulted in a considerable increase of the sensitivity of the system. The acquired signal did not enable the identification of all characteristic cardiac waves. However, it was possible to identify clearly a signal pattern, characteristic of the QRS complex. The properties of the acquired signal restrict their use in rigorous electrocardiographic studies, allowing, however, its application in heart rate variability monitoring and biometric identification without the disadvantages usually associated with conventional electrodes. This makes it specially useful for man-machine interfaces and automated identification.

1. Introduction

Devices to acquire ECGs are certainly not new. To do it without contact, making use of capacitive electrodes is also not new [1, 2]. Interesting enough the acquisition on a single arm, specifically the left arm, is also not new [3,4]. What the authors claim to be new is the approach and localization. The paper in [3] and application notes found on this topic [4] show that near ECG signals can only be acquired from the arm, being the forearm and wrist locations with too weak signal to noise ratio, even using normal Ag-AgCl wet electrodes. In our approach we were able to acquire near ECG type signals on the forearm while avoiding any contact whatsoever with skin on any other part of the body. This makes this an extremely convenient approach for acquiring ECG signals, confortable to wear for long periods of time. Furthermore we were able to use it over long-sleeved shirts and pullovers. Our goal with the development of such a device was to increase the viability of applications like biometric user validation, sports physical effort monitorization, game-play control system, prosthesis control and fatigue and drowsiness detection for drivers.

2. Biomedical considerations

The biosignals in the human body are a resultant of electric potential propagation caused by variations in the distribution of ions. The skin is composed by different layers, which present different electrical properties. The stratum corneum, external layer of the epidermis (dead cells), presents some

semipermiability to the body ions and consequently acts as an insulator. Its capacitance is not well characterized, since its value is dependent of the skin properties, especially the thickness and humidity of this layer [5]. Taheri et al. estimated that the stratum corneum capacitance can have minimum values of $100nF.cm^{-2}$. On the other hand, Rosell et al. obtained values for skin bioimpedance that varied from 10 $k\Omega.cm^{-2}$ till 100 $M\Omega.cm^{-2}$ at 1 Hz, 220 $\Omega.cm^{-2}$ at 100 kHz and 120 $\Omega.cm^{-2}$ at 1MHz, which implies amplifier circuits with a very high input impedance in order to avoid the voltage conversion from the common-mode to differential-mode [6]. The skin-clothing-electrode interface can be approximated by a conductive layer (dermis), which for the usual cases acts as a pure resistor, in series with parallel RC elements (stratum corneum, clothing layers, etc.) and in series with the electrode equivalent circuit. In the case of capacitive electrodes, this circuit can be approximated by a capacitor. On the other hand, the usual ECG electrodes are well described by a half-cell potential, in series with two parallel RC elements (electrolytic paste and stratum corneum) and in series with a conductive layer (dermis) [10].
In this way, when the conventional electrodes are used, a better signal can be obtained if the values of the stratum corneum resistance, capacitance and half-cell potential are reduced. A usual technique to decrease these values is to rub the areas under the electrodes. However, the regeneration capability of this layer is high, which results in variations of these values in time and consequently in the measured signal.

3. Methodology

While the main goal is the assessment of an ECG, we wanted the setup to easily permit determination of bioimpedance. This would allow arterial plethysmography information to be assessed simultaneously. As such the electrode configuration we designed has this dual capability but the focus in this paper will be on the biopotential recording.
The system consists in an armband with four capacitive electrodes positioned according to the schematic shown in figure 3.1.

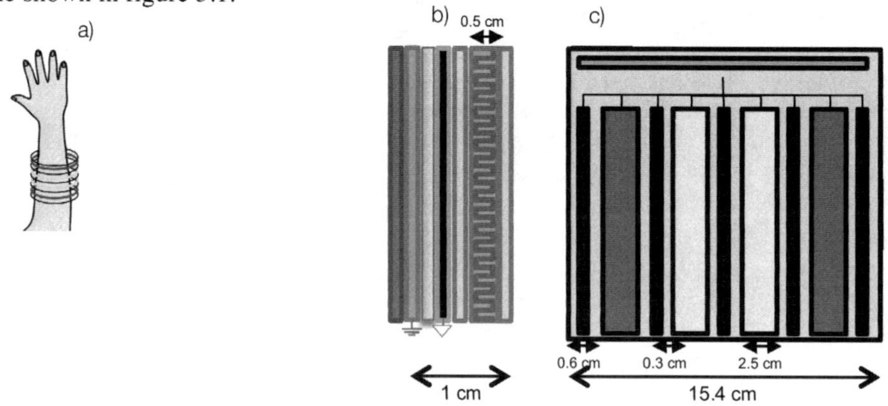

Figure 3.1 – Schematic representation of the developed electrodes: a) Body location ;b) lateral view – cloth support (blue), ground layer (orange), dielectric - paper (violet) guard layer (black), silicone sheet (pink), conductor electrodes/guards (red) and silicone gel (green); c) top view – guard electrodes (black), injection electrodes (red), measurement electrodes (blue) and Velcro (purple)

The first and fourth electrodes are responsible for creating an ion current in the body, which will be used to measure the plethysmographic impedance. The remaining electrodes will alternately measure both ECG biopotential and bioimpedance. The use of different electrodes for current injection and signal acquisition is preferable, since this configuration results in a more uniform current density, avoiding problems related with higher density currents in the areas near the injection electrodes and problems in the discrimination of the impedance of the skin-electrode interface and the area of interest. In order to guarantee the decoupling of both reading and injecting electrodes, three guard electrodes are displaced between these as presented in Figure 3.1 c). In addition two guard electrodes, which are connected to the circuit ground, are displaced in the extremities of the system, ensuring the shielding of the measured signal from the external noise sources.

Two more conductive layers are added to the system. An external layer, which is connected to the ground, allows the shielding of signals from parasite noise sources, such as power lines sources and biosignals originating from other body parts. A medial conductive layer connected to the guard common was also used. The inclusion of this layer resulted from some experiments that showed that the addition of it led to a better signal-noise ratio (SNR). Thin sheets of dielectric material separate all these layers.

4. Electrode design

The proper design of the circuit implies the previous study of the properties of the electrodes. The interaction electrode-skin can be approximated by a capacitor in which the subject arm and the electrode are the conductors separated by a dielectric (clothing and other insulators). Furthermore, since it was intended that the system wraps the entire arm, the interaction skin-electrode can be approximated by a cylindrical capacitor. The capacitance of a cylindrical capacitor can be estimated using both the definition of the capacitance (4.1) and the calculus of the electric field (E) to an infinite cylinder by using the Gauss's Law (4.2):

$$C = \frac{Q}{V} \cong \frac{\lambda L}{\Delta V} \tag{4.1}$$

$$E = \frac{\lambda}{2\pi\varepsilon_0 r} \tag{4.2}$$

in which C is the capacitance, Q represents the ratio of charges between the conductors to the voltage V, L is the cylinder length, λ and ε_0 represents respectively the charge per unit of length in the conductor and the vacuum permittivity and r is the radial distance to the center of the conductor. The voltage difference between the conductors can be computed by integrating E along a radial line:

$$\Delta V = \int_{r_1}^{r_2} \frac{\lambda}{2\pi\varepsilon_0 r} dr = \frac{\lambda}{2\pi\varepsilon_0} \log\left[\frac{r_2}{r_1}\right] \tag{4.3}$$

where r_1 represents the internal conductor radius and r_2 is the radial distance between the medial side of the external conductor and the center of the internal conductor. Substituting equation (4.3) in equation (4.1) and considering a dielectric with the dielectric constant ε_r, the capacitance for a cylindrical capacitor is derived: $C \cong \frac{2\pi\varepsilon_r\varepsilon_0 L}{\log\left[\frac{r_2}{r_1}\right]}$.

The developed system should be as robust as possible, enabling its use over a non-determined set of clothing layers. Thus, in order to enhance the signal acquisition it is important to consider the different parameters that can affect it.

A schematic representation of the relation between the impedance of the electrodes (Zelec), the impedance of the body part in measurement (Zbio) and the acquired signal (Vout) is presented in Figure 4.1. Simplifying, this relation can be approached by a voltage divider referenced to the ground between both impedances.

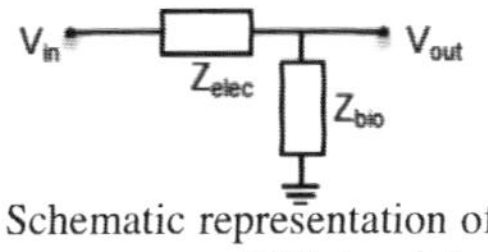

$$V_{out} = \frac{Z_{bio}}{Z_{bio} + Z_{elec}} V_{in} = \frac{Z_{bio}}{Z_{bio} + 1/j\omega C_{elec}} \tag{4.4}$$

Figure 4.1 – Schematic representation of the relation between the impedance of the electrodes (Zelec), the impedance of the body part in measurement (Zbio) and the acquired signal (Vout)

Therefore, the value of Zelec should be minimized as much as possible in order to achieve high values of Vout. Considering the relation between the impedance and capacitance of a capacitor (Z=1/jωC), the minimization of the value of Zelec implies the maximization of the value of electrode capacitance (Celec). Bearing in mind these issues and considering the equation 4.4, two main parameters can be

Journal of Physics: Conference Series **459** (2013) 012065 doi:10.1088/1742-6596/459/1/012065

varied to guarantee high values of Celec – Electric area and electric permittivity. We improved both through appropriate design. To increase the area the electrodes were wrapped like and harmonium, effectively doubling the electric surface. To optimize the permittivity we used high permittivity materials like silicone.

5. Results and discussion

We have performed an extensive set of tests with the purpose of evaluating the performance of the design according to different parameters. These tests include sensitivity analysis, frequency response, lift-off performance and electromyographic interference.

5.1 Sensitivity analysis

Being this essentially a capacitive system, both electrode configurations should be able to detect electric field variations caused by the subject movements. It is expected that the different geometry of the electrodes lead to different sensitivities. These can be determined simply through the amplitude of signals obtained by a simple hovering hand movement at a given distance from the electrodes. This distance was kept the same. Since everything was kept the same except for the electrode geometry, amplitude differences can be attributed to geometry. Figure 5.1 shows the difference of sensibility between both systems, when the user's hand is passed over them, approximately at 6 cm. As expected the corrugated system presented a high sensitivity to hand passages. In fact, this system presented a high sensitivity to a large number of motions, detecting some movements at distances higher than 50 cm.

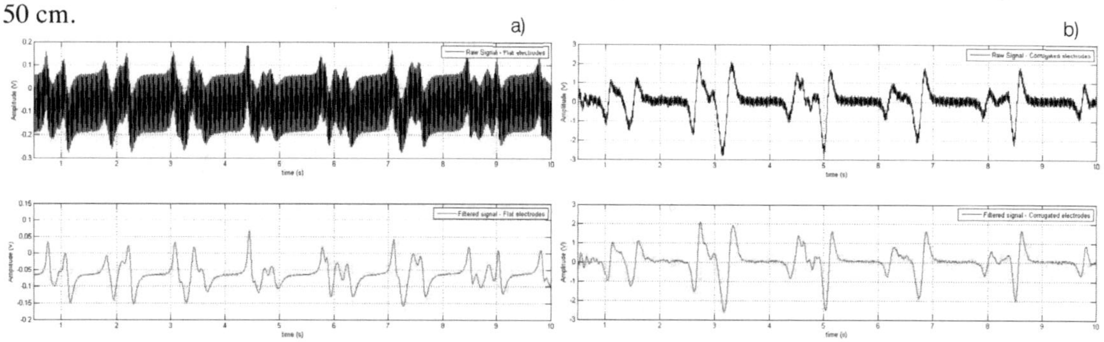

Figure 5.1 – Response of the system to the subject's movements. Each peak corresponds to one passage of the hand, approximately 6 cm over the open band: a) flat electrodes; b) corrugated electrodes.

In particular in figure 5.4.1 can be seen the actual ECG that was obtained by the final device previously described.

While this biopotential cannot be considered an ECG, and thus cannot be used for clinical evaluation of the heart condition, it can be used for heart rate variability assessment, physical strain and biometric identification.

5.2 Frequency response and lift-off performance

A frequency analysis was performed to understand the differences between sensitivities of both systems to different frequencies. A cupper band (2.5 cm x 15 cm) connected to a function generator was placed 4 mm above the open prototype in the middle position. Different frequencies (0.1 Hz - 100Hz) with the same amplitude (3 Vpp) were tested. Figure 5.2 shows the amplitude response of both systems for different frequencies. The corrugated system presented a higher sensitivity, especially to high frequencies. This fact is a direct result of the capacitance increase of the corrugated electrodes.

A similar experimental setting was used to study the effect of the distance in both prototypes. However, in this case, the frequency was maintained constant (5 Hz), varying the distance between the copper band and source. No significant differences were observed between prototypes (see Figure 5.2). It would be expected to observe a higher sensitivity in the corrugated prototype. There is a general trend for the corrogated design to exhibit higher amplitude but the difference was lower than we had anticipated. However, in practice we did observe much better results from the corrogated design than with the flat design as depicted in figure 5.4.2.

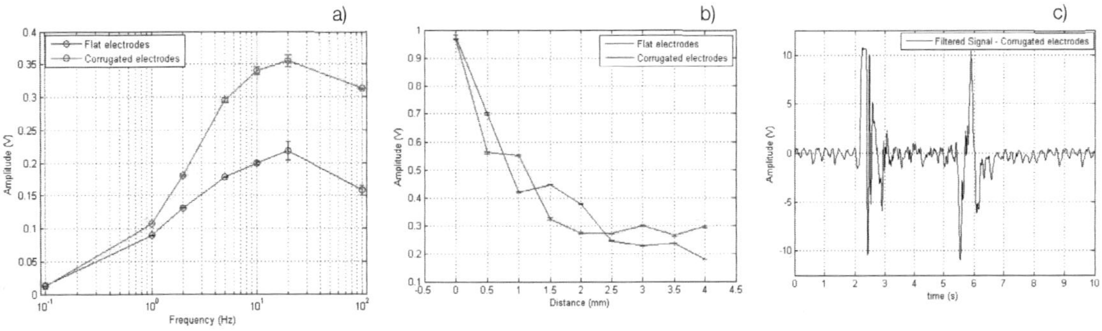

Figure 5.2 – a) Amplitude response of both systems for different frequencies (d=4 mm, Vpp=3 V). b) Amplitude response of both systems for different distances (f=5 Hz, Vpp=3 V). c) EMG interference: 1st peak (~2s) – fingers extension; 2nd peak (~5.2s) – fingers flexion

5.3 Electromyographic interference

One of the issues to consider in this type of systems is the interference of the electromyographic signals (EMG) from the forearm and fingers muscles. The amplitude order of these biopotentials is substantially higher, overlapping the ECG signal. For instance, for a gain of 100 and ±12V supply voltage, the amplifier went into saturation for some of the performed movements. Therefore, a methodology to subtract these signals should be considered in the next developed prototypes. Figure 5.2 c) shows the influence of the flexion and extension of the fingers in the measured signal. In this figure, it is possible to observe differences in the signals according the type of movement performed – flexion and extension of the fingers. The system enables also to detect movements of the hand and of each finger. However, further studies have to be performed in order to understand signal patterns for each type of movement (hand and motion of each finger), enabling the application of this type of prototypes in EMG controlled systems with high potential for man-machine interfaces (MMI).

5.4 Electrocardiogram

The ECG experimental results were obtained in a static sitting position, with the sleeve wrapping the left forearm. A first prototype, without the third layer connected to the ground, was tested. In this case the system became very vulnerable to undesired body noise sources. A clean signal was obtained when the arm was deviated from the body, however when the arm is placed near the body, the QRS complex was more or less visible, presenting the same amplitude of the noise. The introduction of the third layer enabled to route the electric field lines from noise sources to ground, which enabled to acquire ECG signals with the arm near the body. In Figures 5.4.1 and 5.4.2, the ECG reading of a 24 year old male subject with no history of cardiac disorders is shown. Inspection of the graph enables clearly to identify the QRS complex. The obtained results are quite interesting, since besides the fact that these had been acquired from the forearm, no other electrode references were used. The obtained results do not allow identification of other characteristic waves, such as the P wave and T wave. This fact restricts the application of this type of systems in the rigorous study of the electric activity of the heart. However, this type of systems could be used as an ECG monitoring system both in clinic and in ambient assisted living scenarios. Moreover, it could also be applied in the analysis of biometric parameters, without the disadvantages of the usual AgCl/Ag electrodes. Once more, significant differences could be observed between both prototypes. Despite being possible to identify also the QRS complex in the flat conformation, the peaks are not so well defined.

6. Conclusions and future work

One of the main objectives was the development of an efficient yet unobtrusive system to acquire ECG biopotentials in the forearm using capacitive sensing. In order to increase the sensitivity of the system, a new design using corrogated harmonium like electrodes was proposed. The prototype enabled the measurement of ECG signals, resulting in a higher SNR, when compared with a similar prototype with flat electrodes. The obtained signals did not enable their application in rigorous studies of the electric activity of the heart; however, this can be applied in ECG monitoring and biometric systems without the disadvantages of the usual ECG electrodes (Ag/AgCl) or even dry electrode approaches, and without requiring wires to cross the body or to get undressed.

One interesting particularity we observed while making tests to the system was the capability to isolate EMG activity from each finger. This could be used for man-machine interface systems namely the control of prosthesis, robotic arms or game play control. More work will be devoted to increase the electrical surface to physical surface ratio.

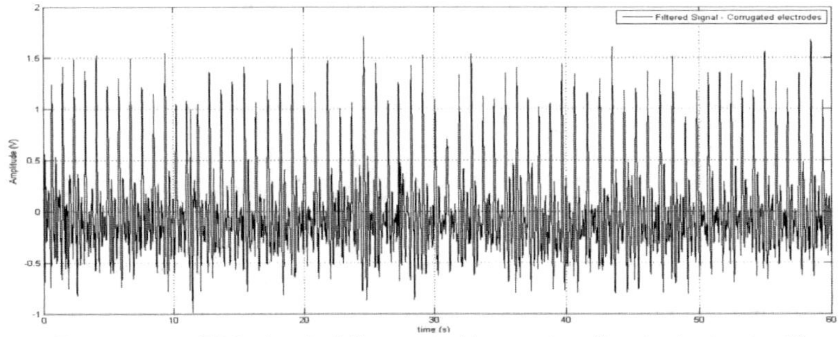

Figure 5.4.1 – ECG signal: a) Corrogated harmonium like electrodes t = 60 s

a) b)

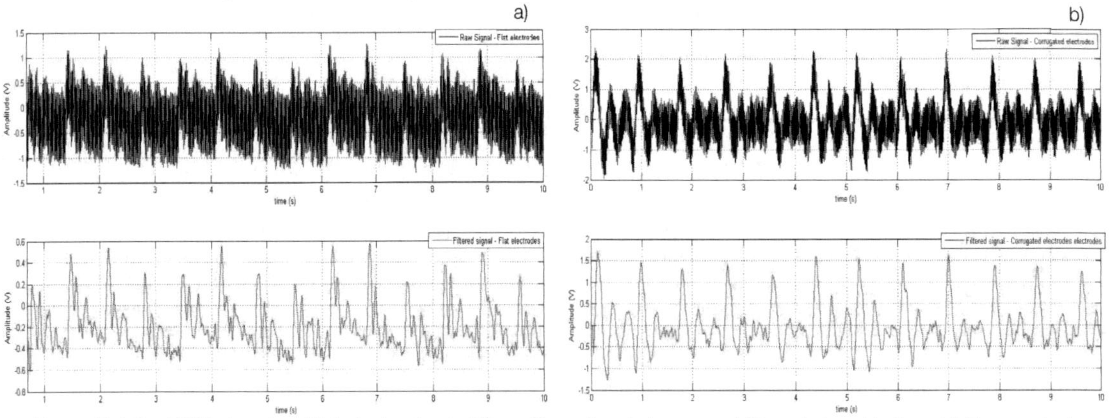

Figure 5.4.2 – ECG signal: a) Flat electrodes t=10 s – Raw signal above and filtered signal below; b) Corrugated electrodes t=10s – Raw signal above and filtered signal below.

Acknowledgments

The authors would like to acknowledge the Foundation for Science and Technology which has funded this work through the grant EXPL/EEI-ELC/0930/2012.

7. References

[1] Oehler M, Ling V, Melhorn K, Schilling M, 2008, A multichannel portable ECG system with capacitive sensors, *Physiological Measurement*, vol. 29, 783.

[2] Ueno A, Akabane Y, Kato T, Hoshino H, Kataoka S, Ishiyama Y, 2007, Capacitive sensing of electrocardiographic potential through cloth from the dorsal surface of the body in a supine position: a preliminary study, *Biomedical Engineering, IEEE Transactions on*, vol. 54, 759-66

[3] Hannula M, Hinkula H and Jauhiainen J, 2008, Development and evaluation of one arm electrode based ECG measurement system, *IFMBE Proceedings*, vol. 20, 234–7

[4] Plessey Semiconductors' Application Note # 291491 Single arm ECG measurement using EPIC

[5] Griffith M, Portnoy WM, Stotts LJ, Day JL, 1979, Improved capacitive electrocardiogram electrodes for burn applications, *Medical and Biological Engineering and Computing*, vol. 17, 641-6.

[6] Rosell J, Colominas J, Riu P, Pallas-Areny R, Webster JG, 1988, Skin impedance from 1 Hz to 1 MHz, *Biomedical Engineering, IEEE Transactions on*, vol. 35, 649-51.

[7] Chi, YM, Tzyy-Ping J, Cauwenberghs G, 2010, Dry-contact and noncontact biopotential electrodes: Methodological review, *Biomedical Engineering, IEEE Reviews in*, vol. 3, 106-19

AUTHOR INDEX

Abdulla, T 55
Accardi, L 3
Affanni, A 71, 275
Agus, M 111
Akiyama, J 245
Araújo, M C C 182
Aristovich, E 153
Armani, F 257
Bacelar, L 175
Belotti, V 345
Benoit, E 5
Berglund, B 1, 100
Bezruczko, N 94
Boffitto, M 59
Boscolo, A 257
Bovio, L 161
Bruna, O 45
Caldwell, D G 169
Camargo, F R 33
Cannella, F 169
Carvalho, F 175
Casaccia, S 76
Ciardelli, G 59
Coghe, F 263
Corazza, A 275
Crenna, F 10, 161, 345
Debei, S 315
Demichelis, A 280
D'Errico, G E 227
Dittrich, P-G 39
Divieto, C 59, 286
Dobiliene, J 251
Düntsch, E 39
Elkarous, L 263
Emardson, R 100
Ercoli, I 76, 355
Esposito, G 275
Esward, T J 215
Fabbiano, L 209
Fernandez, A 82
Filho, J B P F 182
Fisher Jr, W P 117, 123, 129, 321
Fitkov-Norris, E D 105
Fogolari, F 275
Forbes, A B 175
Fujiki, T 245
Gärtner, C 39
Gasparetto, M 161
Giannetti, R 82
Giaquinto, N 209
Golinval, J C 263
Goloni, C M 361
Gonçalves, S 367
Granovskii, V A 141, 303
Guàrdia-Olmos, J 111
Harris, P M 175
Henson, B 33
Hernández, C 203
Hofmann, D 39
Holub, J 45, 51, 333
Irribarra, D T 21, 27
Kaspar, P 298
Khan, S H 153
Kirschen, D 135
Klemm, R 39
Kowalewski, M 339
Kraus, D 39
Laby, D M 135
Lekkala, J 65
Lorenzini, E 315
Luff, R 55
Mambelli, E 350
Marchionni, P 76, 355
Mari, L 2, 21, 221
Marinushkina, I A 15
Martins, L L 309
Martins, R C 367
Massof, R W 135
Maul, A 21, 27
Meadows, D 135
Memeo, M 169
Mencattini, A 221
Meskuotiene, A 251
Miekina, A 147
Mlejnek, P 298
Mokhtar, N A B 245
Monteiro, E C 361
Morawski, R Z 147
Mortati, L 59, 280
Muravyov, S V 15
Nilsson, M E 100
Olivieri, E 169
Otomanski, P 327
Pacini, G 350
Pavarelli, S 280
Pavese, F 197, 233
Peltokangas, M 65
Pendrill, L 100
Pendrill, L R 321
Penna, M P 111
Peró-Cebollero, M 111
Pertile, M 315
Pirlot, M 263
Polano, M 275
Raudiene, E 251
Ravazzani, P 350
Rebordão, J M 309
Reis, M L C C 182
Revel, L 286
Ribeiro, A S 175, 309

AUTHOR INDEX

Rossi, G B 10, 161, 345
Ruhm, K H 239
Sáenz-Nuño, M A 82
Santoro, A 350
Sartori, S 59
Sassi, G 280, 286
Sassi, M 59, 280, 286
Savino, M 209
Sbrignadello, S 350
Scalise, L 76, 169, 355
Schmidt, K M 135
Siraia, T N 303
Slavata, O 51, 333
Soucek, P 45, 51, 333
Sousa, J A 175
Specogna, R 71
Stenner, A J 117, 123
Stephanou, A 129
Summers, R 55
Sundling, C 100
Takayama, S 245
Temporão, G P 361
Testa, R 161
Thomas, P J 88, 292
Tomasini, E P 76
Trevisan, F 71
Tura, A 350
Tutsch, R 203
Vacca, G 209
Valderrama, J M 82
Vehkaoja, A 65
Warsza, Z L 191
Wilson, M 2, 21, 27
Wright, L 215
Yeghiazarian, A 105
Zappa, E 161
Zazula, P 327

Institute of Physics
Dirac House, Temple Back
Bristol BS1 6BE UK

CURRAN ASSOCIATES INC
proceedings.com